FOUNDATIONS FOR THE FUTURE IN MATHEMATICS EDUCATION

Edited by

Richard A. Lesh
Indiana University

Eric Hamilton
U.S. Air Force Academy—Colorado

James J. Kaput

LEA LAWRENCE ERLBAUM ASSOCIATES, PUBLISHERS
2007 Mahwah, New Jersey   London

Senior Acquisitions Editor: Naomi Silverman
Editorial Assistant: Joy Tatusko
Cover Design: Tomai Maridou
Full-Service Compositor: MidAtlantic Books and Journals, Inc.

Copyright © 2007 by Lawrence Erlbaum Associates, Inc.

All rights reserved. No part of this book may be reproduced in
any form, by photostat, microform, retrieval system, or any
other means, without prior written permission of the publisher.

Lawrence Erlbaum Associates, Inc., Publishers
10 Industrial Avenue
Mahwah, New Jersey 07430
www.erlbaum.com

CIP information for this volume can be obtained from the Library of Congress.

ISBN 978-0-8058-6056-6—0-8058-6056-8 (case)
ISBN 978-0-8058-6057-3—0-8058-6057-6 (paper)
ISBN 978-1-4106-1482-7—1-4106-1482-4 (e-book)

Books published by Lawrence Erlbaum Associates are printed on
acid-free paper, and their bindings are chosen for strength and durability.

Printed in the United States of America

10 9 8 7 6 5 4 3 2 1

CONTENTS

Preface — Foundations for the Future in Engineering and other Fields That Are Heavy Users of Mathematics, Science and Technology vii
Richard Lesh

PART I — What Changes are Occurring in the Kind of Problem-Solving Situations Where Mathematical Thinking Is Needed Beyond School? 1
Eric Hamilton

Chapter 1 — The Meanings of Statistical Variation in the Context of Work 7
Celia Hoyles and Richard Noss

Chapter 2 — Problem-Solving and Learning in Everyday Structural Engineering Work 37
Julie Gainsburg

Chapter 3 — Modeling Without End: Conflict Across Organizational and Disciplinary Boundaries in Habitat Conservation Planning 57
Bruce Evan Goldstein and Rogers Hall

Chapter 4 — Mathematical Modeling 'in the Wild': A Case of Hot Cognition ... 77
Wolff-Michael Roth

Chapter 5 — Learning in Design 99
David Williamson Shaffer

Chapter 6 — The Cognitive Science of Mathematics: Why Is It Relevant for Mathematics Education? 127
Rafael Núñez

PART II — What Changes Are Occurring in the Kind of Elementary-but-Powerful Mathematics Concepts That Provide New Foundations for the Future? 155
Richard Lesh

Chapter 7 — Models, Simulations, and Exploratory Environments: A Tentative Taxonomy 161
Judah L. Schwartz

| Chapter 8 | Technology Becoming Infrastructural in Mathematics Education 173
Jim Kaput, Richard Lesh, and Steve Hegedus

| Chapter 9 | Why Build a Mathematical Model? Taxonomy of Situations That Create the Need for a Model to Be Developed 193
Maynard Thompson and Caroline Yoon

| Chapter 10 | Cultivating Modeling Abilities 201
Caroline Yoon and Maynard Thompson

| Chapter 11 | Discrete Mathematics in 21st Century Education: An Opportunity to Retreat from the Rush to Calculus 211
Joseph G. Rosenstein

| Chapter 12 | Formalizing Learning as a Complex System: Scale Invariant Power Law Distributions in Group and Individual Decision Making 225
Thomas Hills, Andrew C. Huford, Walter M. Stroup, and Richard Lesh

| Chapter 13 | Systemics of Learning for a Revised Pedagogical Agenda 245
Andrea A. diSessa

| Chapter 14 | The *DNR* System as a Conceptual Framework for Curriculum Development and Instruction 263
Guershon Harel

| Chapter 15 | Aspects of Affect and Mathematical Modeling Processes 281
Gerald A. Goldin

| PART III | What Kind of Instructional Activities are Are Needed to Develop New Levels and Types of Understanding and Ability? 297
Richard Lesh and Jim Kaput

| Chapter 16 | Beyond Efficiency: A Critical Perspective of Singapore's Educational Reforms 301
Mani Le Vasan, Richard Lesh, and Mardiana Abu Bakar

| Chapter 17 | John Dewey Revisited—Making Mathematics Practical versus Making Practice Mathematical 315
Richard Lesh, Caroline Yoon, and Judi Zawojewski

| Chapter 18 | The Use of Reflection Tools in Building Personal Models of Problem-Solving 349
Eric Hamilton, Richard Lesh, Frank Lester, and Caroline Yoon

Chapter 19 Diversity-by-Design: The What, Why, and How of Generativity
in Next-Generation Classroom Networks 367
Walter M. Stroup, Nancy Ares, Andrew C. Hurford, and Richard Lesh

Chapter 20 When the Model Is a Program 395
Fred G. Martin, Margret A. Hjalmarson, and Phillip C. Wankat

Chapter 21 Uncertainty and Iteration in Design Tasks
for Engineering Students 409
Margret A. Hjalmarson, Monica Cardella, and Phillip C. Wankat

Chapter 22 Teacher Development in a Large Urban District
and the Impact on Students 431
Roberta Y. Schorr, Lisa Warner, Darleen Gearhar, and May Samuels

Chapter 23 Directions for Future Research 449
Richard Lesh, Eric Hamilton, and Jim Kaput

Author Index 455
Subject Index 463

PREFACE

Foundations for the Future in Mathematics Education

Richard Lesh
Indiana University

In fields ranging from aeronautical engineering to agriculture, and from biotechnologies to business administration, outside advisors to future-oriented university programs increasingly emphasize the fact that, beyond school, the nature of problem solving activities has changed dramatically during the past twenty 20 years. For example, powerful tools for computation, conceptualization, and communication have led to fundamental changes in the levels and types of mathematical understandings and abilities that are needed for success in such fields. These observations raise the following questions.

> *What is the nature of typical problem-solving situations where elementary-but-powerful mathematical constructs and conceptual systems are needed for success in a technology-based age of information? What kind of "mathematical thinking" is emphasized in these situations? What does it mean to "understand" the most important of these ideas and abilities? How do these competencies develop? What can be done to facilitate development? How can we document and assess the most important (deeper, higher-order, more powerful) achievements that are needed: (i) for informed citizenship, or (ii) for successful participation in the increasingly wide range of professions that are becoming heavy users of mathematics, science, and technology?*

Authors in this book believe that such questions should be investigated through research—not simply resolved through political processes (such as those that are used in the development of curriculum standards or tests). We also believe that researchers with broad and deep expertise in mathematics and science should play significant roles in such research—and that input should be sought, not just from creators of mathematics (i.e., "pure" mathematicians), but also heavy users of mathematics (e.g., "applied" mathematicians and scientists). This is because the questions listed above are about the changing nature of mathematics and situations where mathematics is used; they are not simply questions about the nature of students, human minds, human information processing capabilities, or human development.

The chapters in this book evolved out of a series of meetings that were held during the spring, summer, and fall of 2004. The first meetings were sponsored by Purdue University's School Mathematics & Science Center; and, the later meetings were sponsored by Indiana University's Center for Learning & Technology. The chief goal of these meetings was to develop key elements of a research agenda aimed at investigating the questions posed above. However, the following questions also were considered.

Why do students who score well on traditional standardized tests often perform poorly in more complex "real life" situations where mathematical thinking is needed? Why do students who have poor records of performance in school often perform exceptionally well in relevant "real life" situations?

These latter questions emerged because many participants shared the experience of encountering our former mathematics students when they appeared several years later in courses or jobs where the mathematics that we tried to teach them would have been useful. In some cases, we have been discouraged by how little was left from what we thought we had taught. On the other hand, we often were equally impressed that some students whose classroom performances were unimpressive went on to develop a great deal from seeds that we apparently helped to plant.

Upon further reflection and research about the preceding issues, most of us gradually developed the opinion that, for most topics that we have tried to teach, the kind of mathematical understandings and abilities that are emphasized in mathematics textbooks and tests tend to represent only a shallow, narrow, and often non-central subset of those that are needed for success when the relevant ideas should be useful in "real life" situations. For example, in projects such as Purdue University's Gender Equity in Engineering Project (Zawojeski et al., in press), when students' abilities and achievements were assessed using tasks that were designed to be simulations of "real life" problem solving situations, the majority of understandings and abilities that emerged as being critical for success included were not among those emphasized in traditional textbooks or tests. Consequently, when we recognized the importance of a broader range of deeper understandings and abilities, a broader range of students naturally emerged as having extraordinary potential. Furthermore, many of these students came from populations that are highly under represented in fields that emphasize mathematics, science, and technology; and this was true precisely because their abilities were previously unrecognized. Such observations return us to the following fundamental question:

What kind of understandings and abilities should be emphasized to decrease mismatches between: (i) the narrow band of mathematical understandings and abilities that are emphasized in mathematics classrooms and tests, and (ii) those that are needed for success beyond school in the 21st century?

Many people assume that students simply need practice with ideas and abilities that have been considered to be "basics" in the past. Others assume that old conceptions of "basics" should be replaced by completely new topics and ideas (such as those associated with complexity theory, discrete mathematics, systems theory, or computational modeling). Still others assume that new levels and type of understanding are needed for both old and new ideas. Examples include understandings that emphasize graphics-based or computation-based representational media. My own perspectives lean toward this third option—without denying the legitimacy of the other two. But, each of these issues will be addressed later by authors throughout this book. So, no attempt will be made to resolve them here. Let me simply point out that, when such issues were discussed by authors in this book, three levels of students were given special attention.

Undergraduate students preparing for leadership positions in fields, such as engineering, where mathematical and scientific thinking tend to be emphasized.

Middle-school students who, with proper educational opportunities, could have the potential to succeed in universities such as Purdue or Indiana University, which specialize in a variety of fields that are increasingly heavy users of mathematics, science, and technology.

Teachers (as well as professors and teaching assistants) of the preceding students.

For K–12 students and teachers, questions about the changing nature of mathematics (and mathematically thinking beyond school) might be rephrased to ask: *If attention focuses on preparation for success in fields that are increasingly heavy users of mathematics, science, and technology, how should traditional conceptions of the 3R's (Reading, wRiting, and aRithmetic) be extended or reconceived to prepare students for success beyond school in the 21st century?*

The book is partitioned into three sections. The first focuses on naturalistic observations aimed at clarifying what kind of "mathematical thinking" people really do when they are engaged in "real life" problem solving or decision- making situations beyond school. The second section shifts attention toward changes that have occurred in kinds of elementary-but-powerful mathematical concepts, topics, and tools that have evolved recently—and that could replace past notions of "basics" by providing new foundations for the future. This second section also initiates discussions about what it means to "understand" the preceding ideas and abilities. Finally, the third section extends these discussions about meaning and understanding—and emphasizes teaching experiments aimed at investigating how instructional activities can be designed to facilitate the development of the preceding ideas and abilities.

Overall, what the chapters in this book suggest is that, if our goal is to create a K–16 mathematics curriculum that will be adequate to prepare students for informed citizenship—as well as preparing them for career opportunities in learning organizations, in knowledge economies, in an age of increasing globalization—then it is not likely to be sufficient to simply make incremental changes in the existing curriculum whose traditions developed out of the needs of industrial societies.

Throughout this book, the challenge that was given to authors was not so much to state conclusions from our research, but rather to use results from our research to describe promising directions for a research agenda related to the questions posed in this introduction to the project.

REFERENCES

Zawojewski, J. S., Diefes-Dux, H., & Bowman, K. (Eds.) (in press). *(under development). Models and modeling in engineering education: Designing experiences for all students.*

PART I

What Changes Are Occurring in the Kind of Problem-Solving Situations Where Mathematical Thinking Is Needed Beyond School?

Eric Hamilton
United States Air Force Academy

The chapters in Part A present ethnographies describing several workplace settings in which significant types of mathematical thinking are needed. This collection of ethnographies clarifies the nature of mathematics-rich thinking and problem-solving that these settings tend to emphasize.

Ethnographies specialize in highlighting and analyzing structural relationships of a social system, in contrast to other approaches that focus on isolating and testing single variables within the system. The environments or systems under study in the following chapters involve factory statistical process control (Hoyles and Noss); architectural design (Shaffer); structural engineering (Gainsburg); a fish hatchery (Roth); and conservation biologists (Goldstein and Hall). Education requirements in each of these venues vary from high school level to advanced degree training. The mathematics most useful in modeling and solving problems varies as much as the environments themselves.

Each author employs a different analytic framework. Hoyles and Noss and Roth anchor their respective studies in activity theory: Hoyles and Noss seek to expose multiple layers of statistical thinking involved in minimizing factory outputs variances; Roth documents the mediating roles of emotion and identity in a technician's fish hatchery care. Gainsburg's and Goldstein and Hall's cognitive ethnographies trace structural engineers and conservation biologists' mathematical modeling journeys when solving real life problems. Shaffer's structural ethnography describes and analyzes the induction of students in MIT's Oxford Studio into mathematics-rich norms and practices of the architectural design community. In each study and more generally, one significant aspect of workplaces is that individuals must master new technological tools and their interfaces. For example, successful preparation for the workplace typically includes mastering simple but profound tools such as spreadsheets, databases, and system-visualization software; mastering more sophisticated tools associated with specific professions; and finding new ways to model and analyze mathematics-related tasks. Part A studies also illustrate how understanding the limitations of modeling tools are now in each successful modern problem-solver's repertoire. Examples in these chapters include software systems providing stress tolerance estimates (Gainsburg) or statistical systems that report production variances (Hoyles & Noss). Knowing when and why tools such as these are *not* useful is as important as knowing how to use them.

In general, facility with complex tools, understanding and adapting to rapid changes in those tools, and communicating with them, are all part of what Hoyles call "techno-mathematical literacies" et al (2003). New tools imply modeling problems or mediating problems in new ways. Problems themselves change as rapidly as the professions and social structures in which they are embedded change. No occupations in these ethnographies involve the same mathematical modeling characteristics required 10 or 20 years previously. Across virtually the full spectrum of occupations in early 21st century society, the paradox holds that jobs are both easier and more complex than 10 or 20 years ago—provided they even existed then. More accurately, they are fundamentally different.

MULTIPLE MULTIPLES

In addition to changes directly related to new technologies, teams often replace individuals as problem-solving entities. More than ever, this requires increased sophistication among team members' contributions in formulating, expressing, and testing solutions. In particular, both Hoyles & Noss and Goldstein and Hall discuss the increasing need for individuals to develop broad conceptual models of full systems of operations, and to be knowledgeable about the priorities, emphases, tools and metrics of other parties in a team. This in turn entails sophisticated representational and "cross-model" communication competencies. Distributed problem-solving and effective communication skills are one trend in incorporating multiple perspectives into mathematical models, creating models reflecting multiple constraints and priorities (mostly non-mathematical), trying multiple paths and test iterations, and arriving at any of multiple potential solutions. Each "multiple" is a problem-solving feature that is increasingly prominent in real-world settings where tasks become more complex and interconnected and require distributed expertise, accountability and collaborative structures.

ALL MATHEMATICS IS LOCAL

Ethnographic approaches carry an especially compelling methodological value documenting mathematics needed beyond school. The chapters in Part A converge around the finding that mathematical problem-solving character in a real-world setting is inextricably intertwined with that setting's structure. The five reported workplaces are distinctive. They collectively stress what might be considered a parallel to the maxim that "all politics is local." The best description of situated and context-dependent mathematical modeling nature outside schools may be: "All mathematics is local." That is, Part A studies make clear that problem-solving success beyond school and workplace involves seeing and manipulating mathematics within the unique systems and representation of objects, relationships, operations, and analytic tools of a specific workplace setting. Mathematical modeling and problem-solving beyond school is context-specific.

Of course, not all mathematics is local or context-specific. The abstract systems that comprise formal mathematics and that are structured by definitions, axioms, and theorems are anything but local or context-specific. Formal or pure mathematics occupies a mythic place; its aesthetic features, modeling power, and position as arbiter and language of physical science give it a unique role in human knowledge. These characteristics have engendered lively and enduring debate about whether mathematics was invented or discovered. Rafael Núñez, the final author in this section, does not directly enter this debate in his chapter, but he does provide a provocative analysis of mathematics and its mythic character. That analysis prefaces his argument that understanding the nature of mathematics requires understanding mathematical cognition and the human activity systems contributing to its development. Núñez elaborates on something of a rubicon between abstract mathematics and mathematical cognition, arguing there is a striking difference between static formulations of formal mathematics and the dynamic nature of mathematical cognition and problem solving as it unfolds or develops in an individual. That is, abstract mathematical formalisms necessarily become decontextualized from the mathematical cognition systems originally giving rise to the formalisms.

It is not difficult to argue that decontextualized structures of formal mathematics have become the foundation for modern mathematics curricula. The steady and universally recognizable pattern of introducing mathematics concepts at the beginning of a textbook section, followed by word problem "containers" embedding those concepts in tailor-made situations is a sizable wager on adequately preparing learners for mathematical success beyond school. It is a wager we appear not to have won. There is little evidence suggesting, for example, that building textbook word problem solving proficiencies leads to improved proficiencies using mathematics to solve problems beyond schooling. We have assumed that by decomposing formal mathematics into 1,000–2,000 sequential stand-alone lessons concepts we can additively recompose those structures through 12–16 school years to produce mathematics-proficient adults. Perhaps the alluring aesthetic and power of formal mathematics as the end-point for mathematical development masks the crucial deficiency of formal mathematics as the starting point for nurturing mathematical cognition in youngsters.

There is a subtle irony about the relationship between formalisms and cognition: on one hand, as an individual's mathematical cognition and modeling competencies evolve upward they necessarily gravitate toward stability and power, successively subsuming formal mathematical structures. On the other hand, simply decomposing mathematics downward—and modern arithmetic to calculus curriculum sequences are indeed decompositions from calculus down to arithmetic—does not appear in itself to optimally yield effective instructional building blocks. It is quite possible, for example and as argued throughout this volume, to expose youngsters to problem situations before giving them all the mathematical tools they need to solve them, instead expecting them to invent their own versions of those tools after determining they are needed. Sherra Kerns (2005), a keynote panelist at the 2005 Annual Meeting of the American Society for Engineering Education, was one of many reflecting on the competencies required of the so-called "Engineer of 2020" (NAE, 2004). "Why," she asked, "do we assume that we have to teach every concept before a student can use it?" The question exposes two critical fault lines in mathematics education. The "concept-then-word problem" sequence of most curriculum approaches constrains problem-solving contexts to those that often artificially contain and highlight the "concept of the day." More importantly, this approach precludes allowing youngsters the opportunity to develop new constructs themselves, out of the necessity of solving a real situation. Instead, students know by experience that the constructs needed to solve textbook word problems don't need to be invented or developed at all—they are instead the ideas that have just been introduced the student's main challenge is the detective work in teasing those ideas out of the text of the problem. This is an unnecessarily impoverished approach to mathematical experience. The need to develop or invent new mathematics out of necessity is precisely what historically has driven the expansion of frontiers of applied mathematics. More modestly, it is precisely the kind of problem-solving experience studies in Part A demonstrate are needed to succeed beyond school.

MATHEMATICS IS PERSONAL

Each of this section's ethnographers note sizable differences between school mathematics and how the individuals they studied practiced mathematics. Among the contrasts is how each reported problem solver formulates a succession of conceptual models solving complicated problems, drawing on their own unique experiences and competencies. That is, they often create or invent mathematical structures their personal conceptual models require to solve problems, structures necessarily unique to their own experience and understanding.

Julie Gainsburg identifies the "co-development" of mathematical ideas and their use in solutions to new problems as a salient characteristic of successful structural engineers. Michael-Wolff Roth's "Hot Cognition" study also highlights person-specific mathematical modeling. Considering the fuller human experience wrapped around the cognitive systems needed to solve problems dimensions is unorthodox. Through Roth's prosodic and text analysis and observation, however, it becomes clear that cognitive processes of mathematical modeling behavior are mediated by personal aspirations, identity, motivation, and affective states.

Shaffer describes one critical norm that the Oxford Studio's architectural design community seeks to impart—to absorb critique and then engage in reflective judgment of design iterations—as a means of highly personalizing professional practice. By analyzing how the same process control requirements are conceptualized and acted upon by different players, team leaders, managers, and statisticians in a factory, Hoyles and Noss stress how layered models of the same phenomenon reflect different roles and conceptual systems of individuals who hold those models. Goldstein and Hall emphasize how broad consequence models are almost never self-contained mathematical puzzles with single authors and docile readers. What is in/out of a model is an organizational question as much as a question for individual cognition. Therefore, model negotiation across organizational boundaries is an increasingly common scientific and technical work form. In other words, each author, in a different way, stresses mathematical modeling and problem solving beyond school is both person-specific and socially shaped.

PREPARING FOR MATHEMATICAL SUCCESS BEYOND SCHOOL

There is no shortage among policy-makers' calls for mathematics education to better prepare learners for the workplace and life in a complex and technologically charged society. Yet despite heated controversies about mathematics education over the past decades and impressive curriculum improvement efforts, classroom practice in mathematics is almost intransigently consistent and resistant to change (Hiebert, 1999) while the world around mathematics education is changing rapidly. Given that mathematics classrooms have proven highly resistant to change for decades and society and workplaces are changing rapidly, it is not surprising school mathematics and mathematics needed beyond school are diverging rapidly.

It is beyond the scope of this short introduction to suggest comprehensive alternative curriculum practices that might legitimately contend for the trifecta, for example, of satisfying proponents of both sides of the so-called math wars of the past 10 years and better preparing for success in mathematics outside of school. Yet the chapters advance an intriguing possibility. They suggest when mathematical problem solving is vital or meaningful or carries high stakes, as is the case when done "in the wild," it cannot be separated from the fuller enabling dimensions of affect, identity, and social structures. Mathematical cognition does not separate well from the objects, operations, or relationship contexts it can model. Of course, formal mathematics can be and is separated from context, but translation to mathematical abstractions loses important meanings and context, as Núñez argues.

When abstractions rather than mathematics context drive problem solving, the context dynamics are necessarily lost. In each setting, whether the goal is assuring a building's structural integrity, reaching a three part per million factory error rate, designing a living space, or raising a healthy coho stock, the context and its meaning are far more important to the actors than whether they possess the mathematical skills to reach the goal. It is through the unique, powerful human nature of adaptivity that actors develop, invent, or otherwise acquire the mathematics needed to reach problem-solving goals. Quite to the contrary of frantic newspaper polemicists who warn that letting youngsters invent mathematical ideas encourages "fuzzy" or "pseudo-math," mathematics is quite safe from subversive danger. When humans invent mathematical constructs or systems to model real-world problems, they cannot create false systems that will withstand testing. That is part of mathematics' enduring power and mystique. It must also be noted, though, that when individuals develop their own mathematics to solve a problem, they do not necessarily create fully stable or comprehensive systems. Instead, they create for the needs at hand. This requires responding to real-world constraints, working and re-working solutions involving mathematical adaptations, engaging others in complex reporting or collaborative relationships, and arriving at solutions that usually are one of countless possible outcomes that fit well, but not perfectly, their constraints. The paradox is when context rather than mathematical concepts drives modeling, mathematical modeling and adaptivity flex their muscles, and new concepts arise with great potency.

Of course, the dynamics of building mathematical understandings and adapting to real world needs are not new to twenty-first century mathematical problem-solving. But the emergence of pervasive and accessible computational and communication tools has dramatically altered virtually every sphere of life. That is, an increasingly complex society possessing rapidly evolving tools has created a fundamentally different context of mathematical thinking beyond school. And context matters. Modern context requires new adaptations and new ways of thinking about mathematics curriculum. These chapters suggest that in structuring the curriculum of the future, there is wisdom in re-thinking the traditional formulation of school mathematics as means to prepare for mathematical problem solving outside of school. In asking how schooling may best prepare youngsters for the workplace, we may actually have the school/workplace relationship incorrectly positioned, or at least miss the most salient contrast between school and workplace mathematics. Mathematical problem solving *beyond* schooling may give singular insight into preparing mathematical curriculum *for* schooling.

This chapter was supported in part by National Science Institution grant 04-33573. This support is gratefully acknowledged. The views in this chapter are those of the author and do not reflect those of the National Science Foundation.

REFERENCES

Hiebert, J. (1999). Relationships between research and the NCTM standards. *Journal for Research in Mathematics Education, 30,* 3–19.

Hoyles, C., Noss, R., Kent, P. & Guile, D. (2003). *Techno-mathematical literacies in the workplace.* UK Economic and Social Research Council (ESRC) grant L139-25-0119.

Kerns, S. (2005). *Innovation in engineering education.* On ASEE board of directors panel, "Highlighting the Engineer of 2020." (D. Giddens, moderator). Annual Meeting of the American Society for Engineering Education, Portland.

National Academy of Engineering. (2004). *The engineer of 2020: Visions of engineering in the new century.*

CHAPTER 1

The Meanings of Statistical Variation in the Context of Work

Celia Hoyles and Richard Noss
University of London

The changing nature of working practices and the ubiquity of computationally based systems has brought with it the need for many employees to develop understandings of the IT-based models that are increasingly part of their working practice. Our goal in this chapter is to sketch some of the categories of mathematical knowledge that characterize working practices in an era of ubiquitous technology, and to explore the meanings of this mathematical knowledge—specifically statistical variation—at the boundaries between the different communities of actors within them.

In his stark analysis of modern working practices, Reich (1991) classifies working practices into three types; symbolic analysts, who solve, identify and broker problems by manipulating symbols, *in-person* services (jobs based on interaction with people) and *routine* producers. His thesis is that only a relatively few workers will contribute materially to the knowledge economy, and he paints a picture of the kind of education that is appropriate for such workers:

> Symbolic analysts solve, identify, and broker problems by manipulating symbols. They simplify reality into abstract images that can be rearranged, juggled, experimented with, communicated to other specialists, and then, eventually, transformed back into reality (Reich, 1991).

We leave aside Reich's broader economic concerns here, as they fall squarely outside of our expertise. Reich's thesis is, however, broadly convergent with other analyses (see, for example, Zuboff, 1988). The implications of this thesis for the kinds of knowledge required by operatives within such workplaces are relatively clear-cut. A classic case is reported by Raizen (1994) who cites Lesgold's work with USAF technicians: the more highly skilled mechanics differed from their less skilled colleagues, chiefly in their ability to hold better "mental models of the systems they were working with (how components work, what their functions are, and how they relate to the system as a whole)…" (p. 73).

The key point about symbolic analysts is that they model the processes of work into abstractions that can be communicated to other specialists. Whether or not Reich is right in the detail, his central contention that it is the manipulation of symbols rather than things is beyond doubt. The question, however, is not only to understand what the symbol-analyzers need to know to develop useful symbolic representations of the workplace, but what the implications are for those who 'consume' the results of the symbol-analyzers. In particular, in the years that have elapsed since Reich proposed his analysis, the relationships and intersections between different layers of the workforce have shifted in significant ways. Part of our agenda in this chapter is to seek to reshape our understandings of who needs to know what, and the ways in which the knowledge takes on a different character in relation to the sub-communities of practice within workplaces.

THE INVISIBILITY OF KNOWLEDGE AT WORK

First we consider a major epistemological and methodological difficulty, which concerns the *invisibility* of knowledge at work. From a socio-cultural point of view, the invisibility of knowledge required in modern work has been commented on by many authors (see, for example, Nardi and Engeström, 1999). From this perspective, invisibility is seen to apply to entire swathes of working practices, such as cleaning or maintenance, rendering the individuals and communities concerned unnoticed and ignored, their place in the chain of work-related activity disregarded. For example, Bishop (1999) argues that distinguishing between visible and invisible work affords an understanding of the ways employment relations are emerging in post-industrial societies, and calls into question a set of implicit assumptions that all invisible work 'should be made visible'. From a rather different orientation, Star and Strauss (1999) consider instances that exhibit a range of indicators of what counts as work and how it appears in different situations. These include the creation of a 'non-person' (in which the employee but not the work is invisible), taken-for-granted work that they characterize as 'disembedded background work' in which the workers are visible, but their work is not, and the abstraction and manipulation of indicators—a category that comes closer to our own concerns, in which both work and people come to be defined as invisible. This last class emerges in two cases:

i. Formal and quantitative indicators of work are abstracted away from the work settings, and become the basis for resource allocation and decision-making. When productivity is quantified through a series of indirect indicators, for example, the legitimacy of work may rest with the manipulation of those indicators by those who never see the work situation first hand.

ii. The products of work are commodities purchased at a distance from the setting of work. Both the work and the workers are invisible to the consumers, who nonetheless passively contribute to their silencing and continuing invisibility (Star & Strauss, 1999).

While questions of legitimacy and silencing are important and challenging facets of the invisibility of work, our concerns are primarily with the visibility of *knowledge* held by persons or communities (rather than of persons themselves or the products of their labor). Knowledge, of course, does not come out of nowhere: it is embedded in people, in contexts, and in situations. Nevertheless, there are, we believe, important aspects of the problem that cannot be captured without a corresponding focus on the ways that knowledge in general and mathematical knowledge in particular—not just people—has been transformed in the modern workplace.

Concerns with the ways that mathematical knowledge is used in workplaces predates the demographic and organizational shifts engendered by technology and has tended to pay more attention to the visibility or invisibility of the knowledge, in contrast to research in the socio-cultural tradition, as mentioned above (for an overview of research in mainly non-technical settings, see Bessot & Ridgeway, 2000). Mathematical knowledge is judged as invisible as it tends to be deeply embedded within the representational infrastructures of the models, tools and artifacts of the workplace. Our own work with bank employees, nurses and engineers has evidenced this invisibility and has shown how mathematical knowledge in use is characterized by fragmented and pragmatic strategies intertwined with meanings of the mathematical knowledge and situational "noise"; and that mathematical knowledge is transformed when it crosses boundaries between different situations or settings (Hoyles & Noss, 1996; Hoyles et al., 2001; Noss et al., 2002; Kent & Noss, 2002; for a summary see Noss, 2002). Others have reported similar outcomes following research in technical workplaces, such as the automotive industry (Smith & Douglas, 1997; structural engineering, Hall & Stevens, 1995, and Hall, 1999; and industrial chemistry laboratories, Wake & Williams, 2001, and Williams & Wake, 2002). It must also be noted that the invisibility of mathematical knowledge is compounded by the general perception that mathematical and technological competencies consist of sets of decontextualized skills or techniques, disconnected from each other, and from their context of application (see, e.g., Smith & Douglas, 1997; Clayton, 1999), an issue that may be generational differences in interpretation (see Zevenbergen, 2004).

All these studies have had to face the methodological challenge of making visible the embedded mathematics of the practice in order to study and analyze it. Most have undertaken ethnographies, often focusing on "disruptions" in the routines of work or on communication across different representational infrastructures (Hall, 1999; Noss et al., 2002). Yet despite its invisibility, Hall et al. (2002) have argued that what makes subject-matter knowledge systems powerful is also what makes them difficult to study; it is powerful knowledge that is both widely distributed and massively influential in shaping the ways people interpret their working activities, while at the same time, profoundly embedded in the representational infrastructures that permeate working practices. From a methodological point of view, Hall et al.—like us—regard enhancing the visibility of subject-matter knowledge as a crucial analytical challenge.

We therefore now provide a brief overview of the methods used in our previous and ongoing studies, which we use to provide the illustrative empirical data in this chapter.

METHODS

In each new study of a particular work sector or factory, we have established a modus operandi that consists of five main components: interviews (which may be telephone, audio-recorded) with site or technical managers, site visits (involving work shadowing, and impromptu interviews where appropriate), iterative analyses (ongoing throughout the site visits), cross-factor analysis to draw out common components, and validation of analyses (with stakeholders within the work sector). Initially, a list of companies in a particular sector of work is drawn up following consultation with and advice from the relevant professional organization, from which a sample for case study is selected by the project team to represent as far as possible the spread within the sector and geographical locations. It is worth noting that there are inevitably some problems in obtaining and retaining access to companies. Each case study then comprises an initial semi-structured telephone interview with a key person in the company, following a set of agreed questions. The aim of the interview is to obtain a broad picture of the work and the range and profile of the employees in terms of skills and expertise, but most crucially to try to identify one or more people for observation and interview on subsequent site visits. Such people might be the Technical Manager or somebody who is in charge of several teams of people. Once identified, data from the telephone interview provide information on the job title of the identified person, what work is involved and where he/she fits into the company structure.

Following the interview, site visits are arranged. These are similarly structured according to sets of pre-defined questions, but with the flexibility to follow up interesting avenues for further investigation, including interviews where possible. This is particularly important as companies give different titles to jobs that might appear the same to the outsider (team leader, shift manager, group leader, process technician), and describe what we might see as mathematical processes in a different language from that of mathematical discourse. After general orientation in the company, we seek to tease out through questions and observations (some gleaned from work-shadowing) how mathematical knowledge might play a role in the workplace and specifically, how artifacts are used and how they might mediate the knowledge required. If possible, we try to document the extent to which employees are expected to display mathematical problem solving skills, for example, appreciate links between disparate data sources or representations, base decisions on underlying structures, models or logical analyses alongside immediate concerns. We similarly probe for specific examples of when a 'skill' might be needed. In this endeavor we often use critical incident techniques and ask for a description in as much detail as possible of an occasion when something critical happened, something did not work out, or there was a dispute as to appropriate action, how and why it arose, and how it was resolved. We paid particular attention to the roles of artifacts, especially computational objects and systems.

Case studies are then drawn up, and provisional versions sent to each company to check for accuracy and to safeguard confidentiality. The next step is to draw out common themes or issues that span case studies from one sector, and could be seen as 'characteristic' of a typical type of work in the sector. Finally, we attempt to involve professional associations in organizing presentations of research findings to appropriate employers and sector representatives allowing for wider validation and if necessary, modification. We are sensitive to the fact that the case studies do not represent a random sample, either of a sector or even of a type of business within a sector. These validation exercises help us see limits to the generalization of our findings as well as assisting us in developing a language to communicate to the relevant work communities.

We are well aware that asking managers to talk about the knowledge requirements of their employees' work is not at all the same as asking the employees, observing the employees at work, or studying longitudinally the activity system and the employees' actions within it. Nonetheless, this is always the first step of our studies and we will present many extracts from interviews with managers. We ask the reader to bear in mind our objective: it is to elaborate our theoretical framework, derived from a substantial corpus of data emanating from several studies in a variety of workplaces—and we use these extracts as illustrations of the theory as it materializes. Viewed in this light, the way that managers describe their knowledge requirements for example does, we think, provide an illustrative backdrop for the elaboration of our emergent theoretical position, by contrasting, for example, their perspective with that of other workers.

INFORMATION TECHNOLOGY AND THE MODERN WORKPLACE

In many manufacturing companies in modern workplaces, huge amounts of data are gathered daily to measure different aspects of the work process. For example, in the manufacturing sectors considered in the Mathematics Skills in the Workplace project (Hoyles, Wolf, Kent, & Molyneux-Hodgeson, 2002), it was reported that the companies surveyed were hugely concerned with improving efficiency and production, and data were gathered and manipulated to provide (potentially) an evidence base for action planning and decision making. Although company practices varied (with the larger companies proceeding more 'scientifically'), it was found that a common feature across sectors was the expectation that most employees, regardless of level, were to be involved in the process of increasing productivity or producing efficiency gains. Therefore the study concluded that all (or nearly all) employees needed relevant and interrelated mathematical and IT competencies.

With increasing emphasis on process control and improvement, and the need to deal with more complex contractual arrangements with customers, data—often represented on a spreadsheet (see, e.g. Figure 1–1)—were used to monitor and analyze a factory's performance at different levels (e.g., on single production lines, or divisions of the factory) and over different timescales (ranging from dealing with day-to-day technical problems, to long-term planning of production improvements up to several years ahead). Two points are worth making at the outset. First, the columns of the spreadsheet represent the variables of the process, some of which are "Key Performance Indicators," KPIs, such as downtime for machine (that has to be minimized), or number of products per hour

crewed hrs	overtime hrs	total hrs	make ready hrs	no. make readies	run hrs	lost time hrs	no. good sheets	av. make ready (min)	av. run speed	% lost time	av. lot size	sheets per manned hour
4302	0	4302	2,444.80	1,276.00	1,076.30	780.60	10,816,550	114.96	10,050	18%	8,477	2514
3930	0	3930	2,270.50	1,295.00	858.60	800.90	8,518,020	105.20	9,921	20%	6,578	2167
4342	0	4342	2,409.10	1,501.00	1,085.70	847.50	10,593,195	96.30	9,757	20%	7,057	2440
4479	0	4479	2,774.50	1,767.00	807.70	897.00	7,692,195	94.21	9,524	20%	4,353	1717
4407	0	4407	2,945.10	2,053.00	724.70	737.40	7,319,290	86.07	10,100	17%	3,565	1661
2722	0	2722	1,892.50	1,048.00	345.50	484.00	2,327,476	108.35	6,737	18%	2,221	855
24182	0	24182	14,736.50	8,940.00	4,898.50	4,547.40	47,266,726	98.50	9,649	19%	5,287	1955
	0%		0.6		20%							

crewed hrs	overtime hrs	total hrs	make ready hrs	no. make readies	run hrs	lost time hrs	no. good sheets	av. make ready (min)	av. run speed	% lost time	av. lot size	sheets per manned hour
784	0	2899	1174	1,792	1,154	442	7,116,944	39.31	6,167	15%	3,972	2455
856	0	3018	1214	1,862	1,179	624	6,967,254	39.12	5,909	21%	3,742	2309
898	0	3414	1437	2,035	1,481	496	8,539,228	42.37	5,766	15%	4,196	2501
880	0	2847	1244	1,082	1,065	539	5,562,340	68.98	5,223	19%	5,141	1954
602	0	1948	769	721	816	363	4,329,169	63.99	5,305	19%	6,004	2222
1005	0	3763	982	1,356	1,959	732	10,859,334	43.45	5,543	19%	8,008	2886
5025	0	17889	6820	8,848	7,654	3,196	43,374,269	46.25	5,667	18%	4,902	2437
	0%		38%		43%							

crewed hrs	overtime hrs	total hrs	make ready hrs	no. make readies	run hrs	lost time hrs	no. good cartons	av. make ready (min)	av. run speed	% lost time	av. lot size	cartons per manned hour
621	0	2608	738	1,521	1,595	226	38,319,552	29.11	24,025	9%	25,194	14693
550	0	2220	660	1,323	1,205	355	33,494,313	29.93	27,796	16%	25,324	15088
551	0	2154	719	1,511	1,169	265	43,375,005	28.55	37,104	12%	28,706	20137
552	0	1540	461	735	759	260	18,945,932	37.63	24,962	17%	25,777	12303
553	0	2744	709	650	1,817	219	14,293,112	65.45	7,866	8%	21,989	5209
522	0	2203	653	1,409	1,177	374	31,757,755	27.81	26,982	17%	22,539	14416
617	0	2837	256	232	2,181	401	35,419,915	66.21	16,240	14%	152,672	12485
701	0	2567	342	755	1,753	472	116,383,850	27.18	66,391	18%	152,151	45338
652	0	2603	524	925	1,636	443	100,011,258	33.99	61,132	17%	108,120	38422
5319	0	21476	5062	9,061	13,292	3,015	432,000,692	33.52	32,505	15%	47,679	20117
	0%		24%		62%							

Av. 60,000; Total no. of Av. batch size; No of good sheets; No. of job; No. cartons; No. of jobs; Having a *job* average; Productivity per manned hour; Overall index.

FIGURE 1–1. Spreadsheet of process data.

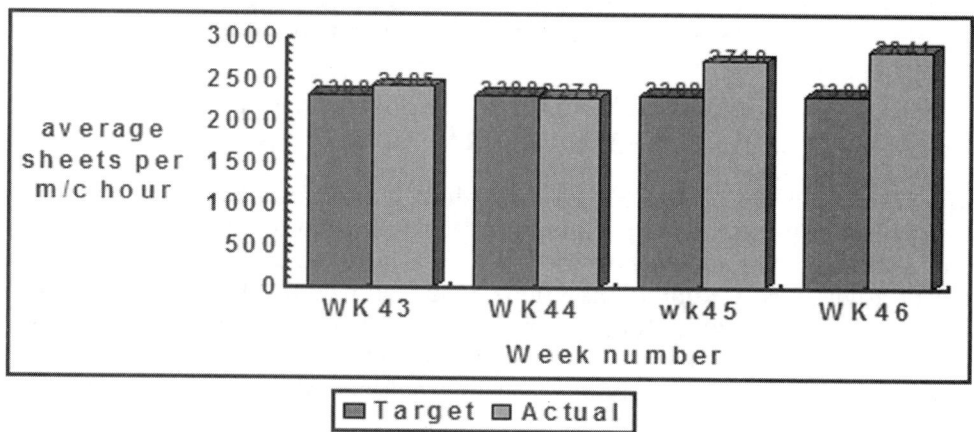

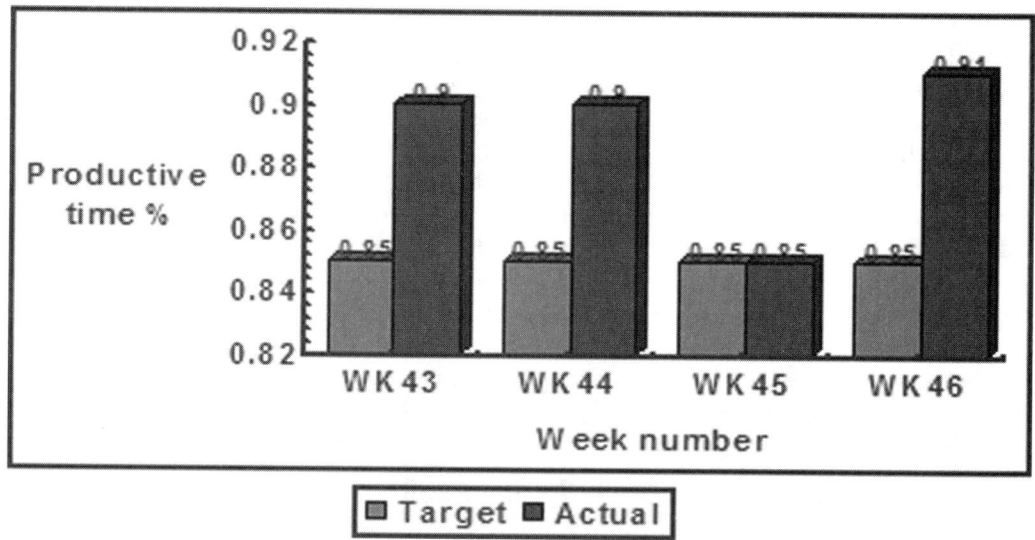

FIGURE 1–2. Bar charts derived from process data.

(which, of course, normally has to be maximized). A second point is that these KPIs are seldom independent and many are connected by a specified (but usually opaque) mathematical formula.

In the Hoyles et al. (2002) study, team leaders were identified as playing a pivotal role in the operations of the factory, "reading in" information to the abstract model (such as that underlying the spreadsheet) for upward communication to management, and "reading out" information from the model for dissemination to operational workers. To fulfill the latter purpose, data related to KPIs were often displayed on factory floors, in the form of spreadsheet-derived graphs, which also depicted output related to given targets (see Figure 1–2). Team leaders also often had to identify poor performance, and recognize anomalies and erroneous input data.

At the same time, managers were increasingly required to take decisions based on output from these quantitative models and to extrapolate trends, while accounting for other, less-easily-quantifiable information, such as the production capacity of the work force and the current "industry climate" (see Hoyles et al., 2002). Apart from the vertical metaphors of upward and downward communication, such information also had to be channeled *horizontally* within a project team to communicate among people with different skills, and *outwards* to customers.

We now proceed to elaborate the new literacies required in the workplace, by presenting some examples using some of the case study extracts that appeared in Hoyles et al. (2002). We begin in the company within which the spreadsheet shown earlier was used. It is reasonably typical: a large packaging company employing about 220 persons that made printed cartons on a contract basis. Contracts could be stand-alone but increasingly there were multinational contracts for corporate clients covering many countries and different types of cartons. Cartons are printed, cut out and creased on very high-speed, computerized printing presses. Typically, carton design is done by the client, but the company has to work out (nowadays, using computer software) how optimally to 'fill' the individual carton designs onto the raw cardboard sheet. Cartons leave the factory as flat sheets, which are then formed into box cartons at the client's packing plants.

First, it was reported that a change had occurred in that key managers of information in many companies were now the team leaders who supervised shop-floor operations, a crucial change instigated by the introduction of IT-based technology. Thus team leaders needed to manage and communicate information much more than in the past. A senior manager told us:

M1: We're data-driven much of the time, and it is clearly our strategy to push a data-driven approach right down through the organization. Now, team leaders have to come and present to me a lot of analytical data about what happened in their shift. Five years ago it didn't happen...

We asked M1 if he could describe an employee who had found it difficult to cope with this complexity in the light of the introduction of massive computerization:

M1: We had somebody, X, who had been in the section a long, long time. He could not manage the six machines, the complexity of it.... He found it very difficult to be able to juggle the dynamics of what was happening.... He could do the mechanics of it because he knew how to do that...but he was not able to interpret it, to transform it into a trend or a problem solving analysis. He wasn't able to really use the information in a well-constructed argument. He could not present an argument which was fact-based, by using the numbers and the information that we do collect...not really understanding what that data means or how it impacts on the business.

The second change was in the marketplace that was more competitive, with profit margins very much leaner than they used to be, and clients demanding higher quality and more complex contractual arrangements. As M1 commented:

M1: If you look back 10 years, the variability in our products would have been huge. And the customers didn't care so much, because their supply chain was not that sophisticated, and waste was not very visible; the drive for profit and cost reduction was nothing like what it is now.

Managers thus needed to determine realistic performance criteria that formed part of complex models of production, including for example, product output as a function of 'man-hours', productive machine time, and relative costs of storage, design and transport. They then needed to use these data to assess performance against criteria and targets. Fluent manipulation of variables, some of which were quite invisible, was therefore crucial. Thus management needed to see data both as a concrete representation of what the company produced, and, at the same time, as an abstract representation that identified the implications of the data in terms of general trends that had both historical and predictive significance. As part of this process, they needed to be able to communicate complex information from interrelated data sets, spot errors and troubleshoot, as illustrated by a second manager we interviewed in another manufacturing company:

M2: We record everything on a log and it actually tends to come to me as well as everybody else and we keep our own Excel spreadsheets which are followed over time and it is obvious if say you have a bad measurement.

"It just leaps out at you. Because obviously the real problem is that if you've got bad measurements and you don't realize that there is something funny about the numbers then you go to the next stage in the analysis. Well the software doesn't know it is a bad measurement so it just throws it in."

So how might we describe what team leaders and managers must 'do', and what, for example, X could not do? It is not helpful to describe this purely in terms of 'competences', of say, understanding variables and trends. This does focus on knowledge, but misses a crucial aspect of what it means to use knowledge *in situ*, where any understandings are necessarily articulated within the workplace discourse, and supported by contextual cues within it. The knowledge mobilized in use is not codified knowledge (Eraut, 2004), in the sense that it is readily available in published sources. It is knowledge that is mobilized in the practices of working cultures, and its realization in activities is achieved in intimate connection with practices and artifacts of the workplace, specifically with the technologies that perform and monitor the work process.

Given the need to characterize the kinds of knowledge involved, we have coined the description *techno-mathematical* knowledge, to highlight the fusion of IT and mathematics. It is obvious that there is huge variation in the depth of this techno-mathematical knowledge required by different elements within and between working communities. From the perspective of Reich's symbol analysts, the kinds of knowledge required are strongly associated with academic knowledge taught, for example, in university departments of mathematics or theoretical physics. It would certainly be worthwhile to explore the extent to which these curricula are actually aligned with the requirements of modern working practices: but this is not our concern here. Instead, we focus on individuals and communities within working practices that are relatively uneducated in these domains, yet who have to work with their products. For this reason, we focus not on techno-mathematical knowledge *per se*, but on techno-mathematical *literacies*, what needs to be appreciated, interpreted, and communicated horizontally and vertically within working communities. The term 'literacies' is appropriate, in that the kind of knowledge required is analogous to that required in relation to reading and writing—not to write books but to read them, not to construct literature but to talk about it, and think about its meanings. *Techno-mathematical Literacies* (TmL) are technically-orientated functional mathematical knowledge, grounded in the context of specific work situations and employed as devices for making sense of the systems that pervade and operationalize them. Part of the research endeavor in our current project, *Techno-mathematical Literacies in the Workplace*[1] (henceforward referred to as the *TmL Project*) is to characterize the different meanings associated with TmL as articulated by the different communities who use them in a variety of working practices.

CONCEPTUALIZING TECHNO-MATHEMATICAL KNOWLEDGE: BOUNDARY OBJECTS AND SITUATED ABSTRACTION

We have argued that the work process has increasingly become mathematized through models instantiated in different technical artifacts, with the outcome that huge amounts of quantitative data have to be interpreted as a basis for action and decision-making. The question then arises as to what happens to individual and collective knowledge under these conditions and how it is mediated by the technology. This in turn requires a more analytical description of the knowledge, who uses it, and for what purposes.

We suggest, following on from the ideas of 'disruption' mentioned earlier, that one method to obtain insight into how TmL is used at work, is to consider how the relevant knowledge is perceived across *boundaries* within the workplace, as it is at these boundaries that ideas may be contested, disruptions may occur, and different languages of description have to be aligned. We conjecture that there are at least two major trajectories of what it means to cross boundaries; the first concerns people and the second objects.

The first phenomenon, which we will call *devolution*, describes how many individuals need now to understand aspects of the system that previously would not need to have been understood by an individual within that community (it would doubtlessly have had to be understood by *someone* even if they were never physically part of any community within the workplace). This is, perhaps, a surprising spin-off of computerization, given that the initial impetus for the introduction of digital technologies was often unashamedly that of deskilling the workforce both quantitatively and qualitatively. Yet, as one manager in a packaging plant recently commented to us: "We replaced six operatives by two, but then we found out that the two remaining people needed to understand a lot more than before!"

The second trajectory, *communication*, names the way in which people need to share knowledge with others across boundaries of space and time. This *horizontal* dimension of knowledge—and therefore learning—has been theorized by Engeström (1996), who argues that a major challenge for individuals and social groups is learning how to cross the social and cultural borders between different activity systems (see Guile & Young, 2003). For Engeström, these 'boundary crossings' point to a *horizontal development* that is to be seen as radically different from the traditional *vertical* notion of mastering a skill or a hierarchical body of knowledge. The second aspect is thus about objects that populate the boundaries and facilitate communication across it, and it is these objects that we now discuss.

---

[1] The research project "Techno-mathematical Literacies in the Workplace" is funded by the UK Economic and Social Research Council as part of the Teaching and Learning Research Programme [www.tlrp.org], Award Number L139-25-0119. Some of the preliminary work reported later in this chapter was conducted in collaboration with Phillip Kent, whose contribution we wish to acknowledge.

Objects at the Boundary: Windows on Devolution and Communication

A boundary object (Star, 1989; Star & Griesemer, 1989), names an important class of knowledge artifact shared between different communities of practice and can be used differently by the communities.

> Boundary objects are objects that are both plastic enough to adapt to local needs and constraints of the several parties employing them, yet robust enough to maintain a common identity across sites.... Like the blackboard, a boundary object 'sits in the middle' of a group of actors with divergent viewpoints (Star, 1989).

Thus boundary objects can be shared between different levels of an activity system or between different activity systems. Each community might view the boundary object differently, talk about it in distinct ways, express its functions diversely or see different parts of the system in it, but nonetheless the boundary object can provide a means to think and talk about an idea and negotiate its meaning. Boundary objects might, for example, comprise a symbolic map of the work process thus opening it up for discussion (devolution), or sit at the intersection between team leaders and managers (communication).

Our assumption is that in the modern workplace, many boundary objects include symbolic representations of mathematical relationships and thus make it possible for them to be interpreted mathematically. Although representations create a layer of invisibility with respect to the mathematical knowledge underpinning them, they simultaneously—and contradictorily—afford a common visible framework that allows diverse communities to act and think as if they had a common purpose. We suggest that this common symbolic discourse can be a starting point for negotiation of meanings. For example, the workplaces described earlier were rich in mathematical representations of process, abstractions of the systems that characterized production, portrayed through spreadsheets or graphical charts. If we consider these representations as forming elements of boundary objects, the techno-mathematical knowledge required to reason with them depends on three factors: first on the way they are represented, whether as lists of figures (as in a spreadsheet) or as a dynamically-represented graph; second, on the uses to which they are being put in the workplace; and third on the experience of the individual and the extent to which appropriate mathematical knowledge might be mobilised. For example, if we consider the spreadsheet illustrated earlier, we know that some columns of data are related mathematically to others. This implies that the variables in the dependent column cannot be manipulated independently of their parent. While most employees would not be required to identify the mathematicalrelationship involved—and certainly would never be required to express it algebraically—they *may well* be required to understand the basic shape of the relationship (is it direct or inverse?), the effect of a change in one variable on others, and the identification of any anomalies on the basis of the underlying relationship—but of course only in so far as these understandings are meaningful in the workplace and not simply as abstract entities.

Individual and Collective Knowledge at the Boundary

From an activity-theoretic perspective, boundary objects facilitate the achievement of an object (goal) in the face of competing interests from different communities (Bowker & Star, 1999), and in contrast to our approach, the knowledge within each community and the way it is recontextualized is not a central concern. Similarly in using activity theory in the context of employee training, knowledge is not explicitly considered: as Guile and Young (2003) put it, "while activity theory stresses how activities are needed that enable students to re-locate what they already know as a step to acquiring new knowledge, it has little to say about the form or content of this knowledge."

For our research agenda, a major interest is precisely the study of the meanings attached to boundary objects, which will include the meanings associated with any mathematical representations. It is worth noting that these representations may not be expressed using the official symbols of mathematics, but rather with *ad hoc* representations of some specific aspect of factory production. Thus we need a language that catches what is simultaneously situated and abstract about the mathematical knowledge involved. We employ the idea of *situated abstraction* to pay equal attention to the notion of knowledge and to the language in which it is expressed: a situated abstraction is the expression of a mathematical abstraction framed by the artifacts and the discourse of a community. Any mathematical models involved may be different from those that are conventionally understood: not necessarily mediated by formal mathematical symbol systems and artifacts, but rather by 'situated' techno-mathematical artifacts (see Hoyles & Noss, 1992; Noss & Hoyles, 1996). Thus we would expect workers using technological tools as part of their practice, tools that mediate the relationships that structure production, to express and communicate their knowledge in ways shaped by these tools.

Situated abstractions are the means by which mathematical ideas can be communicated to self and to others—without necessarily having recourse to conventional mathematical symbolism. Such representations are not to be conceived as an optional extra, an illustrative form that embellishes or gives voice to an already-formed object of idea: on the contrary, we take it as read that artifacts and symbol systems are *constitutive* of meaning and—by implication—of thought (see Sfard, 2001, for an insightful perspective on this question). Thus, the meanings that people hold of mathematical ideas are framed by the representational infrastructure with which they are expressed. So for the spreadsheet shown earlier, the column of numbers stands for a variable in a process which itself stands for a rate, a quantity or even a relationship. Moreover, the meaning of the mathematics is inextricably connected to work process activity. For example, a variable which 'represents' the runtime of a machine, is not to be thought of just as 'time', but understood in relation to the actual machine, its history and its idiosyncrasies.

The traditional view is that workplace knowledge is pragmatic in character, and that there is a natural antagonism between pragmatic and theoretical knowledge, in terms of its purposes and its forms of representation. The notion of situated abstraction calls into question this distinction, suggesting that the seeds of the theoretical are present in the pragmatic, especially in situations where the tools and artifacts involve symbolically represented objects.

In order to give shape to these ideas, we now elaborate our example of TmL using some extracts taken from our ongoing work in the TmL project, derived from case studies of factories involved in mass production and process improvement.

The Meanings of Variation at the Boundaries

A significant number of companies across the sectors studied in Hoyles et al. (2002) were beginning to invest in operating methodologies concerned with quality and continuous improvement. These methodologies go beyond 'conventional' efficiency improvements of the type described earlier. They seek through a strict testing schedule based on statistical theory to establish higher absolute levels of quality, with smaller margins of error and tolerances. In the past, any such testing regime was both designed and interpreted by senior managers and engineers/technicians, with team leaders simply being required to collect information and report problems to more senior colleagues. This trend is only a statistical edge of the current we outlined earlier, but it has a very clear connection with a body of explicitly mathematical theory, and we are concerned to understand just how much of this—if any—is relevant, in what form, and for whom. So the specific example of TmL we now consider in more detail is that which is called into operation in companies that are striving to use a statistical process control (SPC) methodology such as "Six Sigma."[2]

In one such company, a pharmaceutical company, we were told in an interview with a manager that the main task was to reduce variability. Using as an example the target weight of a tablet, he explained:

M3: The whole ethos of manufacturing is to eliminate variability, then you won't have defects and your customers will be happy. When you have poor process control, you have to compensate for it, maybe by increasing the average weight of material in the product (in this case the tablet) [*so the normal distribution lies always sufficiently above the lower specification limit—CH/RN*], but if you can reduce the sigma, you can use less material and make the product more cheaply. So, all our efforts involve asking 'why is the variability like this?', 'how can we close the gap?'

M3 stressed that all layers of the workforce, not just managers, must become SPC-literate and this meant improving understanding of *variation*, since it is this understanding that is thought to play a crucial role in improving practices at work. As Joiner puts it in his book on Fourth Generation Management:

> Our ability to produce rapid, sustained improvement is tied directly to our ability to understand and interpret variation. Until we know how to react to variation, any actions we take are almost as likely to make things worse, or to have no effect at all, as they are to make things better (Joiner, 1994).

But what does 'understanding variation' in the context of work mean? Wild and Pfannkuch (1999) have argued that: "variation is omnipresent; variation can have serious practical consequences; and *statistics* give us a means of understanding a variation-beset world (our emphasis)." But can people without any background in or knowledge of statistical thinking gain understanding of variation from experiences in their workplace insofar as they are required to characterize this variation, quantify it and seek to reduce it?

Konold and Pollatsek (2002) suggest that understanding data means appreciating the existence of a 'signal in noisy processes', a stable value, the central tendency of a data set, with (inevitable) variation. They argue that "implicit in our description of central tendency is the idea that even as one speaks of some stable component, one acknowledges the fundamental variability inherent in that process and thus its probabilistic nature". Because of this, they claim that "the notion of an average understood as a central tendency is inseparable from the notion of spread" and that "average and variability are inseparable concepts". If, as Konold and Pollatsek

---

[2]Six Sigma seeks to control manufacturing processes so tightly that only three products per million will be defective; that is, in the distribution of any measure of the product, there are three standard deviations (sigmas) either side of the mean value that fall within the quality specification limits.

suggest, rather little attention is paid to these ideas in school, is it possible that workers can appreciate them, let alone come to use them effectively?

We seek to illuminate these issues by presenting a further series of extracts that start with the notion of a boundary object, and follow with descriptions of the meanings that different communities have expressed about these objects.

The SPC Chart as a Boundary Object

The first step factories take when moving to process control, is to select key performance indicators, KPIs, of productivity and efficiency, along with a set of standardized in-process control measures to assess them. KPIs are the output of a process, and the data used in SPC are the measures of these KPIs over time. What do workers need to know about data analysis in order to monitor performance? What are the meanings they assign, for example, to the central tendency of a particular KPI process and the variation that might be evident? Fundamentally important in SPC is the drive to control variation and central to achieving this control is to distinguish between common cause variation and signals or 'special cause' variation.[3] Common cause variation is ever present in any process or interconnected set of processes. Special causes arise for reasons *outside* the usual process. They can contribute either small or large amounts to the total variation, but typically have a much larger impact on variation than the common causes—and of course special causes affect the central tendency of the data set:

> The important point about common cause variation is, it is argued, that there is no single cause: it is the result of a set of interacting factors. So, to change common cause variation requires reconfiguring the entire system. Special cause variation, on the other hand, demands immediate investigation with—if possible—an immediate remedy. In order to reduce common cause variation, a much more subtle and in-depth appreciation of the process is required" (Joiner, 1994).

If an SPC-literate workforce is needed to implement process improvement, how do companies proceed? First, our case studies indicate that they seek to make the variation in a data set *visible* in the form of SPC charts, which are produced automatically for each KPI. SPC charts have three major features: data are plotted in time order; a centre line—typically the mathematical average or central tendency of the process (as characterized by Konold & Pollatsek, 2002)—is drawn; and finally statistical control limits are added that indicate the width of the common cause variation, the historical expectation of the normal variation. These upper and lower control limits are based on what the process is capable of doing, calculated by statistical formulae used on historical data sets. Depending on the nature of the data, other features are sometimes added, namely a target production figure and the upper and lower limits of a customer specification (the limits of quality that are acceptable for a particular product).

A simple example SPC chart is shown in Figure 1–3 showing the 'historical mean', the upper and lower control limits, the customer specification and the target line just above the mean. The chart shows one point outside the lower control limit that signals a special cause.

In the factories we visited that used SPC methodology, SPC charts could be seen on the shop floor and in management offices. They are familiar 'mathematical'

---

[3] Signal seems to be used in the context of work rather differently from Konold and Pollatsek (2002), here we use the term as we heard it used in the workplace.

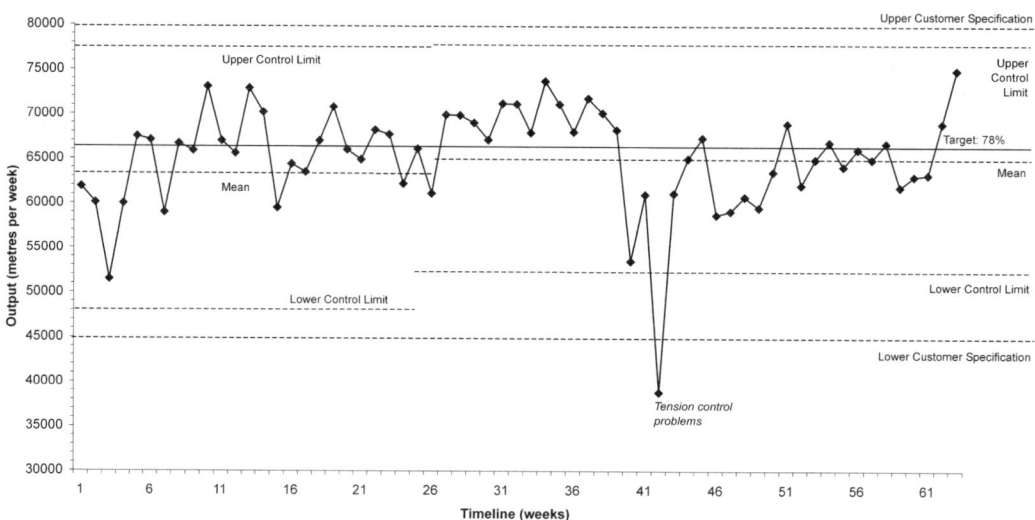

FIGURE 1–3. Illustrative SPC Chart.

objects, regarded as important and taken as the focus of discussion in regular team meetings. The data are meant to be transparent in that they are agreed measures of process and they are input by operators in the factory—although the chart itself and the control lines on it are produced automatically. An SPC chart can be thought of as a boundary object comprising a knowledge artifact based on a mathematical and statistical representation, which might be interpreted differently depending on one's position in the activity structure or degree of mathematical knowledge. What are the meanings that different groups of workers in the factory have of these charts? How far are they aligned with each other and with statistical interpretations? How are they mediated by the technological tools that create the charts? How do the charts provide insights into the factory process?

We present below some different descriptions of the meanings of the SPC methodology and of the SPC chart as related to us by key people in process improvement: first, a company statistician; second, by company managers; and finally, by team leaders working on the shop floor.

The Company Statistician

The statistician's (S) point of view sets out the rationale of SPC. His view, expressed in no uncertain terms, was that everybody working in the company needs to understand the meanings of variation and uncertainty in the context of SPC, which he argued could be achieved rather simply but rarely was evident:

S: What I'd like people to do, whether they're a new manager, a new scientist, a new operator or whatever, is to arrive here understanding variation and issues of uncertainty...there's no reason that I can see why everybody shouldn't arrive here with that knowledge. It's very simple knowledge. But hardly anybody does.

S argued that it was crucial that employees distinguished work outcomes along two dimensions: "what we *want* to get?" (the customer specification) and "what we *expect* to get?" (data varying between the two control limits). Taking effective action demanded that workers compared and acted upon actual results along *both* these dimensions with the desired state in any process being that the expected variation was within the variation set by the customer specification. At the same time no action should be taken when the data points were simply the result of random variation.

S went on to argue:

S: ...There is a great temptation for people to see a number that's lower or higher than the previous one and feel they have to make a change. Deciding this is not really evidence for change therefore I don't need to alter the process.... Anybody going into any industry should believe that this variation issue is important for them and it's important in all their decision-making.

So the first key production technique was not to act on changes that were simply the outcome of common cause variation, even if this meant the customer specification was not achieved:

S: ...A key part of using these charts is you need to know which strategy for improvement to use. A huge proportion of the world *investigates a point outside the customer specification line (a disappointment) and tries to improve* [italics added]. And that is very wasteful and inefficient.

The second technique was to investigate signals, deviations from expectations, explain them and then either correct problems designing them out of the process or planning to incorporate them into the process. In both cases, the aim must be to stabilize process as soon as possible. Reaction to signals was important but still more so was improving the process that is reducing variation. As S put it:

S: Variation has costs, and reduction in variation constitutes an improvement. Thus once the process is stable, it is important to start proactive improvement.

To improve, that is to ensure the control limits are within the customer specification *and* variation is reduced, requires understanding common cause variation, analyzing the multiple factors that give rise to it and then seeking ways to reduce it. As S explained:

S: You'd have to get much more involved in fully studying your process and understanding where that variation is coming from. You'd use all of the data points and lots of information, with expert people to facilitate the work. It could be anything to do with the variation in the raw materials. It could be variation in the environment, it could be differences in the machines you use, it could be differences in the way operators work, it could be differences in the set up of the machines, anything that could vary in the whole process not just in the machine. Any interaction between materials, methods, people, environment, any of those things will have to be thought of as part of the process.

To sum up, there is an unexpected complexity in understanding process control. Although some facets might be judged as "standard" statistics (the notions of mean and variation for example), other facets emerge in the interaction between artifacts and tools, routines, targets, ambient environment and working practices. Clearly statisticians have been engaged in teaching SPC techniques and have addressed some of the problems. Yet on the whole, statisticians have focused their efforts on managers and engineers. Many of this group are paying lip service to improvement methodologies without understanding what they are doing, according to Greenfield (1993), who argued that process improvement would therefore not happen: "without understanding they will surely fail as surely as they would if they had never heard of TQM (total quality management)" (p. 291). Greenfield has asserted that statisticians must make an effort to ensure their techniques are not obscured by unnecessary jargon. We are seeking to take a further step in this direction. In one way, our task may look simple, as the ideas underlying SPC are not hugely sophisticated. But from another perspective, our task is complex, as our audience might have rather limited mathematical—let alone statistical—background.

We now view the same issue of developing SPC literacy, this time from the perspective of company managers who have introduced SPC methodologies.

The Company Manager

The major impression we obtained from interviewing one company manager, familiar with the SPC methodology, was his rejection of the SPC chart as a mathematical object. M4 was adamant that there was little, if any need for anybody to understand the charts in a statistical way—he certainly did not and did not see that others needed to, since there were simple 'rules' that could be applied to read the charts:

> **M4:** With, for example SPCs, we'll pick on that because that's the normal format we're looking at, that obviously as you all know there's some statistics behind them—and that totally loses me never mind others. However in another sense SPCs are very simple and you only need to have a fair grasp...just a very small number of key principles. You don't need to understand all about standard deviations and how they control them, it's where they end up being where they are.
>
> So basically what we're providing people with is this format and the basic understanding that they need to know that there's this difference between common cause variation and special variation. If your data points appear outside of the limits you've got a big signal that you need to investigate. The concept that you need to be looking at...the width of your control limits, the gaps between control limits and one of the aims should be to reduce the variation, reduce the gap between the limits. So they're very easy principles to grasp, you don't need a great deal of mathematical training to understand this.

These ideas (repeated several times during the interview) might be an example of the invisibility of the mathematical underpinnings of the charts. But, is it possible that SPC methods could be successful if they were simply used, as M4 wished, 'like a recipe book', with little if any attention to 'why'? If so what would be the meaning assigned, for example, to the lines on the charts for the average and the upper and lower controls?

> **I:** Does nobody ever ask you by the way what those lines are or why they are way down there (referring to the mean and upper and lower control lines)?
>
> **M4:** There's always one or two. Mainly it is because [name] says it is!... Generally we are expecting people to use these charts to show them...outcomes. We use them very publicly every week in our weekly review meeting to show the teams how they are doing.

Despite being convinced that one did not need to understand the basis of the SPC charts in order for them to be used effectively, a position at odds with the statisticians' perspective, M4 was also convinced that the SPC methodology was not working well in the factory. SPC charts, he maintained, were simply being used to 'track the system', proved useful in displaying data, 'to show outcomes', but were not being exploited to improve the process:

M4: The thing is, somehow we're not utilizing SPC charts to their full potential... we're potentially not optimizing that process, the improvement opportunities on that process.

There's using and using them isn't there? They've been around for a long time, again to be perfectly honest I would say in a lot of ways we are only just starting to get to grips with really using them. In other words for a long while we've used them as a way of presenting information so instead of looking at them in accounting terms or as you say looking at a table of numbers which is dead boring and hard to read; it's much easier to look at profit development or cost development of whatever else on a SPC. But there's a whole different ball game actually using that information as a SPC is intended to be used.

I: For productivity?

M4: Exactly, rather than as a better way of presenting data than a table of numbers. And that's the challenge.

Yet despite this anxiety about the effectiveness of the way the company used SPC, M4 still adhered to his belief that failure to progress along the improvement agenda was not due to any lack of understanding of the 'why' underlying the charts. He and several members of the management team and been through a six-day training session and he complained that the effects had not been noticeable in terms of an upturn in productivity:

M4: I personally don't think the issue is in so much understanding the maths and statistics of whatever behind it. I think it's more to do with commitment and follow through on making continuing improvement happen. During that training we spent a lot of time with the graph paper out drawing SPCs ourselves, so rather than just going into Winchart and typing the data in and saying "hey presto!" and it all works for you we went right behind and did it all manually.... You would like to think having done six days' training and gone into some detail on this that a few weeks time, a few months time, the quality is going to improve, that people will just get on and do it. It doesn't happen.

M4 also did not appreciate any need for others 'lower down' in the factory to understand the charts and thought the weakness was solely in management:

M4: It's got to be supported by management...by the whole supporting infrastructure that enables it, whatever 'it' is that you actually want to happen.... So how do we make this next leap from not just looking at the data but actually using it for continuing improvement? I honestly don't believe that the big stumbling block is because people don't understand the number side of it. I think it's that in the past we've been failing to put enough pieces of this supporting jigsaw in place to make sure that continuing improvement actually happens.

Clearly the meanings of the charts in terms of their role in process improvement were very different as expressed by the statistician and the manager.[4] M4 was, however, beginning to explore other ways to present data, as a way to catalyze the next step in improvement:

---

[4] We intend to explore how these different views might be discussed in boundary crossing scenarios we plan to organize in our ongoing project.

M4: We have, I believe, been putting an over-reliance on SPCs. A while ago we said SPCs were the way to go and there's much better looking data with SPCs. What then happens is everything you look at now is on SPCs and it isn't always the best way to look at it. Especially if you are just wanting to use it to present data because people then start thinking it's just a tool to present data; it's not there to improve processes. Well they may just say it shows us how we are doing. Which is not bad but it's not really what you want because you want it to be a catalyst for action.... Well some parts of the industry use EXCEL spreadsheets actually. Maybe you need multiple ways of doing it?

M4 is correct in that many factories use spreadsheets to display data, as described earlier in this chapter. His suggestion of using multiple forms to display performance on KPIs is interesting, not least because to read data presented in different ways and to appreciate the commonality underlying the representations in terms of properties of a data set (the representations highlight or suppress different aspects) demands quite a sophisticated understanding of the data set and the way the tools mediate the information about it. In fact, Wild and Pfannkuch (1999) have argued that forming and changing data representations of aspects of a system is a fundamental part of a statistical approach to understanding a system more effectively.

M4's position thus seems a little contradictory, if we note that the kind of statistical insight he is advocating—multiple ways of visualizing the data—is 'mathematical', yet he has rejected the need for mathematical understanding. He sees there is a problem, otherwise he would not have spent time showing the operatives how to 'do it manually'; but he is ambivalent as to whether this is effective, or whether the utilization of SPC charts "to their full potential" involves a mathematical component or not.

We will return to this point in the conclusions. Now, we turn to some extracts from another manager, who was in charge of process improvement and thus perhaps more closely aligned with a statistical point of view. M5 was equally concerned that the SPC charts were not being used effectively, but in contrast to M4, was concerned about problems in understanding them:

M5: They (team leaders and operators) are used to seeing SPC control charts and understanding variation and not getting too carried away with it. They could probably give you a fairly brief overview of what it actually means. But whether that is 100 percent accurate or not I would doubt. And that's not for lack of trying because we have held countless workshops over the years. Certainly down to Team Leader level on things like SPC and things like...

So why did M5 see a role for understanding and not merely skill? He argued that it is only if readers of the charts 'engaged' with the data that they would be able to use them purposefully, to look again at them and question familiar procedures, as a first step in systemizing them:

M5: ...to make decisions, need to move from opinions based on what they *observe* hopefully, rather than basically their perception of what they think is happening. But it's just taking it a level beyond it, and taking it away from observations and putting it into fact, which data is—fact—unless you've measured it wrong.

I: And do you find there are some people more reluctant than others to do that?

M5: Definitely.

I: Is that because they don't understand it? (group looks at chart—several unclear comments)

M5: Yes to some extent: they need to know that that's the mean of that process [pointing to the average line in the SPC chart]. But what it tells us, it takes into account that variation as well because some weeks are better than others, that's a fact of life. So we have a control—there it is 4.1 percent. So what that tells us is that that process under normal common-cause variation is capable of producing up to 4.1 percent.

M5 also suggested that *"management* needed to take any data point down a level" (our emphasis), that is, identify the key input variables that created any output KPI, tease out the most crucial influential ones and reduce variation in them by introducing standardized procedures. It was only through these steps that process improvement could be achieved: that is by explicitly recognizing the multiple and interrelated factors that acted together to account for the common cause variation. He explained by reference to an analysis of flawed products:

M5: Yeah say for example, what you could do is you could probably break that down and say we keep 60 percent of our raw material on average for silicon-related problems, 20 percent because we get spots in the adhesive, 10 percent is for creasing in the backing paper, or whatever…so let's say mostly it's the silicon problems. So then what you can do is you can take that down a level and say OK well…what are the things that contribute towards silicon cure, for picking up, a silicon cure? And it might be the temperatures of the silicon ovens, obviously the running speed, it might be the modifier level in the silicon, etc., etc., etc. So then you can take like real process measures then, if you look at the process measures on this type of chart I would expect to see where you've got high degrees of problems in siliconizing you would expect to see a chart for some of these variables that are not in control. So these are the ones that you can then start to do things about. Let's put a standard operating procedure in place for setting up the silicon drying oven for this particular type of product and that should be what people are interested in on an operator level.

So what were M5's understandings of the lines drawn on the SPC charts? When M5 first looked at the SPC chart shown in Figure 1–3, he explained the target line:

M5: Now this is meter square per man hour. There is a target in there but its not something that we really…the target is there basically because we know these are the market demands for this year. Well, we don't know what the market demands are for the coming year but we've got as good information as we can get on that sort of situation. So in order to meet those market demands we know that we should produce at that target.

He then explained the meaning of the upper control limit and also how the control limits were calculated, and changed on the basis of previous data:

M5: That's telling us that we are actually capable of producing up to 7700 square meters per man hour. After the change it's telling us we haven't increased that, but what we have managed to do is reduce the amount of variations so we've actually increased the minimum because before we were actually capable…if we produced only 4800 that would have also been pretty much within the capability of the processor, that's now been increased about 5200.

I: Who made that decision, to increase the lower control limit to that (pointing at graph, when LCL was raised after 26 weeks)?

M5: Well the control limits are not done manually; they are based on the data.

So to appreciate the process, M5 suggested that there was a need to recognize that points outside the control limits have to be investigated as the likelihood of them occurring when the process was working as normal was just very low. He, unlike M4, was beginning to articulate some simple explanations with examples of why the 'recipes for action' had been put in place.

The Team Leader

We now turn to the perspective of team leaders who, as we mentioned earlier, have in recent years taken up a more important role in monitoring production, alongside their role in managing teams of operators. When asked about SPC charts, one explained:

TL1: So the operators will input a data point—their result. And if it's within the natural processing limits there is no issue. They continue with their operation. If the result is outside of the processing limits then they will come to either myself or the technical officer and we work out…we do an investigation to find out what has caused us to have a rule breach [a signal]—something that is outside of the processing limits. So although it is simple for the operators the fact they input the data, the real skill comes in investigating why we've had a data point that is then sat outside of the processing limits. The critical thing for these charts to work is that we do it live. If we have a rule breach then we deal with it there and then.

It is these investigations of signals that TL1 saw as crucial to 'revealing' inefficient practices:

TL1: It's amazing since we've been doing the graphs how many rule breaches are operator error. There was a time we used to blame when a machine stopped working or…the operators would always blame the engineers and the engineers would always blame the operators and now since we have started to take this structured approach with the charts and everything they're starting to realize how easy it is for them to make a mistake.

Thus TL1 mainly saw his role as reacting to and investigating signals, rather than seeking to reduce variation. He read the SPC charts as about identifying special causes. But, in companies moving to SPC, the TLs need to read more information from the chart and to appreciate that remaining within the control limits is a necessary but not sufficient condition for the process to be deemed satisfactory.[5]

We now present extracts from an interview with another team leader (TL2), who was being work-shadowed on the shop floor of a factory that was part of an international group involved in the mass production of large lengths of adhesive paper, cut to various widths. The factory produced huge roles of paper (jumbos) on which was applied a coating adhesive, with paper and 'stickiness' monitored according to the customer specifications. Speed in responding to customer demands, provided quality criteria were met and wastage reduced to a minimum,

---

[5]Sometimes the process changes and moves from something that is basically stable and capable (either by design or by chance) and starts to run along, say, a higher mean (see Figure 1–3). If this occurs, two procedures should be set in motion. First there should be an investigation as to what happened around eight data points earlier (the probability of eight consecutive points above (or below) the mean is extremely low: so the process would have to be investigated at the start of this 'run'). Second, a 'step change' can be initiated, mean is extremely low: so the process would have to be investigated at the start of this 'run').

FIGURE 1–4. A factory floor with example data displays.

was of paramount importance. TL2 was highly experienced and had been at the factory for nearly 20 years, working his way from operator to team leader. He was in charge of one 'web' in the factory: the stream of paper that went through the factory during which time it was coated with adhesive, rolled into jumbos, slit into various widths, and finally packed off on a palette for transport to the customer. TL2 was in charge of about 10 production processes for his particular web, which he controlled and monitored by observing and manipulating dozens of dials and data windows shown on six computer screens, as illustrated in Figure 1–4. The web also ran past the window to his 'hut' so he could see its progress directly as well as through the data.

We wanted to know TL2's interpretation of the SPC chart in Figure 1–3, with the KPI taken to be an output measure for his web, with the amount of product produced reported as 'square meters per manned hour'. TL2 was shown the chart and asked to describe it:

TL2: So they're looking for however many square meters are going out the door in return for the manpower that they've got.... So for the five people working we're averaging from there round about 48,000 to about 78,000 square meters per manned hour. And the average is about 78 something like that, the average. Well, not the average, the optimum…efficiently at full speed constantly, I think it's 78 which is about right, that's right.

First note that TL2 was immediately reading into the SPC chart data that were not presented; for example, the five people working on his web. He read the chart (correctly from a mathematical point of view) as an average with variation, but then insisted that the Upper Control Limit was the maximum length of paper his web could produce. He made a quick calculation as to the length of paper the web could produce, adding that he knew that the machine ran at 700 meters per minute:

TL2: UCL that's the maximum. It can't get any more than that because speed vs time, if you like. So obviously you're running at 700 meters per minute, times that by obviously 60 per hour, times 24, times 144 which is year full capacity and it won't go any higher than that unless you increase the machine speeds or alter the shift patterns.... That line indicates 78,000 square meters per hour. *And you can't get more than that. It's not physically possible* [our emphasis]. You can achieve more if you do something to the machine to make it run faster to produce more or put more hourly capacity into the working week which has 24 hours spare. Say we were running at optimum speeds with no web breaks.

TL2 went on to work out the maximum output more carefully, but then noticed something was wrong and the calculation gave a result that he knew from experience was too large. He then found reasons for the smaller number:

TL2: But it's not is it. It should work out. 700 times 60 is 42,000 linear meters a minute, times 2 is 84 square, times 144 = 12.096 million a week. But then again you've got your start-up, shut-downs, clean-downs and…obviously. That'll be why 84 is your optimum. Your upper control limit is 78 because they've obviously got to take out your wash-downs, clean-downs…

It was hard to follow TL2's calculations as there was so much embedded knowledge there (for example, the 42,000 is multiplied by 2 to obtain the area of the paper since the paper comes in 2 meter widths). Also the shop floor is not conducive to close questioning. Suffice to say that TL2's interpretations of the control limits were largely deterministic, based on an assumption that they were determined only by the speed of the web 'minus special causes', rather than as a property of the data set. TL2 did however have a good grasp of common cause variation:

I: But why is there all this sort of going up and down like this?
TL2: Lots of reasons. It could be lack of orders, lack of raw materials, machine breakdowns, web breaks. Like I said, a lot of things. It could be a process problem, a quality problem, raw materials, lack of work. It could be all sorts, you know.

Clearly TL2 was unsurprised by the common cause variation as displayed in the chart and put it down to a range of multiple and interconnected factors—just as the managers did—but based this recognition on his knowledge of the data and the whole work process rather than of statistics. TL2 could also explain why the data stayed high over some weeks, by reference to the situation—in this case the particular product on the web:

TL2: We're probably running the same adhesive with massive runs of the same material with a less amount of changes and breakdowns.
I: That causes why we've got more output?
TL2: The less change you've got the less waste you've got the more you produce.

For a dip outside the control limit, a signal, TL2 also could immediately provide a simple causal explanation.

TL2: A particular mechanical breakdown ... tension control problems, so the web must have broken...

Like TL1, he saw the main function of the charts as monitoring production and describing the reasons behind such signals, not about explicitly distinguishing the multiple factors in common cause variation mentioned earlier, and reducing it. However, TL2 did talk about his work as part of a process improvement team on waste, where again, SPC charts were being used, again not in his terms to reduce variation, but to bring down the output of waste:

TL2: To get a tighter grip and lower, you know, if it's waste. You want to get it down, don't you?

He explained—just like M5—the way the team set out to identify the main factors in waste, one of which had been at the point of changing materials:

TL2: It's because basically the machine when it's stopped, it's not doing anything, is it. You tick it over it's doing 10 meters a minute. All that 10 meters a minute for however long you're ticking it over is wasting the relay because, it's just waste. Because you're not putting glue on it. We're putting silicon on but we're not putting glue adhesive on and we can't sell it to anybody. The machine needs to be running at a certain speed, tensions and certain coat weights in which ... to make the pressure sensitive labels work correctly.

I: Oh I see.

TL2: See for 10 meters a minute it's no use. So what we'd need to do is to ramp the machine up to say 300 meters a minute or whatever speed it runs at with everything fully in place. And when, after a minute or so, when the adhesive goes through to the reeling or the laminator then we flag it and anything below that flag is waste. So we need to try and do things better, quicker and reduce waste.

I: So you've got to try and do this takeover faster.

TL2: Exactly.

I: How will you go about trying to think how to do that?

TL2: There's certain ways, and certain reasons why realistically what you'd want to do is start the machine up, put everything in place, i.e. bring your ovens down and then go to run and get up to optimum speed the quicker you can. There's certain things you can do like make the ovens get up to temperature quicker, get the technicians involved and make the machine run faster because it only runs at certain speeds.

When talking about process improvement, TL2 talked about actual working procedures: reducing the speed of the machine, increasing oven temperature more quickly, rather than considering 'abstract' data. He was aware that the different factors interacted on the shop floor but was unlikely to have been able to represent how they interacted with any precision or indeed quantify and compare the interactions. TL2 was also clear that although he saw the benefits of moving to more standardized practices, his knowledge of the conditions for his web to work could *not* be codified; there were just too many interacting variables to take into account to decide the best setting for any process:

TL2: Well, we run obviously different grades of paper, well I mean grades—thicknesses of papers, 59 gram, 78 gram, 88 gram, so obviously the reason that we're heating that up, putting temperature into the paper is to cure the silicon. The thicker the paper the more temperature you need. Also you've got to take into consideration the machine speed. The faster you run the more temperature you require so ... . It's not just that, that's just a very, very small fraction of the whole process. You've got to remember things like that for various different products—we have 65 different products—I'm not saying we have 65 different settings, but they do vary. And in this product range we do like laid flat, reverse curl, backprintings, synthetics, you know.

TL2 also summarized all the reasons for faults, again arguing that there were too many to record:

TL2: It's the whole...there's quite an array of things, like you know, there's all sorts of things you could be...the machine could just be ticking over and the paper could just break because you get a raw material fault. If you look at a faulty machine overall there's thousands of variables and all of these variables can have an impact on things that matter.

All this expert knowledge of the process was in his head and could not be written down:

TL2: It's all in my head now.
I: Don't you think that's odd? I mean, you've got all these values here and you have to know that the oven temperature is such and such...
TL2: Yeah, we have discussed this at team leader's meetings. And they would love us to document it, but because you know, I say there's not a forever changing process but it does change and it also changes I believe at the norm for ambient temperature. In the summer, we don't need the ovens so high in the summer because the ambient temperature has an affect upon the product, believe it or not. So you could write down a whole load of settings for a load of everything, but it could change inside a month.

So, in summary, TL2—like TL1—tended to read the SPC charts as a series of causes that were not quantified, that interacted over time, again in ways that were not measured or even thought to be measurable: "went up as no breakdowns, good temperature and raw materials...went down as had an operator error or bad release results or silicone ovens working erratically". The control limits provided a demarcation for signals, but the role of the lines themselves and how they were derived remained unclear. Why were particular data points signals? The team leaders could certainly interpret common variation on the basis of their knowledge of the whole work process. This knowledge was shared with management to help explain variation, but the next step, to control it, may need some way of aligning meanings with a statistical view.

TOWARD A THEORETICAL FRAMEWORK

This chapter started from the assumption that the advent of digital technology as a pervasive force at work necessitates a reappraisal of the nature of mathematical knowledge used at work and how it is communicated. We have sought to illustrate how the models of the workplace, particularly in relation to data modeling and statistical process control, play a range of roles for the different actors in the workplace activity system. We now try to draw these ideas together with a theoretical framework with which to interpret them.

Questions that we began to isolate during our case studies of the use of SPC charts were: What are the meanings ascribed to the line indicating the historical mean of a process? How are signals interpreted? How is a signal conceptualized in relation to the whole process? How are the upper and lower control limits on the SPC chart interpreted and what are the meanings ascribed to the variation within and outside these limits, especially in relation to any targets or customer specifications?

The question of control limits has turned out to be suggestive of a more general phenomenon. For a company to improve, it must ensure that the control limits are within the customer specification but in times of tight profit margins also that variation is reduced, that is, the gap between the control limits is narrowed. For some people in the activity system, but not all, this meant that at some level, the mathematical object represented by the gap between the control limits has to be understood. We say *mathematical* in the sense that there is no concrete instantiation of this gap (or indeed, of either control limit) in the process itself. Like any statistic, say, the mean, this gap is an abstraction, with no straightforward referent in the work system. In this respect, therefore, it is relevant to ask who, among those viewing and working with the charts, needs to view the control limits as mathematical objects, and for what purposes?

Consider the manager who noted that certain phenomena instantiated in the chart would necessitate taking things 'down a level'. Regardless of whether his metaphor of up/down is the right one (or in the right direction) his meaning is clear. The chart's structure would point to complexities and/or interactions between variables, and this would necessitate closer investigation of individual values—either readings on other charts, or readouts of individual machines reported in the form of abstracted data. The operators and the team leaders need, similarly, to look at particular elements of the process—down a level from the aggregated structure of the chart. However, the information they use to effect this level change tends not to be abstracted data, but actual factors they know affect the work process.

In contrast, we conclude that managers have to conceive of variation as a factor to be manipulated, and they necessarily have to see mean and variation as inseparable, i.e. they have to see the charts as mathematical—and many may not. In order to appreciate the sources of common cause variation, some of the managers see the need to analyze the multiple factors that give rise to it and then seek ways to reduce it. Others, as we saw, regard their role more as 'following recipes'. In terms of investigating special causes, this appears a reasonable strategy—they are, after all, deterministic. Where such a management approach falls down is in seeking to reduce common cause variation, which requires engagement with the data and distinguishing its interconnected components. Team leaders, on the other hand, appreciate the 'system on the ground' although they tend to see the charts merely as a way of displaying information—which does the job fine! What is lost is the stochastic dimension, and the meaning of the mean in relation to variation that depends on the historical data set.

Managers in charge of process improvement do, of course, appreciate the complexity and stochastic nature of charts. They understand the importance of the interaction of factors leading to common cause variation *and* the importance of interactions *between* KPIs, although they may not have tools to quantify random variation within *interacting* variables. This latter appreciation requires quantification and analysis at different levels of the process and represents a multifaceted, multileveled TmL. Yet we have seen that there are managers who reject the need for more quantification and the use of statistical thinking throughout the workforce, point to the importance of management, but are dissatisfied with outcomes of process improvement.

Making sense of this ambivalence is an important aspect of the problem we seek to address. In this respect, the role of the chart as a boundary object is particularly relevant, as it draws attention not only to what is different between the various communities in making sense of them, but also to what is the same. This sameness derives precisely from the fact that the charts are abstractions. For one group, the team leaders and operators, the charts are essentially 'triggers' for pinpointing (usually) human error; for another, (some of) the managers and statisticians, they are a source of data for further analysis: they, unlike the operators, are aware that remaining within the control limits is a necessary but not sufficient condition for the process to be deemed satisfactory. The fact that the charts are mathematical abstractions from the production process allows them to become shared territory. At the same time, the kinds of knowledge about the objects and relationships represented in the charts, what the variants and invariants are, the abstractions derived, differ among the actors who use them.

The roles of the charts as boundary objects gives rise to particular forms of knowledge that characterize their use, deriving from their status as abstractions from the work process. Unsurprisingly, we encountered nobody (except perhaps the statistician) who required an explicitly mathematical view of the charts, in terms of statistical theory. We did, however, find persons at all levels, who had derived ways of talking about and conceptualizing what they thought the chart represented, and a language with which to express them. We know, for example, that team leaders now have to come and present—in the words of one manager—"a lot of analytical data about what happened in their shift": they have to know more of the system (devolution) and articulate what they know (communication). These data are not a narrative of what happened, but a situated abstraction of what happened, in the sense that the quantitative readouts of the statistics and charts—encapsulating all manner of relationships between the variables concerned—are one level removed from the actual operation of the machines. We do not yet have enough data to characterize this form of quasi-mathematical expression explicitly, but we know from one of the managers that it is necessary (recall the team leader who was "not able to interpret [data], to transform it into a trend or a problem solving analysis").

The different ways to think about the charts as boundary objects accounts for the ambivalence of the managers themselves as to the role of the chart. M4's apparently contradictory view, particularly, in which he simultaneously asserts there is no need for a mathematical view of the chart, and yet works to induce his trainees to produce statistics by hand or wants them to cope with multiple representations of data sets, reflects this ambivalence. However, the question still remains as to how far is it important for the workforce *as a whole* to engage in statistical thinking to make effective decisions on process improvement.

Our initial thoughts are that negotiation of meanings of the SPC chart may provide a fruitful way forward in providing a window on this question and the work process as a whole. We describe a telling example. We know that SPC software in common use eliminates special causes from any calculation of the mean and control limits as they are not part of the stable process. One manager we interviewed was clearly confused about this important aspect of the charts. After demonstrating the software that generated the charts, pointing out the upper and lower control limits and the special cause points which operators were expected to annotate with reasons, she added "it (the special cause point) is taken out of the calculations of mean and upper and lower controls. The process automatically re-calculates without the points outside limits...not sure why... I suppose as we want to make sure we reduce the variation in process." She is right of course: it would certainly reduce the variation, but not the common cause variation which is the aim of the whole SPC initiative!

Here is the nub of the issue. This manager, and perhaps the workers more generally, need to come to see data sets both as a concrete representation of what the company produces, and as an abstract representation that identifies trends with both historical and predictive significance, which points to strategies to reduce variation as well as special causes, and simultaneously provides theoretical (mathematical) and pragmatic indicators. The distinction between parts of the workforce is *not* as clear-cut as we may have thought: it simply does not seem to be the case that one sub-community "needs" one view, while another "needs" a different one. Instead, it is a question of balance, in which the mathematical element of the charts comes into and out of focus while in use for different purposes, different sectors of the communities, and at different times. We are led to concur with Greenfield's assertion that "understanding" of management is necessary for the success of SPC techniques, and we may go further in suggesting that those who are being managed need some elements of understanding as well.

We conclude by revisiting the overview of modern production processes with which we began. The unidirectionality that Reich describes—from symbol analysts downward—is no longer the only dynamic: increasingly, communication is a two-way process between those who develop the symbolic frameworks that drive machines and systems, and those that operate within them (for a recent discussion of new configurations of work, see Engeström, 2004). We should stress that *nobody* in the factories we visited could be described as a symbol analyst: they were firmly behind the scenes and we have no data even to assess how many there are. Neither, incidentally, do we know how many levels there are between our senior managers and the symbol analysts. It is, entirely possible (probable even) that those responsible for installing the computer systems in the factories were using commercially-available software, or at most, configured versions of off-the-shelf systems. Somewhere along the line there are symbol analysts, but it is quite likely that not one of them has any first-hand (or any) knowledge of the factories into which the system has been adopted.

This point is not only of passing interest. It means that in terms of the techno-mathematical knowledge of the actors in the workplace, we may be dealing with a few slices of a highly complex hierarchy, rather than two or three main delineations. Reich's more coarse-grained classification of the workplace is entirely appropriate for his economic and social perspective, but it may not be adequate for our purposes. The implication is that the knowledge that characterizes the boundaries, across which our team-leaders, managers and statisticians are communicating, may be different more in nuance than in real substance.

Nevertheless, some real differences between elements of the workforce are emerging. There are—as we have seen—surprising differences in the ways that the different sub-communities who engage with, for example, the SPC charts, use them, and the apparent meanings of statistical variation that are shaped by them. These meanings tend to mask the complexity of the system, and render invisible the interactions between a wide range of variables that become 'wrapped up' in the data points—only to be opened up by considering all the day-to-day specificity of the shop floor. Thus the work of the floor must be made visible as a first step in appreciating variation. The formal and quantitative indicators of work no longer 'silence' workers, as Star and Strauss put it, but require workers for their interpretation. Our conjecture is that it is explicitly recognizing these very facets and observing how they interact that could be the key to aligning meanings across communities.

ACKNOWLEDGMENTS

We would like to thank our colleagues, Dr. Phillip Kent and Dr. Arthur Bakker for their contributions and comments on an earlier draft of this chapter.

REFERENCES

Bessot, A., & Ridgway, J. (Eds.) (2000). *Education for mathematics in the workplace*. Dordrecht: Kluwer.

Bishop L. (1999). Visible and invisible work: The emerging post-industrial employment relation. In B. Nardi & Y. Engeström (Eds.), *Computer Supported Collaborative Work. Special Issue: A Web on the Wind: The Emerging Post-Industrial Employment Relation. vol 8* (1–2), 115–126.

Bowker, G. C., & Star, S. L. (1999). *Sorting things out: Classification and its consequences*. Cambridge, MA: MIT Press.

Clayton, M. (1999). Industrial applied mathematics is changing as technology advances: What skills does mathematics education need to provide? In C. Hoyles, C. Morgan, & G. Woodhouse (Eds.), *Rethinking the Mathematics Curriculum* (pp. 22–28). London: Falmer Press.

Engeström, Y. (1996). Interobjectivity, ideality, and dialectics. *Mind, Culture, and Activity*, 3(4), 259–265.

Engeström, Y. (2004). The new generation of expertise: Seven theses. In H. Rainbird, A. Fuller, & A. Munro (Eds.), *Workplace Learning in Context*. London: Routledge Falmer.

Eraut, M. (2004). Transfer of knowledge between education and workplace settings. In H. Rainbird, A. Fuller, & A. Munro (Eds.), *Workplace Learning in Context*. London: Routledge Falmer.

Greenfield, T. (1993). Communicating statistics. *J R Statist Soc 156*, part 2, 287–297.

Guile, D., & Young, M. (2003). Transfer and transition in vocational education: Some theoretical considerations. In T. Tuomi-Grohn, and Y. Engeström (Eds.), *Between School and Work: New Perspectives on Transfer and Boundary Crossing* (63–81). Pergamon Press.

Hall, R., & Stevens, R. (1995). Making spaces: A comparison of mathematical work in school and professional design practices. In S. L. Star (Ed.), *The Cultures of Computing* (pp. 118–143). London: Basil Blackwell.

Hall, R. (1999). Following mathematical practices in design-oriented work. In C. Hoyles, C. Morgan, & G. Woodhouse (Eds.), *Rethinking the Mathematics Curriculum* (pp. 29–47). London: Falmer Press.

Hall, R., Stevens, R., & Torralba, T. (2002). Disrupting representational infrastructure in conversations across disciplines. *Mind, Culture, & Activity*.

Hoyles, C., & Noss, R. (1992). Looking back and looking forward. In C. Hoyles & R. Noss (Eds.) *Learning Mathematics and Logo* (pp. 431–468). Cambridge: MIT Press.

Hoyles, C., & Noss, R. (1996) The visibility of meanings: modelling the mathematics of banking. *International Journal of Computers for Mathematical Learning 1*, 1, 3–31.

Hoyles, C., Noss, R., & Pozzi, S. (2001). Proportional reasoning in nursing practice. *Journal for Research in Mathematics Education*, 32, 4–27.

Hoyles, C., Wolf, A., Molyneux-Hodgson, S., & Kent, P. (2002). *Mathematical Skills in the Workplace*. London: The Science, Technology and Mathematics Council. [Download: www.ioe.ac.uk/tlrp/technomaths/skills

Joiner, B. (1994). *Fourth Generation Management: The New Business Consciousness*. New York: McGraw Hill.

Kent, P., & Noss, R. (2002). The mathematical components of engineering expertise: The relationship between doing and understanding mathematics. *Proceedings of the IEE Second Annual Symposium on Engineering Education*. London: Institution of Electrical Engineers.

Konold C., & Pollatsek, A. (2002). Data analysis as the search for signals in noisy processes. *Journal for Research in Mathematics Education, 33*, p. 263.

Nardi, B., & Engeström, Y., (Eds.) (1999). A web on the wind: The structure of invisible work. *Special Issue of Computer Supported Cooperative Work: The Journal of Collaborative Computing.* Vol. 8, Nos. 1–2.

Noss, R., & Hoyles, C. (1996). *Windows on mathematical meanings: Learning cultures and computers.* Dordrecht: Kluwer.

Noss, R., Hoyles, C., & Pozzi S. (2002). "Abstraction in expertise: A study of nurses' conceptions of concentration". *Journal for Research in Mathematics Education, 33*(3), 204–229.

Noss, R. (2002). Mathematical epistemologies at work. *For the Learning of Mathematics, 22,* 2, 2–13.

Raizen, S. M. (1994). Learning and work. *The Research Base in Vocational Education and Training for Youth: Towards Coherent Policy and Practice* (pp. 69–113). Paris: OECD.

Reich, R. B. (1991). *The work of nations: Preparing ourselves for 21st century capitalism.* London: Simon & Schuster.

Sfard, A. (2001). There is more to discourse than meets the ears: Looking at thinking as communicating to learn more about mathematical learning. Special Issue, *Educational Studies in Mathematics, 46,* 1–3, 13–57.

Smith, J., & Douglas, L. (1997). Surveying the mathematical demands of manufacturing work: Lessons for educators from the automotive industry. In R. Hall & J. Smith (Eds.), *Session 10.39, AERA Annual Meeting.* Chicago.

Star, S. L. (1989). The structure of ill-structured solutions: Boundary objects and heterogeneous distributed problem solving. In L. Gasser & M. N. Huhns (Eds.), *Distributed artificial intelligence* (Vol. 2, pp. 37–54).

Star, S. L., & Griesemer, J. (1989). "Institutional ecology, 'translations,' and boundary objects: Amateurs and professionals in Berkeley's Museum of Vertebrate Zoology, 1907–1939," *Social Studies of Science, 19,* 387–420.

Star, S. L., & Strauss, A. (1999). Layers of silence, arenas of voice: The ecology of visible and invisible work. *Computer-Supported Cooperative Work: The Journal of Collaborative Computing, 8,* 9–30.

Wake, G. D., & Williams, J. S. (2001). *Using college mathematics to understand workplace practice. Final report to the Leverhulme Trust.* Manchester: University of Manchester.

Wild, C. J., & Pfannkuch, M. (1999). Statistical thinking in empirical enquiry. *International Statistical Review, 67,* 3, 223–265.

Williams, J., & Wake, G. (2002, September). *Metaphors and models that repair communication breakdowns and at disjunctions between college and workplace mathematics.* Paper presented at the BERA conference.

Zevenbergen, R. (2004). Technologizing numeracy: Intergenerational differences in working mathematically in new times. *Educational Studies in Mathematics 56,* 97–117.

Zuboff, S. (1988). *In the age of the smart machine: The future of work and power.* New York: Basic Books.

CHAPTER 2

Problem Solving and Learning in Everyday Structural Engineering Work

Julie Gainsburg
California State University, Northridge

In recent years, the education, policy, and business communities have stressed the enhanced problem solving requirements of the modern workplace. The new conventional wisdom is that higher levels of mathematical proficiency are required of a wider range of workers. Reformers of K–12 education have responded with initiatives to provide high-level math education for all students and for greater alignment between school and "real-world" problem solving, i.e., learning (math) by doing (what adults do when they solve quantitative problems). There are two problems with this formulation. First is the authenticity problem: our knowledge about adult problem solving activity is limited, especially regarding the heavy-math use professions about which math educators are specifically concerned. Second is the effectiveness problem: it is uncertain how effectively adult problems (or classroom simulations of them) serve as sites for math learning. Because this knowledge is unavailable, educators understandably adopt a primarily pedagogical focus when developing problem solving tasks for K–12 students, giving less consideration to how authentically the tasks replicate the actual behavior of any group of adults. For example, teachers might design a modeling exercise expressly to promote learning about direct variation, with little regard for whether the exercise reflects real-world modeling activity. This approach helps ensure that students learn the desired math-course content (or state-level curriculum standards), but it leaves open the question of whether these curricular experiences bear any relation to the activities of adults in the modern workplace—the future for which we hope to prepare students.

To address the problems of authenticity and effectiveness, research is needed in both arenas: the classroom and the math-intensive workplace. A strong example of the kind of research necessary in the classroom is Lesh and colleagues' studies of the learning of students as they engage in *model-eliciting activities* (MEAs). MEAs are classroom tasks that require students to interpret realistic and complex situations mathematically, by applying, modifying, or extending constructs and conceptual systems—essentially, developing mathematical models (Lesh & Doerr, 2003). MEAs are often designed to promote *local concept development*: the refinement or extension of a mathematical concept or construct (such as proportion) in the context of a brief problem solving session (Lesh & Harel, 2003). What makes this research so potent is that it goes beyond evaluating the pedagogical value of modeling tasks (where other studies usually stop) to using those tasks as a means to investigate children's learning trajectory. Lesh and Harel hypothesized that classroom activities that afforded students the open exploration of concepts—learning through solving complex problems, as opposed to guiding the learning down a single, constrained path (as in traditional math teaching), would reveal the natural processes of local concept development. The unfolding processes Lesh and Harel observed appeared to be microcosmic versions of the processes of the long-term development of the corresponding general concepts, as delineated by learning psychologists such as Piaget and the van Hieles. More studies of local concept development though modeling activities are critical for resolving the effectiveness problem. Increasing our knowledge about children's learning trajectories should enable us to design classroom tasks that take advantage of natural concept-development processes rather than working against them.

This chapter in some sense parallels the work of Lesh and Harel, investigating the high-tech workplace rather than school. Using episodes from an ethnographic study of the mathematical behavior of structural engineers in practice, I illustrate the nature of their everyday problem solving. Further, like Lesh and Harel, I take the view that solving problems is, in effect, learning, and I contend that through observations of everyday problem solving activity we can gain an understanding of how and what engineers learn in the course of practice. This perspective is supported by Torraco (1999), whose meta-analysis of several workplace ethnographies revealed that "the distinction between learning and working has significantly eroded in today's workplace" (p. 35–34). My observations and interviews convinced me that structural engineers, to be successful, must engage in two main kinds of intellectual activity: solving the problems of their daily work (design and analysis) and building their expertise—in other words, learning. These are done simultaneously; each problem solving episode is a potential learning experience. The expertise of an engineer is not static—cannot be; rather it increases with accrued experience. Structural engineering expertise appears case-based, reminiscent of the practice of law, and the engineers I observed constantly made reference to past projects when solving current problems. But structural projects are extensive and complex, always presenting new problems and often demanding novel solutions. To observe structural engineers solving everyday problems, then, is to observe their learning process, or at least snapshots of it, as each problem solving episode is a building block in the structural engineer's accruing knowledge.

The heart of the intellectual work of structural engineers is the application of mathematical representations and procedures to solve design problems, which usually requires the selection, adaptation, or creation of a model (heretofore called modeling for brevity) (MAA, 2000; Gainsburg, 2006). Developing structural-engineering expertise is largely a matter of becoming increasingly adept at modeling. Over time, engineers build a collection of increasingly sophisticated modeling "tools"—mathematical/theoretical procedures or structures—while recognizing a widening range of applications for each tool. Like students engaged in MEAs, structural engineers learn in an open, uncertain problem solving environment that offers an infinite number of paths along which to develop expertise. Thus, it is reasonable to assume that engineers' ordinary work reflects fairly natural processes of problem solving and learning. In other words, analyzing the everyday problem solving process of structural engineers in practice should reveal something about how adults learn, and more specifically how heavy users of math become increasingly adept at modeling.

Studies like this one, cognitive ethnographies in high-tech workplaces, serve two purposes for education. First, they enhance our understanding of the kinds of problem solving for which we must prepare students, especially those headed into math-intensive professions. Second, they illuminate the relationship between adults' everyday problem solving and the paths by which adults develop concepts that are central to their work. This information could inform efforts to create K–12 classroom tasks and environments that promote learning as well as offer ideas for university educators about how to prepare professionals for the continual learning that must take place on the job.

Below, I present synopses of two problem solving episodes drawn from an extensive ethnographic study of the mathematical behavior of structural engineers (Gainsburg, 2003). The larger study comprised 70 hours of observations of engineers at work in two structural engineering firms, along with 24 hours of interviews of the engineers and the collection and analysis of artifacts of their work. Of the dozen-odd major problem solving episodes I documented in the field, the two selected for this chapter typify the more intellectually challenging ones; they were also selected because the engineers involved were fairly articulate about their learning. I also chose these episodes for their differences. They took place in different firms. One features a fairly new engineer working alone and the other a long-time veteran collaborating with a junior colleague, illustrating that learning continues throughout an engineer's career. Also, these cases represent two major but different purposes for modeling: in one case its purpose is to generate a design solution, while in the other it is to explain an analytic result. Despite their differences, these episodes paint a fairly convergent picture of problem solving activity that holds over the other episodes I observed.

THE CASE OF THE DISCREPANT GIRDER

Lynn,[1] a junior engineer, is designing one of several scheme options for the floor of a new casino. Having just recalculated the size of a floor girder (a major beam), she notices that it is quite a bit larger than her supervisor, Kevin, had originally estimated. This particular girder is designed as a *composite beam*: a system comprising a factory-rolled steel beam, a width of concrete floor slab atop the beam, and many small steel studs protruding from the top of the steel beam and embedded in the slab, to anchor slab to beam so that the slab can help the beam resist bending (Figure 2–1).

Gravity loads will cause the beam to deflect downward in the center, placing its top flange and the floor slab in compression. The point of composite construction is to supplement the compression capacity of the steel beam with capacity provided by the concrete. But with three components acting in concert—the steel beam, the slab, and the studs that connect them—no straightforward method exists to calculate the unique set of optimal component sizes for a given load. Instead, Lynn's company has adopted a simplified calculation method: to design each component to handle the full demand independently. This method instantiates a theoretical model of the composite beam that assumes no interaction among the components; the concrete does not help the steel. This model errs on the conservative side, since the actual capacity of the composite system exceeds that of any of its components,

---

[1] All names for engineers in this chapter are pseudonyms.

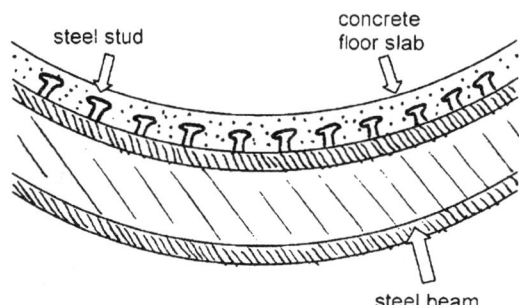

FIGURE 2–1. Lengthwise cross-section of a composite beam.

but it simplifies matters enough to permit the calculations to be done by hand. Given a particular gravity load and beam depth, Lynn can calculate an internal compression demand using a standard formula. Once that demand is known, Lynn can determine the necessary size for the steel beam. Then, in order to "develop" or take advantage of the full capacity of this steel beam, she will size the studs and slab to match the beam's capacity.

The episode begins as Lynn calls Kevin over to discuss the discrepancy between her new beam size and his earlier one. Though the concrete part of the composite system is usually unproblematic, Kevin, for reasons he cannot explain later, wonders if Lynn's concrete slab is adequate to handle the compression; he asks, "Do you have enough slab for that?" To check, he and Lynn perform two routine calculations with memorized, established formulas. First, to find the slab's capacity to handle compression, C, Kevin uses a formula (Equation 1) based on properties of the concrete:

Equation 1
$$C = 0.85 f'_c A_c$$

A_c is the cross-sectional area of the portion of the floor slab that counts as part of the composite beam—the *effective flange*. Kevin uses 3.25" as the slab's thickness and assumes the effective width to be 15'. Early in the project, the strength of the concrete, f'_c, was established to be 4 ksi (kips per square inch).[2] Multiplying these factors, Kevin arrives at a capacity of 1989 k (kips). Next, Kevin and Lynn calculate the compression capacity of the steel beam—the value they will subsequently take to be the demand on the slab. Here, Kevin uses a memorized rule of thumb: the weight of a beam (in pounds per lineal foot, or plf) is 3.4 times its cross-sectional area. Kevin divides the weight of Lynn's beam, 329 plf, by 3.4 to find an area of 96.8 in^2. Finally, he uses a memorized formula (Equation 2) for steel capacity:

Equation 2
$$V = A_s F_y$$

or the product of the steel's area and the steel's strength. (V is commonly used to represent horizontal forces, a category more general than but including compression.) Kevin and Lynn know the steel's strength, F_y, to be 50 ksi, and they calculate the steel's capacity, V, to be 4840 k. Comparing the results of these two calculations, they see that the concrete capacity (1989 k) is far below that of the steel (4840 k), thus, the slab cannot take the same compression as the steel.

---

[2] A kip is 1,000 pounds.

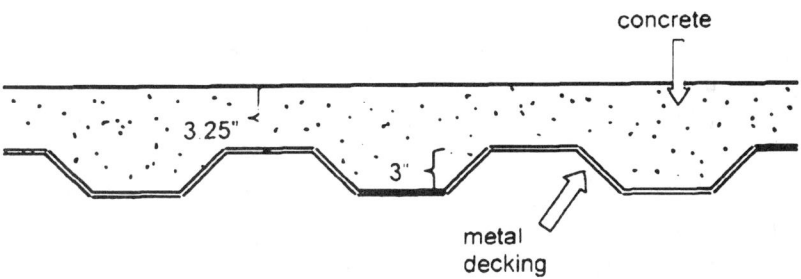

FIGURE 2–2. Concrete slab poured into metal decking.

Though these calculations invoke established, general formulas, they are not completely straightforward. This slab will be formed by pouring concrete into a sheet of ribbed metal decking with 3" deep troughs, thus giving the slab a fluted, or ribbed underside. On top of the "flutes" is poured a solid 3.25" layer of concrete, so that the depth of the slab alternates between 6.25" and 3.25" in stripes across the floor (Figure 2–2).

For this first round of calculations, Kevin and Lynn have used a simple, conservative assumption: that the slab has a uniform thickness of 3.25", which is really only the case for the thinnest parts of the slab. Now that the results of this initial estimate indicate structural inadequacy, Kevin takes a problem solving tack typical of the engineers I observed: use a more complicated but less conservative estimate that more accurately represents the situation. To this end, Kevin uses a new slab thickness of 4.5". He has no intention of actually adding more concrete to the design, even if this new thickness were to solve the problem. Instead, he is simply experimenting with a different model of the slab's cross-section, now taking into account the strips of concrete molded into the troughs. Calling the thickness 6.25" would overestimate the slab's capacity, since only half of the concrete floor is actually that deep. Kevin's new estimate of 4.5" is a sort of average that presumes the ribs and troughs of the decking are symmetric and therefore that the total floor areas that have the two depths, respectively, are equal. This conceptual model assumes that the average of the alternating thicknesses can be used as a single depth for the concrete, in some sense remodeling the fluted cross-section into a rectangular one with the average thickness.

Unfortunately, replacing 3.25" with 4.5" in the calculation only increases the slab's capacity to 2754 k, still well below the 4840-kip capacity of the steel girder. Still, Kevin and Lynn do not give up trying to justify the girder size she has calculated. Kevin recalls that, days earlier, they had sized this same girder using an analytic software program, which had recommended a much larger size. Lynn reminds Kevin that he had advised her to ignore the program's recommendation and use the smaller size he had calculated from codebook specifications. Kevin explains to me later that he justified this override with his understanding of how the program models a composite beam. Experience has shown him that the program makes an overly conservative assumption that causes it to recommend needlessly large beams. Specifically, it assumes the use of narrow-ribbed decking, which permits only a single row of studs along a beam. Wide-flanged beams can actually take wider-ribbed decking and, subsequently, use a side-by-side stud pattern that fits more studs per unit length of rib (Figure 2–3).

However, even when analyzing wide-flanged beams, the program defaults to the single-file stud configuration. The small number of studs the program thus assigns reduces the potential effectiveness of the composite beam because it limits

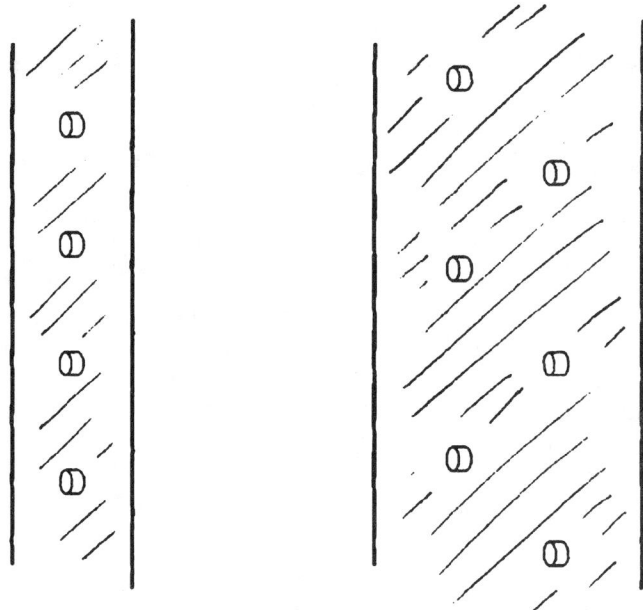

FIGURE 2–3. Studs in narrow vs. wide-ribbed decking.

the transfer of forces between steel and concrete. The program compensates for this reduced effectiveness by recommending a very large steel beam. Kevin, however, knows he can fit more studs into wide-flange composite beams than the program will, and so he habitually either ignores its recommendations or manually inputs a different stud configuration.

Now Kevin turns his attention back to the program's calculation that he had rejected days earlier, but not for the purpose of using its recommended design. Just as he and Lynn had used the hand calculations described above, they intend to use the software's calculation to help them make sense of the current problem: why Lynn's girder, which has been sized by the same simplified, by-hand method they have always used, is not working. Kevin and Lynn quickly retrace the history of Kevin's override. When, days ago, Lynn had reported the large beam recommended by the program, Kevin had wondered whether it was assigning enough studs to "develop" the beam, in other words, so that the studs' capacity to transfer force matched the capacity of the steel beam, thereby enabling the composite system to use the full capacity of the steel. Lynn had checked and discovered that the program had not assigned enough studs for that, and, of course, Kevin had drawn his usual conclusion: that the program was assuming a narrow-ribbed deck with single-file studs. Presuming they could squeeze in more studs, Kevin had advised Lynn to use a smaller beam, sized to handle the known gravity demand. But now Kevin and Lynn are finding the concrete slab inadequate for that demand.

An inadequate slab is not, by itself, a structural problem. Just as the concrete in a composite beam helps the steel girder resist compression, the opposite also applies: the steel girder helps the concrete resist compression. Even if the concrete's capacity were to fall short of the entire compression demand, the composite system could still conceivably work, as the steel could supply the extra capacity. However, an inadequate slab spells trouble for Kevin and Lynn's calculation method, which takes the simple, conservative shortcut of sizing steel, studs, and concrete to each handle the full demand. In past projects, the concrete slab has always been more than adequate—the steel has usually been the weak link—so they are not in the habit of checking the slab during preliminary design. If the slab were ever found to need help from the steel beam, the calculations required to size the components would be extremely cumbersome, as Kevin explains to me later:

> All of my approximate numbers go out the window. Everything that I do, my rules of thumb, are all based on there being enough concrete to develop the steel. I don't have any rules of thumb for this condition.... We don't have a rough way of doing—we don't even want to think about doing this by hand. We can; it just takes—it's an iterative type of procedure and it takes a little while to do it.... But I'm into not spending five pages doing one of these calculations; I'm into doing five calculations in one page and getting to what I consider to be the best design.

But now their calculations are indeed showing that the concrete slab, as currently sized, will not handle the full demand alone. And increasing the slab's thickness just to permit an easier calculation method is out of the question; the building's weight and cost would rise and usable occupant space would decrease.

Kevin now hatches a theory about what might be causing the trouble. He suspects that perhaps the computer program is not limiting the number of studs according to the deck, as it usually does, but instead according to the capacity of the concrete. He now conducts an experiment to determine which factor the program is taking into account. He directs Lynn to rerun the program to see how many studs it assigns to this large beam; when she does, they see it recommends 219 studs with a capacity of 22.7 kips each. Kevin uses these numbers to back-calculate the compression capacity the program must have presumed to determine these studs. He uses the formula (Equation 3):

Equation 3 $$V = (stud\ capacity)(\#\ studs/2)$$

to get (22.7)(219/2) or 2485 kips. Seeing how close this value matches the capacity of 2754 k that he had calculated for the fluted concrete slab, he pronounces this "about right." This supports his new theory that the program is using the concrete capacity to size the studs.

At this point, Kevin's problem solving aims at multiple goals, on different levels. On the most short-term and situation-specific level, Kevin wants to make sense of the calculations that indicate the inadequacy of Lynn's beam to help him decide if the beam can be justified or if they must find a new size. In order to achieve this goal, he must carry out a somewhat higher-level and more general mission: to figure out why the program has recommended the particular beam it has for the current situation. And to accomplish that, through his experimentation he aims for a far-sighted and widely applicable goal of learning about the program in general and the underlying model that explains how it produces its results.

To make sure his new theory is correct, Kevin continues his experiment. He asks Lynn to rerun the program with a new concrete strength (f'_c)—6 ksi rather than 4 ksi. When she does, the program reports that it has upped the number of studs to 281. Hearing this, Kevin responds, "So it's just giving you as much as you can get out of the slab." Here, Kevin has forgone actual calculation and has simply applied his knowledge of functional relationships. Since he is not currently interested in a specific design solution from the program but only wants to discover its stud-calculation method, he just needs to test whether changing a concrete property affects the stud calculation. If the program is sizing the studs to match the compression capacity of the concrete rather than of the steel beam, it must be calculating the compression as a function of concrete properties, using the formula (Equation 1):

Equation 1
$$C = .85 A_c f'_c$$

(as opposed to the formula [Equation 2]:

Equation 2
$$V = A_s F_y$$

which expresses the capacity as a function of steel properties). Changing the value of either A_c or f'_c would change the compression and therefore the stud number. Since changing f'_c to 6 ksi produced a new number of studs, Kevin now has proof that the program is designing the composite beam according to the capacity of the concrete rather than the steel.

Kevin continues to try to justify the use of Lynn's smaller beam. His persistence has both mathematical and practical explanations. Aware of the program's history of questionable assumptions during composite-beam analysis, Kevin is going to need a lot of convincing, through mathematical proof, before he will accept its large recommended beam. But he also knows that a beam as large as the one the program is recommending is impractical and expensive; any design scheme that uses it will not be selected. Consequently, he considers it worth the time to experiment with calculations that might "save" the smaller beam.

Continuing to draw on his knowledge of functional relationships, he now considers another factor that would affect the beam size: the *effective width* of the slab, the portion of concrete floor presumed to contribute to resisting the compression in the composite beam system. The compression capacity of the concrete is a function of its effective width, and Kevin realizes that if a larger effective width could be assumed, the slab's capacity might increase enough to match that of the steel. In that case, Kevin and Lynn could keep their simplified method of sizing all components and, subsequently, prove the adequacy of the system with Lynn's beam. So they turn to a codebook to check whether they have used the correct effective width. On one page, they find three methods for calculating half of the effective width of the concrete slab, i.e., the portion on one side of the beam. These read:

1. One-eighth of the beam span, center-to-center of supports
2. One-half the distance to the center-line of the adjacent beam; or
3. The distance to the edge of the slab.

The code directs them to select the method that produces the most conservative, or smallest width. Kevin immediately identifies the first method as the most conservative for their specifications. Unfortunately, when they apply this formula, they arrive at the same effective width they have been using all along: 15'. While looking at this page, however, Lynn takes the opportunity to learn general information for future reference. She asks Kevin to confirm that the second method would almost never be the most conservative when dealing with girders, because, she reasons, girders (unlike other kinds of beams) are typically spaced quite far apart relative to their lengths.

At this point, Kevin resigns himself to the conclusion that they have encountered the rare situation in which the concrete is the limiting factor in a composite beam's capacity, and that they will have to trust the program's analysis and recommendation. Kevin writes to me later:

> In the normal span and loading situations that arise on our projects, the concrete strength *never* controls. This was an unanticipated development based on our (my) past experience. Because it never controls, we never check it during preliminary design. This is the first project in my memory where it actually controlled. Henceforth my heuristics will tell me that I have to check concrete strength on long-span girders.

Upon finally accepting the program's recommendation, Lynn and Kevin immediately do what I frequently observed among engineers: play out the implications of the design solution in order to judge its feasibility. Lynn makes a nonmathematical assessment, checking a couple of manuals for the availability of the large girder. Girders this large, she discovers, are unavailable domestically. Kevin runs through some calculations to determine the amount of reinforcing steel (rebar) needed to help the concrete handle the demand. The steel girder's capacity that they have been using is 4840 k, while the slab's capacity is only 2485 k. Again, using the formula (Equation 2):

Equation 2 $$V = A_s F_y$$

Kevin quickly calculates that the difference, a V of about 2000 k, would have to be made up with 50-ksi-strength (F_y) rebar with a total cross-sectional area (A_s) of 40 in². This is an untenable amount. Between the need to import the girder from overseas and the unfeasible amount of rebar required, Kevin and Lynn know that the owners will never opt for this particular floor scheme.

What may seem remarkable about this episode is the amount of time and effort Kevin and Lynn devote to purposes other than directly calculating the size of the girder they need. In fact, that calculation was in some sense unnecessary today: the computer program had found the size days ago. But in my observations, problem solving activity for the purposes of proving, making sense of, and justifying results and methods (rather than simply generating them) is the rule, not the exception. Competent engineering prohibits the thoughtless acceptance of results, whether produced by software or by hand. Such competence necessitates understanding the models underlying computer programs and by-hand algorithms. Taking the time for the problem solving required to tease out these underlying models has both immediate and long-term benefits. It helps prevent errors and generates the best solution for the project at hand. Long term, this kind of problem solving leads to more general learning that improves future practice. As Kevin reflects,

> I've disproven my prejudice that the folks who wrote the program are blithering idiots. They are correct in this instance and I am wrong. History does not treat engineers who are wrong with any kindness. It is well that we discovered my error when we did . . . . Had we blindly followed the program we would have gotten the right answer but missed all the lessons. Learning, as it turns out, is a never-ending series of arrogant statements of fact—postulates, if you will—that are either proven wrong or right. Either way we learn. The key is knowing how, when, and what to check.

Most of the problem solving in this episode was directed at discovering and understanding an existing yet hidden model. But structural engineers also adapt and create conceptual models. The next episode demonstrates that creating model can be at least as problematic as uncovering one but just as rich a site for learning.

THE PROBLEM OF DISCRETIZING TIEDOWNS

Tim, a junior-level engineer, is designing the *lateral system* of a four-story, wood-framed apartment complex. The lateral system is the subset of structural elements that work together to resist a horizontal load—in this area of the county, an earthquake. In this particular building, the lateral system comprises extra-strong, or *shear* walls, about 100 on each story. This is only Tim's second wood project, and he uses a spreadsheet he wrote for the previous one. Tim inputs into this spreadsheet the expected seismic load and various properties of each wall, and the spreadsheet reports the internal forces that develop in each wall as a result.

When a lateral force hits a wall aligned in the same direction, the wall has a tendency to overturn, or cartwheel over. This overturning causes the far edge of the wall to compress and the near edge to pull away from the floor (tension). To resist these forces, devices called *tiedowns* are installed in the edges of the shear walls. Tiedowns are constructed of wood posts, to resist the compression, and steel rods, to tie the wall to the floor (Figure 2–4).

These wall forces, reported by the spreadsheet, dictate the size of each tiedown, but these size decisions are not really Tim's to make. Normally, the structural engineer specifies the sizes of the structural elements in his design. But commercially produced tiedowns are built to stock specifications, with each vendor offering a different size run. It is the contractor's responsibility to select the vendors on a project, and if the engineer were to specify tiedown sizes, it might unduly constrain the contractor's choice of vendor. Instead, Tim will only provide the contractor with the wall-force demands and the contractor will select and purchase the tiedowns adequate to handle them.

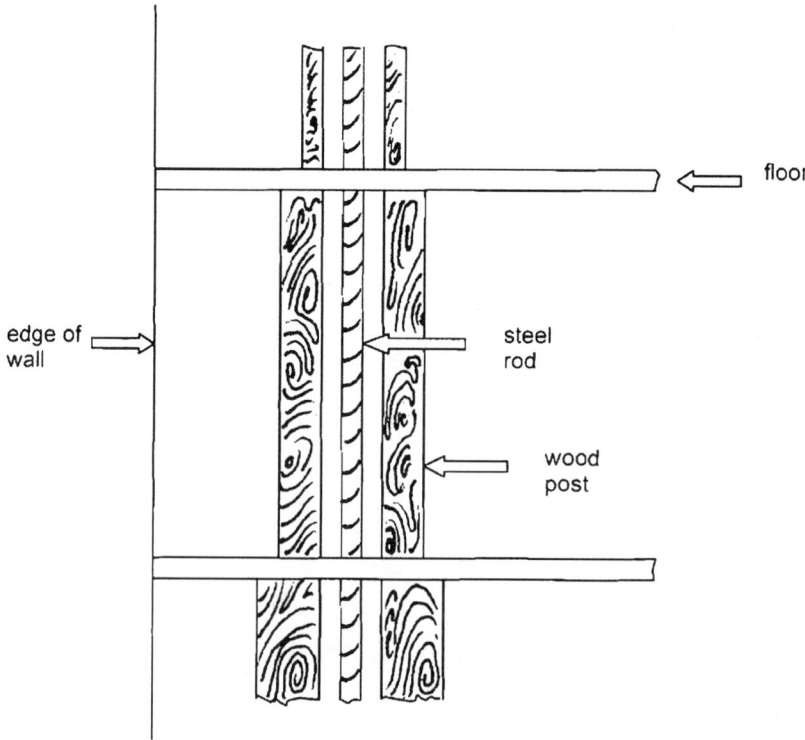

FIGURE 2–4. Tiedown device.

LEVEL	FORCE	T1	T2	T3	T4
4th	Tension	7	5		
	Compression	9	7		
3rd	Tension	16	13	3	
	Compression	20	17	5	
2nd	Tension	29	25	14	3
	Compression	33	29	18	5
1st	Tension	43	40	25	7
	Compression	47	44	29	11

FIGURE 2–5. A table of tiedown forces Tim generates during the task. (*Note:* this vignette concerns tension forces only.)

With nearly 400 walls in the complex, Tim's spreadsheet reports dozens of different tension forces. So Tim must "discretize" these forces, that is, group them so that only a small number of rod sizes need to be used. This grouping task turns out to be highly complex, as Tim gradually discovers. Tiedowns are multi-story devices, with the rods and posts varying (usually decreasing) in size from lower to upper stories. Tim must specify the forces that each tiedown will be required to handle on each story (Figure 2–5 shows a table of forces that Tim generates later in this episode).

Ultimately, Tim will decree that a hypothetical tiedown, called, for example, T1, must be designed to handle tensions up to a, b, c, and d kips, respectively, from the first to fourth stories. Tim will then identify which walls in the building should receive the T1 tiedown—those whose four tension forces are each below their respective story cutoff force of a, b, c, or d kips, but exceeding what the next smallest tiedown (say, T2) would hold. When the contractor receives these cutoff forces, he is free to shop for any commercial tiedown device to be T1, as long as its rods' capacities exceed a, b, c, and d kips, respectively, at each story. Tim will communicate to the contractor the tiedown type (T1, T2, etc.) assigned to each wall and the cutoff forces for each type, but he will not show the contractor the actual tension forces on any particular wall.

Though Tim may not design the actual devices, he knows that he virtually controls the contractor's decisions. Tim has the manuals from two major tiedown vendors, which list the capacities of the available rod sizes. Tim knows the contractor will assemble the most efficient device, the one with the smallest four rods that handle the cutoff forces. Looking in a manual, Tim can predict exactly which rods the contractor will choose given the four force cutoff points. By simply increasing the cutoff force on a story, Tim can force the contractor to move to the next largest rod.

To guide him in this discretization task, Tim refers often to a tiedown table from his previous wood project. But he also uses a metaphor to help him conceptualize this ill-structured and uncertain process: "I'm kind of like fitting the glove onto the blob of numbers, and I want to make breakpoints rationally." The image is vivid: Tim must stuff a shapeless set of hundreds of numbers into a few fingers of a glove, each finger representing a single rod size, in a way that best fits the data. Discretizing is a process common, in some form, to many design projects. Tim describes an earlier experience grouping piers (foundation columns) for a schedule (table of specifications):

> And so I was thinking, well, what am I doing here? It's kind of like glove fitting or curve fitting. The better I do it, the more money they're going to save. But that's not true either. If a glove fit perfectly, it would be nice in some sense, but it would be a bear to build, because there would be no pier schedule. The schedule would have 150 lengths in it. And so it wouldn't really be a schedule at all; it would just be a nightmare.

Tim begins by looking at the tension forces on his spreadsheet for four-story east-west walls (the complex also has three- and two-story walls). He jots down the range of forces on the first story, where tensions are highest: 43–26 k. He wonders if he should designate just one tiedown, with a rod that handles at least 43 k, which would be overkill on many walls, or if he should "value engineer" by designating two tiedowns, the second using a rod with a capacity of, say, only 35 k. Tim looks on his spreadsheet for the north-south four-story walls and discovers that the maximum tension force in this direction is 49 k. This force is problematic. The largest rod listed in the vendor manual, with a 1" diameter, only has a capacity of 42 k. Using a formula in the manual for finding the tension capacity of a rod given its cross-sectional area, Tim back-calculates the diameter necessary to handle a 49-k tension, which he rounds up to 1–1/8". They are not listed in the manual, but Tim feels sure this vendor sells 1–1/8" rods because he recalls using them on his last project. Now he reverses the process, using the same formula to find the tension capacity of the 1–1/8" diameter rod; this turns out to be 53 k.

Tim grapples with a model for this discretizing task. Noting that the minimum and maximum forces differ roughly by a factor of two, he sketches two overlain bell curves, one wide and one tight, to illustrate possible distributions of the tension values (Figure 2–6).

He reminds himself to take into account the actual rod capacities when determining his cutoff, and he decides to break the range of forces into two groups: 40–50 k and below 40 k. This would force the contractor to choose the 1–1/8" rod for the first group and the 1" rod for the other. This is a clear example of how Tim can manipulate the contractor. Even though the 1" rod is rated up to 42 k, by setting the

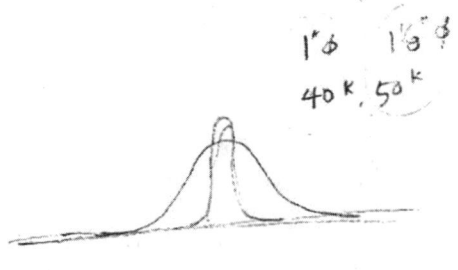

FIGURE 2–6. Tim's sketch of overlain bell curves, representing possible tension distributions.

cutoff point at 40 k Tim forces the contractor to use the 1–1/8" rod for walls with forces between 40 and 42 k. This reflects one of Tim's goals for this grouping task: to build in a margin of safety that he expects the contractor will not. Since Tim will not reveal the actual forces on each wall, he can assign the larger tiedown to a wall with only 41 k of tension on the first story, forcing the contractor to install the larger rod here even though the smaller rod would suffice. But working this way requires a triple bookkeeping system. Tim must keep track of the published rod capacities (in this example, 53 k and 42 k), the actual wall forces (here, the 41 k), and the force cutoff point he will show the contractor (the 40 k).

Deciding to use two sizes of tiedown for four-story walls seems to cause Tim to have a minor revelation about his process. In the prior project—his first experience discretizing tiedowns—he had simply used a single tiedown for all four-story walls (which he called T4), a lighter-duty one for three-story walls (T3), and an even lighter one for two-story walls (T2). He now realizes the inefficiency of that model, saying, "I didn't fit this solution very tightly. I saved time and I just cranked out these points in terms of geometry really, not in terms of load." He now believes his current method will result in more appropriate ties and less overkill.

Tim draws a table on scrap paper, with rows for the stories and two columns headed T6 and T5 and announces that he is abandoning his naming convention from the last project, when the tiedown name matched the number of stories. In the box for the first story of T6 he writes 50 k, for T5, he writes 40 k, and, as an afterthought, he jots the rods' diameters in the boxes, too. Tim now examines the second-story tension forces on his spreadsheet and records their range, 32–17 k, on his scrap paper. He sees in the vendor manual that the next rod size down, 7/8", has a capacity of 32 k. If he lists the maximum force for the second story of T6 as 32 k, which is the truth, the contractor will select the 7/8"rod. But Tim considers this too close for comfort on the 32-k walls. So he simply writes a higher maximum force, 35 k, on the table, forcing the choice of the 1" rod.

As Tim goes on to determine a cutoff for the second story of T5, he realizes a further complexity of the process. He remembers that he has not yet paid attention to the possibility of a wall that might, for example, fit the T5 force ranges on every story but one, which would bump the entire wall up to a T6. Avoiding cases like this would require Tim to consider the walls as ground-to-roof wholes—coherent sets of four forces each—and make sure his discretization scheme aligned with these whole-wall force patterns. He tries to craft a logical approach to this problem with a geometric sketch: a right triangular graph in which the vertical axis represents the stories, the horizontal axis the forces, and the median line his cutoff points at each story (Figure 2–7).

FIGURE 2–7. Tim's geometric representation of the tension forces and the cutoff point.

He explains:

> Wall loads tend to increase linearly from the roof to the ground. A bad wall is going to be bad on every floor and a [good] wall's going to be good on every floor, low and high, so I should see sort of a triangular distribution. Which would mean, I guess, if I'm taking these pairs of numbers, I should try to split them in equal parts at every floor.

At this point, Tim decides to prevent further confusion by creating a second table showing both the actual (secret) forces, which he labels the "Demand," and the ones he will report to the contractor, the "Capacity," as well as the rod diameter that must be chosen to satisfy the capacity. Tim fills in his force decisions so far, but when he reaches T6 on the second story, he sees "a big jump" from the actual demand (32 k, the highest force from the spreadsheet) to the capacity (42 k) of the 1" rod he has forced the contractor to choose. He worries that the contractor might challenge this obviously large rod. Tim reviews his process and now articulates another goal: minimizing construction-worker error, whose probability increases with the number of different-size components specified for installation on each story. Tim realizes that the discretization model used on the previous project minimized potential error by calling, in most cases, for a single rod size per story. With this goal in mind, Tim moves on to the fourth story and selects 3/4" rods for both T6 and T5, though a smaller rod would suffice for most walls.

Tim turns to the tiedown table from his prior project, which lists the component sizes, and is reassured to see that many of those rod sizes match what is emerging for his current project. He takes this as a rough confirmation of his analysis, since both buildings are located in the same seismic zone. Tim mulls over his process and the different possible approaches, including construction feasibility and safety:

> This is the hard thing about design; there are too many options; too many ways to do this. I could bump this, play with these numbers and sizes until they all had some other—I could adjust these from a view of uniform constructability and, like I was just talking about—Right now, I've just taken a first pass at it from another vantage, or from another angle, which is the factor of safety margin, looking at the difference between the demand and the capacity. And as a first pass on design I think that's a rational way to do it. And then when I get done, I guess I can look at: does this mean—? Am I being too fussy in my step size? Stuff like that . . . . So, I think this is good. I'm more comfortable doing it this way. Now, when I did this job the first time, I didn't really understand what I was doing as well as I do now, didn't understand the implications of it.

Referring back to his prior project apparently convinces Tim to reverse the naming system for his current tiedowns, changing T6 to T1, and T5 to T2. At the bottom of his table, he brackets these two columns and labels them "4-story." He begins a column for T3, a tiedown for three-story walls, following the same process as before. He quickly decides to have only one three-story tiedown type, presumably because these shorter walls have lower forces overall, which means fewer savings could be reaped from an additional tiedown. Determining the forces for T3 takes only few minutes; then Tim designates a single tiedown, T4, for his two-story walls.

Tim moves on to other design tasks, some of which alter some of the forces in his spreadsheet. Reviewing the spreadsheet later, Tim sees that the largest tension has popped up to 52 k, just under the 53-k capacity he had calculated for the 1⅛" rod. Giving no reason, he decides to change the 3/4" rod on the top (third) story of T3 to 5/8" and use the same rod for the top (second) story of T4. Perhaps on purpose, the rod sizes in his table now follow a pattern he finds pleasing: the two-story T4 is identical to the top two stories of T3, which in turn are identical to the top three stories of T2. In other words, the rod sizes of T2 through T4 are the same along the diagonals of the table. Later Tim tells me that the emergence of this pattern suggested to him the possibility of some underlying physical logic to his latest model.

Tim phones the vendor with some questions about the devices. When the vendor hears about the 52-k forces, he calls them "ridiculous." The problem is not the rods but the infeasibility of the gigantic posts necessary to resist a 52-k compression force (the compression and tension on each wall are equal). The call motivates Tim to modify his building design, adding shear walls, eliminating a few very short walls with high forces, and altering the location of the rods in the wall. Ultimately, he manages to reduce the highest force to 43 k.

Tim starts to rework his table in light of these new forces. Noticing that he had assigned a rod with a 42-k capacity to the first story of T2, the smaller four-story tiedown, he remarks he could almost get rid of T1, now that the highest force is only 43 k. He recalls that he had chosen the first-story cutoff demand for T2 to be 40 k, to force the choice of a 1" rod. He writes 40 k on his new table, then looks at the second-story forces on his spreadsheet to determine an appropriate tension demand for T2. Because this is the lighter tie, he is not looking for the maximum tension but for a reasonable breakpoint. Though his old table gives a tension demand of 25 k here, he writes 23 k on his new table based on what he sees on the spreadsheet; similarly, he decides on a tension of 12 k for the third story of T2, though his old table says 13 k. Then he remembers he must also pay attention to the whole-wall force patterns, to make sure the breakpoints he chooses align with those patterns and do not exclude most walls just because one story's designated demand is disproportionately low. He considers using a ratio, in other words, adopting a pattern of linearly increasing tensions on a tiedown from story to story. But he drops this idea and resorts to examining some sample walls one at a time. He randomly chooses Wall 19 and looks at its forces on his spreadsheet. Its tension of 26 k on the second story pushes it up to the T1 tie. He calls this a bad example and tries Wall 25, but with a second-story tension of 25 k, it also fails to fit T2. This convinces Tim to increase his cutoff demand on the second story of T2. On the third story, T2 still fails to work for Wall 25, so Tim raises this cutoff as well. He notes that he seems to be recreating his old table, even though those cutoffs had been developed with regard to rod sizes rather than actual tensions. He adds tension cutoffs for the top two stories of T2, then quickly fills in T3 and T4 with the maximum values he sees on his spreadsheet, since these are the only tiedowns for three- and two-story walls, respectively.

In the next two days, design changes again shift the tension forces on the walls. Tim pulls out his newer tiedown table and announces that he needs to make sure it still works for the latest changes, though he confesses he is tired of this process. He starts with the new maximum force, 40 k on the first story, and sees this is easily covered by T1's capacity of 43 k. Tim sees that he had chosen a demand of 40 k on the first story of T2, his lighter four-story tiedown, to force a larger rod size. Now that the maximum force on the first story has come down to 40 k it is within the capacity of T2, and Tim muses that he should eliminate T1. He decides for now to keep it "as a placeholder," in case future changes result in walls with higher forces. Also, he remembers he still has not checked to see if high forces on other stories still put some walls out of the range of T2. When he checks, he sees that the maximum force on every story of the four-story walls is below the cutoffs for T2. Again, he considers going with just a single type of four-story tiedown. He anticipates how two types will complicate the next phase of the task, when he will have to go through every ground-to-roof wall in the project and assign it a tiedown type, T1,

T2, etc. It dawns on him just how unwieldy this procedure could be. There is no easy system for compiling the story forces on every wall. Tim would either have to leaf through the drawings of every story of every wall or tab through every spreadsheet page, comparing the four tension forces to the T2 cutoff forces to see if this lighter tiedown would suffice on every story. Tim is also reminded of the potential confusion of workers carrying multiple sizes of rods to each story. By now, Tim has talked himself into a single four-story tiedown.

Once he removes T1 from his table and is left with only three tiedown types, Tim sees it makes sense to reverse his tiedown names again, calling the four-story tie T4, the three-story tie T3, and the two-story tie T2. Ironically, despite days of work and his earlier claim that "when I did this job the first time, I didn't really understand what I was doing as well as I do now," Tim has come full circle to the same discretization model he had used on his previous project: a "geometric" pattern, where tiedown type is related to the number of stories. Still, the experience has deepened Tim's understanding of the discretization process. He will approach the next such task with a better feel for the appropriate balance point between too many and too few groups of elements. He will be savvier when navigating the odd rules that govern communication with the contractor to attain the desired outcomes. He will know some systems for recording the real and "secret" numerical information and recognize the range of wall forces considered acceptable in the industry. Finally, he will have a better sense of the utility (or lack thereof) of general mathematical patterns as models for grouping. That the old and new models are similar is only superficially significant. Tim knows it is critical to develop a rationale for whatever his final model; like Kevin and Lynn, he will not adopt a model or solution that he does not understand.

THE NATURE OF STRUCTURAL ENGINEERS' PROBLEM SOLVING

These two episodes illustrate common features of the problem solving processes that I observed throughout my study. The crux of most of the significant problems that structural engineers face is to find or develop an appropriate mathematical or conceptual model for a complex system. For Tim, it was to develop a way to systematically group (and communicate) the forces for the tiedown devices; for Kevin it was to discover the mathematical model underlying a computer analysis program as well as to better understand a limitation of a more familiar model. The design solution itself (for example, the size of the beam or force range for a tiedown) is relatively useless to the engineer without the understanding of the logic behind it and the process that generated it. One reason this is true is that the knowledge of the underlying logic is needed to justify the solution—critical in a field where safety is a major factor and proof is required. Another reason is that it is the model, and the understanding about it, that is potentially reusable, that helps the engineer to approach the next project more competently. The direct utility of the specific results of previous work is limited (the case with the tiedown table from Tim's last job), so it is important to emerge from the current experience with increased but more general knowledge.

Problem solving in structural engineering rarely follows a clear path. Many routes are usually attempted and the process involves a considerable amount of testing and revising models, methods, and theories. Neither Tim nor Kevin was able to generate immediately the correct or final model for the system in question; several cycles of trials and revisions were required. Moreover, solving structural engineering problems is not synonymous with mathematical problem solving. Even though developing or understanding a mathematical model may be the goal, multiple resources and constraints—theoretical, practical, and political—shape the process. Tim needed to find a number of tiedown types that would be considered feasible by project partners, and his system of choosing and representing the force ranges had to take into account the norms for communication with the contractor as well as safety and practical matters. Kevin and Lynn knew better than to take their beam to a final design solution; for economical reasons the owner would never select the scheme that included it.

THE LEARNING OF STRUCTURAL ENGINEERS

Kevin, Lynn, and Tim make clear that structural engineers in everyday practice do more than just drive towards the immediate design or analysis solution. Despite the constant pressure to make efficient use of their and their clients' time, engineers will deliberately do extra mental work, in the form of testing, experimenting, and triangulating, to attain general learning that will help them in the future. Even when not expressly engaged in the effort to build their expertise, the work they habitually do to justify solutions, sleuth down the source of errors, or uncover the logic underlying the programs or algorithms that generate solutions is likely to produce general learning with future utility. Structural engineers do not need a "teacher" (or direct training) that guides them along a specified path to greater conceptual development and modeling expertise. They can learn naturally in the course of practice; in fact they must, as there is rarely an omniscient expert with "the answer" (the correct model or method) to real structural problems. Established models and methods are available but can almost never be applied wholesale to an actual project, if their application is even recognized, because projects vary widely and incorporate systems far more complex than those covered by industry-established models and forms. Learning in structural engineering is not necessarily predictable. Tim could probably have expected that he would emerge from this project having learned about discretizing, and his firm could deliberately have assigned him to such a project to boost his expertise in this area. On the other hand, Kevin and Lynn could not have predicted that sizing this girder would teach them something new about their analysis program.

What I observed being learned varied, but it would be difficult to make the case that the engineers became more expert in the use or understanding of the general mathematics they employed in their modeling. Thus, the results of my study shed little light on the question of how effectively authentic modeling activities can teach children general math concepts. It may be that professional engineers have "topped out" in general mathematics, that is, they do not learn general math on the job because they are already proficient in the requisite math skills. What these engineers do learn appears more situation specific or localized: extensions of applications of known models to more situations or new models for specific situations. Yet there is reusable value in much of this, depending on the breadth of the situations. Some of what Tim learned about discretizing was connected to particulars of the tiedown device and the way it was manufactured and vended. Some of what he learned, though, could apply to the grouping of other elements or to communicating with vendors and contractors under similar conditions. Kevin and Lynn learned of a limitation to a calculation method they have used and will continue to use routinely. Further, the information they gained about the model underlying a frequently used computer program will certainly sharpen their future interpretations of its analytic results. These observations speak to a broader debate about learning transfer. Situated theories of cognition connect learning tightly to the specific context in which it arose (Lave, 1988). Critics of these theories find no room in them for the transfer of concepts learned in one situation to another (e.g., Anderson, Reder, & Simon, 1996). What the engineers in this study learned while solving problems at work is deeply colored by the specifics of those problems, illustrating the situated nature of their learning. Yet, some aspects of that learning will be reusable in new problem situations, especially with conscious reflection by the engineer on what new information might generalize to upcoming projects or, as Kevin says, with deliberate effort to not "miss all the lessons." Indeed, the engineers saw the accrual of situation-specific knowledge as the route to advancing their engineering expertise.

Finally, there is probably no endpoint to learning in structural engineering. Modeling and the application of concepts are never mastered; there is no ultimate level of an engineering concept or theory or its application that could be considered to encompass its totality.[3] Individual engineers develop divergent patterns of understanding—patchwork quilts of (situated) conceptualizations based on their personal problem solving experiences. Neither Tim nor any other engineer will ever completely "get" the concept of discretization, nor will they ever accrue the complete collection of appropriate discretization models for every possible engineering situation. Indeed, neither will the industry as whole.

EDUCATIONAL IMPLICATIONS AND FURTHER QUESTIONS

The episodes presented here and the others in my study paint a picture of problem solving activity among workers in a math-intensive profession that resonates in many ways with the goals and methods of model-eliciting activities (MEAs). The way structural engineers solve problems and learn on the job parallels the notion of local concept development in children, a process that is more situated, piecemeal, multidimensional, and unstable than has been depicted by traditional versions of cognitive development (Lesh & Harel, 2003). The open exploration of concepts, within a larger effort to model complex systems, is central to structural engineering work both because it is how rational solutions are generated and because it produces the kind of learning that builds expertise and advances the practice. The specifics of implementation are not clear, but the idea of K–12 curricula that encourage model eliciting and open exploration makes sense, especially if we are concerned about preparing students for math-intensive occupations.

Lesh and Harel also discuss the notion that we can push students beyond the natural development process with enriched environments that "induce" enhanced concept development; indeed, this is the ultimate goal of the designers of MEAs. The episodes in this chapter have shown that structural engineers, at times, will engage in something like the inducing of their own learning. A possible frame for some of Tim, Kevin, and Lynn's behavior is that these engineers were constructing local environments for learning about the concept at hand, engaging in cognitive activity above and beyond what was required just to solve the immediate design problem. This suggests that engineers take deliberate action to enhance the educative potential

---

[3]This point may instantiate a broader issue in discussions about learning and transfer. Given the implausibility of a person's learning all possible applications (projections) of a complex concept (coordination class), diSessa and Wagner (2005) debate what kinds and degrees of partial concept construction should be considered "complete."

of their everyday problem solving experiences. If so, this could be considered an additional component of the intellectual development process of engineers—a metacognitive, self-teaching component—that should perhaps be added to our articulation of the concept development process in children.

The study reported in this chapter raises more questions than it answers about the learning of engineers and what it implies for students, but it points some directions for future research. Longitudinal case studies of individual high-tech workers could follow the trajectories of their personal development of particular models, to see if and how those models broaden or become more reusable and transportable, more flexible across an increasing range of situations. The episodes here and others in my study suggest that learning structural engineering depends on accrued modeling experiences in a variety of contexts, combined with testing, revision, and reflection. But my observation methods would not have detected the role played by sporadic and deliberate teaching events that are not part of everyday work, such as conferences, journal reading, and seminars, in advancing conceptual development; these would be important to study. It also would be enlightening to explore the role of engineers' deliberate efforts to teach themselves general information about models or their applications and, similarly, the role of colleagues' and supervisors' deliberate efforts to teach these things, to determine if such efforts are necessary or whether this learning can happen adequately as a natural byproduct of everyday work. Other case-study research could compare junior- and senior-level engineers' problem solving and facility with models. This could illuminate which kinds of modeling behavior are most efficacious and productive, in other words, what "expert" modeling entails.

Equipped with a better understanding of the modeling behaviors of adult experts (including experimentation, iteration, logical testing, application of abstract mathematical concepts, multiple representations, and seeking underlying assumptions), educational researchers could conduct more targeted studies in the classroom that examined the effectiveness of existing curricular activities, such as MEAs, for promoting those desired behaviors. Conversely, research could search for forms of classroom activities that best promote those behaviors.

ACKNOWLEDGMENTS

This study was made possible by a Stanford Graduate Fellowship and invaluable advising from James Greeno and Diane Bailey. I thank David Miller for his editorial assistance.

REFERENCES

Anderson, J. R., Reder, L. M., & Simon, H. A. (1996). Situated learning and education. *Educational Researcher*, 25(4), 5–11.

DiSessa, A. A., & Wagner, J. F. (2005). What coordination has to say about transfer. In J. Mestre (Ed.), *Transfer of learning from a modern multi-disciplinary perspective* (pp. 121–154). Greenwich, CT: Information Age Publishing.

Gainsburg, J. (2006). The mathematical modeling of structural engineers. *Mathematical Thinking and Learning*, 8(1), 3–36.

Gainsburg, J. (2003). The mathematical behavior of structural engineers. *Dissertation Abstracts International*, A 64/05.

Lave, J. (1998). *Cognition in practice: Mind, mathematics, and culture in everyday life*. New York: Cambridge University Press.

Lesh, R., & Doerr H. M. (2003). Foundations of a models and modeling perspective on mathematics teaching, learning, and problem solving. In R. Lesh & H. M. Doerr (Eds.), *Beyond constructivism: Models and modeling perspectives on mathematics problem solving, learning, and teaching* (pp. 3–33). Mahwah, NJ: Lawrence Erlbaum Associates.

Lesh, R., & Harel, G. (2003). Problem solving, modeling, and local conceptual development. Monograph for International Journal for Mathematical Thinking and Leaning. Hillsdale, NJ: Lawrence Erlbaum Associates.

Mathematical Association of America (MAA). (2000). *Proceedings of the Curriculum Foundations Workshop in Engineering.* Clemson University.

Torraco, R. (1999). Integrating learning and working—A reconceptualization of the role of workplace learning. *Conference proceedings of the Academy of Human Resource Development.*

CHAPTER 3

Modeling Without End: Conflict Across Organizational and Disciplinary Boundaries in Habitat Conservation Planning

Bruce Evan Goldstein
Virginia Tech

Rogers Hall
Vanderbilt University

Practices of modeling that involve real life situations are increasingly seen as both the context for and object of mathematics teaching and learning (e.g., Greeno & Hall, 1997; Lehrer & Schauble, 2004; Lesh & Doerr, 2003). Open-ended, model-eliciting tasks are expected to provide learners with an opportunity to find relevant problems in complex situations, to develop representational tools for describing and analyzing problem structure, and to compare different approaches to solution. Negotiation over model assumptions, interpretation and explanation of model behavior, and model revision in response to evaluations of findings are highlighted as modeling practices that provide both a learning environment and an image of what should be learned. Mathematics classrooms designed to facilitate these modeling practices are expected to engage learners' interests in the real world, to resemble professional practices in ways that are meaningful for learners, and to encourage deeper conceptual understanding of key mathematical concepts for both learners and teachers.

In this chapter, we analyze a case in which modeling becomes "too real" for participating scientists, to the point that modeling goes on, it seems, without end. Participants cannot agree on what are the relevant problems to solve, representational tools are developed and discarded without a clear evaluation of progress, and cycles of model revision do not converge or terminate with findings or explanations that satisfy major stakeholders in the modeling activity. The case is drawn from a four year, ethnographic study of efforts to design a multiple-species habitat conservation plan (MSHCP) for an ecologically sensitive desert region in the southwestern United States (Goldstein, 2004). We focus on work by members of a scientific advisory committee (SAC), locating their efforts within a larger history of plan development. We also look closely at conversational exchanges between scientists, land managers, and plan consultants in a meeting called to review additions/deletions to an existing model of occupiable habitat for an endangered lizard species.

We start by describing the history of conservation planning in the Valley (names of all participants and locations are pseudonyms), leading into the MSHCP planning effort used as a case in this chapter. We interrupt that historical narrative to present two scenes from a SAC meeting convened to identify lizard habitat that should be included in the multiple species plan. Talk-in-interaction from these scenes is analyzed closely to explore differences in professional point of view towards land in the Valley, listed or endangered species that live there, and human activities that increasingly determine the welfare of those species. Land managers in the Valley, regulatory biologists from state and federal agencies that encompass the area, and local biologists with a professional history of studying Valley species (like the fringe-toed lizard) each see the planning process and its outcome differently. A potentially volatile disagreement in the meeting is contained by a decision to fall back on prior planning models for the lizard, but as the meeting ends, this is clearly a tenuous settlement between regulatory and local biologists. We then resume the planning history, describing the fate of this effort to contain disagreement, of the SAC as an organizational entity, and of the MSHCP itself. We conclude by identifying aspects of model construction and negotiation that cross organizational boundaries and may be particularly relevant for cognitive studies of educational practice.

FROM NATURAL PRESERVE TO A "BALANCE OF TERROR" BETWEEN RACKET CLUBS AND LIZARD HABITAT

Table 3.1 shows a narrative timeline of conservation history in the Valley, starting with the region's relative isolation (because of surrounding mountains and desert). This natural isolation was opened with installation of a canal and water supply in the late 1940's (Event 2), which enabled a dramatic expansion of the human population, along with construction of golf courses, tennis clubs, and hotel and related entertainment facilities. Sand-dependent species like the fringe-toed lizard (FTL), found only in the Valley, went into an equally dramatic decline, as habitat was lost to human development. Local biologists, some participating in the SAC meeting analyzed in this paper, successfully petitioned the U.S. Fish and Wildlife Service in the early 1980s to list the FTL as an endangered species. When a new golf course development was proposed in critical lizard habitat, local biologists and developers threatened each other with legal action, creating what one biologist termed a "balance of terror" between the two sides. After a few years of standoff, both sides formed a working group they called the "lizard club" and began preparing a habitat conservation plan (HCP), a regulatory structure newly authorized by the U.S. Congress that allowed regional stakeholders (including cities, land managers, and property owners) to create plans that permit "take" of species and habitat in exchange for mitigation fees, if developers can demonstrate continuing species viability. The fringe-toed lizard HCP was approved in 1986 (Event 3).

With evidence growing that 1986 assumptions about lizard habitat were inaccurate, and in the context of a fierce debate about other listed species in the Valley (Event 4), local biologists assembled a planning group to work on a multiple-species HCP that would include (and revisit) the existing lizard plan. A scientific advisory committee (SAC) consisting of local biologists, regulatory biologists, and area land managers was established and supported by plan consultants in 1996, who expected

TABLE 3.1.
A Narrative Timeline of the MSHCP Planning Process
Running up to the SAC Meeting Analyzed in this Chapter

1500s–1900	(Event 1) Mountains and desert isolate the valley from large-scale land conversion and exotic species that were introduced in other parts of California, creating a natural preserve for endemic valley species.
1948	(Event 2) Construction of a canal makes water available for a substantial increase in human development, including recreational facilities (e.g., golf courses, hotel/entertainment centers). The valley environment is rapidly impacted by human use, and sand-dependent species go into sharp decline.
1980s	(Event 3) Local biologists and developers resolve a "balance of terror" by collaborating in preparation of a habitat conservation plan (HCP) for the fringe-toed lizard. The plan is approved in 1986 by regulatory agencies and local municipalities (the second HCP prepared in the nation).
1990–1996	(Event 4) Amidst controversy over other endangered species and growing uncertainty about the viability of lizard populations in the valley (e.g., assumptions about sand sources), local biologists (several from the 1980s "lizard club") begin discussing a multiple-species HCP. The SAC is convened in 1996, and in their proposed timeline, an approved MSHCP would begin issuing permits in 1998.
1996–2000	(Event 5) Despite guidance provided by eminent external scientists (in 1996 and 1998), there is ongoing disagreement between regulatory and local biologists over assigning conservation value to mapping units in the valley and which areas to include in the plan. After meeting in private for months, regulatory biologists release a 41-page letter, calling for inclusion of new areas and contracting a new hydrology study (2000). Local biologists are furious, some concluding they have been "rejected entirely."
October, 2000	(Event 6, meeting analyzed in this chapter) With new preserve areas and a pending hydrology study still on the table, the SAC meets to discuss lizard habitat as one component of the larger MSHCP.

approval for the new MSHCP within two years. As planning proceeded, the modeling procedures for the SAC habitat preserve design were created with the advice and oversight of a panel of eminent scientists (Event 5), although internal disagreements between local and regulatory biologists became increasingly heated. Regulatory biologists, preferring a plan area that was more expansive than local biologists would allow, began meeting in private, then issued a lengthy criticism of the SAC plan and commissioned scientists from a separate federal agency to study the "hydrology" of wind-blown sand across the landscape, focusing on critical linkages between sand dunes and the mountainous areas that serve as sand sources. Local biologists and even plan consultants felt their work had been undermined by regulatory biologists on the SAC, and this trouble was still underway as they met to revisit the FTL lizard preserve (Event 6, the meeting analyzed in this chapter).

MODELING IN THE WILD: TROUBLE IN SCENES
FROM A SAC MEETING

We now examine two scenes from a SAC meeting (Event 6, Table 3.1) held four years after the committee was established to develop a multi-species HCP for the

FIGURE 3–1. Seating order, participants and active map documents in SAC meeting to discuss conservation plan for an endangered lizard species in the valley. Different professional roles include plan consultants (PC), regulatory biologists (RB), local biologists (LB), and land managers (LM).

valley. As described above, there was substantial disagreement between local and regulatory biologists over which areas of the valley should be included in the plan (Events 4 and 5).

As SAC participants come to the table, they take seats in a pattern that reproduces the growing divide between local and regulatory biologists (Figure 3–1). Local biologists sit together to the left (Nick, Bert, and Dave are active participants), while regulatory biologists cluster at the upper right (Linda, Randy, and Charles). Plan consultants (Mary, Vera, and Edward) take seats that mark a boundary between local and regulatory biologists in the seating order. Land managers and representatives of other stakeholder organizations are distributed across the remaining seats. The purpose of this meeting is to identify preserve areas for a single valley species, the fringe-toed lizard (FTL), which is already protected under a habitat conservation plan established in 1986 (the second HCP approved in the nation, Event 2 in Table 3.1).

Scene 1: Using Satellite Imagery to Model Sandy Lizard Habitats

Dave, a local biologist and manager of the existing FTL preserve, has worked with plan consultants to bring a new map to the meeting. The new map shows soil types by color, and Dave proposes that he has developed a reasonably accurate way to identify areas in the valley that are either good or poor habitat for the lizards. Mary (a plan consultant) has already placed the new soil map at the center of the table, next to a map showing the existing FTL preserve areas. Referring to the new soil map, Dave talks[1] about what a computer can "see" when analyzing data taken from a satellite orbiting the Earth.

Binding reflected light to habitat-species relations and human activity. Dave proposes using reflected light, measured by a satellite, to represent the grain size of soils, associating a particular reflectance range with wind blown sand (i.e., colored red in the new map). Loose sand with this small grain size is good habitat for fringe-toed lizards (a listed species), which "swim" in the sand as part of their daily round. Dave, in his daily round as a local scientist and manager of the existing lizard preserve, has visited these red areas and reports they have wind blown sand inhabited by FTLs.

> **DAVE(LB):** The hot red? Um, Well, these are reflectance values that the computer sees from um, an Ekinos satellite image that's forming a resolution using color, three color infrared...um scanning and it APPEARS based on my field checking and and experience out there, that the RED areas are the areas with the most active and um LEAST compacted sand. And in areas within the preserve boundary that I have checked where it's red, there, it's virtually always occupied by fringe toed lizards.

Three measures are bound together in Dave's extended utterance: (1) A remote sensing measure of light intensity is linked to the relation between (2) grain size and compaction of soils as this provides habitat for (3) confirmed sightings of individuals of an endangered species. Dave speaks as an authority for lizards and the local terrain, and his work in the field and on the computer binds light to lizards and soil quality. The outcome, as proposed and resting on the table top, is a map of the valley with a colorized layer that sorts good ("hot red")" from poor lizard habitat (orange, yellow or brown, described below).

> And the areas that are ORANGE, it's a mix. Um there's some fairly unconsolidated or un-compacted material, but some compacted material as well. And it's sort of intermediate in character and in THOSE areas, so far anyway, I've tended to find LESS fringe toed lizards although some, but um, the flat tails are more common in that particular color type.

HANK(LM): The orange?
DAVE(LB): The orange. The, that's, and I'm just speaking from within the preserve cause those are the only areas I've checked.

> In the yellow area, it's MUCH more compacted, sometimes almost to the extent of feeling like cement, and tends to get coarser and then the brown, which there isn't much there, has got a lot of rocks and gravel and um boulders mixed in with it.

During Dave's visits to the mapped areas, as the sand becomes more compacted (i.e., areas colored orange and brown by satellite data), there are fewer FTLs and other lizard species ("flat tails")" begin to appear. So reflectance values do seem to represent soil type and habitat-species relations accurately. Dave's reflectance layer could provide the SAC with a defensible way to identify current FTL habitat, including areas that have become more or less suitable for lizards since the 1986 plan was approved.

---

[1]Transcripts identify speaker (Figure 3–1) with professional affiliation and show contiguous talk unless otherwise noted. Extended turns are broken at thematic boundaries. EMPHATIC speech is shown in upper case, stre:::tched enunciation is shown with repeated colons, ((*action descriptions*)) are show in italics within double parens, and [overlapping talk is marked with [matching square brackets across speaking turns.

So, one COULD um...figure that if you were just to map the red stuff, that would show you the extent of currently occupied habitat and it, it, at least in the little field checking I've done, which is...just the preserve and the areas I'm familiar with it seems pretty consistent.

You can see over there by Terry's dune area that it's, that's, if it wasn't for the off-road vehicles I'm sure it would be occupied and I'm sure it still has a few lizards in it.

But it also shows the area in front of Stove Top as being bright red and that's an area that just gets hammered by off-road vehicles, so the image may be picking up the fact that it's loose fluffy material because of that and whether or not it would ever be occupied if the um...vehicles were removed, I can't say.

Dave cautions that reflectance values, alone, are not sufficient. Some areas within the selection threshold for fluffy sand (i.e., red areas) do not currently have lizards because of human activity (e.g., off-road vehicles), and that soil quality may even be a result of human activity. In this sense, level of human activity (e.g., terrain "hammered by off-road vehicles")" is added as a fourth measure to the emerging model of suitable FTL habitat.

For a study of modeling in the wild (following Hutchins, 1995), Scene 1 looks like a moment of rapid progress in work by the SAC. The area and relations Dave wants to model are already captured by earlier modeling efforts (the 1986 plan), yet assumptions behind those prior efforts may no longer hold. Dave proposes they displace that earlier model (and its settled agreements) with another, in a way that can be shown to be consistent with the existing preserve but more accurate. To do this, Dave proposes a new representational layer that binds satellites to lizards, linking four measures (light intensity, soil quality, lizard activity, and human activity) to show what has changed since the prior modeling effort.

Regulatory biologists question accuracy and coverage for their anticipated use. Dave's proposal to identify lizard habitat as the correspondence between reflected light and confirmed lizard/human activity seems like a reasonable set of model assumptions. Designated habitat (i.e., what the plan seeks to preserve) would be found by combining objective measures (reflectance imagery) with local, expert judgment in the field (Dave's efforts to "ground truth" soil quality and species sighting). However, regulatory biologists scrutinize the new map (and in-progress model) closely as the conversation continues.

CHARLES(RB): Does the, do the colors comport at all with grain size? Can you tell coarse from wind blown? [Is the red basically wind blown?]

DAVE(LB): [Well, all of those] colors have a degree of aeolian character to them. They all, um the red as I said is the most active, the um, probably has the most consistently, I'm trying to think of the right word, that the the LEAST variation in grain size.

A regulatory biologist (Charles) asks whether the red reflectance class corresponds to coarse or "wind blown" sand, making a distinction between the physical state of soils (fine or coarse) and processes that produce this state (wind sources or human activity). Wind sources are good for sustaining lizard habitat because they require no human intervention. Charles's question cuts through Dave's efforts on theground and local expertise to ask a more fundamental question. Does the reflectance layer tell them anything about sand sources that will operate over longer periods of time?

Another regulatory biologist (Randy) then asks if the entire multiple-species plan area has been covered with satellite reflectance data. Sand sources or even conservable habitat may lie outside the area Dave has mapped, these areas are linked dynamically to what is inside the preserve, and the SAC will want comparable imagery and soils classification for the entire plan area.

RANDY(RB): So its, have you uh, don::e ((*L hand sweeps over white space on map*)) the rest of the sand areas?

DAVE(LB): [No.

MARY(PC): [But just for information, ((*hands sweep over smaller map*)) this is the area that that image covers, overlaid on the whole…um [potential habitat.

DAVE(LB): [I only paid for the preserve and the sand sources to the preserve, so that's all I have, now. I have just recently REQUESTED, and they said it's about a ninety day guaranteed turnaround so it could be sometime next year, an image that would take all the way from Chipper's Finger to the eastern edge of the preserve, so then I can take it all in one image and look [at it that way.

CHARLES(RB): [Um hm.

Dave reports that the larger imagery set will not be available for another three months. Over the history of the SAC planning effort, money to collect mappable data has been scarce, and this creates recurring problems for their planning horizon. Raising money to map the valley, then waiting for additional studies to be completed, pushes a scientifically defensible conservation plan further into the future. Unfortunately, demands from developers (of golf courses, hotels, etc.) and local communities to "take" lizard habitat continue to operate in the present.

As conversation in the SAC continues, regulatory biologists return repeatedly to a distinction between finding areas to preserve, as one function of the plan, and regulating which of these areas land developers might be allowed to "take," as a different function. For example, FTL habitat that is good today (i.e., fluffy sand, populated by lizards) may disappear if regulators allow land developers to take a sand source that maintains that area.

Running Dave's soil classification layer over time. Another local biologist (Nick, a lizard expert) asks if Dave can define reflectance class thresholds consistently so they can compare sand quality over time. He worries that trusting the GIS system to assign color classes will lead them to compare "apples and oranges" across years.

NICK(LB): Can you define:: what frequencies the red um covers so you can compare ((*hands flatten over table*)) between r among years, um [so you're not comparing…apples and oranges?]

DAVE(LB): [In terms of s, its,] Well, you mean in terms of the the the image, ((*hands pull up from table*)) [the reflectance value?

CHARLES(RB): [The year to year image.

NICK(LB): Yeah, so the red this year is the same frequency [range as the red next year.] ((*hands grasp in successive locations*))

Nick's question, completed by Charles, is not just about the validity of computed classes. He proposes extending the reflectance layer from a snapshot of present conditions into a representation of changes in habitat over time. A more dynamic model could show how habitat suitability changes over time, perhaps even addressing the regulatory biologists' questions about the model's capacity to identify FTL habitat that is still maintained by active sand sources. To answer Nick's question, Dave digs into the computational machinery of GIS threshold definitions, contrasting more and less "objective" approaches (below).

DAVE(LB): [Right.] I think you can do that.

> Um what we...we tried it a couple of different ways. One was we just asked the computer to divide the image into ten separate, um, reflectance types. And those are four that were representative of blow sand types or aeolian types of some kind.
>
> The other six were mountainous and vegetation and things like that so we didn't color those, so that's why there's so much gray in that image.
>
> Um, so that's one approach and that is more or less, um, the computer is being more or less objective about that because the computer is making that selection.
>
> ((*describes other technical strategies through which a GIS user can set reflectance thresholds that apply over successive years*))
>
> I mean that would be the best way to make sure that each year you were looking at the exact same thing.

RANDY(RB): USGS might be interested in questions like that.
(3 sec) ((*Dave and Nick looking at Randy*))

DAVE(LB): Well, when I showed this to, I always forget his name =
((*SAC members jointly recall name of a USGS scientist from a neighboring state*))

DAVE(LB): He said he'd been trying to use this kind of imagery to look at grain size for the last ten years and never been successful with it. And I said, well, this may not be grain size per se, but it it's definitely CONSISTENT with grain size. And he was very impressed that it was that consistent.

Dave's new map layer, under questioning by another local scientist, is extended towards a more dynamic model of sand processes that create lizard habitat. A static model of what exists now (the current landscape) might become a model of how things work over time (how sand processes may maintain the current landscape). This will be critical if the plan is to allow developers to take existing habitat that has little future value, to protect areas that are sustainable as lizard habitat, and to protect sand sources for those sustainable areas.

Randy's observation that USGS might be "interested" in Dave's more dynamic model seems out of place, marked (as we hear it) by a pause in ongoing talk among committee members (above). Why would a regulatory biologist, explicitly positioned as a recipient of Dave's in-progress soils model, and even a contributor to insuring its scope and referential adequacy (Randy's earlier questions about areas not yet mapped), allocate the new model to an entirely different organization? Dave, after the slight pause, responds that another scientist, a geologist working for the USGS and not on the SAC, is "very impressed" with his reflected light approach to classifying lizard habitat. But there is no response from Randy.

Why is interest by the USGS a relevant contribution in the SAC meeting at this point? A possible answer, which gains support as the meeting progresses, is that Randy is trying to terminate or defer Dave's model proposal, in light of the fact that he and Charles have recently commissioned a hydrology study (i.e., a map and analysis of sand flow) for the plan area from USGS (Event 5 in Table 3.1). Under this interpretation, Randy withholds assent (or even sustained interest) in Dave's model by referring it to another organization that he and other regulatory biologists have already engaged to provide external scientific advice to the SAC. Several moments later in the meeting (turns not shown), a plan consultant (Edward) follows up on Randy's seemingly off-handed comment, and the regulatory biologists reveal that their external hydrology study will not be delivered for another five months. This is met with incredulity by plan consultants and land managers, who anticipate a significant delay in the SAC planning horizon. What might have been settled in today's meeting is pushed five months into the future.

Scene 2: Conflict and Containment within Coincident Boundaries

What looked like a productive episode of model building at the beginning of Scene 1, where Dave (local biologist) described how to bind satellite imagery to lizard and human activity in different types of soil, now is being pulled apart by the questions and extra-curricular activities of regulatory biologists. Particular mapped areas are disputed, the coverage of Dave's reflectance layer is too small, and his analysis is not dynamic enough both to identify preserve areas and to regulate what developers can "take" when the HCP is actually used. We have selected Scene 2 to show (a) how this tension between model construction and deconstruction spills over in face-to-face interaction during the SAC meeting, and (b) how the conflicts made explicit in this eruption are contained, at least temporarily, within decisions made by the committee.

A plan consultant's effort to bring closure leads to overt accusation and historical retreat. After 40 minutes of further discussion about specific preserve areas, "sand lenses" that supply fluffy sand to lizards, and human structures that act as barriers in these dynamic processes, the lead plan consultant (Edward) tries to bring SAC members to a decision about what should be in the model.

EDWARD(PC): Well, in terms of what we can accomplish uh today, are we all clear now as to what the model is going to represent when finished? [((*looking at Bert*))

BERT(LB): [((*shaking head, negative*))
 ((*general laughter*))

CHARLES(RB): ((*laughs, throws hands up, slumps over table*))

BERT(LB): ((*points to Nick and Dave*)) These guys are. I'll work with them, but I'm not. [Do you understand it?] ((*pointing to Dave*))

DAVE(LB): [Well...] I, I know what they want. But this is, when we put this together we had the same discussion, and it was clear that we weren't of one mind completely. And that's why we created what we did.

HANK(LM): I think what you, what we really need to do is have a written statement that everyone can agree, what the model is intended to represent.

General laughter at Bert's response suggests that SAC members are anything but clear about what will or should be in the model. Charles, in what looks like a show of desperation, slumps over the table. Bert asks other local biologists if they understand what the SAC is proposing, and Dave remembers that the lizard club agreed on the previous FTL model after a similar disagreement over what areas were required to pay the mitigation fee. Hank, who has consistently pointed out that his land management decisions will rest on scientific consensus, asks the committee for a written agreement. But given the circumstances, what would biologists (local or regulatory) agree to write down? Bert and Nick, both local biologists, are next to speak.

BERT(LB): And that's got to come from them. ((*points at Charles*))

NICK(LB): Based on all of the models that we've done [so far.

RANDY(RB): [You know maybe we should back up a little bit and ask the question whether it's worth opening this Pandora's box? It would be far easier and more expeditious to just make a simple assumption and accept the fringe-toed lizard HCP model. And...move on. Because I'm not sure it's going to have LOTS of consequence to the end result.

Randy earlier allocated a local biologist's modeling effort to another organization (USGS), and now Bert allocates responsibility for lack of agreement directly to the regulatory biologists across the table. As Nick (also a local biologist) confirms, the regulators need to agree to some version of their model proposals before the SAC can advance the planning effort. By deconstructing model proposals, withholding agreement, and commissioning external scientific studies, the regulatory biologists have blocked the ability of the SAC to provide scientific advice to valley municipalities just as surely as interstate highways have blocked lizards from dispersing to new habitat.

Randy's proposal to "back up a little bit" (above) pulls the committee back from the brink of what may have turned into a heated disagreement, both at the level of interaction in this moment and the level of their collective advice as scientists. What is in "Pandora's box" is both the current accusation by local biologists and the prospect of reworking the 1986 lizard HCP, which reflects both scientific opinion (at the time) and a long-standing (14 years) set of agreements between local land owners, developers, and regulatory agencies. The valley has literally come to resemble the plan, since regulators have performed its entailments over years of negotiations with property owners. Randy proposes they avoid opening both fights (current and future), since the effort will not have "LOTS of consequence" for the multi-species plan under development.

Consensus by coincident boundaries...political and scientific justification as different orders of work. Randy's proposal, at least for the moment, sidesteps Bert's accusation. As the meeting continues, regulatory biologists and land managers discuss areas in the original (1986) plan that need to be removed because of subsequent development, but leave in other areas like quarries that might collect blowing sand. Once minor edits to the 1986 plan have been identified (the local biologists are largely silent during three minutes of further discussion), Edward, the lead plan consultant again seeks consensus from the SAC.

EDWARD(PC): Okay, so landfill and quarry stay IN. ((*typing*))
 ((*hands spread out to encompass entire table*)) Is our...Linda how do you react to Randy's suggestion?

LINDA(RB): I like it. It's...

EDWARD(PC): ((*looking at local scientists*)) How about you guys?
 (3 sec)((*Nick pushes back from table; all 3 local biologists look back, silent*))

HANK(LM): Well, I think it's workable and defensible.

RANDY(RB): I think we just need some logic. And the logic would be continuity with the past. And there's high acceptance for the existing HCP. It's probably not going to have a big result or ch, make a big difference in the end result.

CHARLES(RB): No.

(3 sec)((*local biologists remain silent*))

Edward tries to get a spokesperson from each side of the conflict, allocating his second question to the local biologists. After a brief silence, Hank (land manager) answers the question, and then Randy calls for "some logic" that could be used to justify their decision to adopt a model that coincides with the 1986 lizard HCP. Local biologists are again quiet, and Hank (below) begins a new topic. As he has done repeatedly during this meeting, Hank asks for advice about a particular land management decision he faces in the valley. Overlapping with the beginning of Hank's speaking turn (below), Bert (local biologist) begins to criticize Randy's historical logic.

HANK(LM): Would it [be permissible at this point to bring up a specific example, another specific example?]

BERT(LB): ((*looking at Edward*)) [Those, those are political (inaudible) That's not the role of...

EDWARD(PC): ((*looking at Bert*)) Well,] but it is a science question. I mean, [do you think that the HCP model is still valid?]

HANK(LM): [And I have all the brains in the industry] ((*looking at Randy, who laughs*))

EDWARD(PC): ((*looking at Bert*)) Given two things. Given, one that as we just said we'll take out stuff that's actually been developed. And two, recognizing that we each DO have the sand source transport, or ecological process OVERLAY, which becomes a part of this.

Edward ignores Hank, instead asking Bert about the scientific merit of going forward with a slightly edited version of the 1986 plan, including new information about sand sources. Hank and Randy stop talking as Bert responds (below).

BERT(LB): And I think the answer to it is yes. Now the justification that, you know, it's historic continuity of the planning process, et cetera. That's not a science question, or an answer. That's political JUSTIFICATION.

DAVE(LB): But you can =

EDWARD(PC): = But you're saying it's scientifically justifiable, too.

BERT(LB): [In my opinion, yes.

DAVE(LB): [Well, all you have to do is...you look at a map like this ((*holds up small map*)), and you look at a map like that red orange and yellow map over there, and you can see that it's pretty defensible scientifically from that standpoint.

EDWARD(PC): Ok.

DAVE(LB): From...if you're looking at historically occupied, occupiable habitat. We're talking within the last hundred years, type stuff.

Bert contrasts two types of justification, scientific and political, complaining that the SAC has adopted the latter. Edward persists, asking if the edited 1986 plan can also be justified scientifically. Bert thinks so, and Dave agrees, pointing to the correspondence between 1986 preserve boundaries and his (currently stalled) reflectance model as "pretty defensible" scientific backing. This correspondence, crafted by local biologists earlier in the meeting, is exactly what regulatory biologists are setting aside in proposing to reuse (with slight edits) the 1986 HCP model.

CONTAINING DISAGREEMENT AS PLAN HISTORY RESUMES

By reusing the 1986 HCP model, edited to reflect undisputed changes in property value (e.g., landfills) over the prior fourteen years, SAC members adopted a tenuous agreement to defer talking about potentially volatile differences between local and regulatory biologists. By allowing the committee to move forward without opening a "Pandora's box" of disputed classifications and scientific uncertainty, the prior plan might provide a "coincident boundary" (Star & Griesemer, 1989) that could contain both the committee's trouble and resolve the "balance of terror" that motivated the 1986 plan approval process. If so, local and regulatory biologists on the committee could continue to coordinate their work in the larger plan without needing to reach full agreement over a model of lizard habitat. But would the container hold into the future and outside the committee?

Table 3.2 resumes the conservation planning timeline that we interrupted to look closely at scenes from a SAC meeting four years into the planning process. As planning continued, regulatory biologists carefully arranged the membership and format of a third external scientific review (Event 7), in which local biologists were not allowed to talk directly with external scientists, and both groups of biologists submitted sets of carefully crafted technical questions for external reviewers. The reviewers' report found use of the 1986 lizard plan scientifically indefensible, on the one hand, but also criticized the SAC for being too conservative when identifying Valley land to preserve in the multiple species plan. In follow-up SAC meetings,

TABLE 3.2.
A Narrative Timeline for the MSHCP Planning Process, Continuing from the SAC Meeting Analyzed in this Chapter

Winter, 2001	(Event 7) Regulatory biologists arrange an external review by a group of scientists, many of whom are unfamiliar with the planning process, and they impose strict rules on interaction between reviewers and SAC members. External reviewers criticize the lizard model (e.g., edits to the 1986 HCP) and caution that the SAC has been too conservative in identifying preserve areas. Regulatory and local biologists reject many of their recommendations.
Summer, 2001	(Event 8) The USGS hydrology survey, contracted 11 months earlier by regulatory biologists, is received. Regulatory and local biologists cannot agree on whether findings in the USGS survey are relevant for preserve areas in the MSHCP.
Fall, 2001	(Event 9) After complaints by municipalities to regulatory agency directors, the SAC is disbanded and the planning process is taken over by these agencies. Regulatory and local biologists continue to give advice, but the SAC plays no further role in the planning process.
2004	(Event 10) A MSHCP for the valley area is released for public review. In planning documents, the SAC is described as giving advice "throughout the planning process," organizing "workshops" for external scientists, and visiting mapped areas in the field. There is no mention of protracted disagreements or disbanding of the SAC.

neither group of biologists accepted the external reviewer's recommendations, arguing these academics were insensitive to local political and financial realities. After almost a year of waiting, the USGS hydrology study commissioned by the regulatory biologists was delivered, but the SAC factions couldn't agree on how (or whether) to use these findings (Event 8). Reflecting a growing sense that the SAC would not be able to reach agreement, agency managers disbanded the SAC, five years after it was organized by local biologists (Event 9).

Neither the coincident boundary of the 1986 lizard HCP nor the committee that adopted it to avoid further trouble survived the planning process in this case. MSHCP planning continued inside sponsoring agencies, and as we write this chapter, a completed plan is finally under public review in the Valley (Event 10), 14 years after the planning process began, which is a longer preparation period than any of the other approximately 50 large-area HCPs approved over the last twenty years (Goldstein, 2004).

PERFORMING DIFFERENCES IN SCIENTIFIC PRACTICE THROUGH MODELING

We began this chapter by describing a case in which modeling, as it is usually understood by design-oriented reformers in mathematics education, became "too real" even for professional modelers. After years of entrenched scientific conflict, with mutual accusations of self interest and bad science traded between local and regulatory biologists, local governments paying for plan development disbanded the SAC. A MSHCP was subsequently developed without formal participation by either group of scientists. How could conservation biologists, whether local scientists or professional regulators, fail to reach agreement when the stakes were so high? How could they abandon the lizards and other endangered or listed species, even as pressures to develop land for human use continued to accumulate? From the perspective of mathematics or science education research, are these people terminally "off task" or resistant to progress?

It is tempting to choose indifference on the part of regulatory biologists (the local biologists' preferred accusation) or overly narrow scientism on the part of local biologists (the regulatory biologists' preference) as explanations, but these simply force us, as analysts, to take sides in a controversy that was not resolved in favor of either professional group (see Latour, 1987, on how to conduct a "tribunal of reason" for settling accusations of irrationality). For example, as evidence against a conjecture that regulatory biologists were insufficiently interested in local species and their habitats, the following excerpt from Goldstein's field notes describes a field trip with the regulators:

> The degree to which the regulatory biologists shared the local biologists' passion for the [...] Valley was impressed upon me in May of 2001 when I accompanied two of the regulatory biologists on a morning visit to the "fault dunes", a series of sand dunes that lay along [an earthquake] fault in the northern end of the valley. We walked over trash-strewn rocky and sandy terrain, looking for a small rare shrub called the Mecca Aster. We were buffeted by thirty mile-per-hour winds, and temperatures soon rose to over one hundred and ten degrees. The dunes were so hot that the sole of one of my Teva sandals melted and separated from the upper part of the shoe. While I was hobbling across the sands, the regulatory biologists were racing around me, identifying the tracks of lizards and flipping over

plywood boards and old tires to find the burrows of pocket mice and ground squirrels. After three hours of this we headed back to [the city] to catch a quick lunch before the SAC meeting that afternoon, where the regulatory biologists engaged in heated debate over the disputed additions to the habitat preserve with the local biologists, who may have been the only other people who shared their enthusiasm for the desert ecology of the Coachella Valley (Goldstein, 2004, p. 320).

Even as Goldstein's plastic shoes were cooling, regulators who had enthusiastically followed lizard tracks in the desert were back at it, tooth and claw, arguing with local scientists who felt that only they were adequately concerned about these species.

From the perspective of mathematics education research, it might also be tempting to decide that the planning effort failed because information needed for the model (e.g., digital map layers with adequate precision across multiple measures) could never be obtained. But conservation planners like those on the SAC are regularly faced with cobbling together information of varied quality from heterogeneous sources, while the plans of developers proceed apace, making claims on the landscape even as the regulatory models are being constructed. The problems of modeling in this case were complex, but not unusually so. Species-habitat relations would never be known with fine precision, and ongoing human projects were, themselves, part of what SAC members were trying to locate in the model. For our analysis, contaminating social interests and inadequate information are poor candidates for explanation.

But what could account for years of intense work in which it appeared modeling would go on without end? It was not, we will argue, bad will or poor communication on the part of either group of professional biologists. Instead, each group participated in the SAC by performing aspects of distinctly different professional practice, and these differences (along with their consequences in the plan under development) led to an inability to construct a model that would be compatible with either group's image of work and professional identity in the future. In a sense, neither group would let modeling end without being able to imagine a viable future for themselves, and through their efforts, the valley itself.

PERFORMING DIFFERENCE BY ENACTING PROFESSIONAL VISION

Chuck Goodwin (1994) argues that professional groups like archeologists bring objects of their work into existence by orienting to the world in ways that reflect years of participation in distinct professional practices. They notice particular objects and relations (and not others), "high light" these in ways that enable coordinated work with others in their field, and encode what they see and talk about in conventional representational forms that (over the history of their practice) structure the intentionality of individual participants. In this sense, practitioners of a discipline actively experience their world of work through historically distinct forms of "professional vision" (Goodwin's term), even as their activity brings that professional world into existence as an ongoing technical practice.

In this case of conservation planning, as we have demonstrated in Scenes 1 and 2, differences in professional vision led to trouble both in ongoing interaction and in the possibility (or not) of creating a scientifically defensible model. In the following paragraphs, we focus on differences in professional vision between land managers, local biologists, and regulatory biologists as the SAC meeting was in progress. We examine an exchange involving participants from each group to explore how different points of view are produced in ongoing talk as SAC members work with map layers. These differences matter, we argue, when setting model assumptions, evaluating proposals for what should be in/out of the model, and for reaching agreement on these matters.

Showing "give and take" in order to "pitch and sell" during plan approval (Land Managers and City Official). In the following excerpt, two land managers (Hank and Kurt) complete a joke about the SAC's planning process as an example of "fuzzy habitat work," referring to the 2000 US Presidential debates, in which Texas Governor George Bush repeatedly accused Al Gore of using "fuzzy math" to explain how he would pay for new government programs. As laughter dies down, Ernie, a local city official, observes that removing habitat areas from the old HCP will show city managers and local land owners ("thousands of people") that conservation biologists are willing to "give and take." In turn, this will allow him to "pitch and sell" the new conservation plan as they seek public approval.

HANK(LM): This is...this looks like fuzzy habitat work. ((*general laughter*)) One percent of the lizards are getting...

KURT(LM): Are getting ninety nine percent of the feed.

ERNIE(CO): It's not a scientific issue, but a, anyway if you see areas that should not be modeled habitat and they're in the old HCP boundary, that shows give and take. And I think, you know, Hank's been talking abouttalking to managers, you know and trying to be able to pitch and sell? And eventually we got to pitch and sell this to...thousands of people. You know as long as moving those boundaries don't get people sideways with the existing HCP? It SHOWS:: a positive effect of this analysis, to have some give and take.

All three SAC members orient strongly to the near term process of plan approval, in which a proposed conservation model will be presented for public hearings and, shortly after, submitted for approval by local governments and state agencies. The joke completed by Hank and Kurt positions land managers as stewards who "feed" habitat to species, but they must do so fairly under public scrutiny. Ernie's contribution is given in a more serious tone, and he identifies specific areas in the plan that will show to the public and city officials that conservation biologists are willing to "give and take" land when it comes to human use. In turn, this will allow him to "pitch and sell" the plan to the public and others involved in the plan approval process. Taken together, the point of view expressed in this excerpt enacts a near term time horizon, concerned with particular areas (and human projects) in space, with contemporary non-specialists as the most significant actors.

Shifting the map and its information into a broader regulatory context (Regulatory Biologists). In the following exchange, Edward (plan consultant) has asked whether there is enough information to preserve habitat in the proposed map, and the response from two regulatory biologists (Linda and Charles) shifts the map into a different regulatory context.

LINDA(RB): But to answer your question Edward. Is there enough information to do the conservation areas? I think...the information is there. But to do the TAKE...that's a diff, ((*leans over, looking at Charles*)) is that, would that be a... Do you agree with that Charles? [Is there enough information?

MARY(PC): [Could you elaborate on that Charles?

CHARLES(RB): I'm not sure what Edward said, but what I said earlier was if we were gonna analyze ta::ke and we had a project, like a dam up here, this is not occupiable habitat but it would still be part of the take issue.

When they look at the map, Linda and Charles see tradeoffs between conserving and taking areas that are valuable to land developers. Their temporal horizon extends out well beyond that of the land managers and county officials (above), to the complexity of managing relations between different organizations as the plan is actually put into practical use (a period of 50 or more years, extending into the future). Mapped plan areas are not simply allocated to needy species (the stewardship approach, above), but enter into a complex managerial relation that balances conservation against the need to "take" habitat and species for human use and development. Charles illustrates this point of view by describing a hypothetical development project ("a dam up here")" that does not involve lizard habitat, but which still should be considered as part of the plan, since developers will be required to apply for a take permit when building the dam. Unlike land managers, time extends well into the future (and their professional career), the relevant terrain is not only soils and species, but also a complex social landscape of competing organizations and commercial interests, and they are central actors who manage this complex natural and social terrain.

Planning with apocalyptic dimensions (Local biologist). In the following excerpt, Hank, a land manager who regularly asks SAC biologists for advice on current land use proposals, begins to explore "the theory" that plan areas might turn into lizard habitat if left alone by humans. When Hank looks at the map, his sense of time is tied to a stack of developers' requests waiting on his desk (i.e., his insistent requests for advice). But Dave's response stretches out over a radically different horizon of activity.

HANK(LM): So, the theory there is what we were talking about in the beginning of the meeting, that if you don't have some structure there, eventually Mother Nature is going to deposit suitable substrate and the lizards could =

DAVE(LB): = Well, our discussion, and Bert brought this to the head was, we're not looking at the next ten years or the next hundred years. From the standpoint of the green line we're looking at geologic processes. So people die out, buildings go away, lizards maintain what happens over the long haul? And so...If we're talking thousands of years, yeah.

When Dave looks at the map, he imagines an apocalyptic future in which "people die out, buildings go away, lizards maintain." Time, which was anchored to plan approval and particular projects for land managers, then at a scale that included careers and organizational life for regulatory biologists, now extends forward and backward into scenes where humans have either not yet arrived or have (for reasons not disclosed) disappeared. In terms of space, what currently shows as human habitat in the map could return to lizard habitat, as sand lenses bring sheets of sand over human structures, these structures crumble, and animals species once threatened by human activity take center stage in an unfolding, apocalyptic narrative. Perhaps surprising, land managers and city officials do not ask who will conserve the golf courses. Most interesting, in the point of view enacted by this local biologist, humans (including authors and users of conservation plans) disappear as part of a broadly encompassing natural order.

The points of view produced in (and used to produce) these conversational exchanges show distinctively different orientations towards time, space, and agency as SAC members who inhabit different professional trajectories look at maps, imagine their reception and use, and judge whether current efforts are adequate for varied purposes. We think it is unlikely that SAC members (or members of any professional group) explicitly orient to these dimensions of experience as they conduct interaction. But differences across professional perspectives may help to explain which proposals lead to agreement or disagreement, and why. For example, when Randy (regulatory biologist) allocates Dave's reflectance imagery model to another organization (USGS), he enacts a version of the ongoing planning process that is made out of and for the work practices of a regulatory biologist. Both at the level of talk-in-interaction (e.g., topic projection) and collective action (e.g., commissioning a parallel study by the USGS), Randy shapes the prospects for a model under development by re-arranging what will count as adequate science, who will give it, and when it will arrive. Dave, in contrast, positions himself (and scientific peers on the SAC) as a spokesperson who binds together satellites, soils, lizards, and off road vehicles without the help or involvement of other organizations. In doing so, he enacts a version of the planning process that is built out of his own scientific expertise and local experience, something that Randy and his professional peers evidently find insufficient. Each biologist contributes to the model in ways that are consistent with his professional vision and expectations about future work.

CREATING DIFFERENT ORGANIZATIONAL FUTURES THROUGH MODELING[2]

Since both regulatory and local biologists felt that no MSHCP at all was better than a dysfunctional MSHCP that undermined their capacity to perform effectively as scientists and conservationists, each refused to give way. Instead, they deadlocked the planning process, causing a controversy that threatened their professional reputations, and endangered the prospects for adoption of the MSHCP. What was at stake in the dispute was their capacity to act effectively as scientists in the valley both now and into the future, and ultimately their ability to realize their conservation vision by setting into motion a natural ecology and a social dynamic that was amenable to the particular way they did science.

For the local biologists, the institutional and ecological setting for the MSHCP would be a kind of "peaceable kingdom". The leadership of the desert cities and surrounding county would predictably abide by the terms of the plan, so the regulatory agencies would never have to intervene to enforce it. Development interests would cooperate with the planning effort, since violating it would only expose them to political and economic turmoil, and the loss of take permits would stop their work altogether. Nature would also obediently play its part in the planning effort, as species lived and died in predictable ways within the habitat preserve. In contrast, for the regulatory biologists the natural and social dynamics of the MSHCP could be characterized as "red in tooth and claw." The cities and counties would have to be closely watched and held to the conditions of the take permits, which they surely would seek to covertly violate. Both

---

[2]Material in this section is drawn from Chapter 10 of Goldstein's dissertation (2004, pp. 317–352, available from the author).

environmentalists and developers would defect from the agreement when it suited them to do so. The natural world would also resist compliance with the terms of MSHCP, as new scientific paradigms undermined the theoretical basis of the plan and predictions based on scanty field data turned out to be false. Fortunately, new opportunities to conserve habitat would also arise, as new development proposals in the valley provided the regulatory agencies with opportunities to modify and adapt the MSHCP.

Both groups tried to bring their different conceptions of the social and natural world into being. The actions of both sides can be understood in terms of sociologist Brian Wynne's observation that, "Validity depends upon whether the world—natural and social—can be restructured and manipulated to accord with and thus 'validate' the tacit models embedded in the technology or knowledge claim" (Wynne 1992, p. 276). For their part, the regulatory biologists chaffed under the organizational regime established by the local biologists, which relied on consensual and trusting social relationships established since the negotiation of the 1986 lizard HCP a decade before. In turn, when the regulatory biologists altered this organizational dynamic to something more harmonious with their understanding of the uncertainty of the natural and social world, the local biologists were bitterly resistant to the change.

DISCUSSION

We have now come a good distance from the optimistic gloss on modeling that we used to open this chapter. Participants indeed could not agree on what were the relevant problems to solve, representational tools were developed and discarded without a clear evaluation of progress, and cycles of model revision did not converge or terminate with findings or explanations that satisfied major stakeholders in the modeling activity. Modeling, in the case of this SAC, did not actually go on "without end," but it exceeded the patience of Valley stakeholders and even Goldstein's fieldwork stamina. Modeling ended because the advisory committee, itself, was broken apart as an organizational container in order to let the planning process reach a conclusion. The MSHCP was eventually delivered for public review, but its local history of production was messy.

Who needs modeling like this, a reader might ask, if we seek empirical images of what should be taught to provide a foundation for the future? Inside the mess, we argue, we can learn a substantial amount that is relevant for teaching and learning when modeling crosses organizational boundaries, something that is now ubiquitous in high stakes technical and scientific work. Our analysis supports several observations:

1. Models of broad consequence are not self-contained mathematical puzzles with single authors and docile readers. What is in/out of a model is an organizational question as much as a question for individual cognition. Model negotiation across organizational boundaries is, we argue, an increasingly common form of scientific/technical work.

2. Model construction and use usually happens against a history of prior modeling efforts. New models, when adopted, displace old models and their negotiated assumptions. Displacements are a disruption to existing representational infrastructure and the work it supports (Hall et al., 2002). As a result, conflict and different perspectives should be expected, and

these are important phenomena for further research in mathematics and science education. The relation of Dave's (local biologist) reflected light model to the 1986 lizard HCP is a particularly clear example of this.

3. Members of different professional groups see models and what they represent in different ways (Ochs, Jacoby & Gonzales 1994). In this case, the underlying ontology of time, space, agency and their relations can be quite different, even among a group of people who self-identify as conservation biologists. Conflicts around model displacement appear to be strongly influenced by these ontological differences (see also Eisenhart, 1996). Distinctly different professional points of view held and enacted by land managers, regulatory biologists, and local biologists appeared to be critical in this case.

4. Anticipating downstream reception and use of models is an important aspect of their design. When modeling and displacement involve negotiation across organizational boundaries, differences in disciplined perception are compounded by different organizational objectives and accountabilities.

5. By studying differences and how they are resolved, we can identify modeling strategies that operate at collective (not only individual cognitive) levels of analysis. Regulatory biologists' successful efforts to re-arrange the SAC's authority and planning timeline provide a particularly clear illustration of this, even if in the negative.

In a larger collection of case studies of math at work among professionals (Hall, 1999), participants (civil engineers, architects, field and conservation biologists) report that learning to work across organizational boundaries is highly valued but rarely taught in school. This has consequences for what students should experience as they move towards professional careers that involve modeling. If negotiation over what is in/out of models is the typical context of using mathematics to model complex systems (i.e., mathematics plays a supporting role in a larger, leading activity), then cognitive studies of modeling as self-contained mathematical problem solving may have little relevance for what work demands of schooling. On the other hand, studies of individual mathematical problem solving in simulated modeling tasks may continue to have great relevance for how mathematical reasoning or application is assessed in schools. This is a larger problem of alignment between schooling and professional practices that this book is organized to address.

ACKNOWLEDGMENTS

This work was supported by a doctoral dissertation improvement grant from the National Science Foundation to Goldstein (SDEST–9987683) and a separate grant to Hall (ESI–94552771). Authors are listed alphabetically. Jim Kaput was an advisor on that project, and his mentoring lives on in this work. We also benefited from lively discussions with members of the Humanities Research Institute Working Group on the History of Quantification and Standards (UC Irvine), in particular Geoffrey Bowker, Martha Lampland, Jean Lave, Mimi Saunders, and Leigh Star. Finally, we thank planners, scientists and stakeholders in the Valley for generously sharing their time and work in the completion of Goldstein's dissertation project.

REFERENCES

Eisenhart, M. (1996). The production of biologists at school and work: Making scientists, conservationists, or flowery bone-heads. In B. A. Levinson, D. E. Foley, & D. C. Holland (Eds.), *The cultural production of the educated person* (pp. 169–185). Albany, New York: State University of New York Press.

Goldstein, B. (2004). *War between social worlds: Scientific deadlock during preparation of an endangered species habitat conservation plan (HCP) and the co-production of scientific knowledge and the social order.* Doctoral Dissertation. University of California, Berkeley.

Goodwin, C. (1994). Professional vision. *American Anthropologist, 96,* 606–633.

Greeno, J. G., & Hall, R. P. (1997). Practicing representation: learning with and about representational forms. *Phi Delta Kappan,* 361–367.

Hall, R. (1999). *Case studies of math at work: exploring design-oriented mathematical practices in school and work settings.* Final report to the National Science (RED–9553648).

Hall, R., Stevens, R., & Torralba, A. (2002). Disrupting representational infrastructure in conversations across disciplines. *Mind, Culture, and Activity, 9*(3), 179–210.

Hutchins, E. (1995). *Cognition in the wild.* Cambridge, MA: MIT Press.

Latour, B. (1999). *Pandora's hope: Essays on the reality of science studies.* Cambridge, MA: Harvard University Press.

Lehrer, R., & Schauble, L. (2004). Modeling natural variation through distribution. *American Educational Research Journal, 41*(3), 635–679.

Lesh, R., & Doerr, H. M. (2003). Foundations of a models and modeling perspective on mathematics teaching, learning, and problem solving. In R. Lesh and H. M. Doerr (Eds.), *Beyond constructivism: Models and modeling perspectives on mathematics problem solving, learning and teaching* (pp. 3–33). Mahwah, NJ: Lawrence Erlbaum Associates.

Ochs, E., Jacoby, S., & Gonzales, P. (1994). Interpretive journeys: How physicists talk and travel through graphic space. *Configurations 1,* 151–171.

Star, S. L., & Griesemer, J. (1989). Institutional ecology, "translations" and boundary objects: Amateurs and professionals in Berkeley's museum of vertebrate zoology, 1907–39. *Social Studies of Science, 19,* 387–420.

Wynne, B. (1992). Risk and social learning: Reification to engagement. In S. Krimsky and D. Golding (Eds.), *Social theories of risk* (pp. 275–297). London: Praeger.

CHAPTER 4

Mathematical Modeling 'in the Wild': A Case of Hot Cognition

Wolff-Michael Roth
University of Victoria

We are in the "wet lab" of a local fish hatchery. Erica, one of the fish culturists and a temporary helper are in the process of "sampling" juvenile coho salmon from the three ponds outside. After sedating the fish with some carbon dioxide gas, Erica takes an animal, places it on a special ruler to measure its length and reads, "one-twenty four." She twists and turns the fish, takes a "close look" checking it for infections and abrasions (Figure 4–1a) and says, seemingly with relief, "Yeah I am not gonna–." There is a 1.53-second pause, before she continues her voice indicating that she is apparently not too convinced, "I mean, his size is really looking good," and then drops the fish into a basin on the scale, previously tared by the helper. The latter reads off the weight, enters it into the computer-based spreadsheet, and suggests, "Drop it from distri–." Erica responds with an outburst, "No, you don't make it up." A little while later, the two are stopping in their task and Erica begins to manipulate the spreadsheet. She produces two histograms, one displaying the distribution of fish weight, the other one of fish length (Figure 4–1b). As the new plots appear, she bursts out, "Oh my god, look at them up there. O::h. I'm there, I am so there. I am not even concerned." (Colons are conventionally used to indicate a drawn out sound, here the "o.")"

Erica had dreaded this moment. This spring has been especially difficult for her. The hatchery management had announced that she would be laid off because of budget constraints. A lengthy strike in her husband's company had led to his temporary layoff. They would not be able to maintain their home, and began to consider selling it. The future looked uncertain, which worried the normally bubbly Erica a lot.

The situation has begun to affect Erica's work. She always has cared for the coho, which she raises from the egg stage to the moment when the fish turn into smolts, ready to leave the hatchery and begin their migration to the ocean. She talks about them as if they were her children: the moment the fish are released has always been filled with a mixture of sadness, for her "babies" were leaving, and joy, for her babies had grown up. This spring was different. She was on an "emotional roller coaster." There were days when she had neglected the coho over her worries about other things; there were days when the fish were not fed despite having been scheduled to do receive food. That is, her emotional ground state mediated how well she was doing her job, both around the hatchery and in her office, where she

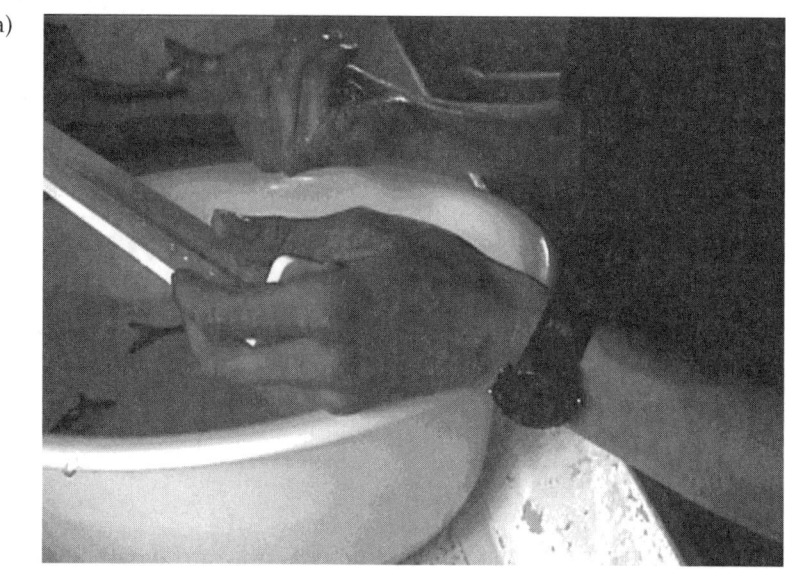

FIGURE 4–1. (a) Erica caringly handles every coho, exhibiting concern for the physical and physiological health of the fish, which she raises from eggs until they are ready to be released; visible in her left hand is the ruler she uses to measure length. (b) Erica has built an extensive database, which constitutes a mathematical model of the fish throughout their life in the hatchery. Emotions are apparent in every action of her work and mathematical modeling.

kept a database that modeled many different aspects of the coho in her care. That spring a fellow worker, manager, or ethnographer watching her everyday actions from the outside might have concluded that Erica was over-rated, that she was not as good a fish culturist as she was said to be. That is, her identity, who she was with respect to others in the hatchery mediated by her emotional ground state that bore on every practical action she produced during the day. Or, knowing of her woes, they might say that the emotions related to her situation affected her work.

We now can understand the emotional expressions during the episode in the wet lab. The emotions made available to others in the room showed that Erica was relieved. While looking at each individual fish, one after the other, she could see that there were no signs of disease or mechanical abrasions. She could see that the fish "looked good," that is, had a perceptual length-thickness ratio that was a sign of good health. There was a loving tone in her voice, a sign that she cared, even though she had neglected the fish this spring. When she did her intermediary check of the size and weight distributions, she immediately saw that the fish as a whole had not suffered. The distributions were "up there," at the average size and weight where she had modeled them to be at this moment of the year, about 30 days before release.

These episodes show how emotions mediate both the content and the process of everyday work. But how well we do something is used by others as evidence for establishing who we are: expert fish culturists, disgruntled employees, just-so workers, or beginners. That is, emotions mediate our every practice-constitutive action, which in turn provides evidence for others who we are, whether we like mathematics, computers, or our jobs and how much we know with respect to each of these. In fact, there is neuropsychological and neurophysiological evidence that emotions do not impinge from the outside on the knowledgeability displayed in practical actions but that they are integral aspect of it (e.g., Damasio, 2000). But because the type and nature of our actions are evidence for who we are with respect to others, our identities are also effects (results) of always emotion-laden practical action. More so, if everyday actions inherently promise a higher emotional valence, we inherently identify with the larger activity in which we participate, we are said to be motivated, which in turn becomes an attribute of our identity. Erica has always given "300 percent," she was recognized not only as an expert but also as a dedicated worker. That is, motivation is also an effect of the fact that emotions constitute an integral aspect of practical action.

A growing body of research, often collectively referred to as ethnomathematics, provides evidence for the numeracy-related competencies people of different walks of life develop (Scribner, 1986). Such studies also show that (a) performance on tasks that require school mathematics and everyday mathematical tasks is virtually unrelated and (b) amount of schooling in mathematics is often unrelated to mathematical competencies in the wild (e.g., Lave, 1988; Saxe, 1991). Levels of schooling, however, do predict performance on school-like problems even when these are structurally identical to everyday work situations (Schliemann & Acioly, 1989). These studies generally provide evidence of the robustness of mathematical competencies in the wild.

More recent studies among banking employees and nurses show that mathematical understandings at work can be enhanced by means of instruction that allows participants to reflect on their use of mathematical representations (Noss & Hoyles, 1996; Noss, Hoyles, & Pozzi, 2002). The development of competencies arises from a dialectical relation between intuitive practical understanding of objects and events that make the workplace, on the one hand, and explanation-seeking modeling activity, on the other hand (Roth, 2004, 2005). Studies among technicians and in scientific research groups showed that activity theory constitutes an important tool for describing and modeling work processes in general and mathematics at work in particular (Roth, 2003a, 2003b).

Cognition is usually considered based on logical relations between language-based concepts. In much of the scholarly literature, emotion and motivation are treated as variables external to but (usually negatively) affecting (mathematical) cognition; and the potential connections between cognition and identity are hardly ever explored. In this chapter, I provide evidence from a four-year ethnographic study of a fish hatchery showing that emotion, motivation, and identity are central aspects of work in general and of mathematical modeling in particular. I begin by articulating a theoretical framework that has arisen from my ethnographic work of mathematical cognition "in the wild," and describe the extensive mathematical modeling and its relationship to the practical understanding of work that I observed over the four-year study in the hatchery. I then provide evidence for the integral nature of emotions to action, and therefore to motivation and identity with respect to mathematical modeling at work. The evidence includes linguistic data and analyses of prosodic features, which are articulations and expressions of emotions that eschew conscious control. I conclude with thoughts about the implications of these results for the teaching of mathematical modeling in schools, activity systems that differ considerably from everyday work.

MATHEMATICAL MODELING OF COHO SALMON AND WORK

This study was conducted as part of a large research project involving more than 60 professors from the natural and social sciences and humanities; it was designed to understand the impact of social and environmental restructuring on environmental and human health in Canada. My part of the project consisted in studying the interaction of local and scientific knowledge, which led me to do two parallel ethnographic studies. One study concerned the practices in an experimental biology laboratory specializing in salmonid vision (e.g., Roth, 2003a); the other study was conducted in a fish hatchery that provided coho salmon to the scientists. Drawing on apprenticeship as an ethnographic fieldwork method (Coy, 1989), I participated in all everyday activities at both sites (collecting and interpreting data in the lab; feeding fish or "doing samples" in the hatchery; collecting and measuring dead salmon in the river). In this chapter I focus on the work in hatchery, particularly with respect to the mathematical modeling of fish and work as exemplified in the practices of one informant, the fish culturist Erica. Whereas most fish culturists have high school education, Erica attended two years of a college business program before dropping out and beginning to work in a hatchery. As her peers in this and other hatcheries, she worked herself up "through the ranks," beginning as a temporary employee and gaining experience, before obtaining the position of a fish culturist. (Nowadays, hatchery managers appear to prefer to hire graduates of acknowledged fisheries and aquaculture programs for any position of fish culturist. According to the old-timers, however, these graduates are "full of book knowledge" and lack a feel for the fish.)

The hatchery where this study was conducted is part of a salmon enhancement program run by the Canadian Department of Fisheries and Oceans and produces smolts (juveniles ready to migrate) of coho and chinook salmon and steelhead trout that are subsequently released into the river for the purpose of maintaining stocks. The fish from all three species generally return to the same sites where they had been chemically imprinted at a particular point in their early development, spawn, and then die (coho, chinook) or return to the ocean (steelhead).

In the fall, the fish culturists and their seasonal helpers collect the fish, which are returning from their ocean migration, in a tank and then, when the females are "ripe" (ready to spawn), take the eggs from the females and fertilize them with the milt from the males. The fertilized eggs are raised until they hatch, at which time the animals are placed in rearing ponds. Here they are fed until they are ready to be released, which takes on the order of several months for chinook and more than a year for coho and steelhead.

Each fish culturist is responsible for a "program," which means a particular species, from egg take to release. They place orders for feed (after consulting with the managers), feed fish or delegate it to a helper, or take a variety of measurements on the fish and environment. Throughout each program, many aspects of the work are recorded and entered into one or more databases. This leads to an extensive mathematical record of the fish and the environment in which they are raised.

Mathematics comes into play from the first moment when eggs are taken. For example, for each female the number of eggs is estimated either by volume or weight. In the former method, for example, a volumetric sample (e.g., 100 milliliter) is taken, the eggs are counted, and then the total volume of a female's eggs is recorded; the total number of eggs is calculated according to the formula

Equation 1 $$\text{total eggs} = \frac{\text{total volume}}{\text{sample volume (e.g., 100 mL)}} * \text{number in sample}$$

by using one of the calculators at hand in the incubation room where this task normally takes place. Episodes of "sampling fish," such as the one featured in the opening of this chapter, includes catching a representative sample, sedating the fish, measuring their lengths, weighing, inspecting each fish for fungal diseases or any apparent abrasions, and then returning them to their pond. That is, data, records, and calculations are produced pervasively as part of everyday work in the hatchery. There are large print notes containing the amount and size of food to be thrown into each pond; those individuals who actually feed the fish will keep scratch papers to note the amounts that they had taken from the feedbags and filled into the feeding buckets. At the end of the day, they add all amounts and transfer the total into a form.

All information produced during a day and entered into the appropriate tables or notebooks is subsequently transferred into a computerized database by the fish culturist responsible for a particular species. The information is then displayed in various forms, assisting the fish culturist to plan future actions, and, when there are anomalies, to better understand some aspect of fish hatchery practices. For example, Erica produces a spreadsheet that contains, in addition to actual average fish weight, ideal average fish weight ("play size") and a scientist's recommended fish weight ("BA Size"), also contains average water temperature in the rearing pond, predicted average fish weight based on initial weight ("Start Size") and weight gain, estimated total number of fish, total (estimated) biomass, and food dispensed per day ("Kg/Day") (Figure 4–2). We also find in the spreadsheet water flow requirements (for maintaining oxygen levels), predicted feed (size depending on the fish size) to be dispensed, actual feed used, and costs of feeding. Finally, there is also a plot of average fish weights: actual, ideal ("play curve"), scientifically recommended ("Brian Anderson's", predicted, and the actual fish weights of a previous program ("2000 Size").

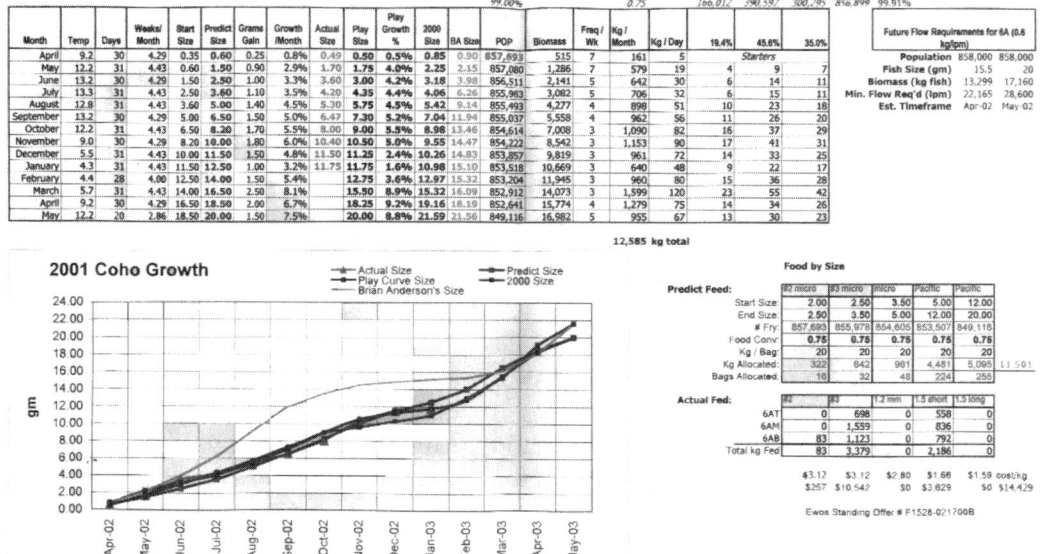

FIGURE 4–2. This printout constitutes a small fraction of the mathematical model Erica has constructed for and about her work. It includes, among others, actual and target average fish weights, food to be dispensed, predicted food costs, and predicted water flow rates to maintain increasing oxygen demand.

My account so far already provided evidence for the fact that the database was not only used as a way of recording various bits and pieces of information, but also to predict future aspects in the program, including average fish weight, recommended weights, desired weights, requirements for water flow through the rearing ponds, and future amounts of food (by size). That is, the database is a resource for mathematical modeling not only of hatchery practices in the (immediate past), that is, descriptive modeling, but also for mathematical modeling of the future, that is, prescriptive modeling. This latter aspect requires knowledge different from the practical knowledgeability that fish culturists develop in the course of working in the hatchery (those I observed had between 10 and 25 years experience). It requires, as I showed elsewhere (Roth, 2005), explicit knowledge *about* fish and fish hatching. The two forms of knowledge, practical understanding of how the hatchery world works and explanatory knowledge used to articulate and explicate work processes stand in a dialectical (mutually presupposing) relationship. That is, the two forms of knowledge arise together through a reflexive elaboration of intuitive understandings. In this, any explanation presupposes practical understanding and knowledgeability, which therefore precedes, accompanies, concludes, and thus *envelops* the explanations (Roth, 2005). But explanation-seeking analysis *develops* practical understanding. This relation is evident in the following quote:

> I usually do the same population every year and there for the most part I can count on certain water temperatures trying to cement it a little closer every year to what works so that you can go, 'Hey, yeah, this time last year, um the fish are this big and I was feeding this much and it's worked.'

Keeping records and modeling the fish population, environment, and the related hatchery practices allows Erica to "cement [the predictions] a little closer every year" and these predictions are based on what she understands as "what works."

Contradictions play an important role in the process, for they lead to explanation-seeking analysis, in the course of which practical understanding is brought to bear and, at least partially, is being articulated. For example, when the average fish weight is off the ideal, Erica wants to both understand what has happened and adjust her feeding schedule in a way to bring the average weight back into line with the ideal. This may sound easy, but in fact involves many variables. I briefly sketch what is implicated but provide a more elaborate account of the complexity of the process elsewhere (Roth, in press). First, the food required to bring about the desired average weight gain is calculated via the biomass. Thus, the total biomass to be gained is calculated using equation 2.

Equation 2   biomass gain = average weight gain * population

This desired biomass gain cannot be obtained in practice simply by adding the same weight in feed. Rather, for each rearing pond, fish culturists keep a record of the food conversion factor (=actual biomass gain/total feed added to pond). This food conversion factor is influenced by particulars of the pond, its exposure to sun and therefore light and temperature, its location with respect to other ponds (a pond upward in the chain of ponds has more oxygen available), water flow rate and water velocity, and other unknown variables. Using the food conversion factor, feed required to achieve the desired biomass gain is calculated according to equation 3.

Equation 3 $$\text{required feed} = \frac{\text{desired biomass gain}}{\text{food conversion factor}}$$

The story is more complicated than calculating required feed and then throwing the feed to the fish—in fact, it turned out that mechanical feeders were both less (cost) effective than human feeders and led to an increase in diseases in part caused by feed decaying on the pond bottoms. Feeding requires watching whether the fish actually feed; and fish may not want to feed but an experienced fish culturist can coax them into feeding. Feeding behavior depends on the weather both in the short and long terms (e.g., pond temperature). Feeding is also influenced by the composition of the fish feed—feeds may contain a "transfer" additive given just before release or a feeding stimulant, which affects other programs as unused feed is dispensed during other periods—or by the structure of the feed—krill, which increases palatability, may be spread throughout a feed pellet or only exist on the outside. Other aspects of hatchery life influence feeding rates, especially during the periods when there are no temporary workers around.

Fish culturists may be required to attend to other jobs within and outside the hatchery (e.g., to implement a program for public viewing in the harbor of the city about 25 kilometers away or release feed into a nearby lake to increase the algae that serve as feed for the sockeye residing there), which may prevent them from distributing some or all feed for the day. Thus, for a variety of reasons to a great extend further mediated by the emotional stress related to being laid off, Erica did not attend to the fish in her program as well as she normally did, which led to additional stress arising from the knowledge that her fish were suffering.

EMOTIONS IN ACTION

Over the course of doing four years of ethnographic work, often closely working under difficult conditions with Erica, it became evident to me to what extent emotions were tied up with every aspect of her work, both concerning the fish and her mathematical modeling activities. It also became evident that her emotional ground state was changing after she received her layoff notice and during those times when she was re-hired as a temporary worker. I begin by providing evidence for emotions at work and then describe a theoretical model explicating the role of emotions in practical action.

Ethnography of Emotions at Work

Erica is a very knowledgeable and successful fish culturist, recognized not only within the hatchery but also to scientists working for the federal Department Fisheries and Oceans in the area. She accounts with pride for the successes of her breeding programs ("dammed, my fish sure look good, now they just had, you see in the two years…there's been incredible survival" "all I know is the vet said my fish look good and they fee—they look good to me").

Erica is very attentive to the needs of the fish: "I try to do what's best for the fish, it's nice to map it all out," "I also think it's important to pay attention to what the fish are doing and let the fish let you know when it's time to go; nature has an uncanny way of knowing what's right," and "my biggest concern is to produce healthy fish." When we looked together at some of the records of past breeding programs ran by others and found out that the fish were made to double their weight in the course of a five-week period starting in April, Erin said deeply concerned, "Scary, I'd be more worried about health problems with that and how much food do you have to throw to get that." However, Erica's concerns and care—expressions of her emotional engagement with the object of the activity—are not merely due to a "feeling for the organism" (Fox-Keller, 1983), a gendered relation to nature. Rather, all fish culturists observed were very attentive to the fish in their care. Male and female fish culturists "listened" to what the fish were telling them, and acted accordingly: "if you look now they sure like to hang at the back screen so that tells me that they're like, 'Okay, well, we're tired of hanging in the in-flow we know it's time to turn around and go the other way" and "here you try to create the ultimate rearing conditions, so 'Oh my flow is always perfect. Oh, I always get food; I never have to worry about starving.' "

In her decision-making, a complexity of issues is in force, all of which are associated with emotional valence. First and foremost, Erica cares for her fish. The care for the fish sometimes comes into conflict with other concerns, such as having to use up existing feed stocks from other programs rather than feeding what she had planned or the costs of a feed that promises health advantages for her fish but that is very expensive.

> I'm kind of ticked about is some of the left over food that I'm feeding right now is something called Transfer and Transfer is touted to be something to feed fish three to four weeks prior to them entering salt water and my issue with it. Either it doesn't do anything or why am I feeding this to fish that aren't going to saltwater for another two months…. Transfer, "well we're not supposed to be going anywhere," I wonder what's gonna do, you know like I don't know and it's sort of hmm.

In the same breath, she also shows concern for the amount of money the hatchery has to spend. Erica keeps in mind the financial costs of her program to the hatchery. Thus, although others in the hatchery used Transfer to assist the fish in making the transition from the (protected) hatchery environment to the competitive environments in the river and estuary, she did not make this available to her fish.

> We're also paying for a lot of research and development, there's been about a twenty percent increase in the price of food um that was a decision making factor for me in ordering food to, to finish off my coho for this year where I had been feeding Moore-Clark ((a feed company)) but because of that twenty percent price increase that they had a little bit earlier than Ewos ((a feed company)) I went with Ewos.... I could could've ordered Transfer say for the end of this program to feed it out prior to saltwater, but they're not going directly into saltwater they'll be going into the river and then they'll be swimming out at there own pace.

In Erica's situation, the deliberations had gone in favor of the hatchery, that is, she did not purchase the Transfer because, as she subsequently legitimized, "till it's absolutely proved to me that yes indeed it makes an incredible difference—and it would have to be substantial—otherwise I'm not I'm not willing to invest in it just yet."

So far, the evidence I provided for the role of emotions may be doubted by some as "anecdotal" or too much subject to interpretive flexibility of everyday actions. However, as linguistic studies show, emotions are made available not only in the content of talking but also, and even more so, in the way people talk, that is, in the prosodic features of their talking (Selting, 1996). Interestingly, people do not have conscious control over prosodic features in everyday actions, and thereby make publicly available core aspects of thinking in real time. Pitch is a typical prosodic feature that provides evidence for the emotional engagement of an individual, particularly when it rises considerably above normal levels (C. Goodwin, M. H. Goodwin, & Yaeger-Dror, 2002). It constitutes an important feature that increases reliability of face-to-face interactions in situated activity, both in mundane settings (M. H. Goodwin & C. Goodwin, 2000) and at work (M. H. Goodwin, 1996). In the following, I provide several pieces of evidence for emotions in action that are of this type.[1]

Erica is exhibits concern for the health and size of her fish not only in the content of her talk but also in the changes of her pitch at particular moments. For example, while doing a sample, which required her taking individual fish, measuring its length, determining its weight, and doing a visual check for its health, Erica greeted particularly beautiful and large specimens with expressions of pleasure:[2] "((Takes fish from bucket)) ↑OUGH (0.66) ↓EH: One Thi:rtee:n." A plot of speech intensity and pitch shows that the interjections "ough" and "eh" were uttered at very high intensity (a function of loudness and frequency), the former also at a pitch much higher than the at the time normal 200–230 Hz, an indication of high arousal emotion (Johnstone & Scherer, 2000). The pitch returns to normal levels at the end of the utterance, which provides a reading of fish length for the helper who immediately

---

[1]The analyses were conducted using PRAAT (v. 4.2), a software package freely available for PC, Macintosh, and Unix environments. It can be downloaded from the URL: http://www.fon.hum.uva.nl/praat/

[2]The following transcription conventions are used: (0.41)—time in seconds; [ ]—in consecutive lines indicate beginning and ending of overlapping speech; RIGHT—capital letters denote louder than surrounding speech volume; °Yea°—degree signs enclose speech with lower than normal volume; *ten*—italicized utterances were stressed; (*away*—down and up arrow followed by underlined words denote descending and mounting volume over the stretch of the underlined text; u:m—each colon indicates an extension of a phoneme by 0.1 seconds; =—equal sign shows latching, that is, two utterances are not separated by the normal pause; (??)—each question mark enclosed in the parentheses represents a word that could not be deciphered; and .,?!—punctuation is used to indicate speech features, such as rising intonation heard as a question, or falling intonation to indicate the end of an idea unit (sentence); ((takes fish from bucket))—double parentheses enclose transcriber's.

and without comment entered the number into the database. Here, the first interjection falls together with the perception of the fish; the high pitch is an expression of the pleasure that she expressed throughout this sampling episode about the fish being in a healthy state despite the previously mentioned lack of attention they had received in the preceding months.

The implication of emotions in work was apparent not only in the everyday practical action in the hatchery, but also in the office where Erica does all of her mathematical modeling. It is not exaggerated to say that Erica is enamored with the mathematical aspects of modeling her work and the computing technology that allows her to do it. Referring to the database, she often says things such as, "I like doing this stuff" or describes herself as "a bit of a geek." In one situation, she guides me through the most recent data and then notes, "I have graphs all over the place...this one is my latest and greatest" or "See, I have too many graphs going at once and I am playing with them all." Erica knows that modeling what is going on in the hatchery gives her an advantage, provides additional room to maneuver when decisions have to be made rapidly: "I think [predicting] is worth while, like I—why wait until your in a panic mode when you have to figure something out then you do it ahead of time." She is well-known and well-respected for this modeling work, for which she stands out in this hatchery and among fish culturists in this part of the province, even if she is being teased, including by her mentor who suggests tongue in cheek that "She is doing it because she worries about making mistakes."

Erica attends with great care to keeping an accurate record, which allows her to improve her predictions with respect to fish weight, future requirements for such things as flow rates and feed, and costs:

> I love details and I've been accused of hanging on to too many details, but I think if you don't take enough initially, you can't get it back if you don't take 'em. So I can weed it out later and pare down later. But like I'm kind'a geeky and the fast thing, well I've been accused of—slow down hmm. I'm not necessarily the best teacher on the computer um, where I can slow down and talk about fish culture things.

Erica vehemently rejects any intimation that she should engage in an action that changes the data with respect to the phenomenon modeled. This can be seen in the following interaction during the same sampling episode from which the opening example was drawn. The episode begins when Erica is reading the length of a specimen off the ruler, and intimates that she was not continuing to worry (line 01), because the size of this fish specifically, and the sizes of all the fish more generally were "really looking good." The helper, who has been silent for much of the day, tarring and reading the scale and entering the data, begins an almost inaudible comment about dropping something from the distribution, to which Erica vehemently responds, "NO YA DON'T TAKE IT OFF!" The analysis of pitch and intensity provides evidence for the vehemence, an articulation of the emotions at work. Thus, compared the early part of the episode (line 01), intensity increase fourfold (every 3 dB means a doubling in speech intensity) and the pitch nearly tripled to level off at about twice the normal pitch height (Figure 4–4).

In this situation, Erica's strong reaction allows us to understand the helper's utterance as an intimation to modify or even falsify the records. Dropping data points to construct the database in general and her mathematical model in particular display in a way that she knows is incorrect is simply out of the question: you just don't take it off. From a purely logical perspective, saying, "No you don't take it off" in a normal pitch and loudness would have sufficed. The fact that both are so far off from the normal range shows that emotional aspect of the response.

MATHEMATICAL MODELING 87

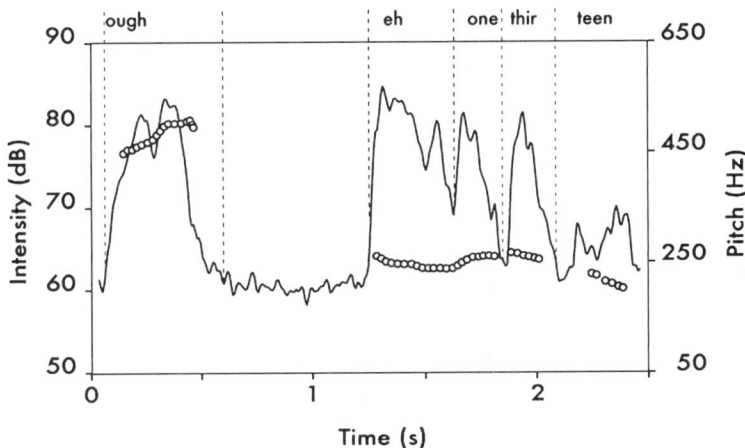

FIGURE 4–3. At the time of the recording, Erica normally has a pitch range between 200 and 230 Hz, which sometimes moves to 250 Hz during emphases. Here, she welcomed a fish of considerable size with emotional expression apparent in speech intensity and much higher than normal pitch levels.

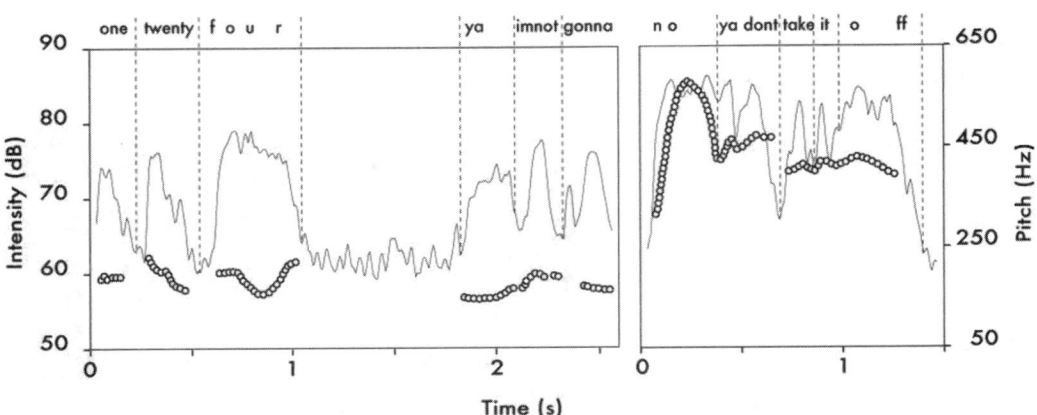

FIGURE 4–4. Erica rejects the modification of data in her mathematical model both verbally and emotionally, as expressed in prosody, much higher than normal intensity and pitch levels. Episodes such as the one displayed on the right are prosodic expressions of emotions, including speech rates, higher speech volumes (intensities), and vastly higher pitch levels.

```
01  E:  One-twenty (0.14) fou::r (0.83) Ya, I'm not gonna—
02      (1.53)
03      °I mean—° His size is really looking good
04      (0.95)
05  H:  °Drop 'em from distri[°
06  E:         [↑NO YA DON't TAKE IT OFF!
```

More so, in this response, Erica does not merely display a general emotional valence toward record keeping and modeling. Rather, her prosody embodies the emotional nature of the (speech) act of rejection.

In the following episode, Erica displays emotions in the act of acknowledging that the weight distribution of a particular sample does not look "bell-curvish" in the way that she has previously described. (See Table 4–1.)

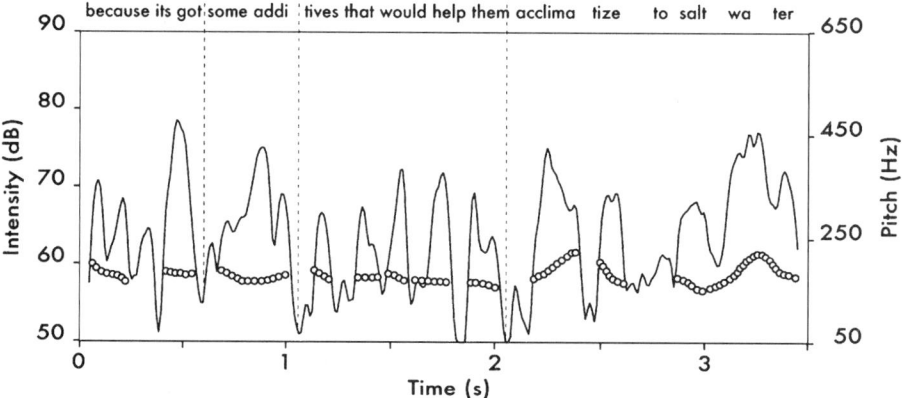

FIGURE 4–5. During the years preceding the layoff notice, Erica's baseline pitch was consistently lower ranging between 160 and 190 Hz, with some higher peaks during emphasized phonemes "clim" and "wa," which, with the higher peaks in intensity, are heard as emphases: ac*clim*atize and *wa*ter.

TABLE 4.1.

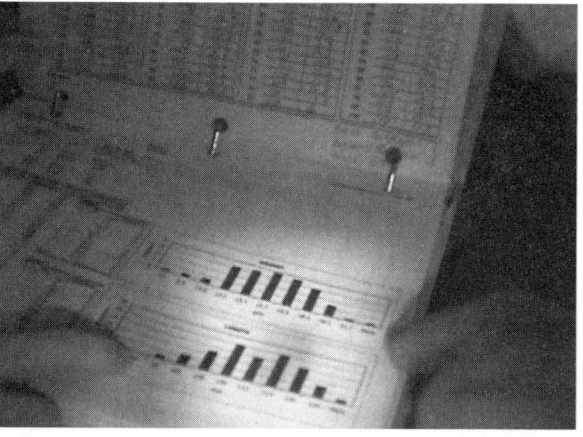

01 M: Can I ask you a question about—you showed me this histogram of the weights and you were saying you wanted it to look in a particular way
02 E: Um I guess bell curve
03 M: but because they don't sort of look bell-curvish
04 E: ↑I know hu hu (0.15) ↓no (0.40) um
05 (1.14)
06 M: Like (0.30) both here in length and weight you ha[ve this
07 E: [Yeah like it looks like I have a nice curve * coming up here and there's a little bit of a drop here.

In this episode, Erica acknowledges the anomaly both in semantic terms ("I know") and the (embarrassed) laughter ("hu hu"). The bell curve therefore is not just an informational item that can be interpreted or constructed in this or that way. It is an object to which Erica relates in an emotional matter, here mediating the difference between an ideal curve and that actually existing in her data.

Higher arousal levels occur with anxiety and nervousness, supported by research that shows increases basic pitch, pitch range, and intensity at higher speech frequency; increases in pitch have also been found for milder forms of the emotion such as worry or anxiety (Pittam & Scherer, 1993). Given the high level of stress Erica has experienced after receiving the lay-off notice, it may not surprise to find out that her baseline pitch levels has increased, she literally is more "high-keyed" than before. My recordings show that over the course of three of the four years, the baseline pitch levels in Erica's prosody were between 160 and 190 Hz during the time prior to the layoff notice (e.g., Figure 4–5), but consistently between 200 and 250 Hz thereafter (Figure 4–4). This therefore is evidence of the different emotional ground state, also available when Erica consciously reflects upon her situation in the hatchery, particularly when she talks about the injustice that temporary fish culturist position has been filled by someone with less experience and competence, someone who has not been in such a position before. In addition to being more high-keyed, Erica is also more "on the edge," evidenced in a higher frequency of sudden pitch changes (Figures 4–3, 4–4).

Emotions and Action in Theory

Recent research in sociology (Turner, 2002) and neurosciences (Damasio, 2000) shows emotions are an integral part of the implementation of practical action. Research with patients that have particular forms of brain damage—lesions between the parts integral for the experience of emotions (i.e., feelings) and those integral for logical evaluation and memory—shows that they consistently take actions that are disadvantageous to themselves, their relations to others, and to their surroundings. That is, although these patients retain full capacities that are normally measured on intelligence tests (including moral reasoning), they no longer *take appropriate actions*, that is, they no longer deal well with the relationship between action and the socially mediated activity (which is characterized by long-term objectives) (Figure 4–6). An individual inherently selects actions on the basis of their potential to reach a state of higher emotional valence, even if this means that in the short run, states of lower emotional valence have to be incurred—no pain no gain (Collins, 2004). Crucially, the bodily ground state always mediates practical action, even if emotional valence is neutral or not conscious at particular moments. This requires us to include emotions in a theory of practical action, always situated in the salient social and material setting. My previous research showed that cultural-historical activity theory is a useful and powerful tool for understanding mathematics at work (e.g., Roth, 2003b).

Cultural historical activity theorists are committed to understanding cognition and consciousness as these arise from practical activity, that is, as these take place in socially organized settings (Leont'ev, 1978). These settings mediate the levels of performances that can be observed, which has *activity* as the unit of analysis. Unlike in normal educational usage, "activity" refers to the entire organization of socially mediated phenomena such as farming, doing research on mathematical cognition, or hatching salmon. In the study of activity, the following, mutually presupposing (dialectically related) entities are helpful to identify structure: (individual, collective) subject, object, and outcome of the activity; community of which the subject is a constitutive part and which mediates the nature of the object; tools in use (e.g., mathematical model of fish); accepted rules of interaction with other elements in the system; and the existing division of labor (Engeström, 1987).

Human activities are analyzed at three mutually presupposing levels: activity, action, and operation (Figure 4–6). Activities, such as fish hatching, are oriented toward conscious motives and are carried out by collectivities (communities); actions are oriented towards conscious goals and are carried out by individuals or groups; and operations are oriented toward conditions and executed in routine fashion without conscious attention. Activity and actions mutually presuppose each other: an activity presupposes the actions that concretely realize it and actions

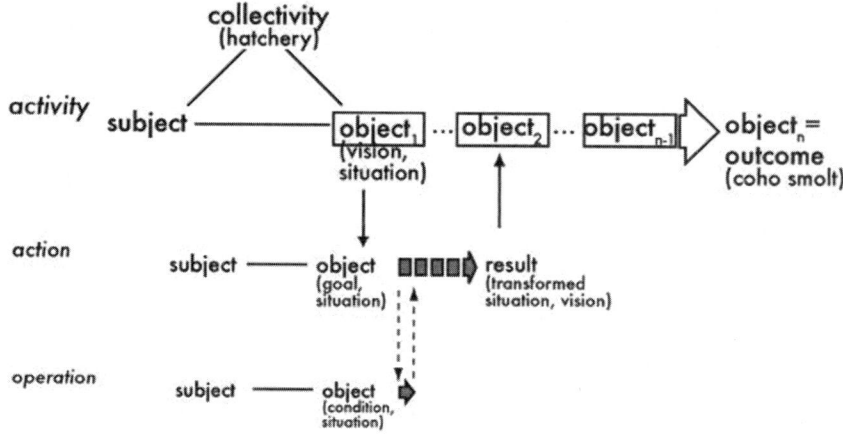

FIGURE 4–6. This activity theoretic model situates each action both in shared social activity and in unconscious bodily operations. Emotions are integral to operations that concretely realize an individual's practical action, which obtains its sense from and is evaluated by the collective.

presuppose the activity that constitutes their raison d'être. Similarly, actions and operations presuppose one another: an action presupposes the operations that concretely realize it and the operations presuppose the action that constitutes, in an unfolding manner, the condition to which each operation responds (leading to a proper unconscious "selection" and "chaining").

Emotions are so important in understanding cognition because they enter this picture in the constitution of practical action (shaded arrows in Figure 4–6), both by setting the emotional background against which decisions are made as well as the emotional valence involved in the decision itself and the anticipated gains or risks to an increase in emotional valence (Turner, 2002). In a variant of development of cultural-historical activity theory, emotion was hypothesized to take such a central place even in pre-human cognition and as early as first active orientation of simple organism in the pursuit of food (Holzkamp, 1983). The data rallied so far provide evidence for the changes in emotional ground state and on Erica's work in general, and the particular expressions of emotions in ongoing (speech) action.

In cultural-historical activity theory, emotions can be incorporated as central features of an account of situated practical action. More so, the framework developed so far also allows us to account for motivation and identity as integrated effects of actions and emotions.

MOTIVATION

One important fact about Erica that has to be explained is her "motivation" not only to do her job but also to do it well ("300 percent") and to do it well even after she has received her layoff notice entailing substantial worries and financial woes for her and her family.

Motivation is a psychological concept usually modeled as part of an affective system that impinges on a separately functioning cognition, normally decreasing its optimal capacities (e.g., Hartman & Sternberg, 1993). From the perspective of cultural-historical activity theory, the concept of motivation has been subject to severe criticism because, the way it is normally used, it is based on circular reasoning

(Osterkamp, 1990). In a reconstructed version of the concept, motivation is integrally related to present and future action possibilities and the related emotional qualities. Thus, an increasing control over one's life conditions (room to maneuver) does not have intrinsic value but constitutes a general, subjectively experienced emotional background (Damasio, 2000; Holzkamp, 1993). The motivational quality of personal reasons for action is a function of the relation of the anticipated gains in action possibilities and the associated higher emotional valence, on the one hand, and the effort, costs, and risks involved, on the other hand (Collins, 2004).

An expansion of the action possibilities loaded with an increase in emotional valence in the context of predictable effort, cost, and risk, constitutes motivation. That is, the promise of an increase in one's personal room to maneuver in decision-making situations is associated with higher emotional valence—no additional variable like "motivation" is necessary to understand why a person is "motivated" or "unmotivated."

Erica always has felt driven to learn more. Using the mathematical model she is building, she predicts fish weights, fish growth, required feed, water flow rate needs, and so forth. Whenever the predicted values do not match with the actual values subsequently measured, there is an occasion for reflecting on the causes of this deviation. This reflection requires an articulation of the hatchery practices (e.g., the nature of feed and feeding practices: "What was the size of the feed pellets?" "When was the pellet size changed?" "Were the fish fed by hand or using a blower?"), environmental conditions (e.g., "What had been the temperatures?" "What had been unusual about the weather?"), or particulars of the fish population (e.g., "Had there occasions of stress?", "Are there any signs of disease and physical damage?", "Any signs of infections?"). Articulating possible causes presupposes a practical understanding of fish culture and fish biology; but it also leads to a better practical understanding of fish biology and fish culture. This understanding allows her to "cement [predictions] a little closer every year to what works." Mediated by her modeling activity, therefore, Erica increases her action possibilities in the hatchery, and a greater room to maneuver in the hatchery leads to an expansion of her modeling activity. Here, Erica increases her room to maneuver, which is associated with a higher emotional valence, and the efforts and costs of getting there do not exceed the ordinary requirements of the job.

It may therefore not come as a surprise that Erica continuously engages in questioning, which leads to an expansion of her possibilities, that is, learning:

> You gotta keep questioning and then, "Well, why does it do that? Why does the formalin work better? Is it because we don't have filtered water?... So, I love learning. It's great and I think if you don't think about things that you become stagnant. And then I think you become, you get into a comfort zone and then you're afraid to get out of the comfort zone and things don't move forward and then you're really shooting yourself in the foot.

This quote shows that there is a potential for risk, which can be avoided by remaining in "the comfort zone." But staying in the comfort zone, in her case, means "shooting [her]self in the foot." Shooting herself in the foot would have been a cost for not engaging in learning (e.g., additional work of keeping detailed records). That is, the risks of not engaging in questioning and learning—with their attendant efforts and costs—are greater than the risks involved in doing just that.

This has not been the case for her mentor (Mike), who, after having greatly contributed to the advancement of hatchery practices—recognized by his peers and scientists alike—no longer engages in experimentation and learning in the way Erica does. After numerous incidences, where the hatchery management has undermined his efforts to mount experimentations and where managers have limited learning of fish culturists (for example, by not allowing them to attend fish culturist conferences), he has shifted from giving "300 percent" to just doing his job. Mike comes to work an eight-hour shift, interacting as little as possible with the managers, and leaves work to seek fulfillment in other pursuits (growing vegetables for commercial purposes, building a house from scratch). For Mike, the risks of being "shot down" and getting into fruitless arguments with his managers are too great to warrant possible payoffs in terms of increased room to maneuver and emotional valence. In fact, the interactions with the managers inherently lead to a decrease in emotional valence at the moment and, cumulatively, decrease the emotional baseline that Mike can experience at work. He sees in his work "just a job," which provides him with the financial resources that allowed him to do those things he really feels rewarding.

Using a traditional concept of motivation, observers certainly would conclude that Erica is motivated, doing more than her job required and learning to become increasingly better in what she is doing but that Mike is unmotivated, merely exchanging his labor for a salary. In Erica's terms, he has "moved into a comfort zone" but also "things [did not] move forward." Received educational and psychological wisdom might suggest that giving Mike some form of "carrot" would increase (externally) his motivation. My exposé should have made it clear that such a conception of motivation is problematic, for although Mike might engage in certain actions to make more money, neither his emotional valence nor his room to maneuver would increase. On the other hand, we see that in Erica's situation, the motive of activity, hatching coho salmon, is associated with gains in emotional valence, with better understanding, greater room to maneuver, and also with knowing that it made her a better fish culturist. Higher emotional valence is compounded in that it encourages her to continue learning, with the promise of even higher payoffs in terms of emotional valence. Each time she engages in action, Erica therefore exhibits her knowledgeability to others and herself. That is, in each action Erica produces and reproduces who she is both for others and for herself—this leads us directly into the concept of identity.

IDENTITY

Each practical action is constitutive of who we are with respect to others and ourselves (Ricœur, 1992). Each time Erica knowledgeably acts—making a deal ordering food (best-buy problem) or feeding fish just enough so that they do not go hungry and that no food is wasted because fish did not snatch it before reaching the bottom—she reproduces herself as a competent fish culturist. Not only does she know herself as a knowledgeable fish culturist but also others can see her to be competent. Each practical action is also a candidate for being included in autobiographical narratives that allow human beings to constitute and maintain their personal sense of identity. Because practical actions cannot be separated from current bodily states, the sense of self is bound up with emotional valences. That is, identity is an effect of practical action and emotion, and is correlated with motivation.

There are therefore two different faces of identity (Roth, 2006). On the one hand, identities are always produced and reproduced anew, in every action, and therefore continuously open to change (Damasio, 1999; Roth et al., 2004). In every action at the job, Erica reasserts herself as a competent fish culturist. More so, because these actions evidently are directed to better understanding and increased room to maneuver, others also recognize her as a "motivated" individual. Each action therefore contributes to allowing others to make attributions and therefore, descriptions of who Erica is. By the same token, others can see in the actions of her mentor that Mike is no longer motivated—not being motivated becomes an attribution to who Mike is with respect to others and his job. However, Mike once has been like Erica, "motivated" and continuously expanding his knowledgeability as a fish culturist. That is, the change from the "motivated" to "unmotivated" attribution has been made available in the changing nature of his actions.

On the other hand, identity has a stable aspect, which allows us to say a person is the same despite the inherent fragility of self arising from the first aspect. That is, although Mike is different now than he has been twenty years ago, we still recognize in him something like a core self. Although Erica's actions have changed after she has received the layoff notice, being at times inattentive to the coho and to her mathematical model, at times even "not her true self," she is still Erica. This stable aspect of identity is both from philosophical and neuroscientific perspectives, not a self-evident quality of human life but a quality constructed by means of (auto-) biographical narratives (Damasio, 1999; Ricœur, 1992). These narratives contain, as their most important component, accounts of a person's interactions with the social and material world. This is quite evident in the following self-description Erica provides relative to her efforts of maintaining a highly detailed database and, therefore, mathematical model.

> See, I'm naughty; I have my own brood binders even though there's official brood binders there. But if you look at, I'll show you, you look at the one's that are there and you look at mine, "Who has the information?"

In this account, one finds different aspects of who Erica is with respect to the hatchery as physical plant and hatchery management. First, she describes the existence of an extra set of "brood binders," that is, the database concerning a particular fish population over its entire time in the hatchery. Not only does Erica spend the additional effort to maintain a separate record but also there is more information than in the official records that the hatchery managers keep. This makes her not only a "naughty" person, a renegade, but also fish culturist "who has the information" that others do not have and therefore cannot use for making more accurate predictions for future populations. By the same token, by not using the information produced, the hatchery managers as also show that they are not interested—a fact that does not go unnoticed among the fish culturists.

Erica, on the other hand, is constructed as a person—especially by the managers—who is "hanging on to too many details" and who spends so much time on establishing and working with the database that the attribute "geek," which she frequently uses in a self-deprecating way, becomes justified. In fact, she appears so enamored with the database that she "ha[s] been accused of" giving them too much attention, when she really should "slow down and talk about fish culture things." That is, she feels to be accused of spending too much time with the mathematical modeling aspects and too little time with the actual fish culture practices that the database models.

The positive emotional valence arising for Erica from her work provides a positive self-image (identity), which feeds back into background emotional valence that Erica brings to work. Furthermore, Erica also knows that others know about her outstanding qualities, as apparent in the following account that pertains to a stint on a research vessel where she has worked for a scientist (Rusty):

> Who was it? Oh, I can't remember, a big name in ocean climate science...he put a call out and I actually got to go out on the trip [ocean research vessel]....Like I'm not a

> scientist, but I know also that any time I want to go back out there Rusty, sure come on out. So I know I've proven myself worthy for, to be involved in his experiment and that's a good feeling, too.

She knows that Rusty has appreciated her contributions to the research project, although she is not a scientist herself, in fact, only has two years of college level education but no certification or degree. Through her actions on the research vessel, Erica has "proven [her]self worthy...to be involved in [the scientist's] experiment." The "good feeling" she has as a result is simply a description of the background emotional valence that arises from being knowledgeable, knowing to be knowledgeable, and knowing that others know her to be knowledgeable.

In a previous section, I show that in each action, Erica makes available to others momentary emotions and a background valence in addition to her knowledgeability. Thus, they establish that she is more high-keyed (evidenced here in the higher pitch ["key"]), which others attributed to the layoff and the financial worries these entailed, and more irritable and "edgy," here exemplified in the substantial pitch shift when a helper intimated to change or omit data from her database. Both kinds of changes are directly available in her (discursive) interactions, and then become attributes of who she is at this point in time, that is, her identity. Similarly, the background emotional valence and the lower emotional valence associated with loosing her job mediate her actions, making her look less competent and less attentive to the fish than she has been in her previous actions. That is, identity is the result of interpretations of practical actions, which cannot be dissociated from emotions; identity therefore is an effect (result) of actions and emotions. But aspects of identity feed back, as the "good feeling" has provided the background emotional valence that mediate her every action.

An interesting aspect of identity emerges from these considerations with respect to differences in motivation. On the one hand, both Erica and Mike not only *are* fish culturists, but also are knowledgeably so. Because they are members of the hatchery, and observable in their every action, both are, for others and themselves, knowledgeable fish culturists (identity). As a consequence, identity is, in part, an effect of the structure of activity in which the individual participates. Erica continues to identify with the object of activity and each action contributes to maintaining and increasing her emotional valence. She "has been motivated" and continues to do so despite the emotional stress related to her layoff. That is, continuously making available to others and herself her knowledgeability as fish culturist and her continuous improvement provides sufficient payoff to keep her going at 300 percent. Mike, on the other hand, no longer envisions sufficient payoff from working more than to rule given the high costs and risks that he would have had to incur. He de-identifies with the motive of activity although continuing to pursue it, given that the income it provides increases his room to maneuver in other, more rewarding contexts (investing in commercial farming, designing and building a home from scratch).

IMPLICATIONS FOR TEACHING MATHEMATICAL MODELING

In this chapter, I provide ethnographic evidence for the integral nature of emotions in the everyday actions of fish culturists in general and the mathematical modeling of one fish culturist in particular. Motivation, rather than constituting an independent construct, is shown to be an effect of changes in action possibility and promise of increases in emotional valence associated with each action and background emotional valence that mediates it. Similarly, identity is shown to be not an independent construct, but the result of interpretations of actions both in their momentary, ever-changing and lasting, continuous dimensions. Especially in the lasting dimension of emotion—associated with the background emotional valence or feeling (Damasio, 1999)—a person's sense of who he or she is feeds back on their situated performance. It also feeds back on the momentary aspects of identity as it arises from action; this momentary aspect subsequently becomes a resource in further, long-term constructions and reconstructions of identity.

These results and the theoretical framework provided have substantial implications for our understanding of mathematical modeling in particular and mathematics education in general. If emotions are central to decision-making, "motivation," learning, and identity, we can no longer theorize mathematical modeling independent of emotions and emotional valence. We can no longer teach mathematics as if only its logical and formal aspects mattered. If emotions are central aspects of decision-making in practical mathematical modeling activity, then mathematics education squarely needs to address this aspect.

An important issue for many students is the promised usefulness of mathematics. In contrast to Erica, who associates her mathematical modeling activity with an increased room to maneuver and emotional valence, students in school and university courses frequently ask, "What do I need this for?" The answers they get are frequently not satisfying and often right out false—students have many examples in their everyday lifeworlds where people live well without knowing mathematics (or science, history, etc.). In addition, there are ways of earning a decent living and coping with the requirements of everyday life without knowing school mathematics—most ethnomathematical studies provide evidence of this. Thus, the payoff, knowing (school) mathematics for its own sake seems low compared to the costs and risks (getting low grades, which may mediate career trajectories). Furthermore, the risks of engagement may be high for the identity of the person—which has had effects especially on the engagement of young women in mathematics. Female students often do not want to do well in mathematics because the emotional risks within their peer groups by far outweigh the future gains they can perceive.

In contrast to the mathematical modeling in the hatchery, which is a tool in the pursuit of a particular object of activity, producing healthy populations of smolt, and the learning that occurs as a result of its use, schooling makes learning mathematics the object of activity. A contradiction is apparent: in the hatchery, learning means an increasing room to maneuver with respect to fish and fish hatching and the relation of payoff and cost. In schools, on the other hand, knowing more is the stated objective, but this objective is neither apparent nor associated with sufficient payoff-cost ratios. The key difference is that in Erica's case, she has reason to engage and learn, whereas in school mathematics, the rationale for particular mathematical activities and content are stipulated from the outside. Problem-based learning may just be a learning environment that sufficiently simulates important qualities of everyday knowledgeability and learning to allow students to identify (rather than de-identify). Students learn because they themselves envision increased action possibilities, and the collaborative contexts in which PBL frequently occurs provides sufficient support for dealing with the costs and risks incurred. It is clear, however, that as long as educational environments control the time for tasks and use tests—the object of activity will remain doing well on tests and other forms of assessments (even in PBL contexts) rather than increasing one's room to maneuver in discipline-relevant aspects.

ACKNOWLEDGMENTS

The data collection was made possible by grants from the Social Sciences and Humanities Research Council of Canada and the Natural Sciences and Engineering Research Council of Canada.

REFERENCES

Collins, R. (2004). *Interaction ritual chains*. Princeton, NJ: Princeton University Press.
Coy, M. (1989). Being what we pretend to be: The usefulness of apprenticeship as a field method. In M. W. Coy (Ed.), *Apprenticeship: From theory to method and back again* (pp. 115–135). Albany, NY: State University of New York Press.
Damasio, A. R. (1999). *The feeling of what happens: Body and emotion in the making of consciousness*. San Diego: Harcourt.
Damasio, A. R. (2000). *Descartes' error: Emotion, reason, and the human brain*. New York: HarperCollins.
Engeström, Y. (1987). *Learning by expanding: An activity-theoretical approach to developmental research*. Helsinki: Orienta-Konsultit.
Fox-Keller, E. (1983). *A feeling for the organism: The life and work of Barbara McClintock*. San Francisco: W. H. Freeman.
Goodwin, C., Goodwin, M. H., & Yaeger-Dror, M. (2002). Multi-modality in girls' game disputes. *Journal of Pragmatics, 34*, 1621–1649.
Goodwin, M. H. (1996). Informings and announcements in their environment: Prosody within a multi-activity work setting. In E. Couper-Kuhlen & M. Selting (Eds.), *Prosody in conversation: Interactional studies* (pp. 436–461). Cambridge: Cambridge University Press.
Goodwin, Marjorie H., & Goodwin, C. (2000). Emotion within situated activity. In A. Duranti (Ed.), *Linguistic anthropology: A reader* (pp. 239–257). Malden, MA: Blackwell.
Hartman, H., & Sternberg, R. (1993). A broad BACEIS model for improving thinking. *Instructional Science 21*, 5, 401–425.
Holzkamp, K. (1983). *Grundlegung der Psychologie*. Frankfurt/M.: Campus.
Holzkamp, K. (1993). *Lernen: Subjektwissenschaftliche Grundlegung*. Frankfurt/M.: Campus.
Johnstone, T., & Scherer, K. R. (2000). Vocal communication of emotion. In M. Lewis & J. Haviland (Eds.), *Handbook of emotion* (2nd ed., pp. 220–235). New York: Guilford Press.
Lave, J. (1988). *Cognition in practice: Mind, mathematics and culture in everyday life*. Cambridge: Cambridge University Press.
Leont'ev, A. N. (1978). *Activity, consciousness and personality*. Englewood Cliffs, NJ: Prentice Hall.
Noss, R., & Hoyles, C. (1996). The visibility of meanings: Designing for understanding the mathematics of banking. *International Journal of Computers for Mathematical Learning, 1*, 3–31.
Noss, R., Hoyles, C., & Pozzi, S. (2002). Abstraction in expertise: A study of nurses' conceptions of concentration. *Journal for Research in Mathematics Education, 33*, 204–229.
Osterkamp, U. (1990). *Grundlagen der psychologischen Motivationsforschung 2. Die Besonderheit enschlicher Bedürfnisse—Problematik und Erkenntnisgehalt der Psychoanalyse* (4th ed.). Frankfurt /M.: Campus.
Pittam, J., & Scherer, K. R. (1993). Vocal expression and communication of emotion. In M. Lewis & J. Haviland (Eds.), *The handbook of emotions* (p. 185–197). New York: Guilford.
Ricœur, P. (1992). *Oneself as another*. Chicago: University of Chicago Press.
Roth, W.-M. (2003a). Competent workplace mathematics: How signs become transparent in use. *International Journal of Computers for Mathematical Learning, 8*(3), 161–189.
Roth, W.-M. (2003b). *Toward an anthropology of graphing: Semiotic and activity theoretic perspectives*. Dordrecht, the Netherlands: Kluwer Academic.

Roth, W.-M. (2004). Emergence of graphing practices in scientific research. *Journal of Cognition and Culture, 4*, 595–627.

Roth, W.-M. (2005). Mathematical inscriptions and the reflexive elaboration of understanding: An ethnography of graphing and numeracy in a fish hatchery. *Mathematical Thinking and Learning, 7*, 75–109.

Roth, W.-M. (2006). Identify as dialectic: Making and Re/making self in urban schooling. In J. L. Kincheloe, K. Rose, & P. M. Anderson (Eds.), *The Praeger handbook of urban education* (pp. 143–153). Westport, CT: Greenwood.

Roth, W.-M. (in press). Motive, emotion, and identity at work: A contribution to third-generation cultural historical activity theory. *Mind, Culture and Activity, 14*(1).

Roth, W.-M., Tobin, K., Elmesky, R., Carambo, C., McKnight, Y., & Beers, J. (2004). Re/making identities in the praxis of urban schooling: A cultural historical perspective. *Mind, Culture, & Activity, 11*, 48–69.

Saxe, G. B. (1991). *Culture and cognitive development: Studies in mathematical understanding*. Hillsdale, NJ: Lawrence Erlbaum Associates.

Schliemann, A. D., & Acioly, N. M. (1989). Mathematical knowledge developed at work: The contribution of practice versus the contribution of schooling. *Cognition and Instruction, 6*, 185–221.

Scribner, S. (1986). Thinking in action: some characteristics of practical thought. In R. J. Sternberg & R. K. Wagner (Eds.), *Practical intelligence: Nature and origins of competence in the everyday world* (pp. 13–30). Cambridge: Cambridge University Press.

Selting, M. (1996). Prosody as an activity-type distinctive cue in conversation: The case of so-called 'astonished' questions in repair initiation. In E. Couper-Kuhlen & M. Selting (Eds.), *Prosody in conversation* (pp. 231–270). Cambridge: Cambridge University Press.

Turner, J. H. (2002). *Face to face: Toward a sociological theory of interpersonal behavior*. Stanford, CA: Stanford University Press.

CHAPTER 5

Learning in Design

David Williamson Shaffer
University of Wisconsin–Madison

In the world beyond school, new technologies are fundamentally changing the mathematical and scientific understanding that practitioners need in fields that use mathematics, science, and technology. This study looks at one such field—architectural design—and at the nature of problem-solving in this real life situation. In particular, I ask: What does it mean to "understand" in the context of architectural design? How does this form of understanding develop? And what kinds of experiences facilitate the development of this form of understanding?

In answering these questions, my focus is not on the development and use of mathematical or scientific understanding *per se*. I look, rather, at what it means to know, to communicate, and to understand in the context of architectural design—and perhaps more important, I look at the systems through which students of architectural design are initiated into these ways of knowing. What I will attempt to show is that these ways of thinking (and ways of coming to these ways of thinking) form a coherent system: a system whose logic differs significantly from traditional school-based learning in mathematics and science.

Such a study is significant in our understanding of mathematics and science education for two reasons. First, it describes the intellectual, practical, and conceptual underpinnings of one of the real world contexts for which students of mathematics and science are being prepared. To the extent that we believe cognitive context matters in education, then understanding the situations to which school is a precursor may be valuable. Second, design and the practices of designers have been used as models for a number of experimental learning environments and progressive curricula in recent years (Cossentino & Shaffer, 1999; Erickson & Lehrer, 1998; Greeno, 1997; Hmelo, Holton, & Kolodner, 2000; Jacobson & Lehrer, 2000; Kolodner, Crismond, Gray, Holbrook, & Puntambekar, 1998; Loeb, 1993; Perkins & Blythe, 1994; Shaffer, 1996). Interest in design as a pedagogical model has produced a number of studies of design practices, particularly of architectural design, which is often seen as a canonical exemplar of the design process. However, much of that work has focused on the practices of professional designers in the workplace (Hall & Stevens, 1996; Hawkins, 1993; Stevens, 2000) or on the psychology of the design process through study of individual designers (Akin, 1986; Branki, 1993; Coyne, 1993; Davies, 1987; Goldschmidt, 1989; Jansson, 1993; Mitchell, 1994; Rowe, 1987; Schon, 1988a, 1988b; Simon, 1996). Such work naturally emphasizes the nature of design activity and understanding, but sheds less light on questions about how design ability and understanding develop.

Here, I take a different approach. Extending the work of Schon (1985; 1987), I look here at the setting in which designers are initiated into their practice. This study describes the Oxford Studio, an architectural design course at the Massachusetts Institute of Technology (MIT) Department of Architecture. My aim is to shed light on both the practices of design and the practices of learning to design. An advantage to this approach is that one of the goals of the Oxford Studio (or any design studio course) is to help young architects understand what it means to be a designer. Activities in the Oxford Studio were thus explicitly organized to clarify as well as to enact design practices and the principles they embody. Design studios at MIT aspire to be model exemplars of this process, and in so doing, they provide a window into the ideals of designing—and the process by which learning to design enacts these ideals in practice.

A number of previous studies have explored particular aspects of design studios in some detail (Anthony, 1987; Craig & Zimring, 2000; Crowe & Hurtt, 1986; Flemming, 1998; Frederickson & Anderton, 1990; Kvan, 2001; Loeb, 1993; Schon, 1985; Uluoglu, 2000). Other studies have examined the social and epistemological implications of studio practices (Dutton, 1987; Heylighen, Neuckermans, & Bouwen, 1999; Ledewitz, 1985; Sancar, 1996; Schon, 1988a, 1988b). The goal of this study is to investigate the ways in which these elements of design practice are constituted as a coherent system of activity: to examine the underlying pedagogical structure of one design studio.

Such a mission suggests an ethnographic perspective. This ethnographic perspective frames both the methodology of the study (observational data collection, interpretive analysis, and descriptive presentation) and its attention to what the participants in one particular studio did, how they understood the significance of their activities, and what we as observers can interpret from that activity about the norms, values, and perspectives that support the practices of design.

As in any ethnographic study, there are not discrete "hypotheses," "results," and "conclusions" to present in the analysis that follows. Rather, I present this study of the Oxford studio in four parts. First is a discussion of prior theoretical and empirical work that frames and informs observations and interpretations of the Oxford studio. Next is a description of the activities I observed in the Oxford studio, including its organization in time and space, the sequence of assignments, and the activities that made up those assignments. This description, which covers activities that took place over a semester in a mid-level graduate/undergraduate design studio, necessarily presents a view of the design process that is both telescopic and technical. The goal is to represent a complex set of relatively specialized activities and interactions in a manner that will be coherent and comprehensible for readers interested in pedagogy but not experts in architectural design.

Following this description of the elements of the Oxford studio is a discussion of the way in which those elements interacted to form a coherent system of activity. Obviously, the elements of the design studio and the interactions among those elements are inseparable in practice; they are distinguished here for rhetorical convenience and conceptual clarity, and only to the extent possible while still reflecting the observed experiences of the participants in the Oxford Studio. Finally, the study concludes with a brief discussion of the implications of such an analysis for our understanding of architectural design as both a target domain and a model for K–12 classrooms.

The analysis looks, in other words, at how design practices were enacted in one design studio, focusing on the relationships between activity and understanding. The study is thus a structural ethnography rather than a hypothesis-driven or micro-genetic account of learning: it attempts to describe phenomena that are local but that operate at a time scale of hours and weeks rather than minutes and seconds—and thus phenomena that appear through observation at a more intermediate level of analysis. This is ethnography closer in spirit and practice to Geertz (1973a) well-known study of conventions for time and identity in Bali than it is to Cobb's (1986) case study of the emergence of abstract mathematical thinking in the concrete activities of one learner.

In a field such as design, where much is already known about specific cognitive and pedagogical processes, such a structural analysis is useful in extending our understanding of the systematic nature of activity. The analysis of the Oxford studio that follows suggests that expression and expressive activity are a significant underpinning of the design studio system. To the extent that elements of design practice are interconnected, the study also suggests key features and relationships of the studio that should ideally be preserved in any adaptation of design to the creation of learning environments in other fields.

BACKGROUND

Research on Design and Design Learning

The design studio can trace its roots back more than a century to the Ecole des Beaux-Arts in France (Chafee, 1977). The focus of a designer's training today still follows the Beaux-Arts tradition of open-ended projects and a variety of structured conversations that culminate in a public presentation of work. Scholars of design education have studied this tradition in some depth. Schon (1985), for example, analyzed a key interaction of the design studio, the *desk crit*: an extended and loosely structured interaction between designer and critic (expert or peer) involving discussion of and collaborative work on a design in progress. Schon suggested the desk crit (or *crit*) is central to the development of a student's ability to design thoughtfully. In Schon's description, the desk crit functions as an instantiation of Vygotsky's (1978) *zone of proximal development*, with development taking place as learners progressively internalize processes they can first do only with the help of others.

Another central tool of the design studio, the *review* or *jury*, similarly mediates the interactions between learner, peers, and experts. The review or jury is a formal group discussion of student work: individuals display their work, present their plans, and get feedback from professionals outside the studio. This model of critical review has been central to architectural training since the foundation of the ateliers at the Ecole des Beaux-Arts in Paris (Chafee, 1977). Reviews were brought to the United States with the founding of the first architecture schools in the 19th century. The problems with the review process—particularly the stressful nature of the experience and the dangers of excessive subjectivity on the part of reviewers—are openly discussed in the architecture community (Anthony, 1987; Frederickson & Anderton, 1990). Nonetheless, external reviews remain a mainstay of the process of architectural design and a powerful tool for connecting work in the studio to professional practice.

Elements of the design process itself have received similar scholarly attention. Schon (1985) and Simon (1996) have discussed the iterative nature of design, in which problems are revisited repeatedly in a generative process. The designer chooses to address a particular issue. A solution is proposed. Strengths and weaknesses of the solution are analyzed (often in a public setting and usually in the form of feedback from others). Based on this analysis, the designer refines the original approach. The new solution is again analyzed, and so the process continues until the analysis of one of the iterations suggests that it is a satisfactory way to resolve the issue. Mitchell and McCullough (1991) discuss the role that media play in this process. They suggest that physical models, digital models, and renderings—as well as more traditional plan, elevation, and freehand drawings—emphasize different aspects of an emerging design. A change in materials produces a change in perspective, and is thus often an important part of the iterative cycle of production and analysis central to design.

Schon (1985) has described the cognitive foundation of design practice as a process of reflection-in-action. In this model, designers make judgments and show skills for which they cannot describe rules or provide explanations. Understanding develops as practitioners refine tacit knowledge through work on subsequent iterations of the design process. This understanding comes partly as Dewey (1934/1958) suggested through overcoming obstacles in the expressive medium, particularly when the obstacles are relevant to the expressive goal and to the underlying structure of the domain. But the norms of a community of practice are also a key element in this process (diSessa, 2000; Erickson & Lehrer, 1998). The demands of producing work for a critical audience provide an important set of constraints. Like overcoming obstacles in the medium of expression, working within the norms of a community demands reflective thinking. Lave and Wenger (1991) argue that learning is always a process whereby an individual comes to participate in the practices of a community: a process that is both inherently social and deeply individual. The norms of the community become a framework for individual thinking and individual identity (Vygotsky, 1978; Wertsch, 1998).

Learning Environments as Coherent Systems

There are, in other words, a number of pedagogical processes and theoretical perspectives that come together in the practices of the design studio. Brown and Campione (1996) suggest that any effective learning environment is not a set of isolated procedures, but rather a coherent system. They argue that such systems depend on a clear articulation not only of "surface procedures," but also of the underlying "principles of learning" that lead to the creation of pedagogical strategies (p. 291). That is, they suggest that in effective learning environments, recurrent patterns of activity or *participant structures* (Herrenkohl & Guerra, 1998; O'Connor & Michaels, 1996) are enacted in such a way as to preserve not only the forms of the individual activity, but the practical and conceptual linkages among different activities. For example, they argue that in the Facilitating Communities of Learners (FCL) curriculum, the pedagogical value of a Jigsaw activity (in which students break into groups to learn about different aspects of a topic and then reform into new groups containing one student with expertise in each aspect) depends in part on the context of a Consequential Task that motivates the interchange of ideas the Jigsaw facilitates. Students merely go through the motions of exchanging information in a Jigsaw if they do not need the information to complete some other meaningful activity—which happens only when the teacher and curriculum designers understand that one of the fundamental principles of learning in FCL is that knowledge is built in the context of meaningful collaborative activity.

In what follows, I extend this basic idea by arguing more explicitly that the pedagogical organization of the Oxford Studio can be analyzed by looking at how the studio's *pedagogical activities* (what Brown and Campione refer to as "surface procedures")" relate, on one hand, to a particular way of thinking, and on the other to the structure of the studio itself. The analysis below looks at:

1. The *surface structure* of the studio: the organization of time, space, persons, and materials in the studio. That is, the physical, temporal, material, and social context of action and interaction—where, when, with what and with whom activities take place.
2. The *pedagogical activities* of the studio: the recurrent participant structures of roles and actions that organize activity in the studio. That is, the kinds of social interactions that characterize activity.
3. The *epistemology* of the studio: the ways of deciding what constitutes a legitimate architectural claim. That is, the conceptual and intellectual warrants that validate activity.

Of course, the challenge is not only to understand how these individual elements function, but also to understand the relationships among them, including:

1. The relationships among surface structures;
2. The relationships among pedagogical activities and the way in which those activities are supported by surface structures; and
3. The way in which these pedagogical activities enact the epistemology of design.

I will argue, in other words, that thinking like a designer (and learning to think like a designer) depends on particular ways of knowing, of acting, and of being in the world. Moreover, these ways of knowing-doing-being are both tightly linked and fundamentally different than the understanding, activities, and structures that comprise a traditional mathematics or science classroom.

METHODS

The Oxford Studio was a one-semester, mid-level architecture course for undergraduate and graduate students, taught by Nigel[1], an experienced architect and studio teacher, and a member of MIT's junior faculty. The Oxford studio had a single professor and two teaching assistants (both doctoral students in the Department of Architecture). There were 11 students in the course. Three were advanced undergraduates majoring in architecture, the remainders were graduate students.[2] All students enrolled in the Oxford Studio course were included in the study, however the portrait presented here focuses on the work of three students in the course (Arnold, Belinda, and Dan) who were studying toward their Master of Architecture (MArch) degree, a program structured to prepare students for professional registration as architects in the United States. These students were chosen because of the proximity of their individual workspaces within the studio, which made it possible to observe the individual work of multiple students, and because of their willingness to be interviewed over the course of the semester. I was present, taking field notes, for roughly one quarter of the studio's teaching hours, and these observations were supplemented by interviews with students and teaching staff. As the semester progressed, one student (Arnold) agreed to participate in a

---

[1]Names reported in this study are all pseudonyms.
[2]Specific demographic information was not collected on the students.

closer examination of his learning process, which forms the centerpiece of the portrait below.

Field and interview notes were analyzed using *case-focused analysis* (Weiss, 1994). Such analysis attempts to understand phenomena by gathering a rich set of data for a limited number of instances to create a thick description (Geertz, 1973b) in which specific examples of experience can be described so as to illustrate how participants understand and organize their activity. Notes on Nigel's, Arnold's, Belinda's, and Dan's activity and their explanations of activity in the studio were coded into thematic categories of surface structure, pedagogical activity, and epistemology, as described above. The analytic descriptions of key elements of the studio and the interactions among those elements that follow were based on these categories, drawing together significant themes from the data collected.

DESCRIPTION: SPACE, TIME, WORK, AND LEARNING IN THE OXFORD STUDIO

In the section that follows, I describe the setting and activities of a semester's worth of work in the Oxford Studio. I begin by outlining the physical and temporal characteristics of the Oxford Studio: where and when activities took place. I describe the sequence of assignments that made up the curriculum, and then, because the assignments all had a similar structure, I describe in more detail the recurring sequence of activities that took place within the assignments. I ground this general description of activity in the Oxford Studio in a progressively more focused explanation of the experience of one student, Arnold, and conclude the description of the studio by describing a fundamental principle of architectural design that this young designer came to understand through his extended design exploration.

Physical Setting

Walking into the Oxford Studio at MIT was quite unlike walking into a lecture hall, seminar room, or classroom in a typical school or college. A well-equipped science lab, with its open plan and benches for student experiments, captures some of the flavor of a studio space, yet still misses certain essential elements of the design environment.

Space

Most apparent in the Oxford Studio was the amount of space allocated to students. In the Oxford Studio, 11 students had more space for their individual drafting areas than many high schools provide for a class of 30 (see Figure 5–1). In addition, the studio offered a meeting space the size of a seminar room for a college course. The students had access to computers and printers within the studio and to woodworking machinery in a nearby room (not shown in Figure 5–1). The studio was connected to an external corridor by rolling garage doors so the hall outside could be used to post students' work for discussion and comment. Finally, the studio had access to additional space for formal presentations of student work. All of these spaces (except the students' individual workspaces) were shared by other studio classes but were available as needed for Oxford Studio activities.

In the Oxford Studio, each student worked throughout the semester in his or her own workspace. Students adapted these work areas to their own needs and

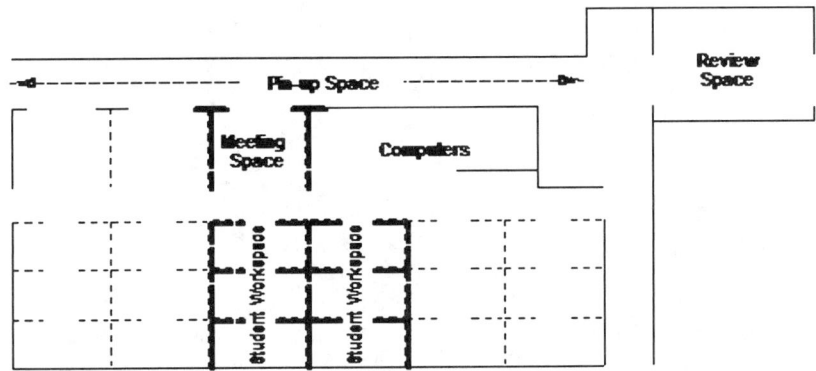

FIGURE 5–1. Diagram of studio space at MIT, in which each student had individual workspace.

working style, and no two studio desks looked the same by the second week in the semester. The low walls of the cubicles were covered with sketches, postcards, inspirational examples of architectural design, and even candy and other junk-food wrappers pinned up as merit badges for work done through the hours of the night. Locked drawers crammed with drafting tools, modeling knives, and rolls of trace paper made each space a workshop; Belinda brought in her computer and secured it with a cable to her desk.

Time

The pace of work in the Oxford Studio was also quite unlike that in a traditional class. Studios at MIT officially met 3 days a week from 2 to 6 p.m. But for the Oxford Studio, this timetable was more a rough guideline than a fixed schedule. Students and teaching staff routinely came to the Oxford Studio before or after 2 p.m. depending on the work they had to do on a particular day. Students and staff often came in at night or on weekends as project deadlines approached. At any given time during official studio hours, the professor and students might be meeting around a seminar table to discuss projects. Or students might be working individually at their desks. Or checking e-mail. Or stepping out for a cup of coffee. Or meeting with faculty.

This informal approach to time in the studio made it difficult, sometimes, to organize activities. Students were not always present for class discussions, and even major events in the semester, like final reviews, started late and had participants drifting in and out. Time management was an issue for students in the Oxford Studio: work was routinely left until the last minute and sometimes suffered as a result. However, the large blocks of time allotted and the flexibility of the routine also made it possible for different studios to share spaces for meetings and presentation.

Curriculum: A Sequence of Assignments

The *project brief* (the design specifications) for the Oxford Studio was taken from a closed competition of prospective plans for a new business school at Oxford University. The proposed site for this new school was on the edge of the urban development of Oxford, so the relationship of the building and site to the larger context was a central design issue in the project. The project as a whole was quite

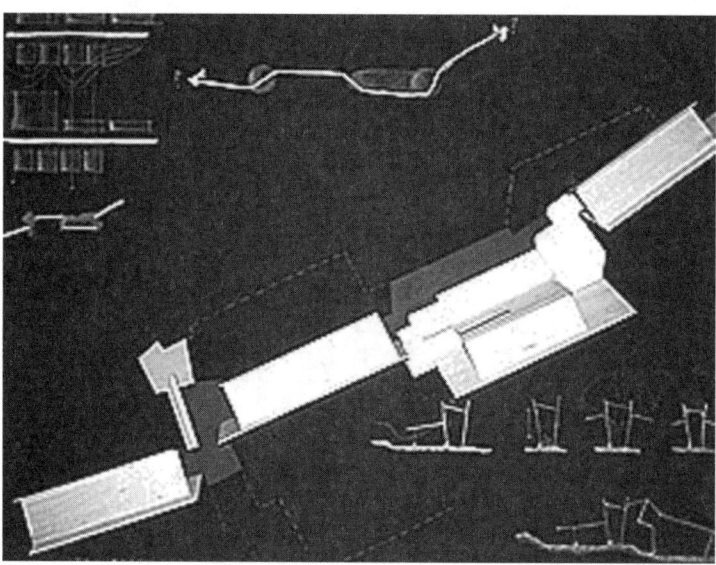

FIGURE 5–2. Belinda's conceptual model of Strawberry Vale Elementary School's main corridor (shown here in aerial view) shows informal gathering spaces (marked by arrows) for teachers and students.

complex and Nigel remarked at the start of the semester that it was "about as difficult a project as students at this level could handle." The project was divided into a series of six assignments, taking students through progressively more detailed examinations of the architectural issues involved in creating a business school in Oxford.

Assignment 1: Precedent

The first assignment asked students to examine a *precedent* for the Oxford project: a building that showed how other architects had dealt with issues of context and community in similar settings. To achieve this, students chose from a list of buildings provided by the professor, then researched their building and represented their findings in a *conceptual model*: a physical representation of some key design principle at work in the building. Belinda studied the Strawberry Vale Elementary School in Victoria, British Columbia, a building notable for the way its architects integrated the building's design into the irregular landscape of the site. Belinda made a model of the school's irregular central corridor, showing how the zigs and zags of the corridor (shown in an aerial view of her model in Figure 5–2) created a network of informal gathering places (marked by arrows in Figure 5–2) for students and teachers. Arnold examined the main hall of Singapore Polytechnic, comparing its hierarchical organization to the decentralized plan of the Oxford campus.

Assignment 2: Urban Structure

The second assignment asked each student to research the city of Oxford and represent their understanding of the city and its surrounding landscape in *diagrams* (drawings that convey design concepts without specific or consistent scale or format). Students also produced a *model* (a three-dimensional representation of architectural forms with consistent scale) showing their general plan for dealing

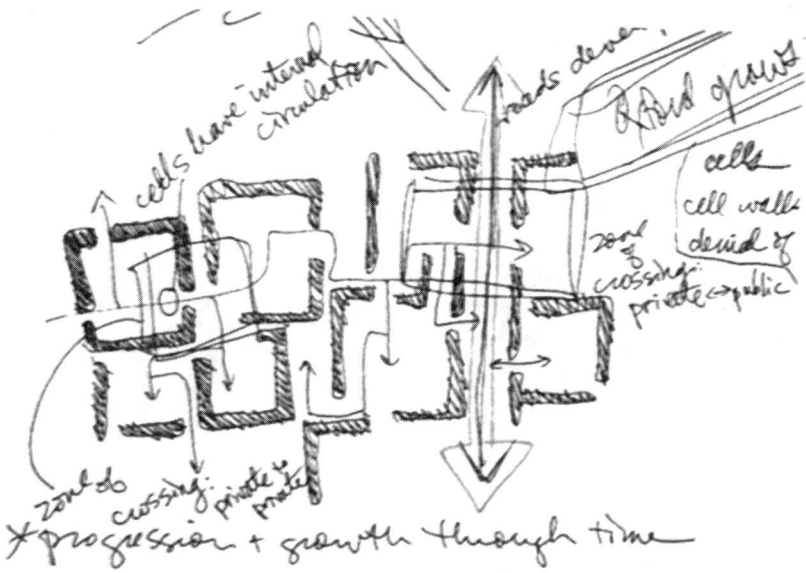

FIGURE 5–3. Arnold's analysis of Oxford's "cellular growth." The thick lines represent the buildings that form the quadrangles of Oxford's colleges, which Arnold suggested function like the semi-permeable membranes of cells. The arrows represent how people move through these membranes and the spaces they create.

with the Oxford landscape in their proposed design for the business school. Arnold's research focused on the history of Oxford's growth, looking at the courtyards formed by the university's many residential colleges as architectural descendants of Medieval monasteries: the academic equivalent of the isolation of the cloister. Arnold's diagram (see Figure 5–3) depicts the courtyards as "the cells of the Oxford quadrangles." Arnold's analysis of this "cellular growth" shows thick lines representing the buildings that form the quadrangles of Oxford's colleges, which Arnold suggested function like semi-permeable membranes. The arrows represent how people move through these membranes and the spaces they create. Arnold identified his site as an opportunity to "bring a graceful end" to Oxford's urban development. Another student's analysis, in contrast, focused on the social fissure between the community and the university (the "town and gown")"; her proposal was to design a "nonhierarchical building" that would symbolically breakdown the traditional English class structure and integrate the city residents and students.

Assignment 3: Conceptual Strategy and the Brief

In the third assignment, students were given the project brief from the original competition, including the *program* (list of requirements) for the building. They were asked to produce a model of the school, as well as a large-scale[3] *plan* (scale drawing made as a view from above showing structural elements of a building). The plan and model were supposed to show how students would address the

---

[3]The language of architectural practice regarding scale is logically consistent but sometimes confusing to nondesigners. *Larger scales* (such as 500:1) produce drawings and models that are *smaller in size; smaller scales* (such as 100:1) produce *larger images*.

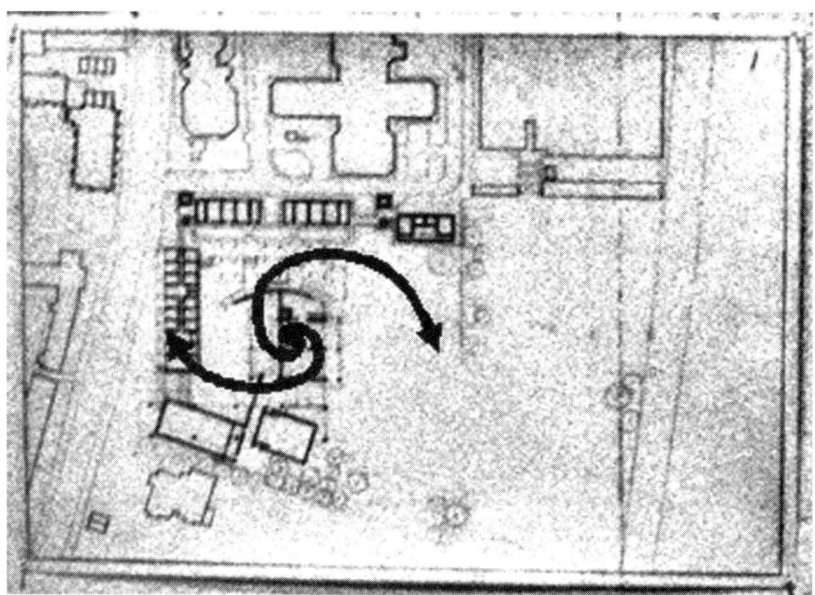

FIGURE 5–4. Arnold's plan shows his conceptual scheme to end the growth of Oxford's quadrangles with a spiraling courtyard (indicated conceptually by the spiraling black arrows, which were not part of the original plan drawing).

school's specific requirements within the broader conceptual scheme outlined in the previous assignment. Arnold, for example, extended the logic of his previous investigations, suggesting that a "spiral courtyard" at the center of his proposed building (see Figure 5–4) would provide the graceful transition from the Oxford quadrangles to the rural landscape by "dissipating" the energy of the city.

Large-scale representations thus give an overview of a structure without much detail; smaller scales are used to focus on specific details of a structure.

Assignments 4 and 5: Integrating Ideas in Three Dimensions

The fourth assignment asked students to provide more detail for the basic models they had proposed. Students were to show how their building would function spatially through *sections* and *elevations* (scale drawings made as a view from the side showing structural details or rendered views of a building). This led to the fifth assignment: a smaller-scale (that is, more detailed) examination of one or more elements of the building, focusing on how these elements related to the overall design strategy. Dan made a model of the school's "learning resource center" (library). Belinda focused on her building's corridors and the informal gathering places they created in her proposed design. Arnold looked at the relationship of his building to its site, exploring how the roof connected his building visually to other buildings nearby.

Assignment 6: Final Review

The final assignment of the studio was to bring these various investigations of the project—from the context of Oxford to the detail of one part of the building—into a presentation showing both a proposal for the project and the thinking process by which the proposal was developed.

Work on Each Assignment

Introduction to the Assignment

For each of the assignments in the Oxford Studio, students received a page of written instructions from the professor, which the class discussed as group. This written description included a summary of the assignment's requirements, an explanation of the reason for the assignment, a description of the professor's expectations, and almost always examples of work for students to use as models. For example, for the third assignment (on site and context), Nigel explained that he wanted students to understand the "inherent characteristics of university space" and show "what it represents" using a series of interpretive sketches. He asked students, through their sketches, to "address an attitude towards the context and site." He suggested that they focus not on details, such as where the entrance or the building should be, but on "the spirit, the opportunities of the site." Nigel explained that for this assignment students needed to integrate three elements: (a) the lessons of the precedents they had examined in the first assignment, (b) their analysis of Oxford with its issues of urban form and growth, and (c) the program of the building. Strategy at the macro scale, he explained, "brings all three together," and "must contain an idea: a creative leap."

Nigel began his explanation by giving an example, sketching on trace paper as he described the "three ugly sisters"—the unattractive buildings adjacent to the business school site—and explaining that students would need to develop a strategy, an "architectural idea," to address issues such as this contiguous architecture. Nigel showed the class sketch diagrams made by noted architect and designer Henry Foster as examples of how to communicate architectural ideas in this way.

Crits and Design Work

After the initial introduction to each assignment, students began work at the leisurely pace characteristic of work on the early stages of the design process in the studio. When students came up with questions, ran into problems in their emerging designs, or finished some coherent stage of their design process, they would sign up for individual conferences with the professor or a teaching assistant.

These conferences, known as *desk crits*, were the heart of the Oxford Studio. Crits were of varying lengths, though they usually lasted somewhere between 20 and 40 minutes. During a crit, a student described his or her work to a *critic*—the professor, a teaching assistant, or another student who had agreed to help the student with his or her design. The student gave an overview of the design process to date, focusing on some area or areas of particular interest or concern. The critic then asked clarifying questions about the design process and design intent. Finally, discussion turned to potential problem areas identified by the critic, which often were not the same areas about which the student had originally been concerned. The goal of the critic was to understand what the student was trying to do with his or her design and then to help him or her develop that design idea. Nigel explained that in a crit with a student he was "trying to get into their head," and "help them flesh out their own ideas, their own perceptions."

For example, in the course of the fourth assignment (showing building details through section and elevation), Arnold met with Nigel for a desk crit, focusing on the roof form of his design. Arnold explained that he was concerned about the relationship between his building, which was modern in design, and the adjacent buildings, which were traditional Oxford buildings with pitched roofs.

"You feel the need to be contextual," suggested Nigel.

"I want to be a good neighbor," replied Arnold.

"What does that mean?" asked Nigel. "How do you define that? Is it pitched roofs?"

Nigel pointed out that that Oxford had a long history of change, suggesting that Arnold did not need to mimic existing building types. But Arnold was still not sure "how to make the building distinctive." This led to an extended discussion about the relationship between design and materials. Arnold wondered whether he should use traditional materials to fit into the existing landscape. In the end, Nigel suggested that Arnold should "let the building be what it wants to be. Don't pander to superficial gestures to adjacent buildings."

"That's reassuring," replied Arnold. But he was not sure what his next step should be. They discussed creating a section drawing of the building, focusing on the structural mechanics of the roof. Perhaps understanding how the roof would be constructed would help Arnold decide what it should look like.

Presentation

Each of the assignments in the Oxford Studio culminated in a presentation. Three of the assignments led to a *pinup*: a group discussion of student work in which individuals pinned their working drawings and models on the wall, described their works-in-progress, and got feedback from the teaching staff and the other students. In a pinup before the midterm review, Arnold presented his idea of Oxford's cellular growth and his plan to use a spiral courtyard to gracefully blend the landscape of quadrangles into the surrounding floodplain. When Arnold finished his presentation, other students suggested that his spiral idea seemed forced and that the current design had too many hard edges to gently end urban growth. Nigel pointed out that the central idea in Arnold's project was clearer at an earlier stage, but was lost when Arnold added a massive overhanging roof in an effort to integrate the building with nearby structures. Nigel suggested that the big roof obscured the spiral form that was supposed to dissipate the growth of the city. He wanted to "rip the roof off" to expose the underlying spiral courtyard more clearly. During the pinup, Arnold tore the roof off of his model, and the other students made a number of suggestions about the design possibilities suggested by this change.

In pinups, each student's work was discussed at a similar level of detail by the professor, the teaching assistants, and the other students. In one case—after the concept and strategy assignment—feedback came from outside the studio. This *guest crit* had essentially the same form as a pinup, but with professors and other professionals in the field of architecture invited to comment on students' work. Two of the assignments (and the studio course as a whole) led to formal *reviews* or *juries*, again with critics from outside the studio.

The essential elements of the presentation process were the same in these reviews as in the pinups, but reviews demanded a higher level of preparation and organization—and the feedback reflected higher expectations for the quality of the designs presented. In the midterm review, Arnold made a formal presentation of the major theme in his design proposal. He described his idea that Oxford's quadrangles function as cell walls that "hold the outside out and the inside in," and explained that the spiral at the center of his building was designed to make a transition between the walls of the city and the open space of the surrounding countryside. The critics asked about details of the proposed design: How would people move through the buildings? How long would the corridors be? Why was the building around the spiral courtyard basically "a long corridor and rows of offices?" One critic suggested that the challenge of the cell model was to show how the "outside can become inside"—to which another critic replied: "Cells! Cells! Cells! But it looks like a dock!" "The central space seems weak," suggested the first critic, and Nigel agreed that "it has weakened over time." The second critic added that in her opinion the problem was that the tension between "hard and soft, inside and out" had not been fully resolved. In the review, the critics thus explored the tension between Arnold's stated design goal and its articulation in his emerging design. In the end, the critics suggested that Arnold was "trying to solve too many perspectives at once" and thus "falling in between" his multiple goals without sufficiently addressing any of them.

Pinups, guest crits, and reviews in the Oxford Studio all brought assignments—and with them, stages of the design process—to a close with pointed criticism that students used to refine their projects.

Lessons Learned

At the end of the midterm review, Nigel explained to the students that "the relationship to landscape came up over and over" in the critics' comments. Moving forward, he asked students to develop this key idea by sketching "one exemplary piece of the building" (the fifth assignment in the studio). Speaking for the students, Arnold suggested, somewhat dispiritedly, that "everyone got taken back a couple of models" in the review—meaning the critics had suggested that students' recent work had lost touch with their original design ideas. Nigel replied: "Steps back don't worry me.... We pushed the envelope. Now we need time to understand how to express those ideas."

After thinking about Nigel's response, Arnold realized that the reviewers had helped him "unhook himself" from old ideas and "invert" his perspective. As a result, he recognized that he needed to "design the space versus design the buildings"—a critical insight about the fundamental nature of architecture and architectural design. This young designer had come to understand that people inhabit the spaces that buildings create rather than the buildings themselves. His next task was to use that insight to inform his work.

A crit with Dan helped Arnold identify the key ideas of this project: the concept of the quadrangle as a cell, the diffusion of the Oxford edge, and the layering of inside and outside (public and private) space. Then Arnold turned to Nigel for help "looking at the space." He wanted to know "how to place the [building] masses to form the spaces" that his design needed.

Nigel and Arnold discussed how space flowed through the center of Arnold's proposed building plan. Arnold said he was "trying to create a pocket, a reservoir" (see Figure 5–5). Nigel suggested that "words like movement and flow are really about the ability to walk." They talked about what Nigel described as "the importance of view" and how "connectivity is key [in] creating an awareness of the space beyond." As a result of the conversation, Arnold created a computer model of his building to help visualize movement through the spaces. Later, Arnold, Dan, and Nigel "walked" through the virtual building to understand the views created by Arnold's building.

Arnold's project began with an exploration of Oxford's architectural heritage of monastic isolation. By the final review, he was working with computer-generated visualizations of the spaces created by his proposed building, exploring how movement through the building could gracefully link Oxford to the surrounding

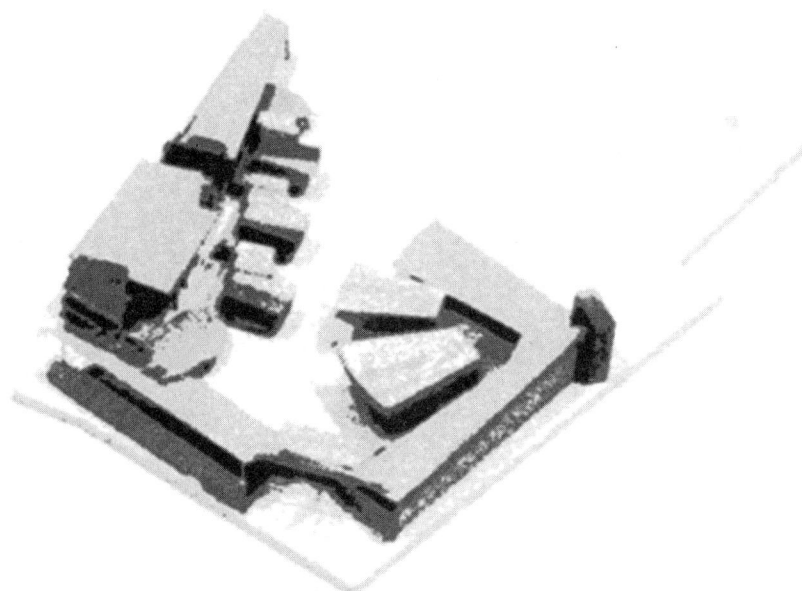

FIGURE 5–5. A model of Arnold's building after the midterm review.

FIGURE 5–6. Professors and design professionals from outside the studio critique a student's work.

countryside. At the final review, the critics suggested that Arnold's organizing metaphor—the structure of Oxford's quadrangles—needed to be "opened up" more than his design proposed if he wanted to express the idea of a quadrangle at the scale of a single building. The space at the center of an Oxford quadrangle is peaceful, they argued. The spaces Arnold had created felt confining.

Through six assignments and presentations, with more than a dozen crits and literally hundreds of sketches, diagrams, and models (physical and virtual), Arnold had created, and now was able to present, a sophisticated architectural proposal (see Figure 5–6). His diagrams and models let jurors at the final review see not just a general design concept, but enough details of a possible building to visualize—and criticize—the "mood" of the spaces he had created.

ANALYSIS: SURFACE STRUCTURES, PEDAGOGY, EPISTEMOLOGY

The Oxford Studio was characterized by (a) a fluid organization of time and space, (b) a series of assignments that revisited a central design problem from multiple perspectives, and (c) feedback from instructors and peers in one-on-one desk crits leading to public presentations of work. As Brown and Campione (1996) have suggested, however, the elements of effective learning environments do not exist in isolation: each helps to create conditions for the success of the others. The challenge is not only to understand how individual elements function, but also to understand the relationships between them.

Surface Structures

Time and Space

As described above, the logistical organization of the Oxford Studio differed substantially from a typical K–12 or college class in a traditional subject. The total amount of space per student was greater in the Oxford Studio, and students were given designated spaces in which they could work on their projects throughout the course. The studio met officially for large blocks of time (4 hours a day, 3 days a week), and unofficially students and staff worked in the studio at all hours of the day and night. These elements were intimately connected. The permanence of individual workspace meant that students could leave work in progress rather than start anew each time they came to class, and thus made it possible for students to work in the studio whenever they pleased.

Personnel

The Oxford Studio had a single professor and two teaching assistants for 11 students, creating a student-teacher ratio of just over 3:1. In addition to the regular presence of these experts, outside reviewers came to give feedback on students' work at three different points during the Oxford Studio course. These reviews were usually 4–5 hours long, taking up most of an afternoon and evening, and demanded from the critics high levels of stamina, architectural sophistication, and sensitivity in giving feedback on student designs.

Ideally, such critics should be thoughtful, skilled, and professional, responding to issues of interest to the student in their comments and criticisms. In the Oxford Studio, students recognized that this ideal is not always met. After a dispiriting midterm review, one student suggested that "sometimes the critics are thinking about Architecture with the capital A," rather than about a student's particular building. The design aesthetics of critics and students were not always aligned in the Oxford Studio, and thus students sometimes experienced criticism as disconnected and harsh. But, as Arnold commented, "guest critics come and guest critics go," and students took their feedback with a grain of salt.

These criticisms notwithstanding, everyone trained as an architect has gone through the review process, and most recognize it as a crucial part of learning to

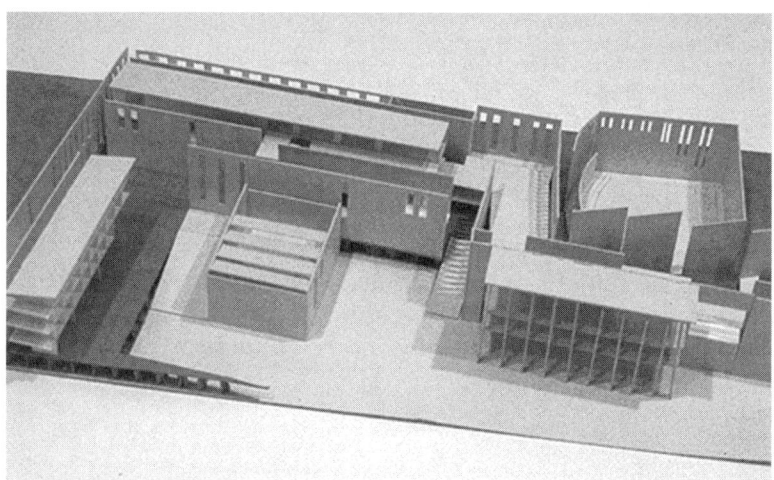

FIGURE 5–7. Models of Belinda's project at 1:500, 1:200, and 1:100 scale. A change in scale focuses attention on different aspects of a design problem.

design. Every studio professor knows that he or she has to serve as a critic in order to get critics for his or her own studio. This tradition of mutual obligation meant that the Oxford Studio had a pool of potential reviewers. Moreover, the open organization of time and space allowed teaching staff to spend a substantial amount of time in desk crits with students and also made it possible to organize extended design reviews with outside critics.

Materials

In the Oxford Studio, assignments asked students to explore the problem of designing a business school at a variety of levels and in a range of media. Each change in materials and scale shifted the students' focus from one part of the problem to another (see Figure 5–7). As Nigel put it, a wall that appears as a thin line on a plan at 1:500 scale has dimension and weight at a scale of 1:100. It is only at smaller scale that one can tell whether it will feel ponderous or playful, whether it will bear a

FIGURE 5–7. Continued from facing page.

load or require structural support. This idea is well developed in the literature on design (Akin, 1986; Mitchell & McCullough, 1991; Schon, 1985) and is central to the practice of good design. "Design means understanding at many scales," explained Nigel. "[It means asking:] how do the scales relate?"

The assignments in the Oxford Studio asked students to investigate their projects at progressively smaller scales—and thus at increasing levels of detail. This approach made it possible for the Oxford Studio to focus on a single project in great depth over an extended period of time. Each assignment asked students to look at the same basic design problem, but at a different scale—and therefore from a slightly different point of view. In part because students were working at different scales and in different media, they were able to focus on different aspects of a single project over a long time—to revisit a central set of design questions—without getting bored. Feedback from outside reviewers could be incorporated in later work in part because a change in medium and/or a change in scale meant that students were *extending* rather than *revising* their ideas. The class was able to move forward even as students circled the same set of fundamental design questions, developing an ever richer understanding of the problem of the Oxford business school and the principles of good design.

Interdependence of Structures

The surface structures of the Oxford studio were thus fundamentally interconnected. The Oxford studio was not simply characterized by a block schedule, permanent personal workspace, a variety of media, and public presentation to external reviewers.

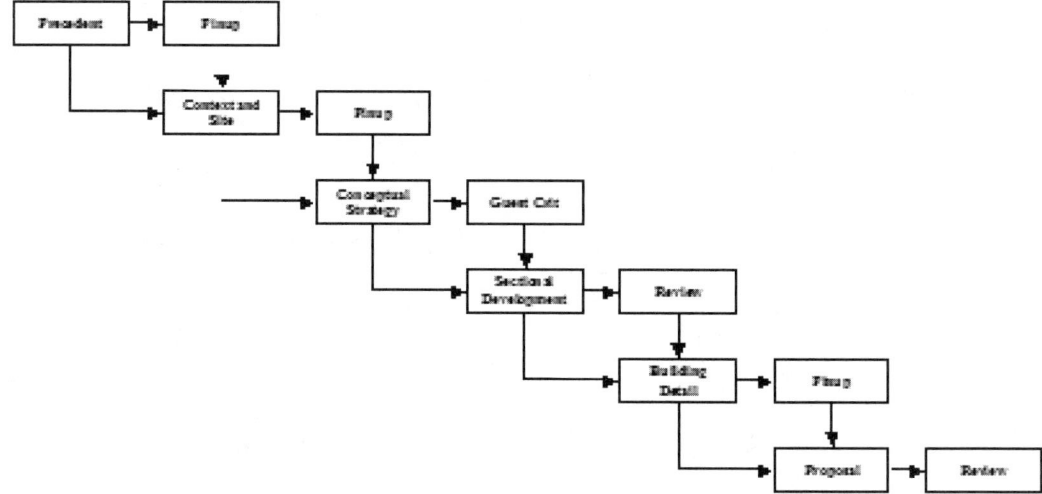

FIGURE 5–8. The assessments of the Oxford Studio as a series of presentations.

Individual workspaces made possible a fluid organization of time...which made possible access to experts in a review process...which was pedagogically useful because extended projects were conducted in a range of media.

Pedagogical Activities

Time, space, materials, and expert feedback were organized in the Oxford Studio so as to support each other. But more important, they were arranged to support two recurrent participant structures: two cyclical processes of learning to design.

Cycles of Design

The design process in the Oxford Studio was a cumulative series of individual assignments and presentations, creating cycles of production and reflection. Each assignment built on the work of the previous assignments, but also on the feedback that came from presenting that work (see Figure 5–8).

    These cycles were made possible, in part, by the nature of the feedback given during presentations. Pinups, guest crits, and reviews were conducted with the explicit idea that students would use the criticism to improve their designs. As Nigel explained, a good critic gains "entry into the thinking of the student"; the critic is able to "lay out the reasoning for criticism" and thus "identify areas for further growth."[4] This feedback, in turn, became the basis for work on subsequent iterations of the design; students addressed the concerns raised by critics as they refined their design work. With each iteration, the scale at which students were working moved from broad questions of context and strategy to detailed questions about individual parts of the building. As their designs moved from larger to smaller scale, students' attachment to their basic design ideas grew. Nigel explained that the midterm review was designed "to get students to commit to a

---

[4]The final review was something of an exception in this sense, and it comes as no surprise that students, professors, and practicing architects feel overall that interim presentations are more useful learning experiences than final reviews (see Anthony, 1987).

design strategy" that would be worked out in more detail in the second half of the semester. In Arnold's case, this meant focusing on his strategy to end the "cellular growth" of Oxford's quadrangles. His commitment to that design goal helped him understand a fundamental principle about the relationship between a building's form and the space it creates. In general, as the presentations moved from pinup to guest crit to review, there was an increase both in their formality and in the expectations for the quality of the work. With each loop around the cycle of production and reflection, students were asked to make deeper and more public statements about their architectural ideas (see Figure 5–8).

Cycles Within Cycles

Alternation of design and feedback was similarly present *within* the assignment-and-presentation cycle in the Oxford Studio. With each assignment, students met with members of the teaching staff or peers for a detailed discussion of their works-in-progress. They received extended and in-depth feedback through these desk crits and then returned to their projects. After working through the ideas generated in the crit, they might sign up for another crit with the professor. They might sign up for a crit with a teaching assistant. They might work out details of the assignment in a crit with another student. Design in the Oxford studio was a process of alternating back and forth between work and feedback within each assignment.

As we saw above in discussions between Arnold and Nigel, desk crits played multiple roles in supporting students' design work in the Oxford Studio. Crits provided what Arnold described as "authorization and validation...confirmation of untested ideas." They also gave students "a sense of where to go from here, including concrete suggestions." Dan described crits as providing "someone to bounce ideas off of, to make sure that they're sensible," but also someone "to bring in other generating ideas," such as (in Dan's case) Nigel's interest in energy-efficient design. Nigel described his role in a desk crit as helping students "focus on a generative idea" and "unlock the door to make the whole thing better." But he also looked for potential problems in students' emerging designs. His experience let him "look at things that you know are going to come home to roost," to see issues that might arise later from current design choices. His role as critic was partly to "ask students to start anticipating [problems] now before they complete their design."

This support for the design process took many forms in desk crits in the Oxford Studio. Critics offered suggestions, pointed out potential problems, or referred to examples of work by other architects that addressed issues similar to those the student was facing. Sometimes critic and student would design together, with the critic sketching a series of design possibilities, showing the student the consequences of possible design choices. In doing so, the critic both offered design ideas and modeled design thinking. In many cases, crits ended with a specific suggestion from the critic, not of a particular design direction to take, but of a way to think productively about the questions raised. Thus, at the end of the extended discussion of the roof of Arnold's building, Nigel suggested not "why don't you cantilever the lecture hall over the courtyard," but "why don't you try drawing that section at a smaller scale to see how the lecture hall relates to the courtyard."

As Schon (1985) described in some detail, desk crits in the Oxford Studio were a venue in which professor, teaching assistants, and peers provided design skills and knowledge that students lacked. With the help of others, the students were able to work, as Vygotsky (1978) suggested, beyond their individual reach. As they became more sophisticated designers, feedback moved to a higher level, always showing the next steps on the path. And so the design process continued: a student expressed his or her design ideas; a critic responded with feedback to those ideas to help the student achieve a higher level of design; the student incorporated that feedback in a new expression; and finally, this epicycle of design and desk crit culminated in a public presentation at the end of each assignment.

Interlocking Rings

Cycles of assignment and presentation, and epicycles of design and desk crit, were related in the Oxford Studio. The tone of desk crits was almost always supportive and nonjudgmental. On the other hand, pinups and reviews, although constructive, were quite blunt and sometimes extremely critical—particularly in the case of formal reviews. Judgment was, in effect, offloaded from the more private desk crits to the more public presentations. This offloading of judgment was explicit. In the last few days before the final review, Nigel repeatedly used the phrase "if I'm on the jury" when commenting on students' work, suggesting that whatever point he was making was intended to help the student present the best possible project for review. Nigel suggested that Dan needed to "develop the concept" for his building—not by making the criticism directly, but by pointing out that "a design is vulnerable to criticism when it doesn't have a compelling idea." Understanding in the Oxford Studio developed as supportive feedback from peers and experts in the desk crits helped students incorporate the norms of the architectural community—personified by the critics—as part of the framework for their individual thinking.

The basic pedagogical organization of the Oxford Studio was this pair of related cycles (see Figure 5–9), in which expression of an architectural idea led to feedback, then to expression, and then again to feedback, eventually producing a design for presentation. The presentation generated more response, leading to the next assignment and a larger cycle of production and reflection that culminated in a final presentation—and in meaningful learning about the process and practice of design. Thus, for example, the Oxford Studio helped Arnold understand a fundamental principle about the design of inhabited spaces in a fundamentally iterative process, in which a series of imperfect solutions to a problem were proposed, analyzed, and refined until they eventually converged on a form that satisfied the initial conditions (Schon, 1985; Simon, 1996).

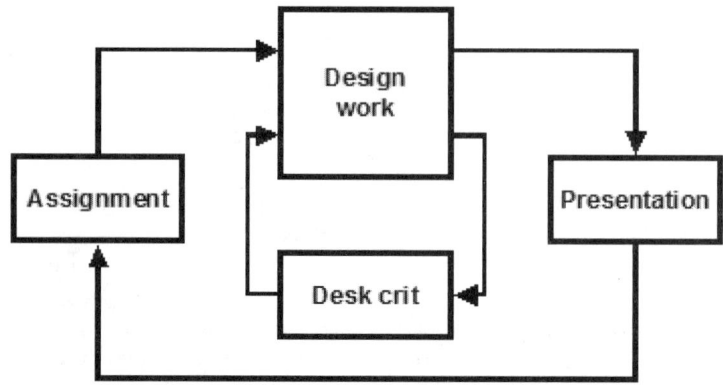

FIGURE 5–9. The cycles of expression and feedback in the design studio.

From Surface to Pedagogy

The surface structures of the Oxford studio made desk crits, cumulative assignments, and public presentations possible. The flexible block scheduling and high teacher/student ratio of the studio made it possible to have extended conversations with students in desk crits. The public presentation spaces shared by multiple studios and tradition of design juries made it possible to critique student work in public reviews. The media of architectural design made possible an extended, iterative project, through which students revisited the same basic design questions on different levels and from different perspectives. Furthermore, the pedagogical activities were related to one another. In desk crits, the professor, a teaching assistant, or a peer supplied skills that students used to prepare their designs for public review. Because they were working on an extended, iterative project, students were able to incorporate criticism they received in reviews into their subsequent work with the help of professor and peers in further desk crits. These cycles of design, in turn, were the means through which students explored (and internalized) the fundamental epistemology of design.

Epistemology

Architectural Ideas

An architectural idea is an understanding about the creation of space that serves as a guide or organizing principle for solving a design problem. Arnold's work, for example, was driven by the goal of designing a building that would "bring a graceful end" to Oxford's urban development by unraveling the "cells" of the monastic grid into the open plains of the landscape surrounding the city. Developing and expressing this central idea in a building plan was the focus of his work throughout the semester.

The existence of such underlying ideas in design is common in the literature on design (Davies, 1987; Hewitt, 1985; Rowe, 1987) and was evident in the language and activities of the Oxford Studio. In the Oxford Studio, professors and critics spoke variously about the need to "develop an attitude," "develop an architectural idea," "find a valid architectural proposition," "decide on a strategy," "take a stand," "take a stance," "develop your criteria"—all expressions of the need to find an underlying idea or ideas to govern the development of a solution to the design problem. The early assignments of the Oxford Studio were deliberately constructed to help students develop such ideas about their projects.

A critical feature of the epistemology of architecture is that these design ideas reflect an individual interpretation of an architectural problem. Students in the Oxford Studio were presented with a design challenge that had an infinite number of potential resolutions. Their task during the semester was to develop a unique solution, to understand that solution, and to convey in words, diagrams, and models how the solution they chose met the demands of the original problem. The idea they developed was to be of their own choosing—as long as they could develop a coherent design based on that idea and defend their rationale. As Nigel said to a student in an early desk crit: "You're in control. Make it whatever size you want. Then I'll ask: 'Why is it that size?' And you'll say: 'Because it's doing this job.' And you'll develop your argument."

The activities of the Oxford Studio were thus organized around the development and articulation of architectural ideas. Conversations in desk crits were about developing these ideas. Representations presented in pinups and reviews were attempts to express these ideas. Criticism and feedback went toward understanding these ideas, refining them, and developing them further. Arnold described the key question of the cyclical design process in the Oxford Studio: "How does this expression of what's in my head inform my underlying concept?" Because the architectural ideas explored in the Oxford Studio were unique to each student, critics could refer to examples from the work of other architects—or even other students—and still leave students free to develop their own thinking. And because the ideas were of the students' own choosing, it was possible for each student to spend a semester refining a vision and turning it into architecture.

In the Oxford Studio, this focus on each individual's architectural idea was supported by the lack of a single canon of design principles. There was no uniform curriculum of skills and concepts that students were required to master, no standardized test at the end of the semester. Individual students were free to focus on developing different design skills and understanding in response to the same series of assignments. Some, like Arnold, worked on developing an understanding of how people move through space; others, like Belinda, looked at the differences between formal and informal space, or, like Dan, examined energy efficiency; still others explored issues of exterior and interior forms, multiple use of spaces, or the relationship between hierarchy and order in a building's organization. The freedom to express was made possible, in part, by freedom from a fixed set of learning objectives.

The Oxford studio was designed to help students learn how to develop and express ideas about architecture through the iterative process of design. The pedagogical activities of the desk crit, cyclical assignments, and generative feedback were means to this end. But these pedagogical activities were also dependent on the individualized nature of design ideas. Eleven students could present their work for mutual critique and review because each student had a unique and valid answer to the common design problem at hand. Because an architectural idea is never right or wrong, but only more or less well-expressed as a solution to a particular design problem, students could work on a single project over a semester to develop and refine an architectural idea and its expression in a range of design media through the iterative process of desk crit and public review. To the extent that the surface structures of the Oxford studio made extended desk crits, ongoing projects, and guest reviews possible, these structures were useful because they were aligned and orchestrated by the expressive and iterative epistemology of design.

DISCUSSION

Prior research has revealed a great deal about design and the processes of designing. This study of the Oxford Studio—which, it is true, represents a particularly well-appointed venue for design activity—reinforces a number of key ideas already established in that literature:

⟨ Design is an iterative process.
⟨ Design is typically conducted in an open environment.
⟨ Design resolves an open-ended problem through a series of intermediate solutions.
⟨ Design learning progresses through a series of exercises that revisit a central problem in progressively more detail.
⟨ This process is mediated by generative feedback from social scaffolds such as desk crits, pinups, and reviews.

We know, moreover, that practices—particularly practices developed to promote learning—are more than just sets of activities in a particular setting. As Brown and Campione (1996) argue, effective learning environments are coherent systems. In effective learning environments, practical and conceptual linkages are preserved in recurrent participant structures (Herrenkohl & Guerra, 1998; O'Connor & Michaels, 1996).

What this study of the Oxford Studio does, I hope, is link these two perspectives to show how the ways of knowing, doing, and being that comprise design activity are co-constructed for students in the context of their design training. In the Oxford Studio, a block schedule, individual workspaces, multiple representational media, and access to external experts made possible social interactions in desk crits and public reviews, which in turn were the vehicles through which students learned to develop and express architectural ideas. That is, surface *features* such as time, space, access to experts, and media of expression were coordinated by a particular set of social interactions to form a coherent physical, temporal, social, and material *structure*. Similarly, social interactions became a *pedagogy* to the extent that they were arranged to convey a particular approach to understanding based on the properties of architectural ideas. The understanding of architectural ideas, in turn, became a coherent *epistemology* when instantiated in the structure and pedagogy of the Oxford Studio. In this sense, theory and activity were inseparably connected in the practices of the Oxford studio.

The goal of this study was to investigate how elements of design practice were constituted in that system of activity. A key component of the coherence of design in the Oxford studio was that students were not merely solving problems; they were engaged in an iterative process of expressing—and thus shaping—their identities. Solutions to design challenges in the Oxford studio necessarily reflected students' individual perspectives. At the same time, solutions were assessed in the context of a larger community of practice. Thus, solving the problems required aligning individual and community—for the students, aligning self and critics—in the context of a set of practices that incorporated a particular approach to understanding.

While the data presented here do not directly address the nature of mathematical and scientific thinking in design, they do suggest that the understandings—including mathematical, scientific, and technical understanding—used in architectural problem solving may be profoundly different than those encountered in traditional mathematics and science class rooms. There are a number of obvious differences between traditional K–12 classrooms and the Oxford Studio:

⟨ In the Oxford studio each student had his or her own workspace; in many science labs and art classrooms, students share workspaces, and a significant amount of time is spent setting up before experiments or projects and cleaning up afterwards.

⟨ The Oxford studio met for large blocks of time, and because students had their own permanent workspaces, they were able to work in the studio outside of these scheduled times; in K–12 classrooms, students have access to the social and material resources of the class for limited periods of time.

⟨ Outside experts played a central and recurring role in the Oxford studio; K–12 classrooms are relatively isolated to interactions among teacher and students.

⟨ A wide range of media for development and representation of problem solutions were central to activities in the Oxford studio; traditional classrooms focus on a limited set of representational tools and forms.

⟨ In the Oxford studio, feedback was generative; in a typical classroom, much of the feedback students receive is *summative*: even if it makes suggestions for improvement, the intent is to evaluate rather than stimulate further work.

While many progressive classrooms incorporate structures and activities similar to those seen in the Oxford studio (see, for example, NCTM, 1991), this study suggests is that the most significant differences between the Oxford studio and a typical K–12 mathematics or science class is that they have different epistemologies. The Oxford studio was based on the development of architectural ideas—ideas that often were deeply mathematical and scientific in nature, but whose salient feature was that they were inherently unique to each individual designer. These ideas were not "right" or "wrong" answers to the architectural problem at hand. Rather, they were attempts to solve a design challenge based on each individual student's sense of what was interesting and important about a building for the proposed Oxford business school. There was no fixed set of content to be mastered here; this was a curriculum focused on the processes of developing, expressing, evaluating, and defending architectural ideas—ideas whose validity and utility were warranted by their ability to generate a coherent and compelling solution to the design problem at hand.

In recent years the design studio has been an appealing model for reformers interested in creating powerful connections among learners, teachers, and intellectual domains. The Middle School Math through Applications Program includes units on designing a dream home, designing quilts, and designing a research station in Antarctica (Greeno, 1997). The Geometry in Design curriculum materials (Jacobson & Lehrer, 2000; Lehrer & Curtis, 2000) use design activities as a context in which students from elementary school through college learn about traditional domains such as mathematics and science through design activities. The Learning by Design project (Hmelo et al., 2000; Kolodner et al., 1998) has demonstrated that design practices can help create effective learning environments for middle school students to learn science concepts and scientific reasoning through design activities.

However, to the extent that such projects are modeled on design, the approach is typically to extract and transform elements of professional practice to fit within the constraints of existing school curricula, organization, and tools. Such adaptations may not take full advantage of the potential of the design studio as a model. The present study suggests that the structures, activities, and ways of knowing of the Oxford studio were highly integrated. Pedagogical activities such as the desk crit and design review depended on both the open structure of the environment and on the expressive nature of the tasks at hand.

An important undertaking—although beyond the scope of the present study—is to understand in more detail how adaptations of design practices function, and the extent to which such adaptations benefit from close integration of elements of design practice such as those found in the Oxford studio. One such study, The Escher's World project (Cossentino & Shaffer, 1999; Shaffer, 1996, 1998, 2002a, 2002b), looked at graphic design projects and practices explicitly modeled on a design studio as a context for learning mathematics. Results showed that elements of the design studio model—expressive and iterative activities with generative feedback in desk crits, pin-ups, and reviews—supported the development of mathematical understanding. As Cossentino and Shaffer (1999) argued, the success of that work was closely associated with the preservation of relationships among features of the design studio, rather than ad hoc adoption of isolated elements of design practice.

This study does suggest, however, that architectural thinking may be fundamentally different than problem solving in traditional K–12 mathematics and science class rooms—and that the ways of knowing and ways of acting in the design studio form a complex and highly integrated system of practice. To the extent that mathematics and science learning should be aimed toward—and based on—practices in the world beyond school, this suggests that we need to consider significant changes not only in the structures and activities of mathematics and science classrooms, but in our ideas about the very nature of mathematical and scientific understanding.

ACKNOWLEDGMENTS

Work on this study was funded in part by LEGO Corporation, the Things That Think Consortium at the MIT Media Laboratory, the Waitt Family Foundation, and the Foundation for Technology and Ethics.

REFERENCES

Akin, O. (1986). *Psychology of Architectural Design*. London: Pion.

Anthony, K. H. (1987). Private reactions to public criticism: students, faculty, and practicing architects state their views on design juries in architectural education. *Journal of Architectural Education, 40*(3), 2–11.

Branki, N. (1993). A study of socially shared cognition in design. *Environment and planning B, planning & design, 20*(3), 295–306.

Brown, A. L., & Campione, J. C. (1996). Psychological theory and the design of innovative learning environments: On procedures, principles and systems. In L. Schauble & R. Glaser (Eds.), *Innovations in learning: New environments for education* (pp. 289–325). Mahwah, NJ: Lawrence Erlbaum Associates.

Chafee, R. (1977). The Teaching of Architecture at the Ecole des Beaux-Arts. In A. Drexler (Ed.), *The architecture of the Ecole des Beaux-Arts*. New York: Museum of Modern Art.

Cobb, P. (1986). Concrete can be abstract: A case study. *Educational Studies in Mathematics, 17*(1), 37–48.

Cossentino, J., & Shaffer, D. W. (1999). The math studio: harnessing the power of the arts to teach across disciplines. *Journal of Aesthetic Education, 33*(2), 99–109.

Coyne, R. (1993). Cooperation an individualism in design. *Environment and Planning B, 20*(2), 163–174.

Craig, D. L., & Zimring, C. (2000). Supporting collaborative design groups as design communities. *Design Studies, 21*, 187–204.

Crowe, N. A., & Hurtt, S. W. (1986). Visual notes and the acquisition of architectural knowledge. *Journal of Architectural Education, 39*(3), 6–16.

Davies, R. (1987). Experiencing ideas: identity, insight and the imago. *Design studies 1987 Jan., 8*(1), 17–25.

Dewey, J. (1934/1958). *Art as experience*. New York: Capricorn Books.

diSessa, A. A. (2000). *Changing minds: Computers, learning, and literacy*. Cambridge, MA: MIT Press.

Dutton, T. A. (1987). Design and Studio Pedagogy. *Journal of Architectural Education, 41*(1), 16–25.

Erickson, J., & Lehrer, R. (1998). The evolution of critical standards as students design hypermedia documents. *Journal of the Learning Sciences, 7*(3–4), 351–386.

Flemming, D. (1998). Design Talk: Constructing the object in studio conversations. *Design Issues, 14*(2), 41–62.

Frederickson, M. P., & Anderton, F. (1990). Design juries: a study on lines of communication. *Journal of Achitectural Education, 43*(2), 22–27.

Geertz, C. (1973a). Person, time and conduct in Bali. In *The Interpretation of Cultures: Selected Essays* (pp. 3–30). New York: Basic Books.

Geertz, C. (1973b). Thick description: Towards an interpretive theory of culture. In *The interpretation of cultures: Selected essays* (pp. 3–30). New York: Basic Books.

Goldschmidt, G. (1989). Problem representation versus domain of solution in architectural design. *Journal of Architectural and Planning Research 42*(3), 17–28.

Greeno, J. G. (1997). Theories and practices of thinking and learning to think: Middle-school mathematics through applications project. *American Journal of Education, 106*, 85–126.

Hall, R., & Stevens, R. (1996). Teaching/learning events in the workplace: a comparative analysis of their organizational and interactional structure. In G. W. Cottrell (Ed.), *Proceedings of the eighteenth annual conference of the Cognitive Science Society* (pp. 160–165). Mahwah, NJ: Lawrence Erlbaum Associates.

Hawkins, R. (1993). The planning process: what goes on behind closed doors. *Architects' Journal, 197*(3), 16–18.

Herrenkohl, L. R., & Guerra, M. R. (1998). Participant structures, scientific discourse, and student engagement in fourth grade. *Cognition and Instruction, 16*, 433–475.

Hewitt, M. (1985). Representational forms and modes of conception: an approach to the history of architectural drawing. *Journal of Architectural Education, 39*(2).

Heylighen, A., Neuckermans, H., & Bouwen, J. E. (1999). Walking on a thin line—between bassive knowledge and active knowing of components and concepts in architectural design. *Design Studies, 20*, 211–235.

Hmelo, C. E., Holton, D. L., & Kolodner, J. L. (2000). Designing to learn about complex systems. *Journal of the Learning Sciences, 9*(3), 247–298.

Jacobson, C., & Lehrer, R. (2000). Teacher appropriation and student learning of geometry through design. *Journal for Research in Mathematics Education, 31*(1), 71–88.

Jansson, D. (1993). Cognition in design: viewing the hidden side of the design process. *Environment and Planning B, Planning & Design, 20*(3), 257–271.

Kolodner, J. L., Crismond, D., Gray, J., Holbrook, J., & Puntambekar, S. (1998). Learning by design from theory to practice. In A. S. Bruckman, M. Guzdial, & J. L. Kolodner (Eds.), *Proceedings of the International Conference of the Learning Sciences* (pp. 16–22). Atlanta, GA.

Kvan, T. (2001). The pedagogy of virtual design studios. *Automation in Construction, 10*, 345–353.

Lave, J., & Wenger, E. (1991). *Situated learning: Legitimate peripheral participation*. Cambridge: Cambridge University Press.

Ledewitz, S. (1985). Models of design in studio teaching. *Journal of Architectural Education, 38*(2), 2–8.

Lehrer, R., & Curtis, C. L. (2000). Why are some solids perfect? Conjectures and experiments by third graders. *Teaching Children Mathematics, 6*(5), 324–329.

Loeb, A. (1993). *Concepts and images: Visual mathematics*. Boston: Birkhauser.

Mitchell, W. J. (1994). *The logic of architecture: Design, computation, and cognition*. Cambridge, MA: MIT Press.

Mitchell, W. J., & McCullough, M. (1991). *Digital design media: A handbook for architects and design professionals*. New York: Van Nostrand Reinhold.

NCTM, C. o. S. f. S. M. (1991). *Professional standards for teaching mathematics*. Reston, VA: National Council of Teachers of Mathematics.

O'Connor, M. C., & Michaels, S. (1996). Shifting participant frameworks: Orchestrating thinking practices in group discussion. In D. Hicks (Ed.), *Child discourse and social learning* (pp. 63–102). Cambridge: Cambridge University Press.

Perkins, D., & Blythe, T. (1994). Putting understanding up front. *Educational Leadership, 51*(5), 4–7.

Rowe, P. G. (1987). *Design thinking*. Cambridge, MA: MIT Press.

Sancar, F. H. (1996). Behavioural knowledge integration in the design studio: And experimental evaluation of three strategies. *Design Studies, 17*, 131–163.

Schon, D. A. (1985). *The design studio: An exploration of its traditions and potentials*. London: RIBA Publications.

Schon, D. A. (1987). *Educating the reflective practitioner: Toward a new design for teaching and learning in the professions*. San Francisco: Jossey-Bass.

Schon, D. A. (1988a). Designing: rules, types and worlds. *Design studies 1988 July, 9*(3), 181–190.

Schon, D. A. (1988b). Toward a marriage of artistry & applied science in the architectural design. *Journal of Architectural Education 41*(4), 4–10.

Shaffer, D. W. (1996). *Escher's world: Learning mathematics through design in a digital studio*. Unpublished master's thesis, Massachusetts Institute of Technology, Cambridge, MA.

Shaffer, D. W. (1998). *Expressive mathematics: Learning by design.* Unpublished doctoral dissertation, Massachusetts Institute of Technology, Cambridge, MA.

Shaffer, D. W. (2002a). The Design Studio: A promising model for learning to collaborate: Thoughts in response to Hall, Star, and Nemirovsky. In T. Koschmann, R. Hall, & N. Miyake (Eds.), *Computer support for collaborative learning 2* (pp. 223–228). Mahwah, NJ: Lawrence Erlbaum Associates.

Shaffer, D. W. (2002b). Design, collaboration, and computation: The design studio as a model for computer-supported collaboration in mathematics. In T. Koschmann, R. Hall, & N. Miyake (Eds.), *Computer support for collaborative learning 2* (pp. 197–222). Mahwah, NJ: Lawrence Erlbaum Associates.

Simon, H. A. (1996). *The Sciences of the artificial.* Cambridge, MA: MIT Press.

Stevens, R. R. (2000). Divisions of labor in school and in the workplace: Comparing computer and paper-supported activities across settings. *Journal of the Learning Sciences, 9*(4), 373–401.

Uluoglu, B. (2000). Design knowledge communicated in studio critiques. *Design Studies, 21*, 33–58.

Vygotsky, L. S. (1978). *Mind in society.* Cambridge, MA: Harvard University Press.

Weiss, R. S. (1994). *Learning from strangers: The art and method of qualitative interview studies.* New York: The Free Press.

Wertsch, J. V. (1998). *Mind as action.* New York: Oxford University Press.

CHAPTER 6

The Cognitive Science of Mathematics: Why Is It Relevant for Mathematics Education?

Rafael Núñez
University of California, San Diego

> Virtually all of basic calculus...achieves its primary meaning through an absolutely essential collection of motion metaphors. These metaphors control the notation. Hence we write limit statements using arrows and use image-ladden words such as "diverge," "converge," "increasing," "constant," and "transform." However, the formal mathematical definitions associated with these notations, being atemporal, are not connected to motion (Jim Kaput, 1979).

One of the main goals of education is to prepare individuals, usually young ones, for a hopefully productive and enriching immersion in the community. This process is normally instantiated through an explicit and systematic effort orchestrated by the society through specific institutions and organizations. The general term "education" is often used along with a noun that refers to a specific domain of human knowledge (which we could call X) such as "music," "language", or "mathematics", thus designating the corresponding subfields of the form "X education": "*music* education," "*language* education," and "*mathematics* education". Interestingly, when confronted to a question such as, what is the *nature* of X? or, what is the essence of that X you are teaching? music and language education tend to provide an essentially different answer from the one given by mathematics education. The former two take their subject matter X to be *inherently* human, while the latter doesn't. In this chapter I'll argue that this situation—seeing mathematics as an essentially pre-given dehumanized body of knowledge—is deeper and more widespread than what we may think, having deep negative implications for mathematics education. I shall defend my position by showing how the study of the cognitive science of mathematics can provide a new understanding of the nature of mathematics education's "X", and help students, teachers, and researchers in the field. My arguments will go along the lines of some deep insights the extraordinary mathematics educator Jim Kaput had more than 25 year ago on the nature of mathematics and the relationship between formalization and mathematical concepts (Kaput, 1979). This chapter is a tribute to him and his well-ahead-of-the-times ideas.

In our society there is a clear folk and academic conception of "music" and "language" as being *inherently* human. Both, music and language are seen as created, developed, and sustained by human beings. Not only are we members of the only species that is able to produce music or full-blown language (as far as we know), but most importantly, music and language *exist* because *we* exist. As an extension, when it comes to teach music or language, these areas of education are taught as human endeavors, bringing into the teaching and learning processes the full extent of human experience, from the most fundamental sensations and perceptions to the analysis of the historical and cultural processes involved. Everything from curriculum development, to drills, to didactical contracts, to teacher education is shaped around the fundamental idea that music and language are human enterprises, manifestations of the ongoing human saga.

The situation is rather different in the case of mathematics education. In the preface of our book *Where Mathematics Comes From*, George Lakoff and I describe the widespread existing folk and academic conception of the nature of mathematics as being essentially independent of human beings (Lakoff & Núñez, 2000). As an extended form of Platonism, this view sees mathematics as being predominantly about timeless eternal objective truths, providing structure and order to the universe (after all it was Galileo, founder of modern science as we know it, who once said "The laws of the universe are written in mathematics. It is our role to learn how to read them"). Lakoff and I called this view the *romance of mathematics*, a kind of mythology that goes like this:

- Mathematics has a truly objective existence, providing structure to this universe and any possible universe, independent of and transcending the existence of human beings or any beings at all.
- Mathematics is abstract and disembodied—yet it is real.
- Human mathematics is just a part of abstract, transcendent mathematics (the concrete and mundane side of it).
- Hence, mathematical proof allows us to discover transcendent truths of the universe.
- Mathematics is part of the physical universe and provides rational structure to it. There are Fibonacci series in flowers, logarithmic spirals in snails, fractals in mountain ranges, and π in the spherical shape of stars and planets and bubbles. Hundreds of books showing how "wonderful" and "magic" mathematics is, continuously sustain this belief.
- Mathematics even characterizes logic, and hence structures reason itself—any form of reason by any possible being.
- To teach and to learn mathematics is therefore to teach and to learn the language of nature, a mode of thought that would have to be shared by any highly intelligent beings anywhere in the universe.
- Because mathematics is disembodied and reason is a form of mathematical logic, reason itself is disembodied. Hence, machines can, in principle, think.

This mythology is sustained by many physicists, mathematicians, engineers, and computer scientists (and even many mathematics educators!). But most importantly, the romance is also in millions of people's lives, including students and parents, in the form of pop-culture books and Hollywood movies such as *2001: A Space Odyssey*, *Sphere*, or *Contact*. These pop-books (some of them written by well-known scientists like Carl Sagan, author of *Contact*) have a basic background assumption that mathematics is the only "truly universal language," which would allow us to communicate with extraterrestrials if we encountered them. In the movie *Contact* this claim is explicitly made by the hero, Dr. Ellie Arroway (played by Jodie Foster), a responsible, professional, and passionate scientist, who, after finding "conclusive" radio proof of intelligent aliens, is discussing with high-level politicians on the urgency of understanding the messages of those aliens. She replies to the mocking question of "why they [the aliens] don't just speak English?" with a serene, solid, and convincing bottom-line argument: *"Mathematics is the only truly universal language, Senator."* Lines like this one are passively and unconsciously swallowed by millions in movie theaters throughout the world, thus adding layer after layer to the solid background sustaining the mythology of the romance of mathematics.

That mathematics is the "only truly universal language" is, unfortunately, a harmful widespread belief. In mathematics education it is often translated into the belief that one is teaching something magic, mysterious, and transcendental, deeply rooted in the unknown structure of the universe. After all, is it not the case that the truth of the Pythagorean Theorem is valid everywhere? Not only in Greece, but also in Easter Island, on the South Pole, at the bottom of the Pacific Ocean, on the moon, in Jupiter, or anywhere in the universe? And is it not the case that the validity of that theorem transcends time, such that its truth was not only valid during the time of the Greeks, but it is still valid now, and it will continue to be valid forever? Is it not the case that the constant π expresses a universal property, namely, the ratio between the perimeter of a circumference and its diameter? Is this not a truth that *any* intelligent being *anywhere* in the universe would be able to grasp? Is it not the case that even today, some scientists, in an effort to see whether there are any intelligent aliens out there in the universe, are spending huge amounts of money sending mathematical codes assuming that these aliens, if they are smart enough, would be able to understand?

Indeed, it is not easy to see that these mathematical truths are in fact the product of human imagination—very peculiar, objective, stable, effective, abstract, well-adapted, and robust ones—but human nonetheless. These extraordinary properties make mathematics a unique form of knowledge, which is gathered not through empirical evidence (as is done in science via experimental or correlational studies) but through formal proof. A piece of mathematical knowledge is accepted by the mathematical community only if there is a proof of its absolute certainty (not just by providing statistically significant robust evidence for it, as it is done in science). And because of these unique properties is that mathematics heavily relies on the power of axiomatic systems and formal definitions. This praxis however, although useful in the context of pure mathematics, is, unfortunately, often carried over to the practice of mathematics teaching. As a result there is an overemphasis on dogmatic oriented forms of education instantiated via formal definitions, reductionistic forms of logic, axioms, algorithms, and so forth, all of which result in forgetting the human nature of mathematics. When did your school teacher, for instance, really *explain* to you, in human *meaningful* terms (not in terms of formal proofs and arbitrary definitions and axioms), why the multiplication of two negative numbers yields a positive result? Or when did your college mathematics professor really explain to you why the empty set is a subset of every set? And what that means? How would *you* explain such simple but profound "truths" to a student or to another colleague? Or if you prefer, you could pick Euler's famous formula $e^{\pi i} + 1 = 0$ and try to explain its *meaning*, that is, not by providing a proof *that* the statement is true, but by providing an explanation of *why* the statement is true by virtue of what it *means*. The answers to these questions are deeply affected in one way or another by the *romance of mathematics*, interfering with a fruitful conception of mathematics as a meaningful human activity and mathematics education as a cognitive friendly enterprise. The answers to those questions are not part of mathematics proper (where formal proof suffices), and they don't exclusively belong to the realm of philosophy or history of mathematics. As I'll claim in this chapter, those answers can be found via the scientific study of how the human mind, with the conceptual systems it creates, makes mathematics possible.

The romance of mathematics, despite its immediate intuitiveness, and despite being supported by many outstanding physicists and mathematicians, is (nowadays) *scientifically* untenable. It is a mythology, and as such, arguing for or against it is a matter of faith, not a matter of scientific discussion. In this chapter, I will argue that Cognitive Science, the contemporary scientific study of the mind, which gathers interdisciplinary efforts from neuroscience to linguistics to cognitive psychology, has important things to say about the nature of mathematics, which in turn can have a positive impact in the enterprise of mathematics education. I will defend the idea that the answer to the question of *what is the nature* of mathematics deeply affects what mathematics education is taken to be, by shaping not only what is to be taught, in what order, through what pedagogical activities and using what methods, but also by prescribing how generations of mathematics teachers should be formed. A scientifically informed view of the nature of mathematics, I'll claim, one that explains the human cognitive everyday mechanisms that make the mysterious edifice of mathematics possible, should help prepare better new generations of mathematics teachers. Teachers who would, not only have a better sense of *how* mathematics is human, but also educators who would be able to design curricula, pedagogical software and classrooms activities in an informed cognitive-friendly manner compatible with how the human mind actually works. The strategy I'll follow in this chapter, therefore, will not be to simply provide suggestions of how to better teach X (mathematics), leaving X untouched. My goal will rather focus on giving an informed account (from the perspective of cognitive science) of *what* is the nature of X such that a cognitive-friendly and meaningful form of mathematics education can follow. In order to illustrate my arguments I will focus on a case study involving limits, infinite series, and continuous functions as they are taught at the college level. The accent will not be put on student's individual performance, or on classroom dynamics, but on the semantic organization of the very concepts themselves.

SO, WHAT IS THE NATURE OF WHAT WE TEACH IN MATHEMATICS EDUCATION?

Mathematics is unique. It is an extraordinary conceptual system characterized by the fact that the very entities that constitute it are imaginary, idealized, mental abstractions. These entities cannot be perceived directly through the senses. Consider, for instance, the simplest entity in Euclidean geometry: a Euclidean point. What is it? How can we find one in the universe? A point, as defined by Euclid is a dimensionless entity, an entity that has only location but no extension. Such a thing doesn't exist anywhere in the entire universe, as we know it. A Euclidean point cannot be actually *perceived* or observed through any scientific empirical method. Humans, however, can create via *imagination* a Euclidean point in a clear, precise, and non-ambiguous manner. They can talk about it and they can create with it more complicated purely imaginary entities, such as segments, planes, polygons, and spheres. A Euclidean point is an idealized abstract entity realized via human cognitive mechanisms. The imaginary (but precise nature) of mathematics becomes even more evident when the concerned entities invoke *infinity*, where, because of the finite nature of our bodies and brains, no direct experience can exist with the infinite itself. Yet, infinity is at the core of mathematics. It lies at the very basis of many fundamental concepts such as limits, least upper bounds, point-set topology, mathematical induction, infinite sets, points at infinity in projective geometry, to mention only a few.

But if mathematics is the product of human imagination, how can we explain the nature of mathematics with its unique features such as precision, objectivity, rigor, generalizability, stability, and, of course, applicability to the real world? How can we give a cognitive account of what mathematics *is*, with all the precision and complexities of its theorems, axioms, formal definitions, and proofs? And how can we do this when the subject matter is truly abstract and apparently detached from anything concrete, as in topics as transfinite numbers, abstract algebra, and hyper-set theory? In the realm of Platonic-oriented philosophies the question of the nature of mathematics doesn't pose a real problem, since the existence of mathematical ideas transcends the world of human ideas. This view, of course, cannot be tested scientifically and doesn't provide any link to current empirical work on human ideas and conceptual systems. The question of the nature of mathematics does not pose major problems to purely formalist philosophies either, because in that worldview mathematics is seen as a manipulation of meaningless symbols. The question of the origin of the meaning of mathematical ideas doesn't even emerge in the formalist world. For many mathematics educators who endorse a social constructivist approach, the question of the nature of mathematics is relatively straightforward: Mathematics, like art, poetry, architecture, music, and fashion, is "socially constructed." Under this account, however, it is not clear what makes mathematics so special. What distinguishes mathematics from, say, fashion or poetry? Any precise enough explanatory proposal of the nature of mathematics should give an account of the unique collection of features mentioned earlier that make mathematics so special: precision, objectivity, rigor, generalizability, stability, and, applicability to the real world. This is what makes the scientific study of the nature of mathematics so challenging: mathematical entities (organized ideas and stable concepts) are abstract and imaginary, yet they are realized through the biological and social peculiarities of the human animal. The challenge then is: how can a bodily-grounded view of the mind give an account of an abstract, idealized, precise, sophisticated and powerful domain of ideas if direct bodily experience with the subject matter is not possible?

In *Where Mathematics Comes From*, Lakoff and I propose some preliminary answers to such questions (Lakoff and Núñez, 2000). Building on findings in mathematical cognition and the neuroscience of numerical cognition, and using mainly methods from Cognitive Linguistics, a branch of Cognitive Science, we asked, What cognitive mechanisms are used in structuring mathematical ideas? And more specifically, what cognitive mechanisms can characterize the *inferential organization* observed in mathematical ideas themselves? We suggested that most of the idealized abstract technical entities in mathematics are created via everyday human cognitive mechanisms that extend the structure of bodily experience while preserving inferential organization. Such "natural" mechanisms are, among others, conceptual metaphors (Lakoff & Johnson, 1980; Sweetser, 1990; Lakoff, 1993; Lakoff & Núñez, 1997; Núñez, 2000; Núñez & Lakoff, 2005), conceptual blends (Fauconnier & Turner, 1998, 2002; Núñez, 2005), conceptual metonymy (Lakoff & Johnson, 1980), fictive motion and dynamic schemas (Talmy, 1988, 2003). Using a technique we called *Mathematical Idea Analysis* we studied in detail many mathematical concepts in several areas in mathematics, from set theory to infinitesimal calculus, to transfinite arithmetic, and showed how, via everyday human embodied mechanisms such as conceptual metaphor and conceptual blending, the inferential patterns drawn from direct bodily experience in the real world get extended in very specific and precise ways to give rise to a new emergent inferential organization in purely imaginary domains.

It is important to keep in mind that Mathematical Idea Analysis focuses on inferential organization and semantic structure rather than on individual cognizing abilities and learning patterns. In order to follow the approach developed in the remainder of this chapter, it is thus advisable to de-emphasize for a moment the actual performances, learning processes, and behaviors of single *individuals*, which usually define the level of analyses in cognitive psychology and those of many important areas of mathematics education, and to focus on the meaning and *inferential organization* of mathematical ideas *themselves* (as accepted within the professional community of contemporary mathematicians). Let us now take a closer look into the study of everyday conceptual mappings and inferential organization.

Conceptual Mappings and Inferential Organization

Consider the following two everyday linguistic expressions: "The spring is *ahead* of us" and "the presidential election is now *behind* us." Literally, these expressions don't make any sense. "The spring" is not something that can physically be "ahead" of us in any measurable or observable way, and an "election" is not something that can be physically "behind" us. Hundreds of thousands of these expressions, whose meaning is not literal but *metaphorical*, can be observed in human everyday language. They are the product of the human imagination, they convey precise meanings, and allow speakers to make precise inferences about them. A branch of cognitive science, cognitive linguistics (and more specifically, cognitive semantics), has studied this phenomenon in detail and has shown that the semantics of these hundreds of thousands metaphorical linguistic expressions can be modeled by a relatively small number of *conceptual metaphors* (Lakoff & Johnson, 1980; Lakoff, 1993). These conceptual metaphors, which are inference-preserving cross-domain mappings, are cognitive mechanisms that allow us to project the inferential structure from a source domain, which usually is grounded in some form of basic bodily-experience, into another one, the target domain, usually more abstract. A crucial component of what is modeled is *inferential organization*, the network of inferences that is generated via the mappings.

In the above examples, although the expressions use completely different words (i.e., the former refers to a location *ahead of us*, and the latter to a location *behind us*), they are both linguistic manifestations of a single general conceptual metaphor, namely, TIME EVENTS ARE THINGS IN UNIDIMENSIONAL SPACE[1]. As in any conceptual metaphor, the inferential structure of target domain concepts (time, in this case) is created via a precise mapping drawn from the source domain (unidimensional space, in this case). In what concerns time expressions, for instance, cognitive linguists have identified two main forms of this general conceptual metaphor, namely, TIME PASSING IS MOTION OF AN OBJECT (which

---

[1]Following a convention in cognitive linguistics, capitals here serve to denote the name of the conceptual mapping as such. Particular instances of these mappings, called metaphorical expressions (e.g., "she has a great future in front of her"), are not written with capitals.

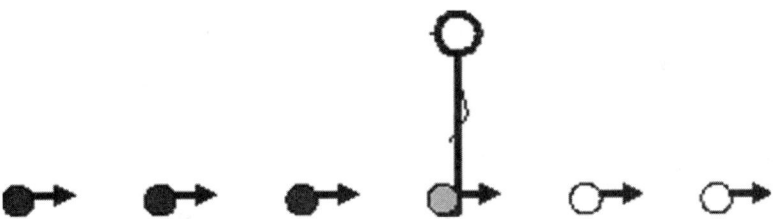

FIGURE 6–1. A graphic representation of the TIME PASSING IS MOTION OF AN OBJECT metaphor.

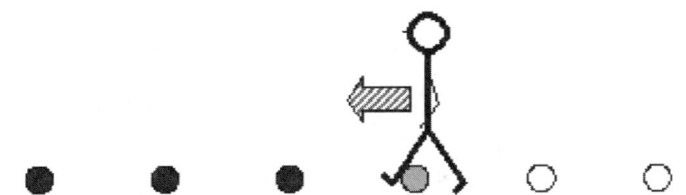

FIGURE 6–2. A graphic representation of the TIME PASSING IS MOTION OVER A LANDSCAPE metaphor.

models the inferential organization of expressions such as *Christmas is coming*) and TIME PASSING IS MOTION OVER A LANSCAPE (which models the inferential organization of expressions such as *we are approaching the end of the month*) (Lakoff, 1993)[2]; The former model has a fixed canonical observer where times are seen as entities moving with respect to the observer (Figure 6–1), while the latter has times as fixed objects where the observer moves respect to events in time (Figure 6–2).

These two forms share some fundamental features: both map (preserving transitivity) spatial locations in front of ego with temporal events in the future, co-locations with ego with events in the present, and locations behind ego (also preserving transitivity) with events in the past. Spatial construals of time are, of course, much more complex, but this is basically all what we need to know here (for details see Lakoff, 1993; Lakoff & Johnson, 1999; Núñez, 1999; For cross-linguistic and gestural studies, Núñez & Sweetser, 2006; For experimental psychological studies based on priming paradigms see Gentner, 2001; Boroditsky, 2000; Núñez, Motz, & Teuscher, 2006). For the purposes of this chapter, there are three very important morals to keep in mind:

1. At this level of the cognitive idea analysis, the primary focus is not how *single individuals* learn how to use these conceptual metaphors, or what difficulties they encounter when they learn them, or how they may lose the ability to use them after a brain injury, and so on. The focus is to characterize (i.e., to model), across hundreds of linguistic expressions, the structure of the inferences that can be drawn from them. For example, if "the spring is ahead of us," we can infer that the summer is not just "ahead" of us, but *further away in front of us*. Similarly, if "the presidential election is behind us," we can infer that the various effects immediately following the election are not only "behind" us, but also *much closer to us* than the election itself.

---

[2]For a different and more recent taxonomy based on linguistic data, as well as on gestural and psychological experimental evidence, see Núñez & Sweetser, 2006, and Núñez, Motz, & Teuscher, 2006.

2. *Truth*, when imaginary entities are concerned, is always relative to the inferential organization of the mappings involved in the underlying conceptual metaphors. For instance, "last summer" can be conceptualized as being *behind us* as long as we operate with the general conceptual metaphor TIME EVENTS ARE THINGS IN UNIDIMENSIONAL SPACE mentioned above, which determines a specific bodily orientation respect to metaphorically conceived events in time, namely, the future as being "in front" of us, and the past as being "behind" us. Núñez and Sweetser (2001; 2006), however, have shown that the details of that mapping are not universal. Through ethnographic field work, as well as cross-linguistic gestural and lexical analysis of the Aymara language of the Andes' highlands, they provided the first well-documented case violating the postulated universality of the metaphorical orientation future-in-front-of ego and past-behind-ego. In Aymara, for instance, "last summer" is conceptualized as being *in front of ego*, not behind of ego, and "next year" not as being in front of ego, but *behind* ego. Moreover, Aymara speakers not only utter these words when referring to time, but also produce co-timed corresponding gestures, strongly suggesting that these metaphorical spatial construals of time are not merely about words, but about deeper conceptual phenomena. The moral is that there is no *ultimate truth* regarding these imaginative structures. In this case, there is no ultimate truth about where, really, is the ultimate metaphorical location of the future (or the past). Truth will depend on the details of the mappings of the underlying conceptual metaphor. As we will see, this will turn out to be of paramount importance when mathematical concepts are concerned: Their ultimate truth is not hidden in the structure of the universe, but it will be relative to the underlying conceptual mappings (e.g., metaphors) used to create them.

3. It is crucial to keep in mind that the abstract conceptual systems we develop are possible *because* we are biological beings with specific morphological and anatomical features. In this sense, human abstraction is *embodied* in nature. It is because we are living creatures with a salient and unambiguous front and a back that we can build on these properties and the related bodily experiences we have to bring forth stable and solid concepts such as "the future in front of us." This wouldn't be possible if we had the body of a jellyfish or of an amoeba.

4. Finally abstract conceptual systems are not "simply" socially constructed, as a matter of convention. Biological properties and specificities of human bodily-grounded experience impose very strong constraints on what concepts can be created. While social conventions usually have a huge number of degrees of freedom, many human abstract concepts don't. For example, the color pattern of the Euro bills was socially constructed via convention (and so were the design patterns they have). But virtually any color ordering would have done the job. Metaphorical construals of time, on the contrary, are *only* based on a spatial source domain. And this is an *empirical* observation, not an arbitrary or speculative statement: as far as we know, there is no language or culture on earth where time is conceived in terms of thermic or chromatic source domains. And there is more: not just any spatial domain does the job. Spatial construals of time are, as far as we know, always based on unidimensional space.[3] Human abstraction is thus not "merely" socially constructed. It is constructed through strong non-arbitrary biological and cognitive constraints that play an essential

---

[3] Although they can, of course, be more complicated. Such is the case of cyclic or helix-like conceptions of time. But even in those cases the building blocks—a segment of a circle or a helix—preserve the topological properties of the uni-dimensional segment.

role in constituting what human abstraction is. Human cognition is *embodied*, shaped by species-specific non-arbitrary constraints. Again, this property will turn out to be very important when mathematical concepts are concerned.

With these morals in mind, we are now in a position to analyze our case study.

A CASE STUDY: LIMITS, INFINITE SERIES, AND CONTINUOUS FUNCTIONS

In the spirit of how the semantics of everyday human language is studied, let's analyze how mathematical ideas are conceived and expressed in the community of professional mathematicians. Keep in mind that the goal is to focus on how mathematical ideas are actually presented, described, characterized, and even formally defined in mathematics books, academic journals, and textbooks (thus de-emphasizing what students may say when they are learning these ideas). A careful analysis of technical books and articles in mathematics provides very good insights into the question of how the inferential organization of human everyday ideas has been used to create mathematical concepts. For the purpose of this chapter, I would like to focus on some concepts known to mathematics education as being particularly elusive, difficult for teachers to teach and for students to learn: limits, continuous functions, and infinite series (Freudenthal, 1973; Tall & Vinner, 1981; Robert, 1982; Núñez, 1993; Núñez, Edwards, and Matos, 1996). But, are these concepts intrinsically difficult to learn? And if yes, why? Or is it that the methods used to teach them are not appropriate? I will claim that "X education" when X is "mathematics" has been too timid in addressing the problem of the nature of X (mathematics) and as a consequence it has been framed for teaching X while leaving X itself largely untouched and unquestioned. Mathematics, as the content to be taught, is pre-given, and it is widely taught as such. In part due to the fact that the praxis of mathematics education perpetuates and sustains the romance of mathematics described earlier, is that mathematics education rarely questions how specific natural everyday ideas such as change, containment, continuity, rotations, and so on, are mathematicized to create what mathematics *is*. In this section I'll try to show that the Cognitive Science of Mathematics can accomplish this goal, and that mathematics education can benefit from it.

Let us start by looking at some classic books and textbooks:

1. While discussing limits, we read in the Russian classic *Matematika, ee soderzhanie metody i znachenie* [Mathematics, its contents, methods and meaning] by A. Aleksandrov, A.N., Kolmogorov, and M.A. Lavrent'ev [1956/1999]:

If a variable x_n may be represented as a sum

$$x_n = a + \alpha_n,$$

where a is a constant and α_n is an infinitesimal, then we say that the variable x_n, for n increasing *beyond* all bounds, *approaches* the number a and we write

$$\lim x_n = a \quad or \quad x_n \to a$$

The number a is called the limit of x_n. (Vol. 1 p. 82; italics are ours).

Strictly speaking, this statement refers to a sequence of discrete and motionless values (real numbers) that a variable x_n takes corresponding to increasing discrete and motion-

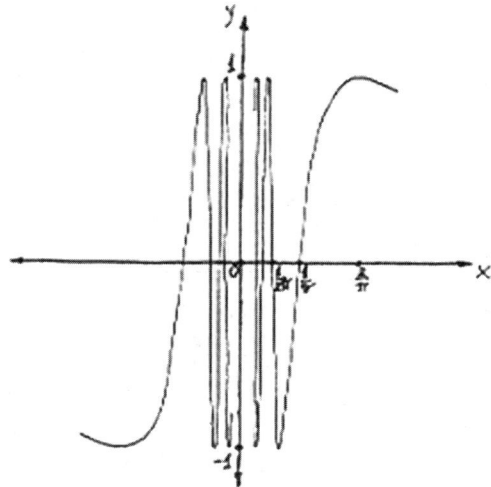

FIGURE 6–3. The graph of the function $f(x) = \sin 1/x$.

less values taken by n. If we examine this statement closely we can see that it describes static facts about numbers. We can observe that there is *no motion* whatsoever involved. No entity is actually *approaching* anything or moving *beyond* anywhere. So, why then did these well-respected Russian authors (or why do mathematicians in general, for that matter) use dynamic language to express static properties of static entities? And what does it mean to say that the "variable x_n *approaches* a number a," when in fact the variable can only have a fixed and distinct value given fixed and distinct values of n?

2. While discussing limits of infinite series, R. Courant & H. Robbins write in their classic book *What is Mathematics* (Courant & Robbins, 1978):

> We describe the behavior of s_n by saying that the sum s_n *approaches* the limit 1 as n *tends* to infinity, and by writing
>
> $$1 = \frac{1}{2} + \frac{1}{2^2} + \frac{1}{2^3} + \frac{1}{2^4} + \cdots,$$
>
> where on the right we have an *infinite series* (p. 64, underlined italics are ours).

This statement refers to a sequence of discrete and motion-less partial sums of s_n (real numbers), corresponding to increasing discrete and motion-less values taken by n in the expression $1/2^n$ where n is a natural number. If we examine this statement closely we can observe that it describes some facts about numbers, and about the result of discrete operations with numbers. Again, *no motion* whatsoever is involved. No entity is actually *approaching* or *tending* to anything. So why then did Courant and Robbins (or why do mathematicians in general) use dynamic language to express static properties of static entities? And what does it mean to say that the "sum s_n approaches," when in fact a sum is simply a fixed number, a result of an operation of addition?

3. Later in the book, Courant and Robbins analyze cases of continuity and discontinuity of trigonometric functions in the real plane. Referring to the function $f(x) = \sin 1/x$, whose graph is shown in Figure 6–3, they say:

"... since the denominators of these fractions increase without limit, the values of x for which the function $\sin(1/x)$ has the values 1, –1, 0, will cluster nearer and nearer to the point $x = 0$. Between any such point and the origin there will be still an infinite number of *oscillations* of the function" (p. 283, underlined italics are ours).

Once again, if, strictly speaking, a function is a mapping between elements of a set (coordinate values on the x-axis) with one and only one of the elements of another set (coordinate values on the y-axis), all what we have is a static correspondence between points on the x-axis with points on the y-axis. How then can the authors (or mathematicians in general) speak of "*oscillations* of the function," let alone an infinite number of them?

These three simple examples illustrate some deep and important issues regarding the semantic structure of mathematical ideas. They show how mathematical ideas and concepts are described, defined, illustrated, and analyzed in mathematics books. You can pick your favorite mathematics books and you will see similar patterns. In all three examples above, static numerical structures are involved, such as partial sums and mappings between coordinates on one axis with coordinates on another. Strictly speaking, absolutely no motion or dynamic entities are involved in the formal definitions of these terms. So, if no entities are really moving, why do authors continue to speak of "approaching," "tending to," and "oscillating"? If mathematical definitions are indeed so precise, why is there still dynamic language when purely static entities are concerned? Where is this motion coming from? What does dynamism mean in these cases? What role is it playing (if any) in the meaning of these statements about mathematics facts?

In order to answer to these questions we will first look at how pure mathematics characterizes real numbers, limits and continuity of functions. We will eventually find that in these cases the logic of formal mathematics of set-theoretic entities and of universal and existential quantifiers is intrinsically static, and that the presence of dynamic content along with its inferential structure is a manifestation of human meaningful cognition that is not captured by mathematics formalisms.

Pure Mathematics and Real Numbers

In pure mathematics, entities are brought to existence via formal definitions, formal proofs (theorems) and axiomatic methods (i.e., by declaring the existence of some entity without the need of proof. For example, in set theory the axiom of infinity assures the existence of infinite sets. Without that axiom, there are no infinite sets). In the case of real numbers, ten axioms taken together, fully characterize this number system and its inferential organization (i.e., theorems about real numbers). The following are the axioms of the real numbers.

1. Commutative laws for addition and multiplication.
2. Associative laws for addition and multiplication.
3. The distributive law.
4. The existence of identity elements for both addition and multiplication.
5. The existence of additive inverses (i.e., negatives).
6. The existence of multiplicative inverses (i.e., reciprocals).
7. Total ordering.
8. If x and y are positive, so is $x + y$.
9. If x and y are positive, so is $x \cdot y$.
10. The Least Upper Bound axiom.

The first 6 axioms provide the structure of what is called a *field* for a set of numbers and two binary operations. Axioms 7 through 9, assure ordering constraints. The first nine axioms fully characterize *ordered fields*, such as the rational numbers with the operations of addition and multiplication. Up to here we already have a lot of structure and complexity. For instance we can characterize and prove theorems about all possible numbers that can be expressed as the division of two whole numbers (i.e., rational numbers). Along a line we can also locate (according to their magnitude) any two different rational numbers and be sure (via proof) that there will always be (infinitely many) more rational numbers between them (a property referred to as density). With the rational numbers we can describe with any given (finite) degree of precision the proportion given by the perimeter of a circle and its diameter (e.g., 3.14; 3.1415; etc.). With the rational numbers, however, we can not "complete" the points on the line, and we can not express with infinite exactitude the magnitude of the proportion mentioned above ($\pi = 3.14159...$). For this we need the full extension of the real numbers. In axiomatic terms, this is accomplished by the tenth axiom: the least upper bound axiom. All ten axioms characterize a *complete ordered field*.

Nothing in the first nine axioms of real numbers helps us understanding the origin of motion in the above mathematical statements about infinite series, and continuity. All nine axioms simply specify the existence of static properties regarding binary operations and their results, and properties regarding ordering. There is no explicit or implicit reference to motion in these axioms. Since what makes a real number a real number (with its infinite precision) is the Least Upper Bound axiom, it is perhaps this very axiom that hides the dynamic secret we are looking for. Let's see what this axiom says:

10. Least Upper Bound axiom: every nonempty set that has an upper bound has a least upper bound.

And what exactly are an upper bound and a least upper bound? This is what pure mathematics says:

Upper Bound
b is *an upper bound* for S if
$x \leq b$, for every x in S.
Least Upper Bound
b_0 is a *least upper bound* for S if
- b_0 is an upper bound for S, and
- $b_0 \leq b$ for every upper bound b of S.

But once again, all we find here are statements about motionless entities such as universal quantifiers (e.g., for every x; for every upper bound b of S), membership relations (e.g., for every x in S), greater than relationships (e.g., $x \leq b$; $b_0 \leq b$), and so on. In other words, there is absolutely no indication of motion in the Least Upper Bound axiom, or in any of the other nine axioms. In short, the axioms of real numbers, which are supposed to completely characterize the "truths" (i.e., theorems) of real numbers don't tell us anything about a sum "approaching" a number, or a number "tending to" infinity (whatever that means!).

Now, let's take a look at the concept of continuity.

What Is Continuity?

What is, according to pure mathematics, continuity of functions? Today mathematics textbooks define continuity for functions as follows:

A function *f* is continuous at a number *a* if the following three conditions are satisfied:

1. *f* is defined on an open interval containing *a*,
2. $\lim_{x \to a} f(x)$ exists, and
3. $\lim_{x \to a} f(x) = f(a)$.

Where by $\lim_{x \to a} f(x)$ what is meant is the following:

Let a function *f* be defined on an open interval containing *a*, except possibly at *a* itself, and let *L* be a real number. The statement

$$\lim_{x \to a} f(x) = 1$$

means that $\forall \epsilon > 0, \quad \exists \delta > 0,$

such that if $0 < |x - a| < \delta,$

then $|f(x) - L| < \epsilon.$

As we can see, pure formal mathematics defines continuity in terms of limits, and limits in terms of static universal and existential quantifiers applied on static numbers (e.g., $\forall \epsilon > 0$, $\exists \delta > 0$), and the satisfaction of certain conditions which are described in terms of motionless arithmetic difference (e.g., $|f(x)-L|$) and static smaller than relations (e.g., $0 < |x-a| < \delta$). Once again, these formal definitions don't tell us anything about a sum "approaching" a number, or a number "tending to" infinity, or about a function "oscillating" between values (let alone doing it infinitely many times, as in the function $f(x) = \sin 1/x$).

A close inspection of mathematics textbooks reveals that often, right before giving this formal ε-δ definition of continuity, a paragraph or two are dedicated to the "informal" characterization of the idea of continuity, one that appeals to an "intuitive" description. Here is, for instance, the famous Russian book *Mathematics, Its Contents, Methods and Meaning* by Aleksandrov, Kolmogorov, and Lavrent'ev (1956/1999) mentioned earlier: "The general idea of a continuous function may be obtained from the fact that its graph is *continuous*: that is, its curve may be drawn without lifting the pencil from the paper." (p. 88; our emphasis).

And here is a quote from the classic textbook *Calculus* by G. Simmons (1985), while discussing the same topic: "In everyday speech a 'continuous' process is one that *proceeds without gaps* or interruptions or sudden changes. Roughly speaking, a function $y = f(x)$ is continuous if it displays similar behavior" (p. 58; our emphasis).

In both texts, we observe a characterization of continuous functions given in *dynamic* terms. In both cases there is something moving: the pencil drawing a curve on the paper in the former, and something unfolding without gaps in the latter. In both cases we have something moving from some position in space towards some other location in an uninterrupted manner. In both books these dynamic descriptions are given as a way of helping the reader by providing some immediate intuitive idea of what a continuous function *means*. The Russian book even characterizes the meaning of a "continuous function" *in terms of* something that *is* "continuous," whose meaning corresponds to what Simmons' Calculus textbook characterize as "everyday speech" (as we'll see later, this meaning corresponds precisely to the concept of *natural continuity* described by

Núñez & Lakoff, 1998). At this point, right after setting this introductory presentation of continuous functions in dynamic terms, textbooks usually make a radical move. In a somewhat downgrading tone they make clear that the "intuitive" examples given so far are merely illustrative, that they are not precise enough, and that a rigorous formal definition is required. Simmons' textbook, for instance, says: "Up to this stage our remarks about continuity have been rather loose and intuitive, and intended more to *explain* than to *define*." (Simmons, 1985, p. 58; our emphasis). This is a remarkable and profoundly informative passage. The choice of the words "explain" and "define" is not random. It characterizes the widespread idea in mathematics education that "explaining" may be a good thing, but what mathematics teaching really is about, is in "defining" entities and properties in a rigorous and precise way. From the perspective of cognitive science this statement goes against most of what we know about how humans learn and make sense of things, from perception, attention, and memory, to categorization and problem solving.

Why is Continuity Difficult to Understand?

An important part of the mathematics education community attributes the difficulties of teaching and learning the concept of continuity to problems related to the use of existential and universal quantifiers. Nowhere is this position clearer than in the work of the famous mathematics educator Hans Freudenthal:

> The difficulties implicit in the continuity concept are quantifiers and the order of quantifiers of different kinds. ... Continuity of f means intuitively: small changes of x correspond with small changes of $f(x)$. Or: if x changes little, $f(x)$, also changes little. Words like 'small', 'big', 'little', 'much', 'short', 'long', may hide a quantifier, but formal linguistic criteria are often insufficient to know which kind. Always, sometimes, everywhere, somewhere—exhibit clearly the universal or existential quantifier, but the linguistic formulation does not unveil that in the continuity definition the second small (or little) hides a universal, and the first an existential, quantifier. To grasp it, a logical analysis is badly needed. The meaning of the quantifiers in 'small' or 'little' is better indicated in the more exact formulation: to sufficiently small changes of x correspond arbitrarily small ones of $f(x)$. Or: if x changes sufficiently little, $f(x)$ changes arbitrarily little. Still from this formulation it is a long step to understand that first the 'arbitrarily little' must be prescribed before the 'sufficiently little' is to be determined. ... The intuitive continuity definition involves two difficulties of formalizing—first, decoding of hidden quantifiers, second, settling the order of quantifiers of different kinds. Good didactics should at least separate these difficulties from each other. (Freudenthal, 1973, p. 561)

In the pages following this citation, Freudenthal provides an insightful analysis of these difficulties and gives helpful recommendations for achieving good didactics regarding the use of quantifiers. While his analysis is precise, deep, and clear, he only focuses on the formal aspects of the ε–δ definition, completely missing the fundamental dimensions of everyday human cognition that may be interfering in understanding such formalization. Perpetuating the common belief that in "X education" where X is mathematics, you can teach X (mathematics) leaving X untouched or unquestioned, Freudenthal (like many in mathematics education) takes the ε–δ–δ definition for granted. This, to the point that for him "Continuity of f means intuitively: small changes of x correspond with small changes of $f(x)$." The statement is indeed a very good linguistic

THE COGNITIVE SCIENCE OF MATHEMATICS   141

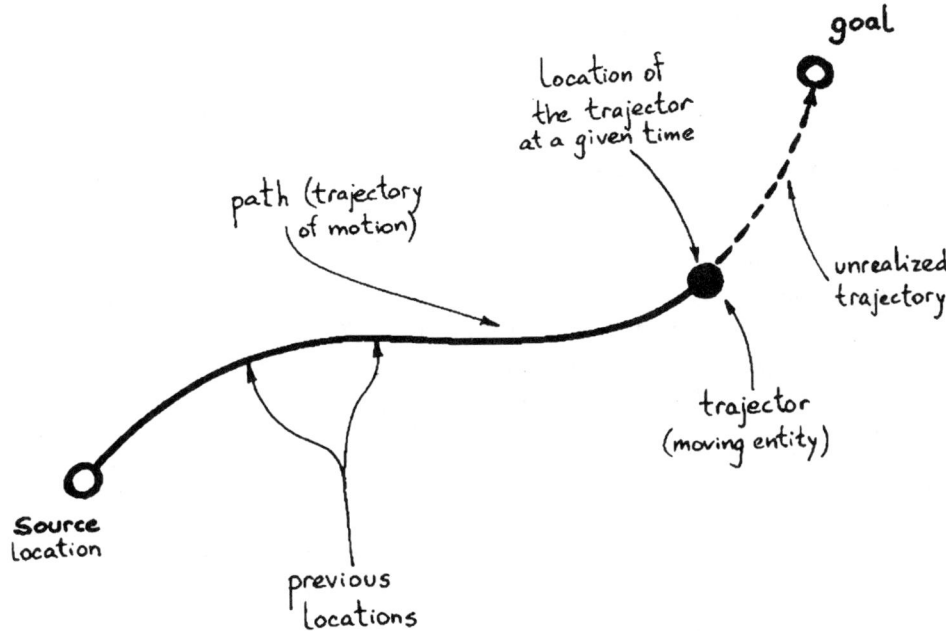

FIGURE 6–4.

formulation of what is intuitive in the ε-δ *definition*. It has the right *static* concepts, and it provides the right relations between them. What is missing, however, is precisely the need of questioning X, that is, the very mathematicization of the everyday notion of continuity—*natural continuity* (Núñez & Lakoff, 1998)—evoked in both our textbook examples. What is missing is the fact that cognitively speaking, the statement *does not* characterize the intuitive meaning of natural continuity—the continuity conceived by the creators of calculus, Leibniz and Newton, in the 17th century (and in fact, all mathematicians up to the 19th century), as well as by the students and teachers who naturally bring it into the classroom today. It is natural continuity that brought Euler to refer to a continuous curve as "a curve described by freely leading the hand" (cited in Stewart, 1995, p. 237), and the great Kepler to measure "an area swept out by the motion of a (celestial) point on a physical 'continuous curve'" (Kramer, 1970, p. 528). Natural continuity—continuity as we normally conceive it outside of mathematics—is based on a *source-path-goal schema*, a fundamental cognitive schema concerned with motion which has the following elements:

A trajector that moves
A source location (the starting point)
A goal—that is, an intended destination of the trajectory
A route from the source to the goal
The actual trajectory of motion
The position of the trajector at a given time
The direction of the trajector at that time

The actual final location of the trajector, which may or may not be the intended destination.

The source-path-goal schema is very general and can be extended in many ways: the speed of motion, the trail left by the thing moving, obstacles to motion, forces that move along a trajectory, additional trajectors, and so on. The schema is topological in the sense that a path can be expanded or shrunk or deformed and still remains a path and it has an internal spatial logic and built-in inferences (see Figure 6–4). For instance, If you have traversed a route to a current location, you have been at all previous locations on that route; If you travel from A to B and from B to C, then you have traveled from A to C; If there is a direct route from A to B and you are moving along that route toward B, then you will keep getting closer to B; If X and Y are traveling along a direct route from A to B and X passes Y, then X is further from A and closer to B than Y is; and so on.

Natural continuity, building on the source-path-goal schema, has the following essential features in its inferential organization (Núñez & Lakoff, 1998):

Continuity, traced by motion, takes place over time.
The trace of the motion is a static holistic line with no "jumps".

None of these features is present in Freudenthal's characterization of what is intuitive about continuity. And the reason is simple. Freudenthal's static characterization corresponds in fact to a radically different human idea with a different inferential organization. It corresponds to the static everyday notion of *preservation of closeness near a location*: being within a given distance from a specific location. Preservation of closeness has static locations, landmarks, reference-points, distances, but no trajectors, no paths, no directionalities, no motion, and therefore no "jumps." As I have argued elsewhere in collaboration with other colleagues (Lakoff & Núñez, 1997, 2000; Núñez & Lakoff, 1998; Núñez, 2000; Núñez, Edwards, and Matos, 1999) "preservation of closeness" is an everyday human concept with a very precise inferential organization, recruited by Cauchy and Weierstrass in the 19th century to carry out the program of arithmetizing analysis (for details see Lakoff & Núñez, 2000, Chapters 12–14). The cognitive science of mathematics shows that the inferential organization of the idea of preservation of closeness is not the same as the one of natural continuity. The two concepts—natural continuity and ε-δ continuity—simply have, cognitively, two radically different logics (Núñez & Lakoff, 1998).

As we saw above, many mathematicians and mathematics educators, thinking that mathematics education can understand how to teach mathematics leaving the very mathematics untouched, believe that the problem of natural continuity is that it is not "precise" enough. They believe that what the ε-δ definition does is to (1) make *precise* and (2) to *generalize* the idea of natural continuity. But this is not true. The idea of natural continuity is indeed very precise, and as we saw earlier, it has a precise inferential organization. The issue is that the inferential organization of natural continuity, which is dynamic and holistic in nature, did not serve the purposes of the arithmetization program which required a reduction of mathematical ideas into static and discrete numeric structures and concepts that later became compatible with static set-theoretic concepts (based on static concepts such as membership relation seen as presence inside a container, and lack of membership relations as presence outside the container, etc.). The inferential structure of the idea of preservation of closeness did fit

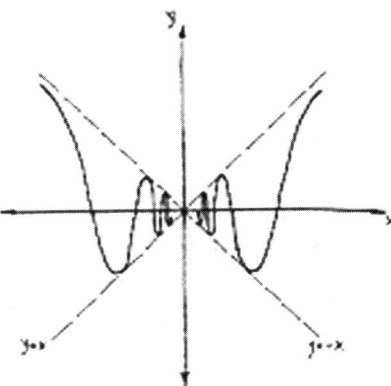

FIGURE 6–5. The graph of the function $f(x) = x \sin 1/x$.

the goals of the arithmetization program, and it is *this* idea (along with others such as *gaplessness*) that was made precise by the ε-δ definition. The moral is that mathematics education, as an endeavor directly dealing with humans and with the question of how they think and learn, shouldn't assume that formal definitions in mathematics (1) make intuitive ideas more "precise", and that (2) they generalize those intuitive ideas. From this it follows that contrary to what Freudenthal (and many mathematics educators) say, "the difficulties implicit in the continuity concept" are not just the quantifiers and their order. The problems are deeper. They start with the false belief (from both the teacher and the student) that formalization necessarily generalizes and makes intuitive ideas more precise. Because the cognitive science *of* mathematics does indeed question the X, that is, the nature of the very mathematics, it is in a good position to show exactly why mathematical formalizations neither make more precise, nor generalize, everyday intuitive ideas. I believe that mathematics education would benefit tremendously by building on these kinds of findings and by acknowledging, in a deep way, that mathematics is indeed the product of human imagination.

The fact that the ε-δ definition doesn't capture the inferential structure of natural continuity shouldn't be a surprise. In *Where Mathematics Comes From*, Lakoff and I showed what well-known contemporary mathematicians had already pointed out in more general terms (Kaput, 1979; Hersh, 1997):

- The structure of human mathematical ideas, and its inferential organization, is richer and more detailed than the inferential organization provided by formal definitions and axiomatic methods. Formal definitions and axioms neither fully formalize nor generalize human concepts (Lakoff & Núñez, 2000).

We can illustrate this with a relatively simple example taken from our book. Consider the function $f(x) = x \sin 1/x$ whose graph is depicted in Figure 6–5.

$$f(x) = \begin{cases} x \sin 1/x & \text{for } x \neq 0 \\ 0 & \text{for } x = 0 \end{cases}$$

According to the ε-δ definition of continuity this function is continuous at every point. Indeed, for all x, it is always possible to find the specified ε's and

δ's to satisfy the conditions for preservation of closeness. However, according to *natural continuity* this function is *not* continuous. The inferential organization of natural continuity requires that certain conditions have to be met. For instance, in the semantics of a naturally continuous line we should be able to tell how long the line is between two points. We should also be able to describe essential properties of the motion of a point along that line. With this function we can not do that. Since the function "oscillates" infinitely many times as it "approaches" the point (0, 0) we can not really tell how long the line is between two points located on the left and right sides of the plane. Moreover, as the function approaches the origin (0, 0) we can not tell whether it will cross from the right plane to the left plane "going down" or "going up." As a result, the function violates two essential properties of natural continuity and therefore it is not continuous. This leads us to think that the reported difficulties students have in learning limits and continuity then, may not be so much due to a lack of mastery in the use of universal and existential -quantifiers as Freudenthal has suggested (1973), but to the fact that the formal ε-δ definition of continuity,

(1) simply doesn't capture the inferential organization of the human everyday notion of continuity (natural continuity), and

(2) contrary to what is claimed in most mathematics books and textbooks, it *does not generalize* the notion of continuity either. The function $f(x) = x \sin 1/x$ is ε-δ continuous but it is not naturally continuous.

The moral here is that what is characterized formally in mathematics leaves out a huge amount of inferential organization of the human ideas that constitute mathematics. As we will see, this is precisely what happens with the dynamic aspects of the expressions we saw before, such as "approaching," "tending to," "oscillating," and so on. Motion, in these examples, is a genuine and constitutive manifestation of the nature of these mathematical ideas. From the point of view of pure mathematics, however, the essential dynamic components of the inferential organization of these ideas are not captured by the ε-δ formalisms and the axiomatic system for real numbers.

BACK TO EMBODIED COGNITION

Let us see what embodied cognition has to say about the origin of motion in the mathematical ideas we have been looking at. In the case of limits of infinite series, motion in "the sum s_n *approaches* the limit 1 as n *tends* to infinity" emerges *metaphorically* from the successive values taken by n in the sequences as a whole. It is beyond the scope of this chapter to go into the details of the mappings involved in the various underlying conceptual metaphors that provide the required dynamic inferential organization (for details see Lakoff & Núñez, 2000). We can point out, however, that there are many conceptual metaphors and metonymies involved. There are conceptual metonymies in cases such as a partial sum *standing for* the entire infinite sum; there are conceptual metaphors in cases where we conceptualize the sequence of these metonymical sums as a *unique trajector* moving in space (*the* sum s_n approaches); there are conceptual metaphors for conceiving infinity as a single location in space such that a metonymical n (standing for the entire sequence of values) can "tend to;" there are conceptual metaphors for conceiving the number "1" (not as a mere natural number but as an infinitely precise real number) as the result of the infinite sum; and so on. Notice that none of these expressions can be literal. The facts described in these sentences do not exist in any real perceivable world. They are metaphorical in nature. It is important to understand that these conceptual metaphors and metonymies are not simply "noise" added on top of formalisms. Indeed, they are constitutive of the very embodied ideas that make mathematical ideas possible. It is the inferential organization provided by our embodied understanding of "approaching" and "tending to" that is at the core of these mathematical ideas.

In the case of the "oscillating" function, the moving object is one holistic object, the trigonometric function in the Real plane, constructed metaphorically from "non-countable" infinite (i.e., $>\aleph_0$) discrete real values for x, which are progressively smaller in absolute terms. In this case motion takes place in a specific manner: moving towards the origin from two opposite sides (i.e., for negative and positive values of x) and always between the values $y = 1$ and $y = -1$. As we saw, a variation of this function, $f(x) = x \sin(x)$, reveals deep cognitive incompatibilities between the dynamic notion of continuity implicit in the example above and the static ε-δ definition of continuity.

Fictive Motion

Now that we are aware of the metaphorical (and metonymical nature) of the mathematical ideas mentioned earlier, we can analyze more in detail the dynamic component of these ideas. From where do these ideas get motion? What cognitive mechanism is allowing us to conceive static entities in dynamic terms? The answer is *fictive motion*.

Fictive motion is a fundamental embodied cognitive mechanism through which we unconsciously (and effortlessly) conceptualize static entities in dynamic terms, as when we say *the road goes along the coast*. The road itself doesn't actually move anywhere. It is simply standing still. But we may conceive it as moving "along the coast." Fictive motion was first studied by Len Talmy (1996), via the analysis of linguistic expressions taken from everyday language in which static scenes are described in dynamic terms. The following are linguistic examples of fictive motion:

- The Equator *passes through* many countries.
- The boarder *runs along* the river.
- The US west coast *goes all the way down* to San Diego.
- After *crossing* the bridge the path *goes through* the forest and then it *reaches* the main house.
- The fence *stops* right after the tree.
- Unlike Tokyo, in Paris there is no subway line that *goes around* the city.

Motion, in all these cases, is fictive, imaginary, and not real in any literal sense. Not only these expressions use verbs of action, but they also provide precise descriptions of the quality, manner, and form of motion. In all cases of fictive motion there is a "trajector" (the moving agent) and a "landscape" (the space in which the trajector moves). Sometimes the trajector may be a real object (e.g., the road goes; the fence stops), and sometimes it is metaphorical (e.g., the Equator passes through; the boarder runs). In fictive motion, real world trajectors don't move but they have the potential to move or the potential to enact or enable movement (e.g., a car moving along that road). In Mathematics proper, however, the trajector has always a metaphorical component. That is, the trajector as such can't be literally capable or incapable of enacting movement, because the very nature of the trajectory is imagined via metaphor (Núñez, 2003). For example, a point in the Cartesian Plane is an entity that has location (determined by its coordinates) but has no extension. So when we say "point P *moves* from A to B" we are ascribing motion to a metaphorical entity that only has location. First, as we saw earlier, entities which have only location (i.e., points) do not exist in the real world, so as such, they do not have the potential to move or not to move in any literal sense. They simply do not exist in the real world. They are metaphorical entities. Second, literally speaking, point A and point B are distinct locations, and in the Cartesian plane no point can change location while preserving its identity since a point's location is uniquely determined by its coordinates and vice-versa. That is, the trajector (point P) can not preserve its identity throughout the process of motion from A to B, since that would mean that it is changing the very properties that are defining it, namely, its coordinates.

We now have a basic understanding of how conceptual metaphor and fictive motion work, so we are in a position to see the embodied cognitive mechanisms underlying the mathematical expressions like the ones we saw earlier. Here we have similar expressions:

- sin $1/x$ *oscillates* more and more as x *approaches* zero
- $g(x)$ never *goes beyond* 1
- If there exists a number L with the property that $f(x)$ *gets closer and closer to* L as x *gets larger and larger*; $\lim_{x \to \infty} f(x) = L$.

In these examples Fictive Motion operates on a network of precise *conceptual metaphors*, such as NUMBERS ARE LOCATIONS IN SPACE (which allows us to conceive numbers in terms of spatial positions), to provide the inferential structure required to conceive mathematical functions as having motion and directionality. Conceptual metaphor generates a purely imaginary entity in a metaphorical space, and fictive motion makes it a moving trajector in this metaphorical space. Thus, the progressively smaller numerical values taken by x which determine numerical values of sin $1/x$, are via the conceptual metaphor NUMBERS ARE LOCATIONS IN SPACE conceptualized as spatial locations. The now metaphorical spatial locus of the function (i.e., the "line" drawn in the plane) becomes available for fictive motion to act upon. The progressively smaller numerical values taken by x (now metaphorically conceptualized as locations progressively closer to the origin) determine corresponding metaphorical locations in space for sin $1/x$. In this imaginary space, via conceptual metaphor and fictive motion now sin $1/x$ can "oscillate" more and more as x "approaches" zero.

In a similar way the infinite precision of real numbers themselves can be conceived as limits of sequences of rational numbers, or as limits of sequences of nested intervals. Because, as we saw, limits have conceptual metaphor and fictive motion built in, we can now see the fundamental role that these embodied mechanisms play in the constitution of the very nature of the real numbers themselves.

Dead Metaphors?

So far in this chapter we have analyzed some mathematical ideas like limits, continuity and series through methods in cognitive linguistics, such as conceptual metaphor and fictive motion. We have studied the inferential organization modeling *linguistic expressions*. But because conceptual metaphor theory, based mainly on purely linguistic grounds, has made important claims *about* human cognition, abstraction, and mental phenomena, some psychologists have rightly questioned the lack of empirical evidence to support the psychological reality of conceptual metaphor. How do we know, for instance, that some of the metaphors we observe in linguistic expressions are not mere "dead metaphors," expressions that were metaphorical years ago but which have become "lexicalized" in nowadays language? How do we know that these metaphors are the actual result of real-time cognitive activity? And how can we find out the answers to such questions? Experimental psychologists have questioned whether there is any psychological reality in people's minds when they listen to, or utter, such metaphorical expressions (see, for example Murphy, 1997; Gibbs, in press). Is it the case that people actually operate cognitively with these conceptual metaphors? Or could it be the case that metaphorical expressions using terms like "approaching" or "getting closer and closer" are simply "dead metaphors," that is, expressions that once had dynamic spatial content but that now have become separate lexical items, no longer with connections with space and motion? Maybe all what we have in the mathematical expressions we have examined, is simply a story of dead metaphors, with no psychological reality whatsoever.

Experimental psychologists have tried to answer similar questions regarding everyday linguistic expressions using priming techniques, a well-known experimental paradigm in which subjects are systematically biased via specific stimulation involving the source domain in order to evaluate whether they carry the corresponding inferences into the target domain. If by priming the source domain of the metaphor one gets systematic variation in the inferences made in the target domain, then one could conclude that individuals do reason metaphorically, or otherwise they would not be sensitive to the priming. In this chapter, however, in order to address the question of the psychological reality of the conceptual metaphors involved, I will use a different form of empirical analysis: real time gesture-speech-thought production. As we will see, the study of human *gesture* provides embodied convergent evidence of the psychological reality of many conceptual mappings, metaphorical, metonymical and others. Gesture studies, via a detailed investigation of real-time cognitive production, bodily motion (mainly hands and arms), and voice inflection, show that the conceptual metaphors and fictive motion involved in the mathematical ideas analyzed above, far from being dead, do have a real-time and very embodied psychological reality.

Gesture as Cognition

In the study of the human mind, gestures have been left out of the picture for a very long time. They constitute the forgotten dimension of thought and language. In Chomskian linguistics, for instance, where language has been seen mainly in terms of abstract grammar, formalisms, and combinatorics, there was simply no room for "bodily production" such as gesture. In mainstream experimental psychology gestures were left out, among others, because being produced in a spontaneous manner, it was very difficult to operationally define them, making rigorous experimental observation on them extremely difficult. In mainstream cognitive science, which in its origins was heavily influenced by classic "disembodied" artificial intelligence, there was simply no room for gestures either. Cognitive science and artificial intelligence were heavily influenced by the information-processing paradigm and what was taken to be essential in any cognitive activity was a set of body-less abstract rules and the manipulation of physical symbols governing the processing of information. In all these cases, gestures were completely ignored and left out of the picture that defined what constituted genuine subject matters for the study of the mind. At best gestures were considered as a kind of epiphenomenon, secondary to other more important and better-defined phenomena.

But in the last decade or so, the field of gesture studies has moved forward dramatically, thanks to the work of pioneers such as Kendon (1982, 2004), McNeill (1992, 2000), Goldin-Meadow (2003) and many others. Research in a variety of areas, from child development, to neuropsychology, to linguistics, and to anthropology, has shown the intimate link between oral and gestural production. Finding after finding has confirmed that gestures are produced in synchronicity with speech, that they develop in close relation with speech, and that brain injuries affecting speech production also affect gesture production. The following is an abbreviated list of sources of evidence supporting (1) the view that speech and gesture are in reality two facets of the same cognitive linguistic reality, and (2) the embodied approach for understanding language, conceptual systems, and high-level cognition:

> Universality: Speech-accompanying gesture is a crosscultural universal (McNeill, 1992; Núñez & Sweetser, 2006; Iverson & Thelen, 1999; Kita & Essegbey, 2001).
>
> Largely unconscious production: Gestures are less monitored than speech, and they are to a great extent unconscious. Speakers are often unaware that they are gesturing at all (McNeill, 1992).
>
> Speech-Gesture synchronicity: Gestures are co-produced with speech, in co-timing patterns which are specific to a given language (McNeill, 1992).
>
> Gesture production with no visible interlocutor: Gestures can be produced without the presence of interlocutors, e.g., people gesture while talking on the telephone, and in monologues; congenitally blind subjects gesture as well (Iverson & Goldin-Meadow, 1998).
>
> Speech-Gesture co-processing: Stutterers stutter in gesture too, and impeding hand gestures interrupts speech production (Mayberry & Jaques, 2000).
>
> Speech-Gesture development: Gesture and speech development are closely linked (Iverson & Thelen, 1999; Bates & Dick, 2002; Goldin-Meadow 2003).
>
> Speech-Gesture complementarity: Gesture can provide complementary (as well as overlapping) content to speech content. Speakers synthesize and subsequently cannot distinguish information taken from the two channels (Kendon, 2000).
>
> Gestures and abstract metaphorical thinking: Linguistic metaphorical mappings are paralleled systematically in gesture (McNeill, 1992; Cienki, 1998; Sweetser, 1998; Núñez & Sweetser, 2006).

In all these studies, a careful analysis of important parameters of gestures such as handshapes, hand and arm positions, palm orientation, type of movements, trajectories, manner, and speed, as well as a careful examination of timing, indexing properties, levels of iconicity, and the coupling with environmental features, give deep insight into human thought.[4] Among many properties, gestures usually have three well-defined phases, called preparation, stroke, and retraction (McNeill, 1992). The

---

[4] An analysis of the various dimensions and methodological issues regarding the scientific study of gestures studies is beyond the scope of this chapter. For details see references mentioned above.

THE COGNITIVE SCIENCE OF MATHEMATICS 149

FIGURE 6–6. A professor of mathematics lecturing on convergent sequences. He explains a case in which the real values of a sequence "oscillate" (horizontally).

FIGURE 6–7. A professor of mathematics talking about an unbounded monotone sequence "going in one direction" (a through e), which "takes off to infinity" (f).

stroke is in general the fastest part of the gesture's motion, and it tends to be highly synchronized with speech accentuation and semantic content. The preparation phase is the motion that precedes the stroke (usually slower), and the retraction is the motion after the stroke has been produced (usually slower as well), when the hand goes back to a resting position or to whatever activity it was engaged in.

With these tools from gesture studies and cognition, we can now analyze mathematical expressions like the ones we saw before, but this time focusing on the gesture production of the speaker (in this case a mathematician teaching a university level mathematics course). The following gestures have been recorded during upper division mathematics classes at a major university in California. Keep in mind that these gestures are abstract (metaphorical) in nature, in the sense that the entities that are indexed with the various handshapes—like points and numbers—are purely imaginary entities.

Figure 6–6, shows a professor of mathematics teaching about convergence of sequences of real numbers. In this particular situation, he is talking about a case in

Figure 6–8. A professor of mathematics in a university level class talking about a constant sequence.

which the values of an infinite sequence do not get closer and closer to a single value as n increases, but "oscillate" between two fixed values. His right hand, with the palm towards his left, has a handshape called *baby O* in American Sign Language and in gesture studies, where the index finger and the thumb are touching and are slightly bent while the other three fingers are fully bent. In this gesture the touching tip of the index and the thumb is metaphorically *indexing* a metonymical value standing for the values in the sequence as n increases (it is almost as if the subject is carefully holding a very tiny object with those two fingers). Holding that fixed handshape, he moves his right arm horizontally back and forth while he says "oscillating."

Hands and arms are essential body parts involved in gesturing. But often it is also the entire body that participates in enacting the inferential structure of an idea. In the following example (Figure 6–7) a professor of mathematics is teaching a course involving notions of calculus. In this scene he is talking about some particular theorems regarding monotone sequences.

As he is talking about an unbounded monotone sequence, he is referring to the important property of "going in one direction" (i.e., taking increasingly large values). As he says this he is producing iterative unfolding circles with his right hand and at the same time he is walking frontward, accelerating at each step (Figure 6–7a through 6–7e). His right hand, with the palm toward his chest, displays a shape called *tapered O* (Thumb relatively extended and touching the upper part of an extended index bent in right angle, like the other fingers), which he keeps in a relatively fixed position while doing the iteration circular movement. A few milliseconds later he completes the sentence by saying "it takes off to infinity" at the very moment when his right arm is fully extended and his hand shape has shifted to an extended shape called *B spread* with a fully (almost over) extension, and the tips of the fingers pointing frontward, slightly at eye-level.

Sometimes, when the sequence exhibits a peculiar property, handshapes adopt specific forms that match the meaning of those properties. In Figure 6–8 we see the same professor talking this time about a situation where the sequence is constant. His dominant hand (the right one, with which he has been writing) curls back, his elbow is bent in 90 degrees and his wrist is maximally bent with the palm oriented down. His fingers are also bent pointing downward (Figure 6–8a). Then while keeping that handshape he extends his elbow (and wrist) producing a small frontward (and slightly downward) motion with his right hand. In the meantime his left hand, with palm toward right, raises slowly, forming a *five* handshape (Figure 6–8b). As he says the word "constant", he abruptly stops the forward motion with his right hand marking a location situated a couple of inches in front of his open five left hand (Figure 6–8c). While keeping his left hand totally fixed holding the same "five" handshape, he iterates a couple of times the same frontward movement with his right hand always stopping sharply at the same location, just a few inches from the open palm of his left hand. These abruptly stopped movements performed with the curled handshape while referring to a constant sequence sharply contrast with the smooth open ended fully extended arm, hand and fingers, of the previous example produced when referring to an unbounded monotone sequence (Figure 6–7).

It is important to mention that in these three cases the blackboard is full of mathematical expressions containing formalisms like the ones we saw in the previous sections (e.g., formalisms, with universal and existential quantifiers, which have no indication or reference to motion). The gestures (and the linguistic expressions used), however, tell us a very different conceptual story. In these examples, these mathematicians are referring to fundamental dynamic aspects of the mathematical ideas they are talking about. In the first example, the oscillating gesture matches, and it is produced synchronically with, the linguistic expressions used. In the second example, the unfolding iterative circular gesture matches the inferential organization of the iteration involved in the monotone sequence, and the entire body moves forwards as the sequence unfolds. Since the sequence is unbounded, it "takes off to infinity," idea which is precisely characterized in a synchronous way with the full frontal extension of the arm and the hand. That motion contrasts with the one in the third example, where a curled shaped hand moves slightly forward but hits repeatedly the same location never being able to go further.

We can conclude from these examples that:

> First, gestures provide converging evidence for the psychological and embodied reality of the linguistic expressions analyzed with classic techniques in cognitive linguistics, such as metaphor and fictive motion analysis. In these cases gesture analysis shows that the metaphorical expressions we saw earlier are not cases of dead metaphors. The above gestures show that the dynamism involved in these ideas has full psychological and cognitive reality, which is enacted in real time while speaking and thinking in an instructional context.
>
> Second, these gestures show that the fundamental dynamic contents involving infinite sequences, limits, continuity, and so on, are in fact *constitutive* of the inferential structure of these ideas. Formal language in mathematics, however, is not as rich as everyday language and cannot capture the full complexity of the inferential structure of mathematical ideas. It is the job of the cognitive science of mathematics to characterize the full richness of mathematical ideas (it is not the job of mathematics itself). These findings should inform mathematics education in order to conceive a more human cognitive-friendly way of teaching mathematics.

CONCLUSION

For many, mathematics is a timeless set of truths about the universe, transcending our human existence. For others mathematics *is* what is characterized by formal definitions and axiomatic systems. And for others it is "just" a social construction, very much like music or literature. These different views of the nature of mathematics produce corresponding views of mathematics education that interfere with a healthy process of teaching and learning mathematics. The first two, which are predominant in our culture, ignore the fact that mathematics is a genuine creation of the abstraction and imagination of the human animal. Under these accounts natural human meaning and understanding are not part of the picture, and therefore they provide inappropriate forms of addressing human learning and sense making. The third one, while acknowledging the human social nature of mathematics, fails to distinguish essential properties of mathematics such as objectivity, formalizability, generalizability, applicability to the real world, and so on. Such view, then, is not in a good position to understand how the highly constrained non-arbitrary peculiarities of human brains and bodies make mathematics (with all its unique features) possible.

From the perspective of the cognitive science *of* mathematics a very different view emerges: Mathematics doesn't exist outside of human cognition. Formal definitions and axioms in mathematics are themselves human ideas (although they constitute a very small and specific fraction of human cognition), and they only capture very limited aspects of the richness of mathematical ideas. Moreover, definitions and axioms neither formalize nor generalize human concepts. A clear example of this is provided by the fact that modern Mathematics (especially after the work by Cauchy, Weierstrass, Dedekind, and others in the 19th century), when dealing with important mathematical concepts such as limits and continuity, is at odds with cognitive mechanisms such as fictive and metaphorical motion that helped to create these very concepts. Anyone who has taught calculus to novices in mathematics can tell how counter-intuitive and hard to understand the ε-δ definitions of limits and continuity are. Many mathematics educators, attributing a higher status to formalisms than to "intuition", have suggested that the problems are caused by lack of mastery of quantifiers. But there may be a cognitively deeper and simpler reason. The ε-δ formalisms are counter-intuitive and difficult to grasp because they characterize a different group of everyday human ideas such as *preservation of closeness* and *gaplessness*, which are at odds with the inferential organization of natural continuity. Furthermore, students are dogmatically (and wrongly) told that the ε-δ formalisms—static and discrete in nature—generalize and make precise the intuitive idea of natural continuity—dynamic and holistic in nature. We have showed, however, that this is not true. Static ε-δ formalisms neither formalize nor generalize the rich human dynamic concepts of natural continuity. When formalisms are not taken for granted, and they themselves are analyzed cognitively, then examples like this one can be found throughout mathematics. The cognitive science of mathematics and mathematics education should work together in order to identify them, study them in detail, and remedy the problematic cases. But this, of course, requires a different look at the edifice of mathematics, one that doesn't see the very subject matter as pre-given.

Mathematics is perhaps the most abstract conceptual system we can think of, but even this it is ultimately embodied in the nature of our bodies, language, and cognition. Conceptual metaphor and fictive motion, being a manifestation of extremely fast, highly efficient, and effortless cognitive mechanisms that preserve inferences, play a fundamental role in bringing many mathematical concepts into being. We analyzed several cases involving dynamic language in mathematics, in domains in which, according to formal definitions and axioms in mathematics, no motion was supposed to exist at all. Via the study of gestures, we were able to see that the metaphors involved in the linguistic metaphorical expressions were not simply cases of "dead" expressions. Gestures studies provided convergent evidence supporting the psychological and cognitive reality of the embodiment of mathematical ideas, and their inferential organization. Building on gestures studies we were able to tell that the above mathematics professors, not only were using metaphorical linguistic expressions, but that they were in fact, in real time, thinking dynamically.

One of the main points of this chapter was to show that the cognitive science of mathematics should matter to mathematics education. The cognitive science of mathematics provides on the one hand, a different philosophy of mathematics for mathematics education—one that embodies the genuine human nature of mathematics and meaning. Such a view should be more compatible with the main goals of education as a human, social and intellectual endeavor, than the predominant disembodied, dogmatic, and platonic views of mathematics. And on the other hand, the cognitive science of mathematics can bring specific empirical findings that can inform mathematics education on how to conceive meaningful, productive and cognitive-friendly learning environments that are compatible with how the human mind actually works. The cognitive science of mathematics can inform mathematics education without leaving the "X"—mathematics—untouched.

ACKNOWLEDGMENT

I would like to thank Laurie Edwards for useful comments and suggestions.

REFERENCES

Aleksandrov, A., Kolmogorov, A. N., & Lavrent'ev, M. A. (1956/1999). *Matematika, ee soderzhanie metody i znachenie* [Mathematics, its contents, methods and meaning]. Mineola, NY: Dover.

Bates, E., & Dick, F. (2002). Language, gesture, and the developing brain. *Developmental Psychobiology, 40*, 293–310.

Boroditsky, L. (2000). Metaphoric structuring: understanding time through spatial metaphors. *Cognition, 75*, 1–28.

Cienki, A. (1998). Metaphoric gestures and some of their relations to verbal metaphoric expressions. In J.-P. Koenig (Ed.), *Discourse and cognition* (pp. 189–204). Stanford, CA: CSLI Publications.

Courant, R., & Robbins, H. (1978). *What is mathematics?* New York: Oxford.

Fauconnier, G., & Turner, M. (1998). Conceptual integration networks. *Cognitive Science, 22*(2): 133–187.

Fauconnier, G., & Turner, M. (2002). *The way we think.* New York: Basic Books.

Freudenthal, H. (1973). *Mathematics as an educational task.* Dordrecht, The Netherlands: Reidel.

Gentner, D. (2001). Spatial metaphors in temporal reasoning. In M. Gattis (Ed.), *Spatial schemas and abstract thought* (pp. 203–222). Cambridge, MA: MIT Press.

Gibbs, R. (in press). Just why should cognitive linguists care about empirical evidence, much less want to go to the trouble of gathering it? In M. González, I. Mittelberg, S. Coulson, and K. Spivey (Eds.), *Methods in cognitive linguistics.* Philadelphia: John Benjamins.

Goldin-Meadow, S., & Mylander, C. (1984). Gestural communication if deaf children: The effects and non-effects of parental input on early language development. *Monographs of the Society for Research in Child Development, 49*(3), no, 207.

Hersh, R. (1997). *What is mathematics, really?* New York: Oxford University Press.

Iverson, J., & Esther Thelen, E. (1999). In R. Nuñez & W. Freeman (Eds.), *Reclaiming cognition: The primacy of action, intention, and emotion.* Thorverton, UK: Imprint Academic.

Iverson, J., & Goldin-Meadow, S. (1998). Why people gesture when they speak. *Nature 396,* 228.

Kaput, J. (1979), Mathematics and Learning: Roots of epistemological status. In J. Lockhead & J. Clement (Eds.), *Cognitive process instruction* (pp. 289–303). Philadelphia: Franklin Institute Press.

Kendon, A. (1980). Gesticulation and Speech: Two aspects of the process of utterance. In M. R. Key (Ed.), *The relation between verbal and nonverbal communication,* 207–227. The Hague: Mouton.

Kendon, A. (2000). Language and gesture: unity or duality? In D. McNeill (Ed.), *Language and gesture* (pp. 47–63). Cambridge: Cambridge University Press.

Kramer, E. (1970). *The nature and growth of modern mathematics.* New York: Hawthorn Books.

Lakoff, G. (1993). The contemporary theory of metaphor. In A. Ortony (Ed.), *Metaphor and thought* (2nd ed.) (pp. 202–251). Cambridge: Cambridge University Press.

Lakoff, G., & Johnson, M. (1980). *Metaphors we live by*. Chicago: University of Chicago Press.

Lakoff, G., & Johnson, M. (1999). *Philosophy in the flesh*. New York: Basic Books.

Lakoff, G., & Núñez, R. (1997). The metaphorical structure of mathematics: Sketching out cognitive foundations for a mind-based mathematics. In L. English (Ed.), *Mathematical reasoning: Analogies, metaphors, and images*. Mahwah, NJ: Lawrence Erlbaum Associates.

Lakoff, G., & Núñez, R. (2000). *Where mathematics comes from: How the embodied mind brings mathematics into being*. New York: Basic Books.

McNeill, D. (1992). *Hand and mind: What gestures reveal about thought*. Chicago: Chicago University Press.

McNeill, D. (Ed.) (2000). *Language and gesture*. Cambridge: Cambridge University Press.

Mayberry, R., & Jaques, J. (2000). Gesture production during stuttered speech: insights into the nature of gesture-speech integration. In D. McNeill (Ed.), *Language and gesture*. Cambridge: Cambridge University Press.

Murphy, G. (1997). Reasons to doubt the present evidence for metaphoric representation. *Cognition, 62*, 99–108.

Núñez, R. (1993). *En déçà du transfini: Aspects psychocognitifs sous-jacents au concept d'infini en mathématiques*. Freiburg, Switzerland: University Press.

Núñez, R. (1999). Could the future taste purple? In R. Nuñez & W. Freeman (Eds.), *Reclaiming cognition: The primacy of action, intention, and emotion* (pp. 41–60). Thorverton, UK: Imprint Academic.

Núñez, R. (2000). Mathematical idea analysis: What embodied cognitive science can say about the human nature of mathematics. Opening plenary address in *Proceedings of the 24th International Conference for the Psychology of Mathematics Education*, 1:3–22. Hiroshima, Japan.

Núñez, R. (2003). Fictive and metaphorical motion in technically idealized domains. *Proceedings of the 8th International Cognitive Linguistics Conference*, Logroño, Spain, July 20–25, p. 215.

Núñez, R. (2005). Creating mathematical infinities: The beauty of transfinite cardinals. *Journal of Pragmatics, 37*, 1717–1741.

Núñez, R., Edwards, L., Matos, J. F. (1999). Embodied cognition as grounding for situatedness and context in mathematics education. *Educational Studies in Mathematics, 39*(1–3): 45–65.

Núñez, R. & Lakoff, G. (1998). What did Weierstrass really define? The cognitive structure of natural and ε-δ continuity. *Mathematical Cognition, 4*(2), 85–101.

Núñez, R., & Lakoff, G. (2005). The cognitive foundations of mathematics: The role of conceptual metaphor. In J. Campbell (Ed.), *Handbook of mathematical cognition* (pp. 109–124). New York: Psychology Press.

Núñez, R., & Motz, B. (2004). Metaphoric spatial construals of time: The psychological reality of the ego- and time-reference-point distinction. Manuscript in preparation, University of California, San Diego.

Núñez, R., & Sweetser, E. (2001). *Proceedings of the 7th International Cognitive Linguistics Conference*, Santa Barbara, USA, July 22–27, pp. 249–250.

Núñez, R., & Sweetser, E. (2006). Aymara, where the future is behind you: Convergent evidence from language and gesture in the cross-linguistic comparison of spatial construals of time. *Cognitive Science, 30*, 401–450.

Robert, A. (1982). L'acquisition de la notion de convergence de suites numériques dans l'enseignement supérieur. *Recherches en didactique des mathématiques, 3*, 307–341.

Simmons, G. F. (1985). *Calculus with analytic geometry*. New York: McGraw-Hill.

Stewart, I. (1995). *Concepts of modern mathematics*. New York: Dover.

Sweetser, E. (1990). *From etymology to pragmatics: Metaphorical and cultural aspects of semantic structure*. New York: Cambridge University Press.

Sweetser, E. (1998). Regular metaphoricity in gesture: bodily-based models of speech interaction. In *Actes du 16ᵉ Congrès International des Linguistes*. Elsevier.

Tall, D., & Vinner, S. (1981). Concept image and concept definition in mathematics with particular refernce to limits and continuity. *Educational Studies in Mathematics, 12*, 151–169.

Talmy, L. (1988). Force dynamics in language and cognition. *Cognitive Science, 12*: 49–100.

Talmy, L. (2003). *Toward a cognitive semantics: Volume 1. Concept structuring systems*. Cambridge: MIT Press.

PART II

What Changes Are Occurring in the Kind of Elementary-but-Powerful Mathematics Concepts That Provide New Foundations for the Future?

Richard Lesh
Indiana University

This book's Part I chapters focus on changes are occurring in: (a) the kind of "real life" situations where mathematical thinking is useful beyond school classrooms, and (b) the kind of mathematical thinking involved in problem solving or decision making situations. Yet these chapters largely accepted traditional views of possible topics that might be useful—even though they often emphasized quite non-traditional conceptions of what it means to understand a wide variety of mathematical concepts and abilities. Part II chapters push these possibilities even further considering new emerging mathematical topic areas which should be considered as possible replacements for "basics" in mathematics education.

If we consider *usefulness beyond school* as primary criteria for choosing content to be emphasized in elementary mathematics curriculum, then emerging new technologies are creating new graphic, dynamic, and interactive ways to think about old topics, such as arithmetic, statistics, geometry, algebra, and calculus. They also are contributing to development of completely new fields of mathematical inquiry. For example, chapters in Part II describe new mathematical topic areas which appear to deserve as much attention in mathematics curricula as topics associated with quaint notions about shopkeeper arithmetic- or calculus-driven curriculum topics designed to prepare students passing through subsequent mathematics courses with little usefulness beyond school for more than a small fraction of students enrolled in such courses. These new mathematics inquiry areas include discrete mathematics, complexity theory, systems theory, and game theory. They also include mathematical modeling drawing on more than a single textbook topic area and often integrating ideas and abilities from more than a single discipline.

Due to technology, modern mathematics is becoming multi-media mathematics. Computer-based computational models are opening whole new ways of thinking about problems involving optimization, stabilization, and other goals that used to require calculus or in other ways go beyond the scope of elementary mathematics. One distinguishing characteristic of a technology-based *information age* is: the same tools providing new ways to help people think about existing worlds of experience also enable these completely new worlds of experience to be designed and created. Thus, increasingly complex systems, ranging from communication systems, to economic systems, to transportation systems, to ecological systems, are among the most powerful "things" impacting both ordinary people and professionals who are heavy mathematics, science, and technology users. However, even before modern technologies began to radically transform the systems that surround us: *What was it that led educators to imagine that most of life's mathematics-relevant experiences can be described using only single, simple, one-way functions—which involve no feedback loops, which involve only a single actor, and which involve no trade-offs among competing factors such as low costs and high quality?* Surely, educators have noticed that everyday experiences of ordinary people involve two-way interactions as much as they involve one-way actions.

A main point in Part II chapters is that new technology-based tools for conceptualization, computation, and communication are inducing significant changes in both: (a) problem-solving and decision-making situations where mathematical thinking is useful, and (b) mathematical understandings and abilities that are useful in such situations. For example:

Systems that need to be understood or explained generally cannot be described with sufficient accuracy using only a single function or a single input-output rule. They often involve interactions among several levels and/or types of interacting agents; and they also involve feedback loops, trade-offs among partly conflicting factors, and goals that often involve optimization, stabilization, or other system-level constructs requiring thinking that goes considerably beyond simple computation.

Problem solvers' mathematical products often involve much more that short answers to pre-mathematized questions. For example, they often involve developing complex conceptual tools designed for some specific decision maker and for some specific decision-making purpose but which seldom are worthwhile to develop unless they go beyond being powerful for a specific purpose to being sharable with others and re-useable beyond the immediate situations in which they were first needed.

The "problem solver" often is not simply an isolated individual whose only tools consist of pencil and paper. Instead, the "problem solver" often is a team of diverse specialists representing different practical and theoretical perspectives, having access to rapidly evolving technical tools, and who may collaborate from remote sites.

The most useful mathematical/conceptual systems often involve non-linear models, discrete mathematical models, graphic media, iterative functions with feedback loops leading to second-order effects that overpower first-order effects, and/or complex, dynamic, and continually adapting systems with interacting agents and emergent properties of the systems-as-a-whole. However, even in situations involving mainly traditional basic understandings related to fractions, ratios, rates, and proportions, elementary-but-deep understandings typically required include many that have been ignored almost completely in paper-based textbooks and tests. In any case, relevant mathematical concepts and abilities seldom fall within a single textbook topic area. Instead, productive solutions to realistically complex problems usually must integrate understandings and abilities drawn from a variety of disciplines and topic areas. Furthermore, they often must be expressed using a variety of interacting representational media. Each emphasizing and de-emphasizing somewhat different underlying concepts and conceptual systems meanings.

Solution processes often involve a series of iterative development→ testing→ revising cycles in which a variety of different ways of thinking about givens, goals, and possible solution steps are iteratively expressed, tested, and revised or rejected. That is, because the solutions often involve developing complex artifacts (or conceptual tools), the development cycles often involve a great deal more than simply progressing from pre-mathematized givens to goals when the path is not obvious. Instead, the heart of the problem often consists of conceptualizing givens and goals in productive ways. Iterative development cycles often involve a variety of different ways of thinking about how givens, goals, and possible solution steps are iteratively expressed, tested, and revised or rejected.

Finally, when humans interpret experiences, they do not simply engage conceptual systems that are purely logical or mathematical in nature. Their interpretations also involve values, feelings, beliefs, dispositions, roles, and a variety of higher-order processes which, in traditional instruction, tend to be treated as "things" that are learned separately from mathematics content.

Situated cognition is a term that is becoming popular for describing socially developed learning-in-context being suggested above. That is, the knowledge and abilities that develop are organized around experience as much as they are organized around textbook-style abstractions. They are created for specific purposes in specific situations by specific communities of people. Taken together, however, the preceding observations suggest that situated knowledge is far more than simply traditional textbook concepts having been developed in specific contexts. It also involves new mathematical topics, multiple integrated topics, multiple interacting media, and a variety of higher-order processes, styles, and dispositions – as well as generalizable knowledge forms developed explicitly to be sharable and reusable beyond immediate circumstances.

Contrary to common sense opinion that "real mathematics" is done in isolation and without any tools except pencil and paper, job interviewers in future-oriented professions consistently emphasize a very different point of view. That is, especially in the case of jobs involving heavy mathematics, science, and technology users, people in highest demand during job interviews tend to be those who are able to: (a) develop useful mathematical explanations of complex systems not coming in premathematized forms, (b) work effectively within diverse specialist's teams, and (c) adapt to rapidly evolving technical tools.

Throughout Part II of this book, it becomes clear that, when groups of specialists use a variety of powerful technical tools to work on complex multi-stage projects, the need for mathematical thinking actually increases rather than decreases. But, in general, "thinking mathematically" is about expression (e.g., interpretation, description, explanation, communication, argumentation, and construction) at least as much as it is about computation or deduction. It is about imposing structure on experience at least as much as it is about deriving or extracting meaning from information presumed to be given in a mathematically meaningful form.

The preceding observations suggest that mathematics educators should try to determine which mathematical concepts and abilities are priorities for students to learn. They should not become so preoccupied with asking what kind of data processing students can do that they neglect to ask what kinds of situations the students can *describe*. In other words, they should entertain notions that mathematics learners and problem solvers are model developers at least as much as they are information processors. They should entertain the notion that models and underlying conceptual systems students develop might be among the most important cognitive objectives of mathematics instruction.

From the perspective of authors throughout this book, two of the most important goals of mathematics instruction should be to help students: (a) develop powerful models, drawing on concepts and abilities associated with a variety of textbook topic areas and which can be used to construct, describe, explain, predict, and intelligently manipulate complex systems; and (b) become proficient model developers, and wise adopters and adaptors of existing models. That is, integrated multi-topic models themselves are among the most important knowledge pieces emphasized in the mathematic curriculum, and modeling abilities are among the most important proficiencies students need to develop.

Of course, discussions about student-developed knowledge are not likely to seem significant to anybody who naively accepts the cliché: *It took brilliant mathematicians hundreds of years to develop most of the most powerful concepts in the elementary mathematics curriculum. It's not realistic to expect average ability children or adolescents to come up with such concepts in a few months or weeks—or during single problem solving episodes.* But, authors throughout this book provide ample evidence students clearly recognize the need for a given type of mathematical description. When "ground rules" are clear students can go through several development cycles during the process of creating final ways of thinking, average ability students often are remarkably able to develop impressively sophisticated and deeply mathematical ways of thinking. We reject the notion that only a few exceptionally brilliant students are capable of developing significant mathematical concepts unless step-by-step guidance is provided by a teacher. Our research is filled with transcripts showing examples of model-eliciting activities in which the models and conceptual tools students develop for making sense of specific problem solving situations also result in significant developments of underlying constructs or conceptual systems (Lesh & Doerr 2003a). In fact, if the goal of instruction is to make significant changes in a student's underlying ways of thinking about important mathematical systems, then virtually the only way to induce significant conceptual change is to engage students in situations where they express→ test→ revise or reject their current ways of thinking. On the other hand, it is not surprising that traditional mathematics tests show when students never go beyond thinking that is based on first drafts; their thinking often appears very superficial, primitive, non-mathematical and illogical.

In order for powerful constructs to emerge, development processes need to be similar to the way people develop useful descriptions of nearly anything. They need to go through multiple drafts. They need to assess for themselves the quality of current drafts. Principles for designing such activities include the following: (1) Students must engage in problem solving activities where they clearly recognize the need to revise or refine their current ways of thinking about the situation. (2) Students must be challenged to express their current understandings in forms they themselves can test and revise multiple times. (3) The conceptual tools students develop should be expected to be sharable with others and re-useable beyond the immediate situation and beyond specific situations for which they were developed (Lesh et al., 2001).

When the preceding conditions are satisfied, our claim is: *If models for making sense of situations involve mathematically significant constructs or conceptual systems, then model development tends to involve significant forms of concept development. Furthermore, such development often is achievable by students who have been labeled average or below-average in ability as measured on traditional school tests and tasks.*

CHAPTER 7

Models, Simulations, and Exploratory Environments: A Tentative Taxonomy

Judah L. Schwartz
Massachusetts Institute of Technology

In the fall semester of the 2001–02 academic year I gave a course entitled "Models, Simulations & Exploratory Environments" to the students in the Technology in Education program at the Harvard Graduate School of Education. It was the first time I had offered the course. Although I was convinced that the subject was an important one, and that I had a reasonable idea of the range of issues that I thought should be covered, I did not have what I considered to be a satisfactory framework for thinking about the subject.... The course description read;

> In many ways the greatest promise of computers is the amplification of human intellectual power by allowing us to explore complexity and to manipulate easily the abstract constructs we formulate. To do this, we often build models and simulations in order to instantiate our theories of real phenomena. These models and simulations are of necessity simplifications and idealizations that attempt to capture the essential features of those aspects of the physical and social world they describe. The widespread availability of inexpensive computation has led to an extraordinary growth of the use of models and simulations for analysis in the natural and social sciences. This course will focus on the exploration of a variety of widely available existing models and simulations. We will attend, in particular, to the ways in which the theories of the model-builders are expressed in the models they build and to the inherent limitations on the fidelity of models.

The course that I offered was a kind of "smorgasbord" that followed the course description touching on topics that, I was certain, were important. I spent a good deal of the time, both during the course and after it was over, reflecting on possible larger frameworks that would allow me to relate the various and seemingly disparate elements of my "smorgasbord" to one another. This chapter constitutes my first attempt at setting down such a framework.

SOME USEFUL CARICATURES OF EXTREME VIEWS ABOUT MODELS AND THEIR ROLE IN EDUCATION

In thinking about a complex problem it often helps to define dimensions whose polar extremes are clear caricatures. One of the virtues of doing so is that one is forced to confront the nuance of complexity by thinking carefully about where on a spectrum between the extremes any given case might fit.... Here are three pairs of polar extremes that have helped me think about models and their role in education.

Models vs. Simulations
Function (process) vs. Structure (constraint) models
Learning by using models vs. Learning by making models

Let us explore each of these in turn.

Models vs. Simulations

The following questions come immediately to mind. What do we mean by models? What do mean by simulations? Are there meaningful distinctions betweens these two terms? If so, how do they resemble and differ from one another? Finally, since both models and simulations, whatever we might mean by those terms, certainly allow us to pose "what if" questions, how do models and simulations resemble and differ from exploratory environments that also support the same sort of inquiry?

Both models and simulations are artificial environments that refer to some external "reality". Both contain a collection of entities. The attributes of (some of) these entities and (some of) the relationships among these entities are incorporated into the model or simulation. Certainly to this extent models and simulations are similar.

In my view an important difference arises when one considers the purpose of fashioning the artificial environment, i.e., the model or simulation. Recognizing the caricature-like nature of the distinction I submit that the purpose of making a model and exploring it is to expose the underlying mechanisms that govern the relationships among the entities. In contrast, the purpose of making a simulation and running it is to provide users with a surrogate experience of the external "reality" that the simulation represents. Thus models seek to "explain" complex referent systems while simulations seek to "describe" referent situations, often by offering rich multisensory stimuli that place users in as rich a simulacrum as possible.

Given this difference in purpose it is not surprising the makers of models seek to limit the complexity of their models so as to make the underlying causal and/or structural mechanisms more salient. In contrast, designers of simulations tend to incorporate as much of the richness and complexity of the referent as possible so as to make the experience of using the simulation as rich a perceptual experience as possible.

In the educational context this difference in purpose points to quite different roles for models and simulations. Consider a natural or social science curriculum. In the elementary grades the emphasis is on having the students exposed to a rich repertoire of phenomena. For these students well-designed simulations can complement direct observation of nature, compensating for spatial and temporal remoteness or mismatch of scale to the human perceptual apparatus.[1] Secondary students, on the other hand, need to learn about correlations and underlying causes. These are best explored in models that make these correlations and causes salient without the overlay of complexity that can so readily disguise them.

---

[1] Simulations of a rain forest allow urban students to overcome one kind of spatial remoteness. Simulations of an urban transport system allow rural children to overcome another kind. Simulations of the night sky that show the trajectories of the planets over several years during the course of several minutes help to overcome a mismatch of temporal scale etc.

Models	Simulations
Refer to a collection of entities, their attributes and relationships among them	
Purpose is to abstract experience	Purpose is to reproduce experience
Models seek to explain	Simulations seek to describe
Complexity of referent deliberately not reflected in model	Complexity of referent deliberately included in simulation
Rules relating entities and their behaviors are explicit	Rules relating entities and their behaviors may be implicit Or may be completely artifactual as in the case of animations.

FIGURE 7–1.

This difference in purpose between models and simulations leads to another interesting difference between them. Models tend to make their underlying generative rules accessible and visible to the user. In contrast, simulations most often hide their generative rules from the user. Indeed in many instances the generative rules of simulations do not at all reflect what is known about the generative rules of the referent situation.[2] These distinctions are summarized in Figure 7–1.

In contrast to models and simulations, exploratory environments refer to *model-building* environments. They may have no particular referent domain as in the case of programming languages—we will describe these environments as *domain general*. Examples include: Modellus, Worldmaker, Link-It, Agentsheets, StarLogo, Stella, etc. On the other hand, they may have restricted domains such as "supposers" of all sorts—we will describe these environments as *domain specific*. Examples include: Geometric Supposer, Geometer's Sketchpad, Cabri Geometre, Genscope, Interactive Physics, Newtonian Sandbox, Tarski's World, etc.

Domain-general exploratory environments allow users to generate models to explore in a wide range of domains limited mainly by the imagination of the user and the input-output affordances of the computing environment. Thus, in Stella, for example, people have written economic models, ecological models, psychological models, sociological models, models of chemical systems, models of physical systems, etc.

Domain-specific exploratory environments allow users to fashion microcosms that they may then explore. For example, the Geometric Supposer allows users to make Euclidean geometric constructions and explore the effects of modifying measures and constraints. Its use in schools has led to many dozens of new theorems discovered by students.[3]

---

[2]Consider, for example, an animation of a running horse or a melting ice cube. Most animations are made by "tweening" i.e., interpolating frames between drawn (or photographed) initial frames. The same algorithm that is used for animating the running dog could be used to animate the melting ice cube. The algorithm that the software uses for interpolating frames has nothing to do with the musculature of the dog or the properties of ice and water.

[3]See, for example, Schwartz, Judah L., "Intellectual Mirrors: A Step in the Direction of Making Schools Knowledge-Making Places", in Harvard Educational Review, Vol. 59, No. 1, February 1989 and J. L.Schwartz, B. Wilson & M. Yerushalmy, (eds.). The Geometric Supposer: What is it A Case Of?, Hillsdale, New Jersey, Lawrence Erlbaum Associates. 1993

Function (Process) vs. Structure (Constraint) Models

The second set of polar extremes that I find useful in thinking about models and their use in education is the contrast between models based on structure and models based on function.

Function-based Models

Much of the excitement in the world of education about the use of using computer-based models and simulations derives from an enthusiasm for having students follow the evolution in time of the system being modeled or simulated. I refer to such models and simulations as function- or process-based.

Independent of the degree of complexity and/or verisimilitude embedded in the model or simulation, function-based models and simulations have the property that the passage of time for the user engaged in the model or simulation is a scaled version of the passage of time in the referent system.[4]

Before proceeding to a discussion of the several type of function-based models a word is in order about the underlying mathematics and its algorithmic articulation. Function-based models are typically expressions of sets of differential equations with time as the independent variable or difference equations with sequence number as the independent variable. Such models have the property that they relate the state of the system at one instant of time to the state of the system at the "next" instant of time[5]. In order to be able to compute the evolution of such a model in time one must specify an initial state of the system so as to be able to begin the computation.

Spatially Distributed Function-based Models

In some ways spatially distributed function-based models are easier to understand conceptually despite the fact that the underlying mathematics is more complex than that of other kinds of function-based models. In a spatially distributed function-based model, one specifies how the values of the variables of interest at a given point in space and a given instant of time depend on the values of the variables at that point and at "neighboring" points in space in the immediate past[6].

Spatially distributed function-based models can also be written in two dimensions. They are particularly useful for the study of predator-prey phenomena, epidemiological phenomena, housing patterns, etc. Consider for example an array of cells each of which has either grass or bare soil. In some of the cells there are rabbits. A very simple set of rules governs the evolution of this system. If a rabbit finds itself on a cell with grass it eats grass. If it finds itself on a cell with a neighboring cell unoccupied by

---

[4]One allows here for the possibility of 1:1 scaling of time, i.e., that model time proceeds at the same rate as referent system time.

[5]Because time, as a continuous variable, is replaced by a discrete variable the ordinarily meaningless locution "the next instant of time" has meaning.

[6]Typically this means that the model is an expression of a coupled set of partial differential equations that are first order in time. Spatially distributed models can be formulated in different numbers of spatial dimensions. In one dimension, for example, one might write a model of the conduction of heat along a metal bar one end of which is maintained at a fixed temperature. The output of the model is the temperature of the bar as a function of position along the bar and the way in which the temperature at every position along the bar evolves in time. Another interesting application of one dimensional spatially distributed function-based models is in the study of traffic flow along a highway. A spreadsheet is often the most convenient tool for the writing of such 1D function-based models.

a rabbit, then there is some probability that it will move to the next cell. If there is a neighboring cell that is unoccupied by a rabbit there is some probability that the rabbit will reproduce and cause the neighboring cell to be occupied by a rabbit. There is some probability that a rabbit on a cell containing soil will die. Cells containing soil that are adjacent to cells containing grass grow grass. This very simple set of rules suffices to produce a model that allows one to explore the relationship between the availability of food and density of animals that consume that food. This is an example of a class of modeling environments called "cellular automata" that lend themselves particularly well to the writing of such models.[7]

Finally, one can write spatially distributed function-based models in three spatial dimensions. Models of the atmosphere used for weather prediction are typically of this form. The atmosphere is divided into cells. The values of the temperature, wind velocity and air pressure in a given cell at a given instant of time depend on the values of these quantities in that cell and neighboring cells in the immediate past. Models such as these tend to be enormously demanding computationally and their utility on the educational scene is somewhat dubious.

Lumped Variable Function-Based Models

A second kind of function-based model ignores the spatial distribution of the variables of interest.[8] The underlying mathematics is that of coupled first order differential equations converted into the appropriately corresponding difference equations in order to allow for numerical computation.

A particularly interesting environment for the writing of such models is called STELLA. It permits users to state relationships among variables and their rates of change graphically. Once the model is thus partially defined in terms of the qualitative relationships among the variables, the user is prompted to supply those quantitative values that are needed by the model in order to compute the time evolution of the system. See Figure 7–2.

The qualitative relationship in the model was defined by establishing a variable called *number of nuclei* (in the diagram), and its rate of change whose magnitude is controlled by a quantity called *decay rate*. The size of *decay rate* is related to the value of the variable *number of nuclei*. Then once an initial value for *number of nuclei* is specified the model may be evolved in time.

It should be pointed out that all phenomena that decay exponentially can be reasonably modeled in this fashion. Moreover, all phenomena that grow exponentially can also be model in this fashion by changing the sign of the value of *decay rate* (We advise changing the name of the variable in this case to *growth rate*).

Mathematically this is a particularly simple example and one that can be solved analytically by any first-year calculus student. However, the modeling environment is far more permissive in terms of the functional forms it allows people to explore. The constraint of analytic solubility that limited exploration of complex systems to simple soluble cases is no longer a serious obstacle to our building and exploring models.

---

[7]I include under this general heading environments that make use of "sprites" such as StarLogo. These environments may be thought of as generalizations of cellular automata in which cells do not have fixed location and are not obliged to interact with near neighbors only.

[8]Indeed the variables of interest may not have spatial extent at all. In modeling immigration policy, it is ...important to define variables like immigration... rate and availability of jobs without concern for how these quantities vary from place to place. Of course, a more detailed model may try to take this kind of variation into account.

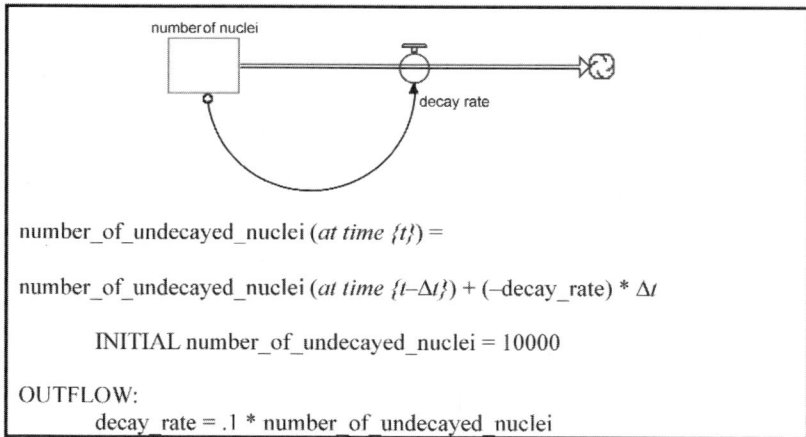

FIGURE 7–2.

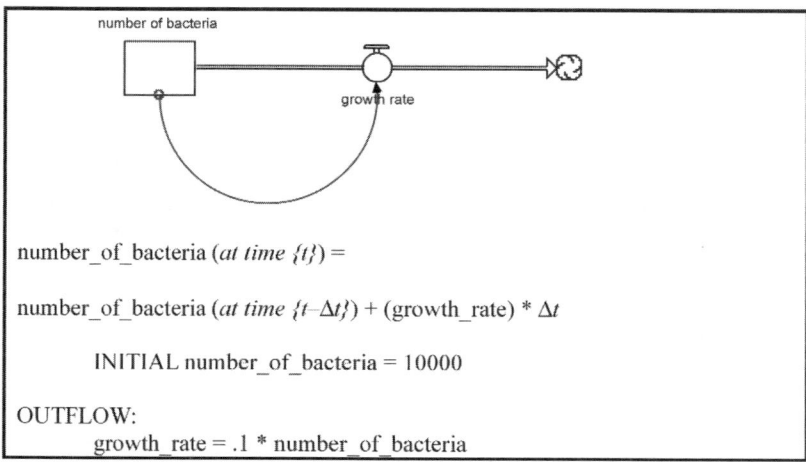

FIGURE 7–3.

Structure-Based Models

Not all useful models involve systems that evolve in time.

Spatial Structure-Based Models

In what sense are these objects models? They are models in the sense that each represents the *form* of a referent object on a grossly different spatial scale. Moreover, their various parts represent spatial interrelationships among the various parts of the referent objects. For example, the model airplane may have a wingspan that is 1.2 times as long as the fuselage of the model airplane, reflecting the corresponding relative sizes of parts in the airplane being modeled.

Similarly, the model of the benzene molecule constructed, say, of colored balls connected by short sticks, is essentially planar (flat) and hexagonal. The real benzene molecule, made up of carbon and hydrogen atoms is also planar and hexagonal.

It is interesting to note that there is a sense in which each of these models is made for essentially the same reason, i.e., to bring into comfortable human perceptual range an object that is much too large or much too small to otherwise study carefully.

That having been said, one is led to ask—just what is it about the referent object (airplane or molecule) that can be studied with the model? It seems to me that models of this sort are primarily useful for studying the spatial properties, both topological and geometric, of their referent objects.

If one is interested in the flow of air around the wings and fuselage of an airplane, one can study the problem using a model only by exercising extreme caution. While one can readily make a 1/100 scale model plane there is no way to scale the size of the air molecules flowing past the airfoil in the wind tunnel. Similarly, the stresses on the fuselage of a model are not related in any simple way to the stresses on the fuselage of a real airplane.

Using models of the very small also presents problems. Carbon and hydrogen atoms are not blue and white spheres and there is no sense in which they can be said to have sharp boundaries. In fact, there is no sense in which they can be said to have color at all!

The primary use of such models based on structure is the study of spatial interrelationships in the referent objects. As the user contemplates a models airplane or model benzene molecule, the passage of time bears no relationship to the system being studied.

Non-Spatial Structure-Based Models

There is a class of models that allow users to explore explicit relationships among a set of related variables. Consider, for example, the phenomenon of the height of a thrown projectile, a subject studied in all introductory physics classes. If we denote the height by y, the initial height by y_0, the initial vertical velocity by v_0, the acceleration due to gravity by g, and the elapsed time since the projectile was thrown by t, one can write a relationship that relates all of these quantities to one another. That relationship is

$$y - y_0 - v_0 t + \tfrac{1}{2} g t^2 = 0.$$

Any one of the quantities y, y_0, v_0, g in this relationship can be written as an explicit function of the time t. In like manner, any of the quantities y, y_0, v_0, g may be treated as the independent variable instead of the time t.

If one holds y_0, v_0, and g constant and systematically varies t while observing the variation of y, one can observe the behavior in scaled time of a real system. Suppose, however, that one holds y, y_0, and g constant and varies t while watching the changing values of v_0. Even though we are systematically varying time in this case, it is not scaled time in some real referent situation. It is rather a depiction of how one parameter of a complex situation depends on another.

To drive this point home consider the relationship among the following three quantities: the pressure of the gas in a container P, the volume of the container V, and the temperature of the gas T. This relationship, studied extensively in all chemistry and many physics classes, is known as the Perfect Gas Law. It can be written as $PV = \alpha T$ where the constant α is related to the amount of gas in the container.

Clearly one can hold V constant and systematically vary T watching the behavior of P as one does so. Similarly, one can hold T constant and systematically vary V watching the behavior of P as one does so. In neither case does elapsed time for the user of the model have anything to do with the temporal behavior of the system. In fact, the model describes a system in equilibrium, and as such, contains no explicit reference to time.

It should be pointed out that a very large fraction of the spreadsheet models used by business analysts around the world fall into this category of non-spatial structure models.

Learning by Using Models vs. Learning by Making Models

The third set of polar extremes is learning by using models vs. learning by making models.

Wilensky[9] argues strongly, in the tradition of Papert[10] and his followers, in favor of students being provided with powerful modeling building environments with an eye toward their building their own models of interesting referent situations. This stance is set in opposition to having students use previously built models as arenas for exploration. His arguments, which are characteristic of this perspective on the use of computers in education, are summarized in Figure 7–4[11].

The contrast drawn between the two perspectives would lead one to believe that they are in profound opposition to one another. In practice, the dichotomy turns out to be a *false* one. Students most often encounter modeling in education through the use and exploration of models written by teachers or other authors of

learning using models	learning making models
passive	active
viewing a "received mathematics and science	constructing mathematics and science
transmission of ideas	expression of ideas
dynamic medium for viewing output of mathematical thought	dynamic medium used as executor of mathematical thought
an expert's question	the learner's own question
an expert's solution	the learner's own tentative solution
learning in a single step	learning through debugging
experts must anticipate relevant parameters for learning	learners can construct parameters relevant to their learning

FIGURE 7–4.

---

[9]See Wilensky, U., GasLab: An extensible modeling toolkit for connecting micro- and macro-properties of gases, in *Modeling and Simulation in science and Mathematics Education*, W. Feurzeig & N.Roberts, eds. New York, Springer-Verlag, 1999, pp.151–178.

[10]See for example Papert, S., *Mindstorms: Children, computers and powerful ideas*, New York, Basic Books, 1980.

[11]See note 8.

curriculum. At first their use of models tends to be exploration by the modifying of the parameters of the models. Next they are often asked to modify the models themselves, thus providing them with a range of related models to work with. Finally, students make be asked to devise models of phenomena *ab initio*[12].

It seems to me that the good sense of teachers is well expressed in this sequence. Students who have little notion of the purpose of a model and the ways in which they enable people to explore their own understanding will probably not benefit greatly from the writing of their own models, if indeed they can learn to do so at all. In my view, pedagogic artistry lies in helping students move through this sequence in ways that are appropriate to the depth and breadth of their understanding of the subject as well as their mastery of the mechanics of the modeling environments available to them. For example, it is perfectly reasonable to ask secondary students to write their own models of motion using simple force laws in simple constraint geometries. Such models as they write might then be explored and provide insights and deepened understanding. On the other hand, if such students are asked to write their own models of incompletely understood complex systems, they are likely to become bogged down in the complexity and learn precious little about either the system or the process of modeling.

CONCLUSION

The central critical ability that a well-educated scientist/engineer can bring to the table is the ability to confront an unorganized, and possibly even inchoate, body of data and formulate an analyzable model of the structure and/or mechanism that underlies the confusing presenting face of nature. This is not to say that we no longer need the skills of the well-trained (as opposed to well-educated) scientist/engineer who can manipulate the formalisms and explore the consequences of well-understood models. The availability of computer-based modeling environments makes it possible to redress the fact that until relatively recently much too little emphasis in school settings is placed on the formulation of analyzable models. For example, central to the ability to formulate models is the ability to decide on an appropriate scale for the elements of the model. It is difficult to recognize macroscopic emergent behavior from atomic scale models. This is a quiet plea for rethinking the rhetoric of micro-macro and replacing it with well thought through approaches that are micro-meso-macro (perhaps even with several levels of "meso").

As an example of such a modeling activity, let me offer an example from the work of the Balanced Assessment in Mathematics project. The project devised a set of tasks called "-ness" tasks (for reasons that will presently become clear). The purpose of these tasks is to see how well students can model a relationship that they are aware of perceptually but probably have never attempted to describe in any formal, not to mention quantitative, fashion. For the sake of specificity, Figure 7–5 is an example:

These tasks require students to identify and describe formally a geometric property of some two or three dimensional shape. It is important to stress that properties such as "squareness" are not formal geometric properties. There are no academically correct or universally accepted answers to these questions. On the other hand, there are sensible (and non-sensible) answers.

---

[12]It is interesting to note that in his paper describing GasLab (note 8) Wilensky describes his students as proceeding through this same sequence from learning using models to learning making models.

Elements of performance on this task include:

a. Choosing the most and least square figure
b. A verbal description of the geometric property being modeled (in this case what one means by "squareness"
c. Identifying the geometric elements that combine to form the measure of "squareness"
d. Forming an algebraic relationship among these elements
e. Computing values of the measure for various figures
f. Discussing the advantages and disadvantages of the measure under varying circumstances

This task can clearly be extended to include defining measure of "squareness" for all parallelograms, or for all quadrilaterals, or for that matter for all closed convex curves in the plane.

Other such tasks involve defining measures of "sharp-ness" of bends in roads (not as obvious as one might think), "crowded-ness" of collections of dots, "smooth-ness"

"Square-ness"

Below is a collection of rectangles.

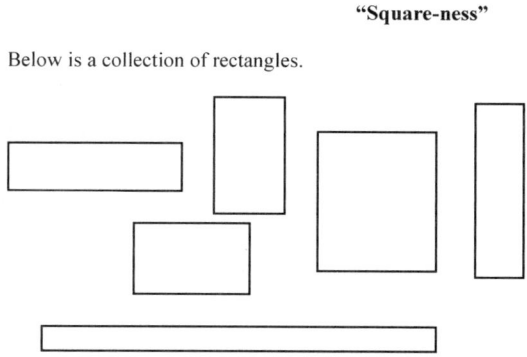

1. Which of the rectangles is the "squarest"?
2. Arrange the rectangles in order of "square-ness" from most to least square.
3. Devise a measure of "square-ness," expressed algebraically, that allows you to order any collection of rectangles in order of "squareness."
4. Devise a second measure of "square-ness" and discuss the advantages and disadvantages of each of your measures.

FIGURE 7–5.

of "spherical" objects (e.g., golf ball, orange, Earth, etc.), "disc-ness" of cylinders (e.g., tuna fish can, penny, stick of uncooked spaghetti, etc.), "good-ness" of fit of a functional form to a collection of data points, etc.

The essential point in such tasks is this—by asking people to formally model the elements and relationships of a perceptually familiar image, the task focuses on the act of modeling itself, largely unconfounded by the need for specialized domain knowledge.

REFERENCES

Mellar, H., Bliss, J., Boohan, R., Ogborn, J., & Tompsett, C. (Eds.) (1994) *Learning with Artificial Worlds: Computer Based Modelling in the Curriculum*, London: The Falmer Press.
Papert, S. (1980). *Mindstorms: Children, Computers and Powerful Ideas.* New York: Basic Books.

Schwartz, J., Yerushalmy, M., & Wilson, B. (Eds.). (1993). *The geometric supposer; what is it a case of?* Hillsdale, NJ: Lawrence Erlbaum Associates.

Wilson, & Yerushalmy, M. (Eds.). (1993). *The geometric supposer: what is it a case of?* Hillsdale, NJ: Lawrence Erlbaum Associates.

http://phoenix.sce.fct.unl.pt/modellus/
http://worldmaker.cite.hku.hk/worldmaker/pages/
http://www.agentsheets.com/
http://el.www.media.mit.edu/projects/macstarlogo/index.html
http://www.hps-inc.com/Education/new_Stella.htm
http://www.cet.ac.il/math-international/software1.htm
http://www.keypress.com/index.html
http://www-cabri.imag.fr/index-e.html
http://genscope.concord.org/
http://www.interactivephysics.com/
http://webassign.net/pasnew/newtonian_sandbox/newtsand.html
http://www-csli.stanford.edu/hp/Logic-software.html

CHAPTER 8

Technology Becoming Infrastructural in Mathematics Education

Jim Kaput and Stephen Hegedus
University of Massachusetts, Dartmouth

Richard Lesh
Indiana University

THE GOALS AND CHALLENGES OF ACCOUNTING FOR TECHNOLOGY'S IMPACT

The Goals of the Paper and Related Complications

The goal of this chapter is to provide a framework that helps us understand the gradual, manifold evolution of the roles of technology[1] in mathematics education. The underlying idea is that the changes over the longer term amount to a process in which technology is gradually becoming "infrastructural." Our first task is to describe what we mean by "infrastructural," and second, to use this characterization to analyze where we are in the long process of building more productive symbiotic relationships of mathematics and mathematics education with technology.

The situation is both complex and subtle if one takes the point of view, as we do, that *technologies and tools co-constitute both the material upon which they operate and the conditions, particularly social conditions, within which such operations occur.* This means that we do *not* take an essentialist view that mathematics somehow exists independently of tools and technologies, and that technologies and tools infiltrate and transform the mathematics over time. Instead, that technology is an equal player in the development of mathematics education, and can provide a substantial infrastructure for the development of critical, 21st century mathematical problem solving, for all. Our focus is on the affordances of ubiquitous forms of technology in schools and its impact and co-evolution on new forms of teaching, improved learning, new curriculum and more effective problem solving skills.

---

[1] For brevity, unless otherwise specified, we use the term "technology" to refer to electronic, mainly digital, information technologies, both hardware and software, in their many forms—including computers of various sizes and configurations, visual display and communication systems, memory storage systems, and so on (the commonplace use of "technology" in mathematics-education). Of course, it is essential, and indeed this is one focus of this paper, to recognize the greater generality of the term across space and time.

The rate of change of technological tools has become very high in the digital age (from our perspective in mathematics education mid 1970s onwards), and the inherited corpus of shared mathematical knowledge that was produced in interaction with pre-digital technologies is large and stable, particularly that part of it that intersects with schools and schooling—the mathematics of mathematics education. The same can be said of the *practices* of mathematics education, which are embedded in, and part of a very stable system of, social practices and institutions. Hence, despite the fact that we do not take an essentialist view, we find ourselves in the position of analyzing the impacts of new technologies on older mathematics and educational practices and institutions. Consider the impact of dynamic geometry, an innovation in the early 80s, but still impacting curriculum writers, students and teachers 15+ years on, about how such environments can be used to allow students to construct and discover their own mathematics. But in addition, our account must include provision for the rapidly evolving new mathematics that is generated by affordances associated with new digital tools and technologies, and it must deal with the fact that this mathematics influences and helps make possible yet new tools and technologies, compounding the interaction. In total, we will describe a new educational environment, with core examples, which provides a futuristic view of 21st century technology use in classrooms that reconstitutes a new form of mathematical problem solving in modern day classrooms. Thus, we will discuss the impact of such technology use on teaching and learning, new forms of curriculum, and resulting new forms of practice.

Finally, while one can analyze relevant interactions from a collective, historical perspective, in mathematics education we must attend to the matter of how mathematics comes to be known through acts of teaching and learning, within schools, classrooms and individuals, in the short term. Therefore, a full account must deal with multiple time-scales and social scales.

CHALLENGES IN ACCOUNTING FOR TECHNOLOGY'S INTERACTIONS WITH MATHEMATICS EDUCATION

The Epistemic Complexity of Mathematics

One challenge is the complexity of mathematics itself. First, there are many kinds of mathematics, in terms of mathematical domains, whose relationships with digital technologies vary—analysis, algebra, geometry, number theory, the mathematics of data, applied vs. "pure" mathematics, and so on. Second, there are important epistemic dimensions within mathematics. One dimension concerns the kinds of objects and relations that are regarded as mathematical within the various mathematical domains; and, a second dimension concerns the kinds of representations and languages by which the former are represented and acted upon—many of which are shared across domains. A third dimension concerns the forms and modes of justification and truth. Each of these dimensions interacts with technology in different ways.

Two Different Kinds of (Software) Technology Affordances: Representation and Communication

In mathematics education, we have long examined the *representational* affordances of technology: syntactical manipulation of representations, where computationally intense duties (e.g., root-finding, algorithm, iterative procedures) are offloaded to the technology (e.g., Mathematica, Maple), adding new actions to traditional notation systems (e.g., actions on coordinate graphs, dynamic geometry figures), supporting new interactive notation systems which include programming languages and spreadsheets, linking notation systems (e.g., tables, graphs, and algebraic functions), new or old, and so on. Within these affordances, some representational uses are more directly computational (such as symbolic manipulation or iterative computation) than expressively representational or visual. But, we will group these together, and refer to them simply as representational affordances. We also include human-computer-interaction interfaces within this category, most especially the use of "direct manipulation" systems such as those that occur in many new software simulations.

On the other hand, computational power is increasingly used in support of human *communication* in the world at large, both in a hidden infrastructural way (as when computers control the flow of information inside networks, telephones, or less interactive technologies, such as television) or in the visible interactions that occur when we send messages (e.g., AIM) or computationally defined objects to one another (e.g., email, software documents/objects). The Internet and its grand application, the World Wide Web, are conspicuous examples that innovative mathematics education programs increasingly exploit in a variety of ways, including ways that often are not especially domain specific to mathematics (e.g., Flash movies, PDF, On-line services). For example, such shifts often emphasize the social constitution of mathematical ideas and knowledge dissemination rather than communication. How often does a student prefer to Google™ a definition, or an answer to a question, utilizing the extreme power of connectivity and linked services, databases, etc. vs. problem solving within a social communicative structure. Later in our chapter, and in the chapter by Stroup and his colleagues in this book, we will offer core examples and discuss within-classroom communication affordances that are distinctly specific to mathematics education, and can significantly impact *mathematical* problem solving.

A hybrid computational affordance involves the importing of physical data into the computational medium ("MBL"—Microcomputer Based Laboratory), as when motion sensors are linked to computational devices so that position, velocity or acceleration data are represented on the device. Indeed, this process can be reversed, as with "LBM"—Line Becomes Motion as exploited by Nemirovsky and colleagues—where graphs serve as control devices that regulate the motions of external objects. For simplicity's sake, we treat these as communication affordances.

Next, we should acknowledge a third major use of technology databases, that is, the storage and manipulation of data. For simplicity's sake, since this use is usually embedded within the other two uses in mathematics education, we will not define a third explicit category, and will treat it simply as an implicit use of the previously mentioned two.

Finally, even though our description of technology is very different from one that is based on *use-patterns* which classically are captured in the "tool, tutor, tutee" analysis, we should acknowledge this as a fourth category which emphasizes the use of technology in systems where interactivity takes the form of *pedagogically organized assistance*—perhaps in the context of programming or design environments.

Over time, all of the preceding uses of technology have tended to be combined in various ways and seldom occur in pure forms, a trend likely to accelerate with the increasing availability of connectivity, which makes resources, including tools and pedagogical resources, available in new ways.

Three Types of Technology Hardware:
Computational Technologies, Display Technologies, and Network Technologies

It is useful to distinguish three distinct types of technology that are available to mathematics educators: computational devices, display technologies, and network technologies.

- Computational technologies with which people have direct interaction include calculators of various capacities ranging from 4-function numerical calculators with or without memory capacities to scientific calculators, graphing calculators, tablet and PDA-like devices, laptop computers, desktop computers, and sophisticated workstations.
- Display technologies range from the built-in screens of the computational devices to monitors and whole-class display systems.
- Network technologies include both wired and wireless, and each varies in scales, from a few devices within a room or site, to the Internet, and their topologies vary to include server-client relations to point-to-point communication, and virtually any mix of these.

Within each of the preceding types of devices, we have seen relatively clear trends. Except for the simplest calculators, with the processing resources continuing to grow while simultaneously becoming more compact according to Moore's Law, software affordances have trended downward from larger to smaller devices as smaller devices become capable of hosting independently produced software (e.g., graphing calculators with flash RAM capability thereby becoming flexible computational devices). As a result, similar software is increasingly appearing across multiple device types.

Display devices with high resolution are becoming cheaper and more ubiquitous, particularly as visibly shared displays in classrooms. Display devices are increasingly used as input devices via touch-screen on PDA-type and tablet systems, and sometimes combined with large displays as in Smartboards.

Networks have increasingly become more ubiquitous, so that computational devices are seldom used in isolation, and wireless connectivity is moving into schools and classrooms, replacing wired networks (see the last section below). Moreover, not only network systems, but also technology in general, is becoming ever "easier" to use—with increasing deployment of self-configuring systems, user-configurable devices and convergence of interface styles. However, "easier" is a relative term and has much to do with our notions of "infrastructure" as it will be discussed in following sections of this chapter.

Combinations of these trends across device types have made possible ever increasing availability and manipulability of digital objects of all kinds—but perhaps most notably in the context of video and audio objects. To date, these "heavier" objects, in terms of bit-weight, have not been centrally utilized in mathematics education.

Managing the Complexity of Mathematics
Education—Some Arbitrary Category Choices

Inevitably, we must deal with the complexity of the field of mathematics education. So, in this chapter, our strategy for managing this complexity is to describe the field in terms of "dimensions." In a sense, defining such dimensions cannot avoid being arbitrary and *ad hoc*, and must include decisions about how fine grained a decomposition to use. Further, these dimensions are in no way independent or "orthogonal." Indeed, interconnectedness is a basic feature of the field as a human endeavor.

With the preceding caveats in mind, it is useful to sort out the three major activities of teaching, learning, and assessment. Then, there also are activities associated with the organization of mathematics content to be taught and learned—the curriculum—as well as activities associated with teacher education and the political-social administration of mathematics education in schools. Furthermore, there is the economic side of curriculum material and its distribution; and, there is the all-important epistemological dimension, the dimension within which we situate mathematics content itself, which can be further decomposed into the various branches of mathematics. For the purposes this chapter, we keep the categories simple.

MEANINGS OF "INFRASTRUCTURE" AND RELATED ISSUES OF LITERACY COMMUNITIES AND COMMUNITIES OF PRACTICE

Recall the dictionary definitions for *Infrastructure*:

> *American Heritage*: 1. An underlying base or foundation especially for an organization or system. 2. The basic facilities, services, and installations needed for the functioning of a community or society, such as transportation and communications systems, water and power lines, and public institutions including schools, post offices, and prisons.

The word "infrastructure" has been stretched over the years to include not only material supports for activity, but also social systems at different size scales, sometimes referred to as communities of practice in the sense of Lave and Wenger (1991). As we shall discuss further, technology plays multiple roles as a part of the infrastructure supporting other infrastructures, both physical and otherwise.

Our intuitive starting point for understanding the process of technology becoming infrastructural is dual: First from the phenomenological perspective of individuals, the *experience* of technology, transitioning from strange to familiar to invisible. Second, from an external, analytic perspective. From this perspective we employ several analytic frameworks in an eclectic way, particularly as we attempt to keep in mind the multiple time-scales in which technological change can be described, ranging from months to millennia.

The Technology-Driven Evolution of Technology-User Communities

We need to keep in mind the sociological side of "infrastructurality" partly captured in diSessa's notion of technology and literacy infrastructures (in his book, *Changing Minds*), which he links to literacy communities defined by their literacy practices—he uses romance novel readers as an example, who typically read this particular genre while commuting to and from work, and who are predominantly female. However, as a community, they are passive users compared to the active users of most mathematics or mathematics education software, where active participation in a practice is an intrinsic property of membership. This is the case whether one uses the technology as an interactive tool or as a medium in which one designs and builds interactive artifacts (technology as "tutee")". Hence we merge the two notions—literacy community and community of practice—into *technology user community*.

Computer Algebras System (CAS) users comprise an example of a technology user community among technology users in Mathematics Education. For them, CAS software is an important infrastructure, and for analysts, they are a fairly well defined community whose features, especially its induction processes and relations with other communities, are significant factors in an account of technology becoming infrastructural.

Another such community is comprised of Dynamic Mathematics (e.g., Sketchpad, Carbi, Fathom) users. In the past, while the software systems that they took as infrastructural were independent from the CAS systems, the overlap with the CAS community has been defined by professional and educational interests—a particular person might use both systems and identify with each community, but use each system for different purposes. However, now that the respective underlying software functionalities of CAS and Dynamic Geometry are increasingly overlapping, their user communities may merge. But more importantly for our purposes, *the emergence of a common infrastructure uniting two communities is deep evidence of technology becoming infrastructural at the epistemic level*. As a result of the software systems growing together into a new, more integrated whole, the practices of the Dynamic Geometry community, whose mathematics domain embodies certain kinds of objects and relations, as well as certain styles of justification and proof, inevitably interact with those of the CAS community, whose objects and relations become transformed when viewed as geometric ones. Further, the styles of exploration, justification and proof in one domain interact with those of the other. In a sense, the seed planted by Descartes blooms and bears fruit in the computational medium, and the respective technology user communities unify in a more unified mathematical practice.

Necessity vs. Convenience as a Factor in Infrastructurality

Another factor in technology's being infrastructural is the matter of necessity, the fact that when the technology is fully infrastructural for a particular use, its use is necessary and not optional. Thus, for example, the study of the behavior of certain dynamical systems and their visual features requiring millions of iterations of functions (e.g., Feigenbaum Bifurcation Diagram) requires computational technology in ways that plotting the intersection of two low-degree (say quadratic) polynomials for an approximate solution does not. In effect, the objects of interest literally do not exist without the technology in the first case, while the technology is a convenience in the second. Of course, "convenience" shades into necessity and at a certain point technology creates new forms of practice and hence new mathematics.

The Issue of Interactivity and Agency

Over time, thanks to more powerful processing and increasingly "natural" input modes utilizing natural (or at least, widely occurring) human capacities, software has become ever more interactive in the sense of responding to human-user input in ever wider communication bandwidth and ever shortening feedback cycles, approaching real-time feedback in many circumstances, e.g., in most direct manipulation interfaces. Furthermore, what we have broadly described as "tools" are increasingly responsive to context and individual use patterns. While we commonly describe this change as the "increasing intelligence" of the tools, it may be at least as fruitful to think of the tools as *increasing in agency*. The is because the tools we have in mind are not like power saws and other tools that sit waiting to be turned on and used; instead, they are continually active and seeking interactions. Consequently, when we speak of our engagements with such technologies, we regularly treat then as having minds and even personalities of their own.

Once we ascribe agency to tools and technologies, we have an opportunity to rethink our relation to technology in terms of *partnership*. While the notion of intelligence distributed across tool and user is far from new, going back at least to Dewey, by adding the notion of tool-agency, we may be able to examine our relation with technology in a deeper way that helps us understand the impact of technology in mathematics education in more systematic ways.

Recent work by David Shaffer (2004) has taken this stronger form of distributed intelligence to include interacting agency—in effect, the new unit of analysis is the partnership, which he refers to as "toolforthought"—deliberately combining words to emphasize the unification of the entity under consideration. A *toolforthought* may be regarded as a cross between a cyborg, a computer-augmented human and a human-enhanced tool. This orientation may reveal patterns in the evolving relations between increasingly powerful tools and the body of mathematical knowledge and skill that are taken as the educational agenda of mathematics education. It may also help us understand more fully the meanings of "powerful" in the preceding sentence. The next section is intended to take a step in this direction.

Regularities in the Changes in Distribution of "Skill" Across Tool and User

On one hand, we are familiar with the processes of "handing off" computations and syntactically-defined operations to software, or having the system provide or link to a new representation, and hence the redistribution of operational skill from human to machine. On the other hand, we are as yet unclear on the new skills that this handing-off requires, especially when the software, as with the typical CAS, can provide multiple representations of the operations, their product, or both (Fey, 1989).

Our first point is that, from a longer-term perspective as well as from any of several theoretical frameworks, this tool-user relationship issue is a particular species of a broad research issue that occurs across all areas of education, not merely mathematics. It is sometimes framed as a language or culture vs. individual learning and/or development issue, sometimes as a cultural anthropological issue, sometimes as a history of culture issue, sometimes as an epistemological issue, and so on. We take the view that in mathematics some of the most important and generative advances are representational—the means by which people think, compute, communicate, etc. In particular, across history regularities in procedures come to be encoded (either deliberately or through natural selection-like processes) in notation systems and actions upon these.

As Shaffer & Kaput (1999) argued, extending the work of the evolutionary psychologist Merlyn Donald, the human mind was externally augmented by static records based on writing systems (and, as Moreno points out, human externalization and crystallization of action occurred much earlier, e.g., with tally systems), and then, several thousand years later in mathematics by syntactically structured *operative* notation systems that, in partnership, support human action, e.g., the algebra representational infrastructure, as well as the standard arithmetic representational infrastructure. Then, in the 20th century, operative systems became autonomously executable within the computational medium. Indeed, the more general idea due to Turing and von Neumann is that of a "program" which was applied to encode and extend previously constructed operative notation systems, leading to, among many other things, CAS systems. Such systems call upon a different kind of partnership between human and tool, or, if you prefer the strong version, they generate a new kind of toolforthought object.

Alternatively, as formerly internal cognitive functions are progressively off loaded using tools (e.g., for functions related to information storage, processing, and retrieval), and as attention shifts beyond individual isolated learners and problem solvers toward learning communities (where the "problem solver" is in fact a group of diverse individuals), it is natural to regard thinking and knowing as being distributed across individuals and across media (and tools).

While the broad pattern of externalization and partnering with external tools is evident, the details are very complex and involve changes in cognitive and perceptual activity as learning occurs within individuals, within and across communities, and across history. One needs to learn to work as a partner with ever more agency-laden tools, tools that themselves carry out increasingly complex processes, processes that themselves were once carried out by humans, sometimes requiring considerable mental effort and skill. The same thing happened in the 15th century with the adoption of the base-ten placeholder system for numbers and the algorithms that were built upon it and that continue to dominate elementary school mathematics. In a deep sense, the mathematical experience is continually changing. So, in particular, our task as educators and designers of educational experiences is to design for an ever-changing partnership rather than to worry over the loss of human skill with obsolete tools as new tools increase in capacity. We build on this theme with examples of how mathematical problem-solving is changing in a later section.

The Rate of Change Problem: Changes in Representational Infrastructures vs. Changes in Surrounding Institutional Infrastructures

Thanks to the plasticity of the computational medium and its ability to fuel its own change as its own infrastructure, the rate of change is accelerating—the change is more like $\exp(\exp(x))$ than $\exp(x)$ or polynomial change. Our challenge, then, as mentioned earlier relative to the essentialist view of mathematics, relates to the problem of time-scales of change. If representations and partnerships change at a rate faster than the rate of change of the surrounding institutions, then we have a major problem. Because educational institutions (schools, assessment systems, teacher education systems, and so on) are inherently conservative social systems, this is exactly what we have!

But the problem is more complex because some of the most important changes in technology, especially computational and representational technologies, do not occur in the straight-ahead direction, but appear from unexpected directions. Who saw the WWW coming? This leads us to what we believe is another important change that is about to descend upon us, related to the communication affordance of technology; and, it is likely to interact in unpredictable ways with the representational affordances.

It is not a new theme. But, as future-oriented theorists such as Dewey and Bruner emphasized long ago, in an age of increasingly rapid change, "learning to learn" (i.e., learning to adapt) is one of the most important goals of instruction.

However, whereas educators in the past have taken this observation to mean that mathematics education should shift toward trying to teach content-independent processes that are not connected to any particular context, modern learning science theories are emphasizing views of knowledge and knowing that are both socially and contextually situated. Just as software systems can be designed for to optimize continuing adaptability and backward compatibility, so to can mathematical concepts be learned in such a way that they are malleable rather than rigid and fixed.

The Communication Technology Affordance and Wireless Classroom Connectivity

One major change is underway that is likely to yield profound consequences because it takes place at the communicative heart of the epicenter of mathematics education—the classroom—and because involves the teacher in a central way. Thanks to technologies developed outside education (as is usually the case), the ability to wirelessly link devices of various kinds and, as noted earlier, the ability to scale software and data across device types, an entirely new universe of classroom possibilities is emerging. This is the focus of recent work by Hegedus & Kaput (under revisions); Kaput (2002); Kaput & Hegedus (2004); Stroup, Ares, Lesh, & Hurford (this volume), and other close colleagues such as Roschelle, Abrahamson, & Penuel (2003); Roschelle & Pea (2002); Wilensky & Stroup (2000). In effect, the communication affordance is moving from outside the classroom to inside the classroom.

Key Issues in Classroom Connectivity and Its Infrastructural Role

The major new ingredients that we have been studying involve (a) the mobility of multiple representations as reflected in the ability to pass information bi-directionally and flexibly between teacher and students and among students, using multiple device-types, (b) the ability to flexibly harvest, aggregate, manipulate and display representationally rich student constructions to the whole classroom, and (c) thanks to the at-handedness of hand-helds, to do all of this in ways that respect and build upon naturally occurring social and participation structures. Thus, we are able to engineer and implement a surprising array of new activity structures in concert with the mathematics to be taught and learned.

We see classroom connectivity (CC) as a critical means to unleash the long-unrealized potential of computational media in education, because we see its potential impacts as direct and at the communicative heart of everyday classroom instruction. For this reason, we are now beginning to build insight into how those new ingredients, in combination, may provide concrete means by which that potential may be realized. These new ingredients may, in fact, help constitute the first truly educational technologies, intimately situated within the fundamental acts of active teaching and active learning. This embeddedness may indeed be more profound than we initially recognized, because these ingredients resonate deeply with broader views of learning as participation—and no longer fit within a "learning in relation to a machine" (large or small, in the lab, classroom or even your hand) view of educational technology. Indeed, the paradigm is shifting towards one where the technology serves not primarily as a cognitive interaction medium for individuals, but rather as a much more pervasive medium in which teaching and learning are instantiated in the social space of the classroom. We deliberately choose "instantiated" ahead of "situated" because we have repeatedly seen mathematical experience emerge from the distributed interactions, which are enabled by the mobility and shareability of representations (see below).

By emphasizing new forms of classroom connectivity, we often observe students "being mathematical" as joint experiences, shared in the social space of the classroom in new ways as student constructions are aggregated in common representations—in ways reminiscent of, but distinct from those of a participatory simulation as studied by Resnick, Stroup, and Wilensky (Wilensky & Resnick, 1999; Wilensky & Stroup, 1999). Cognitive activity is distributed in the socio-material space in the sense of Hutchins. Similarly changed is how students interact mathematically with each other and their teacher, and, critically, how their personal identity is manifest in their shared mathematical experience in the classroom.

For *teachers*, changes occur in the nature of teaching by fundamentally altering how participation structures can be defined and controlled, how attention can be managed, how information flows and can be displayed, and how pedagogical choices and moves are made in real time. This is especially true when the mathematical activity completely embeds the infrastructure of connectivity allowing teachers to aggregate students' work in mathematically meaningful ways. Adding to the richness of the picture is the fact that the continuing evolution in the underlying technological-communication affordances affects all these factors in important ways.

For *researchers*, the ground also shifts in a profound way, challenging our categories of phenomena. The fulcrum of the balance between "individual-cognitive" and "social-distributed" has shifted. Semiosis has become a new kind of operation, applying to shared objects in newly sharable contexts.

The Need for Research in Classroom Connectivity

The potential of CC can be realized only if we understand it sufficiently to inform the design and improvement of (a) its technologies, (b) classroom activities, teaching practices and forms of assessment that optimally exploit it, and (c) the preparation and support of teachers to utilize this new constellation of technologies, activities, practices, and assessments. This will require a new, highly interdisciplinary domain of educational research, one that is now in its early stages, uncovering the new phenomena to be investigated, formulating issues, descriptive languages, candidate theories, and research agendas, and building research communities to extend and elaborate the inquiry.

Starting Point Questions for Investigating Classroom Connectivity

A large number of new phenomena need to be studied. Among these, we have defined three Opportunity Spaces generated by CC; and, we have posed central questions associated with each. We believe that focusing on these three spaces can open up lines of inquiry into what types of mathematical problem solving are now possible when computational power is combined with communication infrastructure.

Learning and activity structures. What new activity structures are possible and appropriate that exploit CC across diverse device-types, that increase learning of traditional topics, that render new topics more accessible, and that increase breadth of student participation and intensity of engagement?

Teaching and pedagogy. How does CC impact teacher decision-making and pedagogical options, both positively and negatively, and how can teachers learn to use CC to maximal advantage?

Assessment, classroom management and information flow. How can teachers use CC and related analytic tools to exploit what we know about student thinking and learning in order to actively diagnose and efficiently respond to student thinking on a regular basis? What kinds of tools can facilitate the flow, organization and display of information in the classroom?

Two sets of questions cross-cut these:

Questions of representation. What kinds of representations and uses of such (e.g., on single devices, distributed across devices, aggregated or not, etc.) support optimal student learning and participation, teacher decision-making and activity-design, interpretation of assessment data, and information flow and display?

Questions of technology. Which combinations of technological characteristics within and across devices (e.g., screen and physical interface, communication capacity, portability, processing power, etc.), networks (e.g., peer-peer, server-based, pull vs. push communication, etc.) or software (e.g., common data-structures and interfaces across devices, control at-a-distance, etc.) enable ease, fluency, and effectiveness of mathematical learning, teaching, assessment, and information flow?

We are actively pursuing these issues at the present time; but, even though our investigations are only at preliminary stages, it already has become abundantly clear that the evolution of this new technology infrastructure is driving a need for new theoretical frameworks and constructs as well as new research methodologies. In a sense, what we are seeing is an example of technology infrastructure reflexively redefining the conditions of its own study.

EXAMPLES OF HOW MATHEMATICAL PROBLEM SOLVING IS CHANGING IN CC ENVIRONMENTS

Our work in these three spaces has produced a set of activity structures that exploit the shifting technological infrastructure we are developing. Consequently, this section will offer some prototypical examples to demonstrate possible impacts of CC in providing rich opportunities for mathematical problem solving in grades K–16. We will emphasize examples that not only impact student learning but also have implications for deeper social-cultural phenomenology in mathematics education.

A "Networked" Representationally Rich Infrastructure

Our examples focus on the use of SimCalc MathWorlds software (see http://www.simcalc.umassd.edu for more details). Currently, SimCalc is being used in algebra and pre-calculus classrooms in combination with communication infrastructures (e.g., TI-Navigator's Navigator system) that use representationally "parallel" software to wirelessly connect students' work on hand-held devices with a teacher's host computer. A long-term goal of the SimCalc Project (Kaput, 1994) has been to (a) exploit technology's capacity for interactive visualization tools and (b) emphasize simulations that are linked to mathematical representations, which provide an alternative to the algebraically based prerequisite structure of topics such as calculus. A central goal is to avoid the algebra bottleneck while at the same time democratizing access to big mathematical ideas that are now inaccessible to the great majority of students due to the algebra barrier.

The SimCalc Project has developed strategies that use interactive representational affordances of technology (visualization, linking representations to each other and to simulations, importing physical data into the mathematical realm in active ways, graphically editing piecewise-defined functions, etc.) to energize and experientially contextualize existing algebra courses. We do this in ways that also lay the base for more advanced mathematics, particularly calculus. Recently, we have focused on studying the profound potential of combining the representational innovations of the computational medium (Kaput & Roschelle, 1998) with the new connectivity affordances of increasingly robust and inexpensive hand-held devices in wireless networks (Roschelle & Pea, 2002), which are linked to larger computers (Hegedus & Kaput, 2003; Kaput, 2002).

Prior empirical work of the SimCalc Project investigated the impact of our constellation of technological, curricular and pedagogical innovations on student learning, *especially as measured by independent standard test items* on a pre/post-test basis. Under various NSF-funded grants (ROLE: REC–0087771; 0337710 and IERI Programs: REC–0089094; 0228515), we have run intense teaching experiments aimed at core topics in algebra in MA, CA and TX. Results (Hegedus & Kaput, 2003; under revisions; Roschelle et al., 2003) demonstrate comparably positive outcomes under substantially different instructional and technological conditions, somewhat different curricular targets, and different student demographics.

One particular 5-week after-school intervention taught by a high school teacher in one of our participating schools combined 7th, 8th and 9th grade students of greatly varying backgrounds. The middle school students took the course as enrichment whereas the 9th graders were students who had either failed or nearly failed the state 8th grade state (MCAS) mathematics test, and were identified to be at risk of failing the high-stakes 10th grade state exam. The course made heavy use of the kinds of aggregation activities described below. Our work has illustrated the positive impact that our connected curricular activities had on student achievement on core algebra skills.

Gain from pre (0.427; var = 0.14) to post (0.659; var = 0.14) demonstrated statistically significant increases ($p < 0.001$) in student mean scores (effect = 1.6) but with an even higher effect on the at-risk 9th grade population (effect = 1.9) when we disaggregated the data. *A major finding of our work was that experience in a networked classroom not only improves students' achievement on core algebra questions but also indirectly improves critically important skills such as graphical interpretation not attended to in our intervention.* Consider one item of the test which involves graphically interpreting the geometric relationship between circumference and diameter, $C = \pi d$, as linear. We observed shifts in mean scores from 0.33 (equal to State mean) to 0.875 on the post-test.

We have similar data from a year long teaching experiment involving a class of academically disadvantaged college freshmen (n = 12, whose demographics mirror those of urban high school students, with average SAT-M under 400) and other undergraduate classrooms for pre-service teachers.

Example: Mathematical Performances

The activities we emphasize may focus on either individual student creations, or on small group constructions, or on constructions that involve coordinated interactions across groups—where results may be uploaded from hand held apps to a host computer application, and then displayed (typically with some narration by the originators of the construction).

The instructional strategy of having students describe their work orally (using individual student recitals, or showing their work at the board) is as old as deliberate education. But, in today's classrooms, where teaching-to-the-test forces teachers to emphasize time-on-task (often for exceedingly low-level tasks), such complex and time-consuming techniques often are viewed as being inefficient uses of class time, especially when the students produce complex written or drawn products that are time-consuming and difficult to score.

In contrast to the preceding scenario, teachers in the CC context are able to upload students' dynamic computational objects rapidly, preview these, and "replay" them on their own computers, which may be connected to a public display. An exemplar of such an interaction occurs in a SimCalc activity known as the "Exciting Sack Race." This activity combines CC with the expressiveness of piecewise defined position vs. time graphs which in turn drive a motion simulation where the expressiveness of natural language is used to address the critical idea of slope as rate-of-change (e.g., steeper means faster, horizontal means the object is stopped, negative means the object is going backwards). Students are given a linear position graph for one object "moving at a constant rate of 2 ft/sec" and they are asked to (a) create a graph for a second object so that the two objects enact an exciting race ending in a tie, and (b) write a short story that describes the race and that can be read as the race is "run" on the whole class display. The students' functions are then uploaded to the teacher's computer and the teacher selects individual functions to animate as the students simultaneously "call their race" using their stories. Or alternatively, students might consider a pair of graphs, one as the student might construct it on their hand-held, and another after uploading as it might appear on the classroom display.

Another extension of the activity might involve using motion-detectors (e.g., Calculator Based Ranger, CBR) or other probes (e.g., MBL) to physically instantiate such "stories" into a cybernetic form (so that motions can be re-animated). Or, such motions can be examined with traditional representations such as the kind of graphs and tables that the computational media automatically provide. For example, SimCalc MathWorlds allows students to import position and motion data that the program then interprets into a graphical representation as well as into an animation, which is based on data about relevant positions and motions.

This is a core example from our work that illustrates the potential establishment of new educational environments that can evolve through a combination of representation-rich mathematical software and connectivity infrastructure, as discussed in the earlier section on technology's interaction with mathematics education. In doing so, new forms of problem solving can emerge in the classroom. We continue to outline examples of such activity.

*Example: Participatory Aggregation to a
Common Public Display*

The activities described in the preceding section involve systematic variation, either within small groups, across groups, or both—in situations where students produce functions that are uploaded and then systematically displayed and discussed

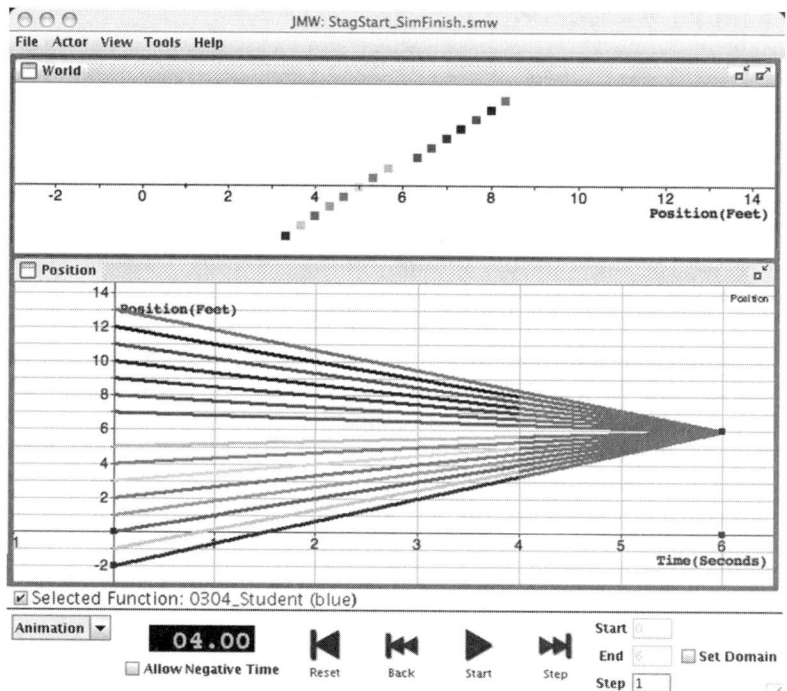

FIGURE 8-1   Staggered Races—varying slope.

to reveal patterns, to elicit generalizations, or to expose or contextualize special cases. Another goal is to help raise students' attention from individual objects to *families* of objects. In a connected algebra environment, the class of students typically is subdivided into numbered groups, where the size and number of groups fit both the given size of the class and the mathematical activity (ranging from the whole class to pairs). Groups are often, although not necessarily, defined "geographically"—students who are physically near each other. The students usually also "count-off" inside the group, so that each student then has a two-number identity that can then serve as a "personal parameter," a Group Number and a Count-Off Number. Students then create mathematical objects—in the cases discussed here, linear Position vs. Time functions, that drive animated screen objects.

One genre of activity that we call "Staggered Races" requires students to use their count-off numbers as varying parameters in polynomial and trigonometric functions. One simple example, based on the equation $Y = MX + B$, requires the students to start at 3 times their Count-Off number but "end the race in a tie"—in a situation where the object is controlled by the target function $Y = 2X$ (the target racer moved at 2 feet per second for 6 seconds and started at zero—see the graph in Figure 8-1). In this activity, students need to calculate how fast they have to go to end the race in a tie. Also, since they start at different positions, the slope of their graphs changes depending on where they start; and, where they start in turn depends on their personal Count-Off Number. Each group is limited to 5 people; and, even though the group number does not affect their constructions, it gives rise to a smaller and more manageable set of functions to discuss. Secondly, and more importantly, the Count-Off Numbers 4 and 5 give rise to two important slopes. The person with Count-Off Number 4 has a graph with constant slope, $Y = 0X + 12$, since he starts at 12 ft, which is the finish line. So, he does not have to move! Whereas, the person with Count-Off

Number 5 starts beyond the finish line (15 ft) and has to run backwards. Thus, this student is forced to calculate a negative slope.

In the preceding scenario, organizing and displaying student work is a strategic pedagogical decision, e.g., in focusing student attention on the underlying mathematical structure. An important question to ask before animating the motions for the whole class is "What will the race look like?"

The aggregate motion in the classroom display becomes a personal reality through the personal link with Count-Off Number; and, by aggregating and displaying the class work, students can observe how their personal construction fits into the "race" and allows them to note how people from other groups had constructed identical motions because of their Count-Off Number. In addition, the shape of the graph and the parity of the slope for those who had to start past the finishing line (i.e., run backwards) was made more realistic and understandable in this motion-based scenario.

These examples typify a genre of activity that creates new learning opportunities and problem solving environments in mathematically meaningful ways. In an earlier section we discussed the affordances of combining hand-held devices for the purpose of creating combinations of users. Here, in the connected classroom, the mathematics of change and variation can emerge through the interaction of multiple students' individual contributions, shifting the focus from isolated, individual learning or problem solving, to broader, integrated, learning communities. Thus, the classroom community is creating an infrastructure that builds upon the technological infrastructure.

Example: Cyclical Upload, Display/Discuss and Broadcast Sequences

In the preceding activity, we harvest students' work, publicly examine and discuss results, and then broadcast some version of results (perhaps a teacher-modified version of an individual's construction) to the class for another cycle of exploration, construction and harvesting. Perhaps harvesting involves homework within the cycle. For example, the assignment might be posed in the following way. "Now that we see this new kind of function that Sarah made, your job is to make your own version of it for homework, and tomorrow we will look at the functions that you built, to see if they have anything in common.")" For example, in the Sack Race examples, Tania's race-function can be sent back to the class so that the students can now "race against Tania" or "dance with Tania."

Using such strategies, we have seen extraordinary excitement and learning occur. Thus, such activities not only investigate new mathematical ideas in new ways but they also frequently attend to a motivation problems in school mathematics—by *deeply embodying the mathematics in the social realities of the students.*

Example: Project Work—Group-Constructed Relay Races Using Point-Slope Forms and Simultaneous Equations

In this genre of activities, students who are working individually or in collaborating groups often produce some mathematical construction, usually to be displayed and discussed by the whole class. For example, sometimes these constructions involve student-designed a whole-class Participatory Aggregation activity of the following type.

Students make linear functions, perhaps using the Point-Slope form, whose starting and ending position and/or time depend on their Group and/or Count-Off numbers. For example, each group might be asked to make an "up and back" swimming pool relay race where students move in a sequence determined by their respective Count-Off Number at a speed equal to some multiple of their Count-Off Number. Here, each student after the first needs to determine when the previous one finishes, a job that increases in complexity as your count-off number increases. This task also requires a sense of which direction to travel in (the sign of the coefficient of X).

Another case involves a unidirectional race where the *duration* of each leg is fixed, but the *speeds* vary according to Count-Off Number. Other variations involve relays across groups, e.g., where the last person in one group "hands off" to the first person in the next group. Critically, the collaboration is deeply mathematical because the objects *must fit together mathematically*—in the same ways that, when a technical worker or team works in a larger project, that person or team's work must fit *functionally* in the project.

Pedagogical Implications

Our recent studies have shown significant improvement in low-achieving students' abilities to solve standardized and applied problems (Hegedus & Kaput, under revisions; Kaput & Hegedus, 2004); and, we also are demonstrating significant shifts in participation structures from non-CC to CC contexts. This is important because, if technology and communication are to become deeply embedded in mathematics education instruction, then the impact is not just on learning but also on pedagogy—as well as on inside-classroom belief structures and norms. So it is significant that we are seeing a variety of shifts in practice as we have studied teachers of varying degrees of exposure to such contexts.

The use of representationally rich software in mathematics education calls for a re-conceptualization of both traditional and applied mathematical concepts. The additional layer of dynamics—e.g., simulating functions with moving objects, utilizing physical motion data, or creating suites of interactive tools that allow users to create, manipulating large sets of operations simultaneously—poses an intellectual challenge for traditional teaching because the degrees of freedom are clearly are immense. But, an increase in freedom can allow broader and more diverse problem solving environments in which discovery and argumentation about the conceptualization of ideas can be more prominent—as well as being in the hands of learners. We have seen that such requirements, which actually call for a shift in agency as described earlier, does take time and growth in teachers' perception of the role of technology in "their" classrooms.

Our studies have contrasted teachers' practices working in both non-CC with CC environments; and, results have revealed distinct differences in fundamental processes such as posture and gesture as well as discourse. Teachers using the activity structures we have described in this chapter appear, in general, to be more positive and effectual, physically, in the classroom, as they use tools to manage the contributions made by a whole class.

In the preceding activity, the teacher's computer often projects results of students work onto a whiteboard where annotations can be made (either using marker pens or electronic media) to highlight and organize information that such an environment deliberately affords. To manage such complexity, the structure of the activity is critical in enabling the teacher to facilitate discussion around general trends in the emerging mathematical objects that the whole class has contributed. For example, consider the staggered races genre, where we have 5 degrees of freedom (one per group) that allows clustering and duplicity across groups within the whole class. So, outliers become more prominent and easier for a teacher to observe and focus discussion on. To support such discussions, we also have built graphical tools to assist in the flow of information, using coloring (by group) and view controls to hide and show individual or whole groups of contributions, including their various representations (e.g., position or velocity graphs, or formula). In essence, this is a form of "pedagogically organized assistance" that was discussed at the start of this chapter.

In non-CC contexts, teachers' roles are critical to success. Yet, they tend to move out of the central workspace; and, the classroom management of ideas becomes a linear process. That is, decisions are implicit in the structure of the activity itself about who speaks one at a time, and where student responses are appreciated, analyzed, or simply dismissed without comments.

Gesture plays an important role, especially, with new forms of an activity structure where personally constructed mathematical objects are now readily contributed and discussed in a CC-enhanced classroom. Students now have an investment in the objects they submit for public examination; so, they need to develop new registers to "talk about them," as well as discuss their properties relative to other students' work.

The objects that students create also are dynamic. So, useful registers also need to be developed to be richer and more applied in nature; and, new language needs to be created which includes metaphors and mathematically-oriented narratives. For example, in asking several groups to create a function to represent an actor, moving at a constant rate equal to their count-off number, but starts at a distance equal to their group number ahead of a target graph (e.g., $Y = 2X$), we observed interesting struggles in students developing a sense of what their group or whole class collection of graphs would look like. We purposefully do not show graphs as they are contributed to enable students to make conjectures about the whole set of contributions. In such an activity, one student held up his hand with fingers spread out, simultaneously saying, "it's going to look like a fan." As new forms of expression, gesture and linguistic, develop, students and teachers are enabled to look at broad sets of mathematical objects, and point, highlight, locate and describe attributes of functions and families of functions, more abstractly defined in mathematics, in later years. It also can help highlight variations in objects and hence access the idea of mathematical variation and co-variation as a new "experience" via a "social-constitution" of ideas.

We also have observed differences in the amount and types of questions that teachers ask in non-CC vs. CC contexts. In both contexts, we observed an interesting phenomenon on numbers of questions being asked when we analyzed several samples of classroom transcripts. In a both a CC and non-CC context, we counted the accumulated number of questions, and plotted these numbers with respect to the question asked. We observed a strong linear relationship across both contexts (an interesting fact in itself), but a higher rate in non-CC contexts to CC contexts. So teachers are in general changing the rapidity of their questioning, but the nature of the questions also appear to change in the CC-context. Therefore, to verify this observation, we have begun to formulate categories of questioning, working with our teachers to unpack the rationale behind why these types of questions are changing, their shifting beliefs and pedagogical practices as well as the impact of such questions on student participation and types of interaction. Through meetings with teachers, we have begun to develop a "community of practice" where we analyze each other's teaching practices (e.g., questioning strategies) and pedagogical beliefs to inform our analyses. Two important results are emerging: (i) the importance of us regulating the types of questions being asked in CC-contexts with an analysis of the expected responses, (ii) the need for question schemas which are intimately bound up with the nature of the activity structures and the new phenomenon of CC. One example we offered in Kaput & Hegedus (2004) is in Table 8–1 below:

TABLE 8.1.
Constructing Pedagogical Actions

	How do your—	
Motion(s)	Look Different as	An Individual vs. the Group
Graphs	Look the same as	An Individual vs. the Class
Formula		The Group vs. the Class
Tables		

In effect, connectivity supports the pedagogical manipulation of student's focus of attention. But the teacher knowledge needed to take advantage of a connected classroom requires extended development. Table 8–1 outlines a simple structure, which can guide the teacher's inquiry. Choosing one item from each column leads to a particular question that can be addressed to individual students, groups or the whole class.

It seems necessary that a co-evolution of pedagogical growth and mathematical cognition is not only evident but imperative in CC-contexts and that they can lead to deeper access to core ideas and an appreciation of technology as infrastructural. Consequently, the kind of multi-tier research designs described by Lesh and Kelly are ideal for structuring our investigations (Lesh, 2003; Lesh & Kelly, 2000).

CONCLUSION

The most ubiquitous form of communication infrastructure, the Internet, fuelled by increases in bandwidth and communication tools such as video-conferencing, will continue to drive outside-classroom interaction and education. But, inside-classroom new forms of communication should lead to similar advances, as technology itself becomes an infrastructural part of mathematics education. Technology becoming infrastructural means a shift to a fundamental yet invisible role—similar to the way electricity, water mains, and telecommunications are given in most modern day homes. Furthermore, this shift will lead to an emphasis on new levels and types of ideas and abilities (such as those related to systems theory, game theory, and discrete mathematics)—as well as new ways to think about traditional concepts and skills (such as those related to algebra and calculus). For example, as technology becomes more infrastructural and as it begins to exploit significant advances in communication power, the interactions will change that occur within learning communities of learners and teachers; and, mathematical discovery and problem solving will take on a more natural and socially situated feel.

In school-based education, these possibilities are only likely to occur due to rapid advances in wirelessly networked classrooms, which facilitate new types of social interaction and thinking—and new ways to make mathematics less abstract and more accessible to a wider population of students. Nonetheless, to realize this potential, new types of pedagogical diversification also will be needed. So, significant advances are only likely if teacher development is not ignored.

REFERENCES

Fey, J. (1989). Technology and mathematics education: A survey of recent developments and important problems. *Educational Studies in Mathematics, 20,* 237–272.

Hegedus, S. J., & Kaput, J. (under revision). Improving algebraic thinking through a connected SimCalc MathWorlds classroom. *Journal of Research in Mathematics Education.*

Hegedus, S. J., & Kaput, J. (2003). The effect of SimCalc connected classrooms on students' algebraic thinking. In N. A. Pateman, B. J. Dougherty, & J. Zilliox (Eds.), *Proceedings of the 27th Conference of the International Group for the Psychology of Mathematics Education held jointly with the 25th Conference of the North American Chapter of the International Group for the Psychology of Mathematics Education* (Vol. 3, pp. 47–54). Honolulu, Hawaii: College of Education, University of Hawaii.

Kaput, J. (1994). Democratizing access to calculus: New routes using old routes. In A. Schoenfeld (Ed.), *Mathematical thinking and problem solving* (pp. 77–156). Hillsdale, NJ: Lawrence Erlbaum Associates.

Kaput, J. (2002). Implications of the shift from isolated, expensive technology to connected, inexpensive, diverse and ubiquitous technologies. In F. Hitt (Ed.), *Representations and mathematical visualization* (pp. 177–207). Mexico: Departmento de Matematica Educativa del Cinvestav-IPN.

Kaput, J., & Hegedus, S. (2004). An introduction to the profound potential of connected algebra activities: issues of representation, engagement and pedagogy. In *Proceedings of the 28th Conference of the International Group for the Psychology of Mathematics Education* (Vol. 3, pp. 129–136). Bergen, Norway.

Kaput, J. J., & Roschelle, J. (1998). The mathematics of change and variation from a millennial perspective: New content, new context. In C. Hoyles, C. Morgon, & G. Woodhouse (Eds.), *Rethinking the mathematics curriculum* (pp. 155–170). London: Springer-Verlag.

Lave, J., & Wenger, E. (1991). *Situated learning: Legitimate peripheral participation.* New York: Cambridge University Press.

Roschelle, J., Abrahamson, L., & Penuel, B. (2003). *Catalyst: Toward scientific studies of the pedagogical integration of learning theory and classroom networks.* Available at: http://firefly.ctl.sri.com/wild/review.html

Roschelle, J., & Pea, R. (2002). A walk on the WILD side: How wireless handhelds may change computer-supported collaborative learning. *International Journal of Cognition and Technology, 1*(1), 145–168.

Shaffer, D. W. (2004). When computer-supported collaboration means computer-supported competition: Professional mediation as a model for collaborative learning. *Journal of Interactive Learning Research, 15*(2), 101–115.

Shaffer, D. W., & Kaput, J. J. (1999). Mathematics and virtual culture: An evolutionary perspective on technology and mathematics. *Educational Studies in Mathematics, 37,* 97–119.

Wilensky, U., & Resnick, M. (1999). Thinking in levels: A dynamic systems approach to making sense of the world. *Journal of Science Education and Technology, 8*(1), 3–19.

Wilensky, U., & Stroup, W. (1999). Learning through participatory simulations: Network-based design for systems learning in classrooms. *Paper presented at the Computer Supported Collaborative Learning (CSCL '99) Conference.* December 12–15, 1999, Stanford University.

Wilensky, U., & Stroup, W. (2000). Networked gridlock: Students enacting complex dynamic phenomena with the HubNet architecture. *Proceedings of The Fourth Annual International Conference of the Learning Sciences,* June 14–17, 2000, Ann Arbor, MI.

CHAPTER 9

Why Build a Mathematical Model? Taxonomy of Situations That Create the Need for a Model To Be Developed

Maynard Thompson and Caroline Yoon
Indiana University

In recent years, research into students' thinking and learning has become increasingly aligned with the perspective that students develop powerful conceptual understandings of systemic ideas in mathematics and science by *modeling* meaningful situations (Lesh & Doerr 2003). As a result, many researchers have focused on studying the process of model building. One approach has been to study naturally occurring instances of model-development through ethnographic observations of professionals in fields such as business, engineering and medicine where modeling is part of their daily work routines (Hoyles et al., 2001). Another has been to recreate these situations in the form of modeling activities that imitate these real-world modeling situations, as in the case of Model-Eliciting Activities (Lesh et al., 2000). By having students work on these activities, researchers are able to create situations where modeling is likely to be observed, thus minimizing the risk of protracted study that is unproductive as to model building. Consequently, the development of these activities as research tools for studying modeling has become a vibrant area of research and development in Education (Diefes-Dux et al., 2004).

If, however, researchers want to use these activities to study modeling systematically, then they need to be sure that the activities elicit a broad range of models. Therefore, in order to develop a more representative range of Model-Eliciting Activities, researchers need to understand more about the characteristics of situations that create the need for model building. For example, researchers could benefit from having a comprehensive catalog of characteristics of situations where models need to be developed. This chapter makes a preliminary attempt at outlining such a catalog, by describing characteristics of six situations that may call for a model to be developed. These situations are not categorized according to the contexts that give rise to them, nor the mathematical ideas used in them, although such classifications exist and are useful for other purposes. Instead, these are sorted into categories that answer the question: What aspects of a situation, or what questions about a situation, give rise to the need for a model to be created? Or more simply, why would one want to build a mathematical model?

SIX SITUATIONS THAT GIVE RISE TO THE NEED FOR MODEL BUILDING

If an investigator is interested in a situation, then in general there is a goal or some questions behind that interest. Whether or not a situation calls for a model to be developed is largely determined by the goals or the kinds of questions that are being asked about the situation. A pianist that has been commissioned to perform a newly composed sonata may want to know "Who is the composer of this sonata?" and "What key is this sonata in?" However, neither of these questions will require the pianist to model the situation because the characteristics of the situation relevant to these questions are already immediately available by simply examining the score. Alternatively, the pianist may want to know something much less obvious such as "How long will the performance of this sonata be?" In this case, the length of the performance cannot be known by doing something as simple as counting the number of pages on which the music is written. Indeed, the length of the performance depends on a variety of factors including the number of bars in the piece, the number of beats to each bar, the tempo of the piece indicated by the number of beats per minute, changes in tempo throughout the piece, the presence of other musical embellishments that either slacken or speed up the pace of the piece, and the presence of components where the pianist has considerable discretion. Consequently, the pianist will have to take into account (i.e., model) this system of interacting factors in order to determine the performance time of the sonata.

What differentiates this third question from the first two is that the information sought (the performance time of the sonata) is so much dependent on a system of other factors (tempo, number of bars, beats per minute, musical embellishments) that it cannot be attained without first understanding the system on which it depends. It is only through modeling the system that the information can be found. Most situations that create the need for a model have this characteristic—that is, the perceiver of the situation wishes to know something that depends on a system. Consequently, another way of answering the original question of "What situations create the need for a model to be developed?" is through addressing the related question "What kinds of information cannot be obtained

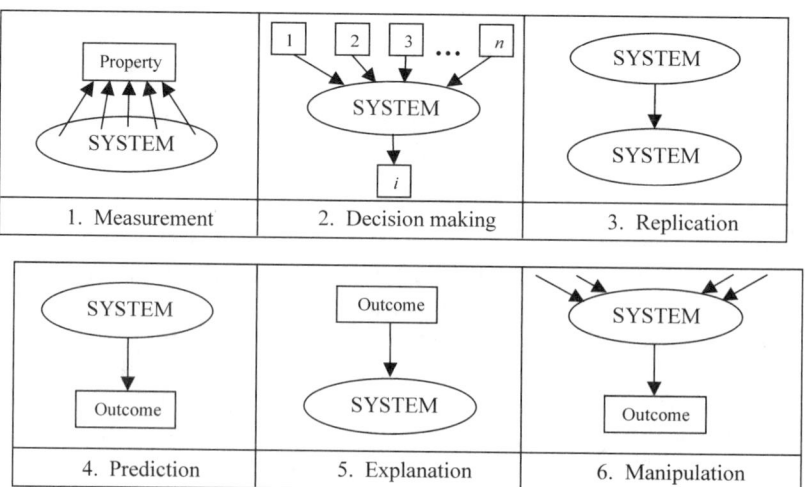

FIGURE 9–1. Summary of the situations that create the need for a model to be developed in order to obtain information that is highly dependent on a system.

directly, but only indirectly through identifying and studying an appropriate system?" Six situations that create the need for model building are described below. These situations occur when the information sought is dependent on a system in the following ways:

1. The information sought is a measurement of some property of a system.
2. The information sought is a way of deciding between alternatives that is justifiable (i.e., based on a system).
3. The information sought is a template that will allow one to replicate a system.
4. The information sought is a prediction of the outcome of a system.
5. The information sought is an explanation of some outcome of a system.
6. The information sought is an understanding of how to manipulate a system so that it produces some desired outcome.

The essential features of these situations are highlighted in Figure 9-1.

Preliminary taxonomies such as this one should be viewed as descriptive and organizational tools that are intended more as general guides than as complete catalogs. To be sure, the situations we describe are caricatures. In contrast, the situations and models that arise in practice rarely fit neatly into exactly one niche. More often, they are blends of several, and the dominant characteristic may be more in the eye of the modeler than an intrinsic feature of the situation. Nevertheless, we consider the task of differentiating between these six caricatured situations as a necessary preliminary step in the process of developing a more complete and nuanced catalog of model building situations in the future.

Situation 1: A Model Is Needed in Order to Measure Some Property of a System

Models are often needed in situations where one wants to measure some phenomenon that is often not directly perceivable, and that is a property of a system (see Figure 9-2). The scenario described above is one such example—the performance time of the sonata is not immediately apparent, but is a property that emerges from the system of relationships between the tempo, embellishments, time signature, and number of bars in the piece. To be sure, if the sonata had been as well known as Beethoven's Moonlight sonata, then anyone could easily find a good estimate of the performance length by referring to the album cover of any number of recordings of the piece. But a pianist trying to decide whether to accept a commission to be the first performer of an unknown sonata does not have this luxury, and instead, must rely on the system of indicators expressed through musical notation.

This tension of wanting to measure some inaccessible phenomenon often arises because the thing to be measured is too small or too big to be directly perceivable, or

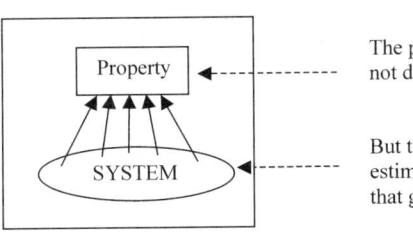

The property to be measured is not directly perceivable

But the property can be estimated by modeling a system that gives rise to the property

FIGURE 9-2. Modeling enables one to *measure* some property of a system.

the operational definition of the measurement involves relationships among several observable quantities. For example, a forester in a department of conservation may be asked to estimate the number of trees in a tract proposed for logging based on aerial photographs of the tract. Because the photographs are taken from 1,000 feet or so in the air, and the trees number in the thousands, it may be impossible to count each individual tree. However, the forester may be able to construct a good estimate by synthesizing related information such as the average height, crown breadth, and crown overlap of the types of trees in the picture, the height and angle to the ground from which the photograph was taken, the area of the forest, and a ground count of the number of trees in a small portion of the forest.

Situation 2: A Model Is Needed in Order to Decide Between Alternatives

Models are often developed when one needs a way of deciding among a set of alternatives (see Figure 9–3). These situations often are a result of pressures such as there being too little time to try out each alternative, the alternatives are costly to test, or a study of the alternatives results in too little or too much information to process. In addition, making a decision means that there is a meaningful definition of the decision criteria, a definition that can be used to compare alternatives in practice. For example, a student may wish to decide which cell-phone plan is most suitable for her lifestyle; an employer may wish to decide who to hire from amongst a pool of applicants; a university administrator may wish to decide which strategy to adopt to attract and retain students from a variety of suggestions from deans and advisors. For each of these cases, there are several ways to proceed. On one extreme, one could make a decision by putting all of the options into a hat and pulling out one at random. However, this method—and other very simplistic decision processes—do not take adequate account the various pros and cons of each option, and a poor choice may have bad repercussions in high stakes situations. The cell phone companies may require the student to sign up for a two year minimum contract; the employer may be required to work closely with the new hire; the university may require a higher enrolment to secure state funding. Such situations call for a more systematic way of assessing the relative benefits and drawbacks of each available option to make the optimal choice. In other words, the strengths and weaknesses of each option need to be modeled.

In many cases the strengths and weaknesses of various options may conflict. For instance cell phone carrier X may be less expensive than carrier Y, but the service area of Y may be larger than that of X. How should these strengths and weaknesses

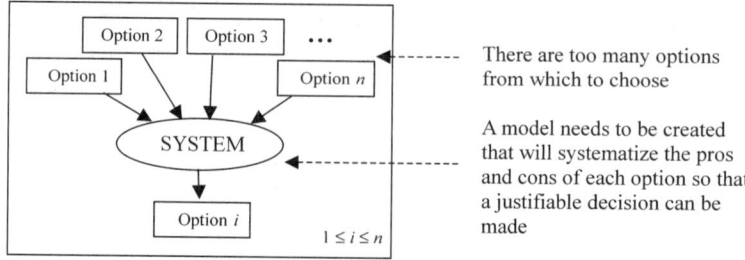

FIGURE 9–3. Modeling enables one to make a justifiable decision from among many alternatives.

be compared? Here, it helps to have a priority ranking of the attributes, which is another modeling problem.

An additional reason why a model may be required to make a decision is that frequently the decision-maker needs to justify the choice to another interested party. For example, the student's parents who are actually paying for the cell phone plan may need to be convinced that their daughter has found the most appropriate choice; the employer may need to explain to a disgruntled and unsuccessful applicant why he/she was not hired; the university administrator may have to convince the university board that the strategy being chosen is likely to yield the highest enrolment in the long run over options that the board believes will be immediately effective. In such cases, the decision-maker needs to explicitly describe the system (model) that was used to arrive at the decision.

Situation 3: A Model Is Needed in Order to Replicate or Modify the System

The need for a model will often arise when a system is to be reproduced or modified (see Figure 9–4). A clothing designer who has designed a particularly flattering cut of pants may wish to profit from her design either by producing pants or selling patterns that will enable others to do so. To do so, the designer will create computer software (or paper patterns) to enable others cut out pieces of fabric to produce pants of various sizes. In addition, she will need to provide instructions for how the pieces fit together. Just having the initial pair of pants (i.e., the system to be replicated), is not sufficient to replicate it. Instead, models are needed that are reusable and portable together with instructions for creating the system. These situations often occur because people want to capitalize on systems that have been shown to be successful.

When producing a template of a system for replication, it is also necessary to understand the new conditions under which the system is to be replicated. Consider the case of a botanist wishing to grow orchids in a greenhouse. In order to produce a workable template of the tropical ecosystem conditions under which orchids naturally thrive, the botanist will also need to know the constraints in the greenhouse environment. For example, in lieu of natural rainfall, plants receive water through artificial sprinklers, and the soil will need to have specific nutrients added to it. Knowledge of these constraints will help the botanist know what aspects of the tropical ecosystem will need to be modeled and adapted, and which can be omitted. In the example of a clothing designer, when preparing patterns for others to use, she may assume certain abilities of those using her patterns in a commercial environment and somewhat different abilities on the part of people who sew only occasionally.

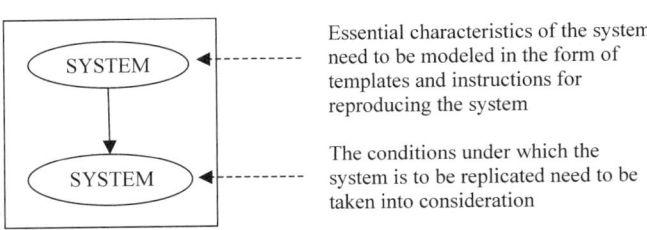

FIGURE 9–4. Modeling enables one to replicate a system.

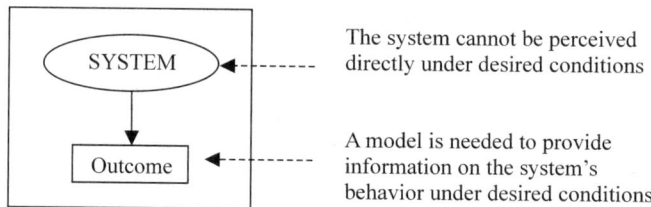

FIGURE 9-5. Modeling enables one to predict what a system might do under specified conditions.

In many circumstances models which enable easy study (or use) of systems which are modifications of the original one are especially useful.

Situation 4: A Model Is Needed in Order to Predict the Outcome of Some System Under Certain Conditions

A very popular reason why models are created is to predict what a system might do under specified conditions (see Figure 9-5). An ecologist may wish to predict the population size for an endangered turtle if nest predation is reduced 10 percent from current levels. In order to make such a prediction, the ecologist may model the turtle population using a predator-prey model, and then see how the predicted turtle population size changes when the predation level changes. Such a study may provide information on how much the predation level would have to be reduced in order to make a significant change in the turtle population size. Another model might be needed to determine how (or whether) such a reduction in predator levels might be achieved.

Sometimes the prediction is retrospective. A bank manager may wish to reconstruct a particularly busy day the previous week when people complained about the length of the queues for teller service. In order to do this, the bank manager will need information on the number of people in the queue, the length of time needed for one person to be served, and the number of servers available.

Predictions are often necessary because you want to make a decision about the system prior to an occurrence (or reoccurrence) that is undesirable—too expensive, dangerous, or impossible to observe directly. In this situation a model needs to be created that provides information regarding the system that can be used safely, inexpensively, and accurately.

Situation 5: A Model Is Needed in Order to Explain How a System Produced a Certain Outcome

Each time a car is taken to a mechanic because of some suspicious behavior—say, the engine starts making an unusual noise when the accelerator pedal is pushed, or the check oil light comes on without apparent cause—the opportunity arises for a model to be created in order to explain how the suspicious behavior is the result of some system. An experienced mechanic will examine the parts of the car and (hopefully) explain to the owner what parts of the car system are producing the observed behavior. The owner hopes that the explanation will lead to recommendations on how to resolve the problem. In situations of this nature, the modeler maps backwards from the unusual results of the system, to the mechanisms in the system which are potentially responsible (see Figure 9-6).

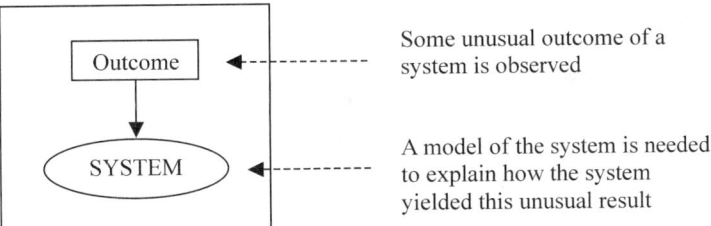

FIGURE 9-6. Modeling enables one to account for unusual outcomes of a system.

People typically seek explanations of outcomes of systems when the results are unexpected or unusual so that they can know how to respond by fixing faults in the system that gave rise to the phenomenon, or anticipating future recurrences. In most cases, an experienced practitioner will have some preliminary ideas about the source of the unexpected outcomes and will suggest tests which provide additional information. The test results may confirm the original diagnosis, or suggest the need for additional tests. The goal is to learn enough about the system so that remedial action can be taken. This process of generating ideas and seeking confirmation through tests is a common attribute of the modeling process. It is important to recognize that frequently one does not know whether positive results are a consequence of having a good model or a consequence of another, possibly completely unrelated, event.

Situation 6: A Model Is Needed in Order to Manipulate or Control the System

Sometimes, instead of wishing to predict what a system will do under certain conditions, one may wish to ensure that the system behaves in a way that yields a particular result or achieves some goal (see Figure 9-7). Accordingly, one would need to understand how to manipulate the system—what options are available—so that the desired outcome occurs, and what other consequences (good or bad) are associated with the various options. A conservation official may wish to decrease the algae in the lake by manipulating the ecosystem in which the algae survives. There are three options: controlling the nutrients in the lake, introducing natural competitors for the nutrients, or direct chemical treatment. Each option has a cost, a likelihood of achieving the desired result, and associated "side effects" that must be understood and evaluated. Rather than a single model, situations of this kind may require a collection of models and a way to establish priorities among disparate aspects of the situation.

These situations often arise when people are trying to optimize some aspect of a situation (maximize benefit, maximize the probability that something happens,

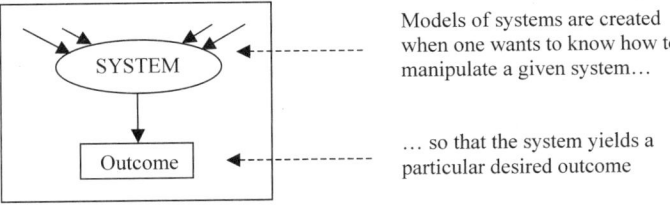

FIGURE 9-7. Modeling enables one to manipulate a system.

minimize cost, minimize risk, etc.). Many applications of mathematics in business have this characteristic, and many effective models have been developed. At the same time, there are many situations where models have been developed, but there are differences of opinion as to their effectiveness. The chairman of the Federal Reserve may wish to control interest rates and inflation, and tries to do so by raising and lowering the discount rate, and dropping comments in different settings to set future expectations. In certain circumstances the model being used seems to lead to desirable results, but there are times when its use seems less effective.

SUMMARY AND FUTURE RESEARCH DIRECTIONS

In this chapter we began with the question: "Why would one want to build a mathematical model?" We pointed out that models are often built because they enable the modeler to access some otherwise unattainable information about the system at hand. Then, we described six situations where attaining the desired information required understanding their associated systems, and showed how each case necessitated the development of a mathematical model of the system.

We have presented this description as an initial step toward a future research goal of developing a more complete catalog of situations that create the need for model-building. Such a catalog could be used by task developers to design a wider range of modeling activities. These activities function like photographic negatives in the sense that they elicit from students certain kinds of ideas that research can witness evolving as they work on the activities, much like one can watch an image slowly appear on a Polaroid picture. Consequently, a wider range of modeling activities will enable researchers of students' thinking and learning to observe and study the evolution of a larger set of students' ideas.

REFERENCES

Diefes-Dux, H., Moore, T., Zawojewski, J. S., Imbrie, P. K., & Follman, D. (2004). *A framework of posing open-ended engineering problems: Model-eliciting activities.* Proceedings of the 34th ASEE/IEEE Frontiers in Education Conference. Savannah, GA.

Hoyles, C., Noss, R., & Pozzi, S. (2001). Proportional reasoning in nursing practice. *Journal for Research in Mathematics Education, 32*(1), 4–27.

Lesh, R., & Doerr, H. (Eds.) (2003). *Beyond constructivism: Models and modeling perspectives on mathematics problem solving, learning and teaching.* Mahwah, NJ: Lawrence Erlbaum Associates.

Lesh, R., Hoover, M., Hole, B., Kelly, A., & Post, T. (2000). Principles for developing thought-revealing activities for students and teachers. In A. Kelly & R. Lesh (Eds.), *Handbook of research design in mathematics and science education.* Mahwah, NJ: Lawrence Erlbaum Associates.

CHAPTER 10

Cultivating Modeling Abilities

Caroline Yoon and Maynard Thompson
Indiana University

Experts in mathematical modeling often point out that the only way to get better at modeling is by doing it (Edwards & Hamson; 1989, Maki & Thompson, 1973). Reading books and guides on mathematical modeling may give insight into the modeling process, but will not by themselves endow the reader with the ability to go through that same process on their own. Also, understanding a wide variety of formally constructed mathematical models may give insight into the kind of products that result from modeling, but does not provide the ability to develop one's own models. In short, most experts will claim that no amount of preparation, no matter how thorough, will substitute for actual time spent modeling.

Although we agree with this sentiment, we know that the solution is not simply to swing to the other extreme, and give students modeling activities to work on without *any* preparatory instruction. Aside from rare exceptions, students' modeling abilities do not usually improve just by virtue of doing more modeling activities. In fact, the unguided repeated exercise may even reinforce bad modeling habits instead of helping them develop productive ones. It seems that cultivating students' modeling abilities requires some form of guidance that both builds on and enhances the students' own modeling experiences, whilst not attempting to take their place.

In this chapter we describe one approach to cultivating modeling abilities that combines teacher guidance with students' own actual modeling experiences. In this approach, the teacher takes on the role of a post-game coach, encouraging students to actively reflect on their modeling experiences after each activity. We do not claim originality for this approach, being aware that many experienced teachers of modeling already use some variant of this form of instruction. What we do offer is a theoretical explanation for why we believe students' modeling abilities can be cultivated in this way, as well as some concrete suggestions as to how to make this approach even more effective.

We begin by defining in very simple terms what we mean by "modeling abilities". Next, we consider an approach that tries to "teach" such abilities by training students to use specific modeling skills. We contrast this approach with the method we advocate, which focuses on getting students to reflect back on the productivity of the modeling skills they used in their own modeling. At the heart of this reflection is a consideration of how the skills enabled (or prevented) the modeler to interpret the situation in new ways, and subsequently develop a new model based on this new way of thinking. Finally, we end the chapter by describing the design of reflection tools that are intended to facilitate students' reflection on their own modeling experiences.

WHAT DO WE MEAN BY 'MODELING ABILITIES'?

In many peoples' eyes, "mathematical modeling" is little more than the act of applying traditional mathematics (e.g., ideas from geometry, calculus, algebra, etc.) to solve real-world problems. Under this perspective, mathematical modelers are viewed as people who, through years of learning and study, have amassed and stored a vast selection of mathematical models in memory and other resources. Then, when they are presented with a new problem, they need only find the right mathematical model from their storehouse to fit the situation, just as a librarian will find a patron's desired book from rows and rows of shelved items. However, this image of the mathematical modeler as curator of ready-to-use models is incomplete. Although most expert modelers certainly know a wide range of mathematics and a variety of successful models, they do not so much fit pre-assembled, intact models into new situations as they create "new" models for each situation they are given. For sure, these models that they develop are not usually entirely original, but will often draw on mathematical tools that they know of and have used before. Nevertheless, it is seldom the case that the problematic situation can be modeled perfectly by an intact available model. Modeling (i.e., model-development) is in large part, a creative activity.

If we consider a more familiar creative activity—creative writing, we notice that writers do not usually pen polished short stories and poems in one sitting, but develop each piece by reworking and revising it through multiple drafts. Similarly, mathematical modelers will seldom develop a new model for a given situation without considering, revising and rejecting several alternative models. An overview of the various stages involved in this process is presented below:

> Beginning in the "real world" with a situation that needs to be modeled, the modeler will try to formulate a mathematical model that describes the relationships and interactions in the situation mathematically. Then, the modeler will obtain the model's results, either by letting the model "run" (if it's in the form of a computer program), or by manipulating mathematical symbols to derive a solution. These results will then be interpreted back into the original real-world situation, whereupon the modeler will choose whether to revise or reject it and go through the cycle again. Typically, a modeler will go through this cycle multiple times, shuttling back and forth between designing a mathematical model and manipulating it in the model world, then testing it out in the real world, before presenting the final solution to the client.

Many discussions and textbooks on mathematical modeling have described the modeling process in a similar manner (Giordano & Weir, 1985; Maki & Thompson, 1973; Maki & Thompson, 2006; Roberts, 1976). Therefore, in this chapter, when we talk about cultivating students' *modeling abilities*, we are simply referring to their ability to go through multiple modeling cycles as described above to produce a mathematical model for the given situation. But how can such modeling abilities be cultivated effectively in students? This diagram is not prescriptive. That is, while it may give students a good overview of the modeling process, it does not show them how to go through it on their own.

In the next two sections we present in turn, two different approaches for cultivating students' modeling abilities. The first one takes a rather didactic approach, whereas the second, less common approach is the reflective method that we advocate. In order to point out clearly the benefits of the second approach, we have chosen to exaggerate both approaches, casting them as idealized opposite poles. In reality, most instructional approaches will be much less extreme, and indeed, the best approach is probably a combination of both. However, our task is not to propose the "best approach", but merely to suggest that some kind of post-activity reflection would be a useful component of any attempt to cultivate students' modeling abilities.

ONE APPROACH TO CULTIVATING MODELING ABILITIES: TEACHING SPECIFIC MODELING SKILLS

One way educators have tried to make diagrams like the modeling flowchart in Figure 10–1 more prescriptive has been to add more detail—specifically, by breaking up each stage of the diagram into more precise modeling skills, such as:

- listing factors
- making approximations and simplifications
- making assumptions
- choosing mathematical structures
- translating into mathematics
- isolating variables
- reducing the number of parameters
(Part of this list was taken from Edwards & Hamson, 1989)

A common approach to cultivating modeling abilities, then, centers on teaching students these individual modeling skills. Under this approach, instructors will demonstrate the use of each skill with portions of a modeling problem that have been carefully selected so as to highlight the usefulness of the skill. After observing the instructor using each of these skills, students will then practice using the skills themselves on similarly selected portions of problems. And through repeated demonstration and practice, students will build up an impressive knowledge of modeling-related skills.

Expert modelers can indeed be observed using such modeling skills in their own modeling. However, while expert modelers use them naturally and productively, students who have been taught modeling skills in this didactic manner are often less adept at using them in their own subsequent modeling activities.

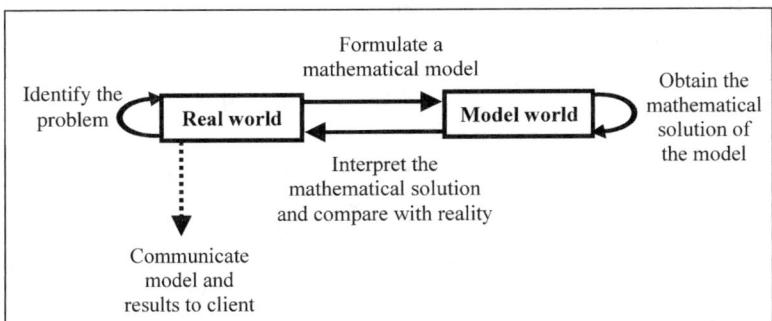

FIGURE 10–1. Modeling flowchart adapted from Edwards & Hamson (1989) and Lesh & Doerr (2003).

Without someone there to tell them when to use the skills at the appropriate stages, students often either forget to use them at all, or may use them in a way that actually interferes with their modeling progress (Zawojewski & Lesh, 2003). For example, they may make assumptions that prevent them from noticing certain relationships that need to be modeled later on. They may not realize that their model could be improved if they revised their assumptions throughout the activity as they notice new details and relationships in the system being modeled. This is because they have not learned how to recognize when each of the skills needs to be used during the modeling process, but only how to perform them when told to do so. This situation can be likened to a dysfunctional restaurant kitchen staffed only by chef's assistants that are highly trained to perform certain specialized skills (slicing, sauce-making, etc.,) but have no head chef to tell them what to do. Without the directions of a head chef that can oversee the production of the meals, the individual chef assistants may just sit around doing nothing. Consequently, none of the requested dishes would get made despite the abundance of talent in the kitchen.

If we were to take this approach in cultivating students' modeling abilities, then, we would need to do the equivalent of introducing a head chef into the kitchen. That is, we would need to give students an accompanying managerial understanding of when to appropriately use each of the skills they have been trained to perform. It is not clear however, that this can be done in the same way that students are trained to use the skills. The appropriateness of any given skill depends on so many different interacting conditions that learning when to use each one appropriately would be impractical and unnecessarily tedious. An alternative approach that is sometimes suggested is to have someone (a teacher, or another student) act as a monitor that observes the students' progress and suggests that they use certain skills at certain times. The problem with this is that although students may learn to automatically start using certain skills at certain times, each situation is different, and no prescribed algorithm is going to work at all times.

AN ALTERNATIVE APPROACH TO CULTIVATING STUDENTS' MODELING ABILITIES: REFLECTING ON THEIR OWN MODELING EXPERIENCES

In the previous section we described how students are often unable to use modeling skills productively in their own modeling activities, despite having successfully learned them in prior practice exercises. In contrast, researchers have pointed out that in some situations, students without any formal training in modeling skills will spontaneously and appropriately use them while engaging in modeling activities. In particular, Zawojewski and Lesh (2003) have given examples of students creating their own specialized skills when they are needed. That is, students naturally develop a sense of when to use certain skills without necessarily knowing the names of the skills they are using, or even without being consciously aware that they are using them at all. However, the skills that they spontaneously use are often primitive versions of the ones formally taught in the approach described above (see Zawojewski & Lesh, 2003). In restaurant kitchen terms, they already have a "head chef" that organizes the assistants productively, but the assistants themselves are mediocre. Thus we are faced with the opposite problem—namely, we need students to integrate more sophisticated modeling skills into their practice—i.e., we would wish the chef to get acquainted with some more highly skilled assistants.

The approach we advocate centers around having students reflect back on the success of the modeling process they engaged in after each modeling activity. In this approach, the instructor behaves like a post-game coach, encouraging students to actively reflect back and identify the modeling skills that they spontaneously used, and assess how productive they were in helping them in their modeling. The word "reflect" is often taken to mean engaging in dreamy reminiscence, but in this chapter, we use it in a very active and focused sense. Students use recall techniques to become aware of the modeling skills they used during the activity, and analyze how these skills helped or hindered their modeling efforts.

This practice of post-activity reflection offers a definite advantage over the alternative approach, by not upsetting the process of modeling as it occurs naturally. As we already pointed out, students (and experts) naturally use these skills in an intuitive and spontaneous manner, rather than consciously and deliberately. For example, students do not usually consciously say "now let us move to the next skill on the list," but instead, do so when the need arises. Indeed, forcing students to consciously think about the skills that they should use *while they are working* on a modeling problem will likely distract them from focusing on the task at hand. Furthermore, excessive analysis may confuse them to the point of being unable to perform what was once natural and spontaneous activity—much like the well-known fable of the centipede becoming unable to walk when asked to explain how it moves its feet. Therefore, instead of asking students to consciously think about why they should use some skill while they are modeling, post-activity reflection gives students an opportunity to think about the productivity of skills they used after the fact, so that they need not be repeatedly disrupted during the activity itself.

But how will a student know whether a skill was productively used or not? At the heart of the modeling process is the desire to revise one's current model in light of new evidence that does not fit into one's current way of thinking about the situation, or is inconsistent with the results of one's model. Therefore, modeling skills like those described earlier can be used productively if they help lead you to a new way of thinking about the situation. For example, the practice of "listing factors" can help focus your attention on identifying deeper relationships between factors in the situation that affect what you want to find out. However, the very same skills can be used in an unproductive way if they serve to lock you into a primitive way of thinking early on (for a similar argument about the productive use of problem solving heuristics, see Zawojewski and Lesh, 2003). To use the same example, listing factors may prevent you from noticing new relationships between items you had not initially listed if you adhere to your original list too closely. Modeling skills are not inherently useful or not useful in themselves—but only insofar as they enable the modeler to express, test, and revise their current way of thinking that is reflected in the model they are designing. Consequently, when we ask students to reflect on how productively they used various modeling skills in their own modeling, we are ultimately asking them to consider how their use of the skill did or did not result in a shift in their thinking.

MODELING CYCLES: SHIFTS IN THINKING/INTERPRETATION

In this section, we take a short detour from describing the approach that we advocate for cultivating modeling abilities, and describe in more detail, how the process of modeling can be thought of as a sequence of shifts in thinking. When describing the conceptual shifts that are the result of repeated modeling cycles, researchers sometimes use the terms "interpretation", "model", and "way of thinking" interchangeably (Lesh & Doerr, 2003). Ultimately, the model that is developed is an expression of one's current way of thinking, or interpretation. Therefore, what really changes first is the modeler's interpretation or way of thinking, which is then expressed in a model. To some extent models are just formalized externalizations of one's interpretation of the situation. The goal then, is to change the person's interpretation so that it will be reflected in the model that is produced. However, once expressed, this model becomes a new window through which the situation is re-interpreted, and subsequent model-reality mismatches become the source of further changes to one's interpretations. Thus, models themselves contribute to helping one change one's way of thinking.

We will illustrate how shifts in thinking occur during modeling by considering a concrete example—the volleyball activity as described in chapter 17, this volume, by Lesh, Yoon, Zawojeski. In this activity, the modelers are given tables of information on eighteen volleyball players' abilities, such as the individual players' heights, jumping ability, speed, ball skills (spike and serve results), as well as comments made by their coach. They are asked to use this information to develop a way of sorting the eighteen players into three fair teams of six players each, as well as a general procedure for sorting out arbitrarily large groups of players into fair teams. Determining what "fair" means is itself part of the task. The following points show

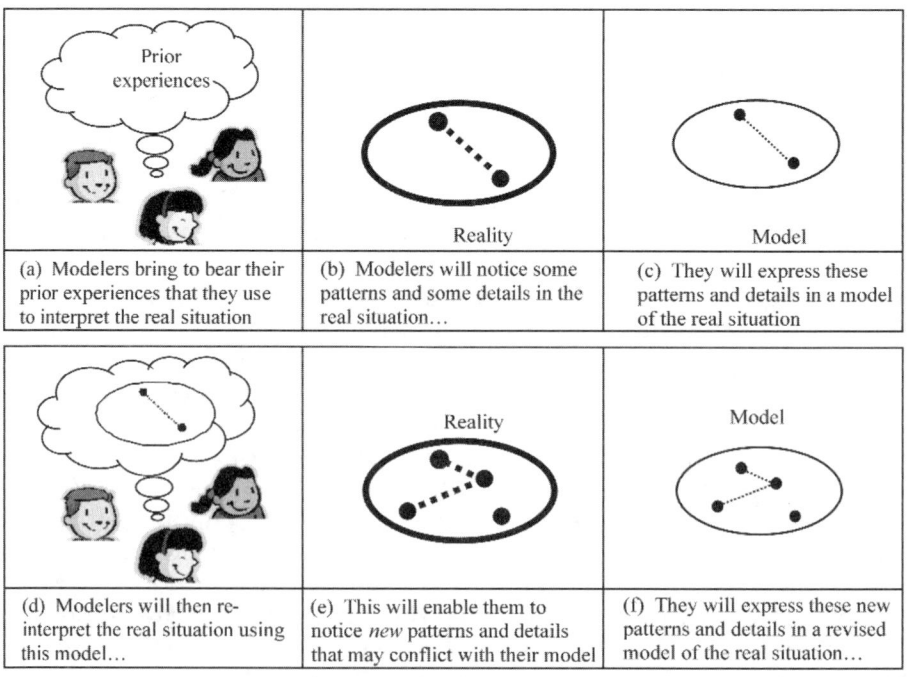

FIGURE 10–2. Diagram showing how modeling is a process whereby one's interpretations get successively refined and extended.

how modelers' ways of thinking undergo significant shifts in the volleyball problem. These points are then illustrated in Figure 10–2.

Modelers bring to bear their prior experiences that they use to interpret the real situation. Although they may not have encountered this specific problem before, the modelers would probably have experienced situations where rankings were used to order and differentiate between people competing in a task. For example, they may have seen judges of singing competitions, or gymnastics, or piano recitals, or seen how grades are given for academic achievement in their own classrooms. So their first thought might be to interpret this situation as a place where the players' scores should be aggregated in some way to find an overall ranking, and then sort the players out according to their ranks. Modeling always begins by interpreting the situation using some prior experiences which help you to organize the information in the situation accordingly.

Modelers express their interpretation of the situation in a mathematical model. Having decided that the way to approach this activity is to aggregate the scores in the categories, the modelers may next try to make sure that the scores can indeed be aggregated. However, it may become apparent that some scores do not count the same way—high scores are preferred in serving ability, whereas high scores in running speed are not as they indicate slowness, and consequent adjustments will need to be made. Additionally, qualitative or descriptive information may need to be converted into quantitative information, and some sets of quantitative information might have to be converted into common units before they can be combined (for example, students' heights may need to be converted into inches before they are added to their vertical leap). In doing so, the modelers develop a mathematical model that structures the information so that it fits into their interpretation.

The model then becomes a lens through which the modeler can re-interpret the situation. Having developed a model, the modelers can use it to interpret the situation again, and will likely notice new patterns and details that were not previously obvious. They may find that the information does not fit in the way they wanted it to, but that it would be better described using some other structure. For example, they may decide that some of the categories of abilities are more important than others (the players' running scores are not as crucial as their ball handling scores), and may consequently assign different weights to each of the scores (so that ball handling scores count twice as much as running scores).

The model will be revised in light of the new interpretation, and so on. Sometimes modelers will only have to make minor adjustments to their model in light of their new interpretation, like having to weight some categories of abilities heavier than others. At other times, the new interpretation may be so at odds with the previous model that it will warrant a complete overhaul. For example, the modelers may decide that instead sorting players into teams based on their overall ranking, a more efficient approach may be to score each player's ability in various things, and sort out the teams in terms of having equal amounts of each type of skill, regardless of the player's overall score. In any case, regardless of the scale of the revisions, the new model will again be able to be used as a lens to re-interpret the situation, and will thereby potentially lead to multiple further shifts in the modelers' interpretations.

REFLECTION TOOLS

A productive area of research in modeling instruction has been in the design of reflection tools for assisting students' post-activity reflection on various aspects of their modeling processes (see the Hamilton, Lesh, Lester, Yoon chapter in this book). In general, the purpose of reflection tools is to preserve a record of the students' modeling process, focusing in on particular areas of interest (like students' engagement, group roles, or modeling skills). Students can then use these tools to reflect back on their modeling process, and make decisions about how to modify their modeling approach in subsequent modeling activities. In this section we give a brief outline of how reflection tools could focus on students' use of modeling skills so as to facilitate their post-activity reflection that we described in the approach of cultivating modeling abilities.

As we pointed out earlier, the productivity of any given modeling skill ultimately depends on whether it helps one interpret the situation, produce results which are consistent with observations, and notice new patterns and details in the situation that one did not see previously. Therefore, in order to reflect on the productivity of the modeling skills used while modeling, the student needs to consider three things: (1) the modeling skills that were spontaneously used, (2) the shifts in their ways of thinking that occurred during the modeling activity, and (3) how the modeling skills either enabled or prevented these shifts to occur. However, since the modeling skills were used spontaneously and without conscious thought, it is often difficult for students to recall what skills were used after they have finished using them. Similarly, by the time students have produced their final model, it is not easy to recall each of the previous ways of thinking that preceded their final way of thinking. Reflection tools for modeling skills would ideally preserve records of fluctuations in both the kinds of modeling skills used and the ways of thinking that were revised and rejected so that students can reflect on how the former might have influenced the latter. Effective design and implementation of these tools will be a next step in research into cultivating students' modeling abilities through post-activity reflection. These issues are discussed in more detail in the Hamilton et al chapter in this book.

SUMMARY AND FUTURE DIRECTIONS

In this chapter, we started off with the question—how can we get students to develop their modeling abilities? We showed why learning modeling skills through teacher demonstration does not necessarily help one develop a better model, and argued instead for an alternative approach where students reflected post-activity on the skills they spontaneously used in their own modeling experiences. Next, we showed how these skills were only useful insofar as they helped the modeler to get to the next cycle of interpretation, and described how the essence of modeling involves refining one's interpretations of the situation. Finally, we suggested that reflection tools could be a productive way of helping students to reflect on the productivity of the skills they used while engaged in the modeling activity during post-activity reflection and debriefing sessions. One of the immediate goals ahead for research on the cultivation of students' modeling abilities will be to develop these reflection tools more fully.

REFERENCES

Edwards, D., & Hamson, M. (1989). *Guide to mathematical modeling.* New York: Macmillan.
Giordano, F. R., & Weir, M. D. (1985). *A first course in mathematical modeling.* Monterey, CA: Brooks/Cole.

Lesh, R., & Doerr, H. (Eds.) (2003). *Beyond constructivism: Models and modeling perspectives on mathematics problem solving, learning and teaching.* Mahwah, NJ: Lawrence Erlbaum Associates.

Maki, D. P., & Thompson, M. (1973). *Mathematical models and applications.* Englewood Cliffs, NJ: Prentice-Hall.

Maki, D. P., & Thompson, M. (2006). *Mathematical models and computer simulation.* Belmont, CA: Brooks/Cole.

Roberts, F. S. (1976). *Discrete mathematical models.* Englewood Cliffs, NJ: Prentice-Hall.

Zawojewski, J.S., & Lesh, R.A. (2003). A models and modeling perspective on problem solving. In R.A. Lesh & H. Doerr (Eds), *Beyond constructivism: Models and modeling perspectives on mathematics problem solving, learning and teaching.* Mahwah, NJ: Lawrence Erlbaum Associates.

CHAPTER 11

Discrete Mathematics in 21st Century Education: An Opportunity to Retreat from the Rush to Calculus

Joseph G. Rosenstein
Rutgers University

One way of characterizing K–12 mathematics education in the second half of the 20th century is that it has been a rush to calculus. Perhaps the first half of the 21st century will be seen as a time of retreat from that rush, when, at the high school level, mathematics is enriched by new topics like probability, statistics, and discrete mathematics; traditional topics that lead to calculus are treated more deliberately and more thoroughly; and these perspectives are reflected throughout the curriculum. This chapter anticipates that discrete mathematics topics will play a larger role in the K–12 curriculum, as more educators learn these topics and recognize their value in achieving the goal of improving students' understanding of mathematics.

THE RUSH TO CALCULUS

The phrase "rush to calculus" captures one significant aspect of the changes in mathematics education at the high school level over the past 50 years.

At the high school that I attended and graduated from in 1957—Benjamin Franklin High School in Rochester, New York—I took trigonometry and solid geometry in my senior year and, at the end of the year, took the New York State Regents Examinations in those subjects. When I enrolled at Columbia College that fall, the honors class that I attended began with analytic geometry; my classmates included graduates of Bronx Science and other selective schools, so my mathematical background was not unusual. By today's standards we would be at least three semesters behind[1]; indeed, 73 percent of the 61 students in my first-semester calculus class in Fall 2002 at Rutgers University had taken at least one year of calculus in high school.

---

[1] Starting college "three semesters behind" did not lead to mathematical disaster for me. Indeed, I earned a Ph.D. in mathematics and continued mathematical research for fifteen years before focusing my attention on improving K–12 mathematics education.

What changed? Certainly a major contributor was the introduction and popularization of the Advanced Placement Examination in Calculus. "Advanced Placement" is an alluring title and an alluring idea. Any parent would love to hear that his or her child is "advanced" and to have that confirmed by the child's acceptance into a course that is nationally recognized as "advanced." Moreover, there is the promise that the child would start college ahead of his or her peers, take even more advanced courses earlier, and maybe even complete college work ahead of schedule.

Here is how the Advanced Placement Program is promoted on its Web site: "AP can change your life. Through college-level AP courses, you enter a universe of knowledge that might otherwise remain unexplored in high school; through AP Exams, you have the opportunity to earn credit or advanced standing at most of the nation's colleges and universities." Moreover, the Web site indicates that by taking AP courses and examinations, students will "gain the edge in college preparation, stand out in the college admissions process, and broaden your intellectual horizons."

Not only will you "gain the edge in college preparation," but you will also "stand out in the college admissions process." I don't know whether this is true in general, but admissions officers at seven major state universities called in May 2005 by Rutgers graduate student Allison McCulloch all indicated that AP Calculus would be considered more favorably that any other high school mathematics course. Interestingly enough, they all indicated that scores on the AP Calculus examination had no weight in their decisions, presumably because test scores would not arrive until well past the time for admissions decisions. Thus, simply taking the AP Calculus course is considered a positive factor by college admissions officers.

This practice is called into question by a recent "major study by researchers at the University of California at Berkeley" that shows that "college-level courses offered in high school, such as Advanced Placement (AP) or International Baccalaureate (IB), do not appear to improve academic performance in college, unless students take the tests at the end of each course."[2] The illusion has been established, however, that simply taking AP Calculus leads to success.[3]

A second major contributor to the rush to calculus is that parents all around the country have somehow come to the conclusion that taking calculus, even if it's not AP calculus, will make a decisive difference in getting their children admitted to college. That conclusion was not supported by the calls to the admissions officers; indeed, the general sense seems to be that taking a non-honors course in calculus would usually not make much of a difference, although taking mathematics for four years was clearly preferred. How important calculus is in the admissions process is, thus, not very clear, and is a question that bears investigation.

Another question that bears investigation is what percentage of students who take calculus actually obtain advanced standing. Nearly 20 years ago, long before college admissions were handled electronically, I physically went through the folders of hundreds of Rutgers College first-year students to see what they had taken in high school. I do not know what happened to the raw data that I collected, but what I remember clearly is that:

- about half of the students who took calculus in high school took an AP course,
- about half of those who took the AP Calculus course took the exam,

---

[2] As reported in the *Washington Post*, December 23, 2004.

[3] This illusion is reflected in President Bush's plan to train 70,000 high school teachers to lead AP math and science courses as part of the "American Competitiveness Initiative," when there already exists an abundance of skilled professionals who can and do teach this material just one year later – college professors.

- about half of those who took the AP Calculus exam received grades of 4 or 5 (sufficient to get college credit), and
- about half of those who received grades of 4 or 5 actually continued on an accelerated path.

That means that about 1 out of 16 students who were accelerated into a calculus course in high school actually took advantage of that acceleration by earning and making use of advanced placement in mathematics. So for the students who took calculus in high school, the main outcome of their acceleration was that when they got to college they could slow down and repeat calculus. That does not make much sense.

I am currently engaged in a more careful study of more recent data.[4] Of the 2,120 students entering Rutgers College in the fall of 2003, exactly 332 (or 15.7 percent) took the AP Calculus exam (or, more precisely, Rutgers received copies of their scores). Of these students, 161 (just less than half) received college credits; 112 received 4 credits and 49 received 8 credits. This data is consistent with the third bullet in the previous paragraph. How many of those 161 students continued on their accelerated track? For those who received 4 credits, an appropriate definition seems to be that they successfully completed second and third semester calculus in their first year at Rutgers; 39 did and 72 did not. For those who received 8 credits, an appropriate definition seems to be that they successfully completed two of the three second-year mathematics courses (multivariate calculus, linear algebra, differential equations) in their first year at Rutgers; 26 did and 23 did not. Altogether, of the 161 students who were able to start college on an accelerated path, only 65 continued that acceleration. These findings are consistent with the final bullet above.

The second part of the study involves examining the high school transcripts of 335 students randomly chosen from the cohort of 2,120 students described above. Of these 335 students, 105 students (29.6 percent) took an AP calculus course in high school and an additional 82 students (24.5 percent) took a calculus course that was not an AP calculus course. This is consistent with the first bullet above; although more schools are offering and more students are taking AP calculus courses, more students are also taking non-AP calculus courses; altogether over half of our incoming students have taken a full-year calculus course, which was certainly not the case 20 years ago. Of the 105 students whose transcripts indicated that they took an AP calculus course, only 51 appeared on the list of students who took the AP calculus exam; the other 54 did not. This is consistent with the second bullet above.

Altogether, we can say that 105/187 of the students who took calculus in high school actually took an AP calculus course, that 51/105 of those who took an AP calculus course took the AP exam, that 161/332 of those who took the exam received credits for calculus, and 65/161 of those who received credits continued in their accelerated path. Combining these fractions, we conclude that fewer than1 out of 18 (5.3 percent) of those who took calculus in high school continued in their accelerated path. These results corroborate the earlier informal findings, but these results still do not make sense.

---

[4]The results of this study will be published as "The Rush to calculus: An analysis of Course-Taking Patterns by Rutgers students in High School and College" (2007).

Now one might argue that high school students who take calculus do in fact "gain an edge in college preparation" even if they take calculus over again in college. The reasoning is that the second time you see the material you are able to consolidate and move beyond what you learned the first time. However, the problems first-year students have with calculus are not generally with the concepts of calculus but with the manipulations of algebra. For example, in evaluating the expression for the derivative of the function $f(x) = 1/x$, the expression $[1/(x+h)-1/x]$ arises. Some students will not recognize that the appropriate step is to combine the two fractions into a single fraction, others will not know how to combine two fractions, and many who are aware of the procedure will implement it incorrectly. Students often make errors that reflect misunderstandings of the fundamentals of algebra (such as inappropriately canceling terms in fractions). Taking calculus a second time will not correct these misconceptions; a more demanding high school algebra course might be a better strategy. It is no wonder that many college teachers would prefer their students to come to college with stronger algebraic skills and understandings than with a background in calculus.

In order to take calculus as seniors, students take an accelerated math program that starts with Algebra I in grade 8, and continues with Geometry in grade 9, Algebra II in grade 10, Precalculus in grade 11, and Calculus in grade 12. We have already seen that only a small percentage of the students who take calculus benefit from this acceleration and that most indeed come to college with insufficient understanding of the topics of high school mathematics. What about those who do not eventually take calculus in the 12th grade? It is true that many students seem to be ready to take Algebra I in the eighth grade; however, many of them will not be ready for Algebra II in the tenth grade. At one school where we interviewed students a number of years ago, we found that students who thought that they were mathematically competent hit a wall when they took Algebra II in the tenth grade. The consequence of their taking Algebra I earlier than necessary was that they ended up dropping out of math.

Thus the rush to calculus has a negative effect on the students who do not get to calculus as well as on those who do. And by making calculus the norm forcollege-bound students, the rush to calculus ends up distorting the entire high school curriculum, so that students chase after calculus rather than learning algebra and geometry well and studying mathematics topics that might be more valuable to them, like probability, statistics, and discrete mathematics. The rush to calculus, and the resulting rush *through* algebra, geometry, and trigonometry, produces students who have a myopic view of those subjects, that is, they may have an ability to carry out narrowly prescribed operations, but do not have the big picture.

This phenomenon is likely to be exacerbated by the current fad, a rush to algebra. More and more students are being encouraged to take a formal algebra course in grade 8, despite the fact that many educators believe that not all students are cognitively ready for a formal algebra course in grade 8. The initiative that all children should take Algebra I in grade 8 has a noble goal, that of equity. It is apparently based on the idea that calculus in high school is a positive goal and, from the perspective of equity, should therefore be a universal goal. Unfortunately, because of the state of elementary and middle school mathematics education in the United States, most students are not prepared to take a formal algebra course in grade 8 and most 8th grade mathematics teachers do not have the mathematical background to provide students with a strong grounding in algebra. For all these reasons, the attempts by states and districts to ensure that all students take a traditional Algebra I course in the 8th grade are, in my opinion, misguided.[5]

We pay a big price for rushing to calculus and rushing through algebra, with very little payoff. We accelerate all students so that a few can earn college credits for calculus, and then only a small percentage of those who are in a position to benefit from those credits actually do so. We develop curricula that focus on calculus, depriving students of the opportunity to learn other mathematical topics and to gain mastery of algebra and geometry.

WHAT ARE THE ALTERNATIVES?

As noted in the abstract, at the high school level, mathematics can be enriched by new topics like probability, statistics, and discrete mathematics, and traditional topics that lead to calculus can and should be treated more deliberately and more thoroughly. At elementary and middle school levels, students should be learning other topics as well as the traditional ones. The rush to and through algebra should be decelerated.

In this chapter I will concentrate on the discrete mathematics option since that has been one of my main focuses for the last 15 years.[6]

Some may ask, using language that emerged from the discussions of the Third International Mathematics and Science Study (TIMSS) Report, whether our mathematics curriculum is already "a mile wide and an inch deep"? That is, are we not already including too many topics in our mathematics curriculum and not going into them at sufficient depth? Certainly the lack of depth and thoroughness is a problem, as noted earlier. However, the argument that we should reduce the number of topics is not convincing. The set of topics covered in the TIMMS assessment were those common to the curricula of all of the participating countries. So topics such as probability, statistics, and discrete mathematics were excluded from consideration. It would not be unexpected for the performance of American students on the common topics to be slightly lower than in other countries, where 100 percent of instructional time was devoted to those common topics. If our goal were success on the TIMSS assessment, then we could argue for reducing the number of topics. But why should we set our sights so low, when we could expect students to achieve a broader and deeper understanding of more topics than what is expected in other countries?

WHAT IS DISCRETE MATHEMATICS?

Unfortunately, there is no simple answer to this question. Unlike algebra, geometry, probability or calculus, discrete mathematics is not a well-defined area of mathematics. One could contrast "discrete" with "continuous": Whereas the functions in calculus (and, more generally, real analysis) typically have domain consisting of all

---

[5]Writing on "Pushing Algebra Down" in the March 2005 News Bulletin of the National Council of Teachers of Mathematics, President Cathy Seely notes that "students should take algebra early only if they are highly motivated to do so, only if they intend to study mathematics through the calculus or statistics level, only if the system is structured to accommodate this advanced study, and only if the student also studies proportionality and the most important components of a good middle school program. Far more important than *when* they study algebra is *what* they study and *whether* they are taught in a way that helps them learn it and use it for the long term."

[6]The efforts over the past 25 years to introduce discrete mathematics into schools in the United States are described in DeBellis and Rosenstein (2004).

real numbers (the "continuum"), the domains of functions in discrete mathematics are typically the natural numbers; their graphs are discrete sets of points rather than continuous curves. But dividing all of mathematics into "discrete" and "continuous" places too many topics in the "discrete" basket. There is no general agreement on which topics should actually be included in discrete mathematics; indeed, an article by Stephen Maurer in *Discrete Mathematics in the Schools*[7] is called "The Many Definitions of Discrete Mathematics."[8]

Instead of trying to *define* "discrete mathematics" we may agree on some areas that it includes. Indeed, the *Principles and Standards for School Mathematics* (PSSM) of the National Council of Teachers of Mathematics (NCTM) recognizes three particular areas—vertex-edge graphs (often referred to as "networks"), combinatorics (that is, systematic counting), and iteration and recursion (modeling change discretely)—as important at all K–12 grade levels, and adds matrices as an important area at the high school level. However, several important areas are omitted from PSSM, like the constellation of topics that are related to fairness and social choice (e.g., apportionment, elections, fair division) and the topics that are related to information (e.g., codes, sorting).

In any case, discrete mathematics is a wide-ranging collection of mathematical topics including coloring maps, finding shortest routes, scheduling tournaments, constructing fractals, conducting elections, sorting, and counting systematically. Here is a list of simply-stated problems that fall under the rubric of discrete mathematics, following the order of the five areas in the preceding paragraph:

Vertex-Edge Graphs

- Which way of connecting a number of sites into a network involves the least cable?
- What's the best way for a robot to pick up items stored in an automated warehouse, or for a courier to collect deposits at all ATM machines in the assigned region?
- What is the smallest number of colors needed to color the 48 states in the continental United States if states that share a border must be colored with different colors (so that all borders can be clearly distinguished)?

Combinatorics

- How many different pizzas can you have if each pizza must have at most three of the eight available toppings?
- How many tickets do you have to buy to make sure that you have a winning ticket in the contest that involves correctly selecting six numbers from 1 to 36?

---

[7]Two important collections of articles about discrete mathematics in K–12 education are *Discrete Mathematics in the Schools* (1997) edited by Joseph G. Rosenstein, Deborah Franzblau, and Fred Roberts, and *Discrete Mathematics Across the Curriculum K–12*, the 1991 NCTM Yearbook, edited by Margaret Kenney and Christopher Hirsch.

[8]Maurer explores and rejects four approaches that attempt to define discrete mathematics by "specifying the properties," including, for example, "discrete mathematics is the mathematics of discrete sets" and "discrete mathematics is any mathematics that doesn't involve limits." He also provides lists of topics for five different textbooks that focus on discrete mathematics. Finally, he discusses the possible ways of grouping discrete mathematics topics by their emphases and by their goals.

Iteration and Recursion

- What should be the daily dose of medication if, to function effectively, the medication must be at a specified concentration and if a given percentage of the medication in the body is eliminated each day?
- If the population of deer increases by 10 percent each year, how long will it take the population of deer to double?

Social Choice

- What is the best system for reapportioning the 435 seats in the United States House of Representatives among the states after each census? What system is actually used?
- What is a good strategy for dividing up a pie among three people so that each is satisfied with the portion he or she receives?

Information

- What is the quickest way of alphabetizing a list of 1,000 names—on index cards, or in a database?
- How are transmission errors detected and corrected when coded versions of pictures are sent from space?

If discrete mathematics is so important, why didn't we learn about it when we went to school? One simple answer is that, when we went to school, many of these problems were computationally too difficult to solve. Consider for example the problems in the second bullet above, variations of what is traditionally called the Traveling Salesman Problem. If the courier has to collect deposits from 10 ATMs, she can do that in any one of 10! possible orders since she can visit any one of 10 sites first, then any one of the remaining nine sites, then any one of eight sites, etc. To determine which route would be shortest (or would take the least time) would require an examination of over three million possible routes. It is only in the age of computers that such a task might be possible. Once computers became available, the problems of discrete mathematics achieved a greater degree of importance, both practically and theoretically.

WHY SHOULD DISCRETE MATHEMATICS BE PART OF THE K–12 CURRICULUM?

One reason that discrete mathematics should be part of the K–12 curriculum is that it is widely used. It is a rapidly growing and increasingly used area of mathematics with many practical and relevant applications, particularly in the technological and information sciences. Vertex-edge graphs are used, for example, to model the networks that underlie telecommunications systems and the flow charts that present a complex project in terms of simpler components. Combinatorics is fundamental to probability and to modern applications of codes. Iteration and recursion are the mathematical ideas that are basic to the study of discrete change in economics and biology. Discrete mathematics is grounded in real-world problems (like those in the list above), and, as noted in NCTM's *Principles and Standards for School Mathematics*, "As an active branch of contemporary mathematics that is widely used in business and industry, discrete mathematics should be an integral part of the school mathematics curriculum" (p. 31).

That discrete mathematics is widely used obviously has implications for those students who anticipate using mathematics in their future endeavors, since it is likely that the applications that they will encounter will involve the use of discrete mathematics. However, the wide use of discrete mathematics also has implications for those students who may not use mathematics in their careers and even for those students who do not anticipate going to college. As informed citizens and consumers, they should be aware of problems like those listed above and how mathematics can help solve them.

A related reason for including discrete mathematics in the high school curriculum is that it provides ready answers to the question that high school students repeatedly ask: "What is all this mathematics good for?" Many of the applications of discrete mathematics are accessible to all high school students. Students can see the applicability of algebra and geometry, but those on the calculus track have to complete two additional years of preparation—algebra II and precalculus—and a semester of calculus before they see their first applications of calculus to optimization problems (maximum and minimum problems). Discrete optimization problems can be introduced much earlier. For example, once you introduce vertex-edge graphs, you can add weights (representing distance or time or cost) to the edges and ask, "What is the shortest path from A to B?"

This brings us to a third reason for including discrete mathematics at the high school level—discrete mathematics problems can be introduced without much preparation. All students can understand the question posed above, and if provided a "map" like that in Figure 11–1, where the numbers represent distances, can find the shortest path from A to B by trial-and-error or by a systematic analysis of all possible routes.

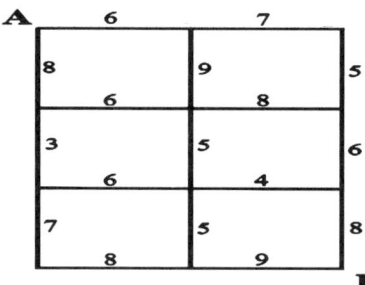

FIGURE 11–1. What is the shortest path from A to B?

Moreover, they can understand and apply (although probably not discover) an algorithm that will indeed provide the shortest path in any such map. What mathematical content knowledge is prerequisite for this activity? Only addition of whole numbers. (Of course, if the distances were given as fractions or decimals, then the ability to add such numbers would be needed; on the other hand, shortest path problems provide a painless way of reinforcing addition skills for such numbers.) That discrete mathematics problems can be introduced with little preparation has two important consequences.

First, it can be used to provide students with a different view of mathematics. Through many years of focus on the mechanics of mathematics, students come to equate mathematics with formal arithmetic and algebra. For them, mathematics is not a domain where big ideas are discussed, or where creativity is exercised. Although students may see applications of mathematics in their science courses, those applications are again likely to involve routine calculations (like balancing equations in chemistry). They are not exposed to mathematical modeling in a variety of arenas or to the variety of tools that mathematicians use to analyze everyday situations. Discrete mathematics is an arena where mathematical modeling is easily understood. If you want to make sure that your yearbook is completed before graduation, you can use vertex-edge graphs to analyze the collection of tasks that need to be done and then use critical path analysis to determine the date by which each task must be completed.

Second, for those students who have not been successful in mathematics, discrete mathematics offers a "new start"—it provides mathematical topics that do not have algebraic skills as a prerequisite, but that rely mainly on reasoning and problem solving. For these students, discrete mathematics offers an arena where they can be successful in mathematics without realizing that they are doing mathematics (since students often equate mathematics with arithmetic and algebra), where they can overcome the stigma of repeated failure, and with newly generated self-confidence can find perhaps that their stumbling blocks to learning traditional topics were not insuperable. Enrichment through discrete mathematics has perhaps more potential than repeated remediation for these students.[9]

Another quality shared by many topics of discrete mathematics is that they are engaging—the problems are more visual than computational, more geometric than algebraic, and they often have a puzzle-like quality—and students easily get engaged in them. Thus discrete mathematics is a very attractive alternative for those students who are moderately successful at traditional mathematics, but are not fans of mathematics. They struggled successfully with fractions and algebra, but did not find them fun. Discrete mathematics provides an opportunity for them to enjoy mathematics again.[10] And students find discrete mathematics challenging because, while many problems in discrete mathematics are easily stated, they are often not easy to solve.

Indeed these qualities of discrete mathematics make it particularly appealing to those students who are really adept at mathematics. From the outset, challenging problems can be presented, including problems whose answer is unknown. Students can quickly learn about the P vs. NP Problem, whose solution will earn $1,000,000 for its solver from The Clay Mathematics Institute of Cambridge, Massachusetts (CMI). This is one of seven "Millenium Problems" named by CMI in 2000 to celebrate mathematics in the new millennium. Given what has been said about discrete mathematics in this chapter, it is not surprising that this problem is the only one of the seven Millenium Problems that is accessible to high school students.[11]

These qualities of discrete mathematics—that it has many applications, that it provides ready answers to the question, "What is mathematics good for?", that many of its topics have few prerequisites, that *all* students find it engaging, and that it presents challenges to the strongest students—provide strong justification for why discrete mathematics should be included in the high school curriculum.

---

[9]For further discussion of the "new start" that discrete mathematics provides for students, see the articles by Biehl, Picker, and Rosenstein (1997) in the volume *Discrete Mathematics in the Schools*.

[10]The opportunity that discrete mathematics offers to make mathematics enjoyable for students can only be realized when they are actively engaged in the learning of the mathematics. This point is discussed further in the penultimate section of this chapter.

[11]For a discussion of the P vs. NP problem, and the other "Millenium Problems," visit the Web site http://www.claymath.org/millennium.

WHY SHOULD DISCRETE MATHEMATICS BE INCLUDED IN THE K–8 CURRICULUM?

Although the qualities of discrete mathematics discussed above apply also to some extent at the K–8 grade levels, there are other reasons as well for the inclusion of discrete mathematics in the K–8 curriculum. Let us reflect for a moment on the statement in NCTM's *Principles and Standards for School Mathematics* cited above. The statement recognizes the value of the *mathematics* in discrete mathematics. Unfortunately, it does not recognize the *educational value* of discrete mathematics. And that is the critical aspect of discrete mathematics that makes it valuable at the earlier grade levels.

Children need to be able to understand and use a variety of concepts and techniques from different areas of mathematics, and to be able to solve problems that draw on their toolbox of mathematical and problem-solving strategies. Learning the basics is no longer enough. Today's children live in a technological age where they will need to think critically, solve problems, and make decisions using mathematical reasoning and strategies. The mathematics standards, both state and national, expect students to go beyond the basics—in terms of additional content and substantially increased problem-solving and reasoning, and in terms of applying school math to daily situations.

A major obstacle to students' going beyond the basics is that their teachers often never went beyond the basics. Many K–8 teachers have taken few courses in mathematics, and have never themselves been engaged in a single real problem-solving mathematical activity or been asked to explain or justify their reasoning mathematically. They use worksheets to provide lots of practice to their students, thinking that they are engaging their students in problem solving, not realizing that a problem is different from an exercise; what makes a problem "a problem" is that you do not recognize immediately how to solve it.

Moreover, since they use only direct instruction (the counterpart of lecturing at the college level), they often do not engage their students in the learning of mathematics; students are expected to recite mathematics (such as memorized facts), manipulate symbols, or apply formulas, but are rarely expected to be actively engaged in figuring out the answers to more difficult questions or in verifying their claims. Those students whose learning style thrives on direct instruction do well; not so, however, for the substantial numbers of students who need to build their own understanding of a concept or technique. So teachers who have never gone beyond the basics cannot reach *all* students, as recommended in the standards.

Discrete mathematics offers an entree into mathematics for teachers who have not been successful at mathematics. It offers them an opportunity to see what mathematics is all about, an opportunity to become engaged in problem-solving and excited about solving challenging problems. It offers an opportunity to understand the standards of problem solving and reasoning, and a path to achieving these standards in their classrooms.

Since 1995 we have offered (initially with support from the National Science Foundation) an institute for K–8 teachers, the Leadership Program in Discrete Mathematics (LP-DM). What we have found is that discrete mathematics provides teachers, particularly elementary school teachers, with non-intimidating access to interesting and important mathematical ideas and strategies that they can use in their classrooms to strengthen reasoning and problem-solving skills for students at all levels and of all abilities. Also, because of its visual nature, discrete mathematics appeals to the learning style of many elementary school teachers; many teachers find, for the first time, an area of mathematics that can be presented in a way that fits their learning styles and the learning styles of their students.

In the very first activity of the LP-DM, groups of participants are engaged in coloring maps. They sit around large maps of the United States, where the interiors of all the states are white, and try to determine (using chips of different colors) the minimum number of colors that must be used if you want to color each state and, at the same time, ensure that bordering states are colored using different colors so that the borders are recognizable. All teachers are engaged in this activity, all find it both fun and challenging, all are exercising their problem-solving and reasoning skills, and all realize that the same activity will engage *all* of their students, independent of grade level or ability level. Not only that, but they also realize that *all* of their students can be successful at such problems.

Here is how this realization comes about.[12] Initially, teachers do not recognize that they are working on mathematical problems, so their own negative associations with mathematics are not activated. They find that they are successful at doing these problems. They learn that they were doing mathematics all along. They realize that they can be successful at mathematics, although they always thought that impossible. They conclude that if they can do it, then so can their students. When they saw themselves as unsuccessful at mathematics, it was reasonable to expect that some of their students would also be unsuccessful at mathematics. But now when they realized that they could be successful at mathematics, then their expectations were raised for *all* of their students.

Since most teachers are unfamiliar with discrete mathematics, questions like this are not even posed in traditional math classes, depriving students of a rich source of problem-solving situations. This activity also introduces them to mathematical applications; techniques for coloring maps efficiently are used in solving a variety of scheduling problems, such as class scheduling, where two courses that share a student (i.e., a student is taking both courses) have to be assigned different meeting times.

These kinds of experiences provide teachers, both practicing and prospective, with an understanding of what their students can achieve mathematically, and the tools for them to become effective teachers of mathematics.

Discrete mathematics is more than just a collection of new and interesting mathematical topics: it is also a vehicle for giving teachers a new way to think about traditional mathematical topics and a new strategy for engaging their students in the study of mathematics.

LOOKING TO THE FUTURE

Although NCTM's *Principles and Standards for School Mathematics* proclaims that discrete mathematics will be incorporated into the discussion of the other standards, the actual references to the topics of discrete mathematics in the document itself are few and far between. Moreover, they are generally accidental—that is, there is no reference to their source in discrete mathematics. The authors of the

---

[12] We hope to conduct a study that will determine the extent to which the anecdotal account that follows reflects reality.

standards, unfortunately, seem unaware of discrete mathematics and the opportunities discrete mathematics provides to implement the standards.[13]

Over time, this will change. Indeed, paraphrasing the concluding sentence of the abstract of this article, "more educators will learn these topics and will recognize their value in achieving the goal of improving students' understanding of mathematics." Accompanying this heightened understanding of the value of discrete mathematics will be a proliferation of research studies that verify the claim that a healthy dose of discrete mathematics, if properly administered, will improve mathematical understanding of teachers (particularly K–8 teachers) and students (at all levels).

"If properly administered..." A few words of caution. Discrete mathematics can be routinized as effectively as any other topic. Memorizing algorithms of discrete mathematics can become as deadly as memorizing arithmetic algorithms, and with less value. An important value of discrete mathematics is that it offers an easy way of accessing the mathematical enterprise of conjecture, exploration, discovery, and proof, and provides students and teachers with "Aha!" opportunities. Focusing on the content of discrete mathematics to the detriment of the spirit of discrete mathematics will be counter-productive, which explains why "if properly administered" appears in the previous paragraph. The key to this is, as always, proper preparation of teachers.

Should there be an AP test in discrete mathematics? Considering the comments that I made earlier about the AP test in calculus, it is not surprising that I would not advocate for an AP test in discrete mathematics. That is not to say that the content of discrete mathematics is unimportant. It is of course important, and those students who are mathematically inclined will benefit from a rigorous study of discrete mathematics. However, in my eyes, the greater value of discrete mathematics is the opportunity it offers to the vast majority of students who would benefit from a glimpse of what mathematics is all about, a glimpse that they will not get from exclusive focus on traditional topics. We must keep in mind that discrete mathematics is an important vehicle for democratizing mathematics.

CONCLUSION

In this short chapter, I have described the nature of discrete mathematics and its dual value in K–12 education—as a collection of important content material and as a vehicle for improving mathematics education. I have indicated that discrete mathematics can be valuable to students of all levels of ability and achievement, and have described the important role it can play in opening a door to mathematics to elementary school teachers. Over time, more people will come to know the subject and recognize its value; over time, I anticipate that the claims that I am making in this article based on our experience will also be supported by research studies. I hope that this process will be accelerated as a result of the publication of the following two sets of materials in the coming years—the NCTM's volume on discrete mathematics in its Navigations Series (*Navigating through Discrete Mathematics*, authored by Valerie DeBellis, Eric Hart, Margaret Kenney, and myself), and a textbook for a mathematics course for prospective K–8 teachers (*Making Mathematics Engaging: Discrete Mathematics for K–8 Teachers*, authored by Valerie DeBellis and myself (2007)).

---

[13]This applies also to the authors of NCTM's recent "Curriculum Focal Points for Prekindergarten through Grade 8 Mathematics" who did not incorporate this author's suggestions for using discrete math activities to enhance the "connections to the focal points.".

ACKNOWLEDGEMENTS

The author acknowledges the support of the National Science Foundation Center for Learning and Teaching Grant ESI-0333753 to "Metromath: The Center for Mathematics in America's Cities," of which he served as Director during the preparation of this article.

REFERENCES

Biehl, L. C. (1997). Discrete mathematics: A fresh start for secondary students. In J. G. Rosenstein, D. Franzblau, & F. Roberts (Eds.), *DIMACS series in discrete mathematics and theoretical computer science: Discrete mathematics in the schools* (Vol. 36). American Mathematical Society and National Council of Teachers of Mathematics. Providence, RI.

DeBellis, V. A., & Rosenstein, J. G. (April 2004). Discrete mathematics in primary and secondary schools in the United States. *International reviews on mathematical education* (Zentralblatt für Didaktik der Mathematik), Vol. 36, Number 2, Electronic-Only Publication, ISSN 1615-679X.

DeBellis, V. A. & Rosenstein, J. G. (2007). *Making Mathematics Engaging: Discrete Mathematics for K-8 Teachers*, Preliminary Edition. Javelando Publications, Greenville, NC.

DeBellis, V. A., Hart, E., Kenney, M. J. & Rosenstein, J. G. (2007). *Navigating Through Discrete Mathematics*. National Council of Teachers of Mathematics, Reston, VA.

Maurer, S. (1997). What is discrete mathematics? The many answers. In J. G. Rosenstein, D. Franzblau, & F. Roberts (Eds.), *DIMACS series in discrete mathematics and theoretical computer science: Discrete mathematics in the schools* (Vol. 36). American Mathematical Society and National Council of Teachers of Mathematics. Providence, RI.

National Council of Teachers of Mathematics (2006). *Curriculum Focal Points for 'rekindergarten through Grade 8 Mathematics*. Reston, VA: National Council of Teachers of Mathematics.

National Council of Teachers of Mathematics. (1991). *Discrete mathematics across the curriculum, K–12*. M. J. Kenney & C. R. Hirsch (Eds.). Reston, VA.

National Council of Teachers of Mathematics (2000). *Principles and standards for school mathematics*. Reston, VA: National Council of Teachers of Mathematics.

Picker, S. H. (1997). Using discrete mathematics to give remedial students a second chance. In J. G. Rosenstein, D. Franzblau, & F. Roberts (Eds.), *DIMACS series in discrete mathematics and theoretical computer science: Discrete mathematics in the schools* (Vol. 36). American Mathematical Society and National Council of Teachers of Mathematics. Providence, RI.

Rosenstein, J. G. (1997). Discrete mathematics in the schools: An opportunity to revitalize school mathematics. In J. G. Rosenstein, D. Franzblau, & F. Roberts (Eds.), *DIMACS series in discrete mathematics and theoretical computer science: Discrete mathematics in the schools* (Vol. 36). American Mathematical Society and National Council of Teachers of Mathematics. Providence, RI.

Rosenstein, J. G. (2007). The Rush to Calculus: An Analysis of Course-Taking Patterns by Rutgers Students in High School and College. (In preparation).

Rosenstein, J. G., Franzblau, D. & Roberts, F. (Eds.) (1997). *DIMACS series in discrete mathematics and theoretical computer science: Discrete mathematics in the schools* (Vol. 36). American Mathematical Society and National Council of Teachers of Mathematics. Providence, RI.

CHAPTER 12

Formalizing Learning as a Complex System: Scale Invariant Power Law Distributions in Group and Individual Decision Making

Thomas Hills, Andrew C. Hurford, and Walter M. Stroup
University of Texas at Austin

Richard Lesh
Indiana University

> *For every complex problem there is an answer that is clear, simple, and wrong.*
> —H. L. Mencken

There is a tendency in education to answer the complex problem of learning, including learning in group contexts like classrooms, with "clear" and "simple" pedagogical methods like direct instruction (Silbert et al., 1997). Rather than fall into the trap outlined by Mencken, we look to move in a quite different direction. We start by acknowledging and respecting that learning is complex, and as a result, we specifically focus on investigating the use of complexity theory in education. An important distinction needs to be made, as there are two broad ways in which complexity theory is engaged in the education literature. One way complexity theory is discussed relates to teaching and learning *about* complex systems (e.g., Resnick & Wilensky, 1998, Wilensky, 1999; Jacobsen, 1998; Hmelo et al., 2000; Wilenksy & Stroup, 1999). The other relates to learning *as* a complex system (e.g., Davis & Simmt, 2003; Ennis, 1992; Hurford, 1998; Thelen & Smith, 1996). Although "learning about" and "learning as" are apt to be mutually informative and learning about complexity and systems analyses is becoming an increasingly important candidate for school curricula, the primary focus in this discussion is on learning *as* a complex adaptive system.

Consistent with the tenor of some of the prior analyses, we begin this chapter with a top-level discussion of why complexity-based approaches might be seen to better fit the phenomenology of learning. What is distinctive is that we then go on to take up one of the more challenging aspects of making the case for the applicability of complexity theory to learning: identifying and formally investigating instances of scale invariance (or self-similarity) in classroom-situated decision making. Using some of the formalisms associated with complexity theory, we identify and investigate learning behaviors that are self-similar across scale from individuals, to groups, and to groups of groups. We believe one of the accomplishments of this line of work is a shift from saying learning and cognition are, in certain ways, *like* complex systems to showing that they actually satisfy some of the more stringent formal conditions for *being* complex systems.

In a sense then, we are responding to the challenge of formalization made by one of the major complexity theorists:

> ...common usage of the term *complex* is informal. The word is typically employed as a name for something that seems counterintuitive, unpredictable, or just plain hard to pin down. So if it is a genuine *science* of complex systems we are after and not just anecdotal accounts based on vague personal opinions, we're going to have to translate some of these informal notions about the complex and the commonplace into a more formal, stylized language, one in which intuition and meaning can be more or less faithfully captured in symbols and syntax. (Casti, 1994, p. 270)

Can we move beyond "anecdotal accounts based on vague personal opinions" and translate some of our informal notions about learning as a complex system into "formal, stylized language...more or less faithfully captured in symbols and syntax"? In order for the efforts to engage learning as a complex system to gain traction, we believe we must respond to this challenge. Accordingly, the data and analyses presented here are, so far as we know, the first time human behaviors have been shown, in formalized language, to be scale-invariant both at the level of the group and at the level of the individual and in ways that are specifically related to underlying cognitive processes predictive of success in a problem-solving environment.

To the extent that the results presented here generalize to learning in other environments, we are inclined to ask in what ways is learning complex? What is the scope of this complexity claim? When, if ever, might learning not be complex? We attempt to provide answers for these questions based on existing theoretical work in other complexity sciences. This work may help us begin to understand the combination of internal and external constraints that are most likely to engage cognition as a complex adaptive system, and what implications for activity design might follow from designing with and/or for complexity. By exploring the formal features of the learning system that would result in the emergence of the observed power law relations we present here, we are further able to address both the developmental features of learning as a complex adaptive system and the kinds of environments that support that system.

We start with a discussion of why we think approaching learning as a complex system rather than as a linear and exclusively individualistic process is appropriate. We then present our empirical results and conclude with a discussion of these results.

PLAUSIBILITY ARGUMENTS FOR TREATING LEARNING AS A COMPLEX SYSTEM

The great majority of theory building in cognitive and constructivist learning theories has been focused on the learning of an individual. Behaviorist perspectives (Skinner, 1954; Stein, Silbert, & Carnine, 1997), information processing perspectives

(Anderson, 1983; Anderson, Reder, & Simon, 2000; Mayer, 1996), novice-expert perspectives (Chi, Feltovich, & Glaser, 1981; NRC, 1999, chap. 2; Reiner, Slotta, Chi, & Resnick, 2000), schema-theoretic perspectives (Derry, 1996; diSessa, 1993), and constructivist perspectives (Cobb, 1994; Ernest, 1996; Piaget[1], 1923/1959, 1924/1969, 1929/1951; Vygotsky, 1987, chap. 6) are all predominantly individualistic views of learning. These efforts have provided many insights and have been very successful in helping researchers build useful models of learning. Although individualistic approaches to learning have been quite productive they also have several limitations.

Limitations of Individualistic and Linear Theories of Learning

The first limitation of individualistic theories of learning addressed here is that they do not "scale" well—that is, the learning of a classroom of students is not very profitably described as the linear combination of a number of individual learners. This type of scaling to whole classrooms of learners does not and cannot take into account the complex interactions and synergetic effects derived from the properties of groups. Very little of what goes on in classrooms can be understood in terms of straightforward cause and effect relationships among simple aggregations of individual learners.

Individualized and linear models of learning tend to be more static than dynamic. Behaviorist models (e.g., Stein, Silbert, & Carnine, 1997, pp. 3–29) assume that learning is the simple accumulation of fixed and appropriately sized knowledge bits that are taken in as given, without much, if any, active adaptation or interpretation on the part of the learner. Another line of (individualized) learning research posits that learners possess relatively static conceptual structures and that teaching and learning constructs knowledge structures and then repairs or replaces "misconceptions" (Reiner, Slotta, Chi, & Resnick, 2000, p. 7) with increasingly "expert" structure. In each of these lines of research, knowledge is envisioned as chunks of information stored in and accessed from static conceptual structures internal to individual "knowers."

Individualized approaches also tend to focus on a learner at the expense of the learner's context and her or his membership in a learning community. There are many aspects of the surrounding contextual situation that influence how learning takes place and what gets learned (Lave, 1988; Lave &Wenger, 1991; Wertsch, del Rio, & Alvarez, 1995). Students and teachers are embedded in a wide variety of social, historical, and cultural systems (cf., Bowers, Cobb, & McClain, 1999; Hiebert, et al., 1996, p. 19; Lave & Wenger, 1991, pp. 67–69) that profoundly affect learning (Cobb, Perlwitz, & Underwood, 1996) and individualized approaches to learning generally overlook these important and complex influences. The sociocultural-historical milieu of the classroom can be seen as part of the environment relative to which adaptation occurs. At any given moment, classrooms and learners are immersed in a wide variety of interconnected and often competing activities and goal structures. Theories of learning that focus on individuals generally do not take these kinds of complex and ubiquitous learning conditions into account.

---

[1] Although there is controversy over whether or not Piaget was actually an individual-constructivist most of the work done in the Piagetian tradition is decidedly focused on individuals.

Finally, individualized accounts of learning do not offer very much to teachers in the way of helping them to make sense of, or design for, whole-classroom activities. Although teachers may develop individualized educational plans, they almost never design classroom activities with a single individual in mind. Classroom activity is inherently a *group activity*, and there is very little in the language and ideas of individualized cognitively-based or even standard constructivism-based learning theories that enables teachers to make sense of the activities of groups of learners.

Affordances of Systems-Theoretic Approaches to Learning

In contrast with individualized approaches, complex systems-theoretical perspectives have much to offer in terms of helping teachers, researchers, and others to focus on and make sense of learning at the level of the group. First, a systems perspective enables thinking about classroom learning in terms of a dynamic, continuously changing "dance" between the group, its members, and the contextual situation. Second, as discussed above, classrooms are much more than a linear sum of individual learners, and a systems perspective enables thinking about the synergetic affordances and "lever points" (Holland, 1995, p. 39) inherent in classrooms. Third, it may well be that the most important affordance of systems-theoretical approaches to learning is in the *language* of complexity itself.

Complex systems terminology and the ideas that terminology represents are increasingly finding their way into the discussions and literature of cognitivist, constructivist, and sociocultural learning theory camps. It seems that the use of the language of complexity is preceding more rigorous and careful application of systems-theoretical tenets—Casti's (1994) "formalization" (pp. 274–276)—to perspectives on learning. The hope is that use of the language will serve as precursor and enabler of more systematic complexity-based modeling in future education research. In any event, the combination of the fundamental ability to address dynamic and complex interrelations, the ability to provide a powerfully descriptive language for talking about what happens cognitively in classrooms, and the potential for formalization of these things into useful models all serve to demonstrate the potential of systems-theoretical approaches for research into learning.

Systems perspectives *do* scale well in terms of considerations of group activity. In one way or another, every complex systems viewpoint addresses both the individual and the aggregate. For example, from the structuralist perspective of Jean Piaget[2] (1968/1970, chap. 2), the group and the elements of the group are mutually constitutive. That is, the dynamic creation of a group, the activity of individual members of the group, and the group's context each influence the others, forming a complex adaptive system[3]. An example of this in a classroom is when students are aware (or quickly become aware) of their status within the larger ensemble, and those status considerations then have powerful effects on the students' and the group's subsequent activities (cf. Empson, 2003). Complex systems analyses (Casti, 1994; Camazine, et al., 2001; Clark, 1997; Holland, 1995, 1998; Stroup & Wilensky, 2000; Prigogine, 1984, 1997) focus on higher-level patterns (e.g., aggregation, flows) that are generated by activity and adaptation at the level of individuals whose behaviors are based solely on the local environment and the individual's own

---

[2]Here, we are considering the systems-theoretical, non-individualistic aspect of Piagetian constructivism.

[3]We would also agree with Chapman's observation that "The kinship between Piaget's theory of equilibration and contemporary theories of self-organizing systems is particularly promising..." (1988, p. 340).

internal models. In contrast to individualized theories of learning, systems-theoretical points of view are fundamentally concerned with viewing learners and groups as mutually constitutive agents whose behavior is to be understood in the context of their larger patterns of activity.

Systems-theoretical points of view tend to be very dynamic—characterizing activity in terms of evolving *patterns*. Rather than focus on static "snapshots" of an individual student's or the group's learning, such as quiz grades or end-of-year tests, we focus our investigations of learning in the context of on-going problem-solving behavior. In the following research study, we recorded the decisions made by all students in a multi-player network simulation as they participated in an open-ended problem concerning the distribution and acquisition of food-like resources. Our results show individuals and groups both exhibited power law distributions in their decision making processes, and that the length of their inter-decision intervals was correlated with overall success in the foraging activity. In addition, both individuals and classrooms showed distinct variation in their fit to power law distributions and our analysis suggests that this variation is informative with respect to learning and individual strategies.

INTRODUCTION TO POWER LAWS

Power law distributions are characterized by self-organized arrangements with many small events and few large events. The distributions are often called scale-free because there is no limit to the largest size event and no natural scale of measurement. The relationship between neighboring elements or events in a distribution is preserved regardless of the scale of observation. Spatial scale-free phenomena that are ordered in this way are called fractal. Temporal distributions of similar arrangement exhibit a behavior known as one-over-f (1/f) noise.

Power law distributions are found in diverse self-organizing systems ranging from forest fires to web links (e.g., forest-fires, Malamud et al., 1998; corporate firm sizes, Axtell, 2001; avalanches, Bak et al., 1988; earthquakes, Johnston & Nava, 1985; biological extinctions, Raup, 1986; stock-market fluctuations, Mandelbrot, 1982; volcanic activity, Diodati et al., 1991; city sizes, Ioannides & Overman, 2000; turning-times in fruit flies, Cole, 1995; and the organization of the world-wide-web, Huberman & Adamic, 1999; also see Bak, 1994). Power law distributions have also been observed in human reaction times when learning procedural tasks, for example, how to roll cigars ("the power law of learning", Neves & Anderson, 1981; Anderson, 1982; Logan, 1988), and in retention ("the power law of forgetting", Anderson & Schooler, 1991; Rubin & Wenzel, 1996; Wixted & Ebbesen, 1991).

While power law distributions have not yet been described for higher-level cognitive processes (i.e., above reaction times) or for social problem solving, power law relationships have been observed for distributions of words in human language. This is designated as Zipf's law (Zipf, 1949) and is mathematically representative of a Pareto power law distribution (Adamic & Huberman, 2002). When words of a given language are ordered by decreasing frequency, the frequency of the ith word, P(i), is proportional to i^{-A}, where A is approximately equal to one; this is mathematically identical to 1/f noise described above. This is suggestive of cognitive constraints associated with human symbolic manipulations, where the use of a new symbol is based on the frequency-dependent relationship it shares with the symbolic population in which it originates. Mechanisms for the evolution of power laws in human communication have been described by Cancho & Sole (2003) and these authors suggest that power law distributions may be "required by [all] symbolic systems" (p. 791).

METHODS

To investigate patterns in human 'thinking' about complex problems, we collected data from five separate classrooms participating in a biology-related foraging simulation (Hills & Stroup, 2004) built in the Hubnet environment (Wilensky & Stroup, 1999; 2000) in which individuals controlled the search patterns of individual foragers (avatars) in a 2-dimensional playing field, hunting for invisible "food" pixels that were grouped in small clumps (Figure 12–1). Students sat at laptop computers and could move their avatars up, down, left, or right one pixel for each individual move. The group space was projected on a screen at the front of the classroom and the simulation was fast enough that avatars responded in less than 500 ms. Each run of the simulation lasted at least three minutes. During this participatory simulation, the students were unaware of the position of the pixel 'resources' but did know the location of their individual avatar as well as the position of other players (Figure 12–1, shows orange food 'pixels' [individual boxes], but students could not see them). The student-avatars are the other icons visible in the figure.

By moving their avatars around the students could construct mental images of the underlying distribution of resources as their computers reported to them when they encountered food, enabling the participants to recognize the size and arrangement of resource clumps. Using this information, they could then adapt their foraging behavior to meet the requirements of the environment and maximize their acquisition of resources.

Students could also listen to the verbal reports and see the actions of other students in the classrooms. All students shared the same resource patterns, so information from other students could be informative about distributions. Separate resources were allocated for each individual, so students were not competing for resources. However, students were asked at 30-second intervals to vocally report how many pixels of food they had acquired (reported on their individual monitor). With this information students could alter their strategies in accordance with their relative success or failure during reporting intervals. Throughout the simulation we recorded all information about the avatars, including position, cumulative resources found, and wait times between individual actions.

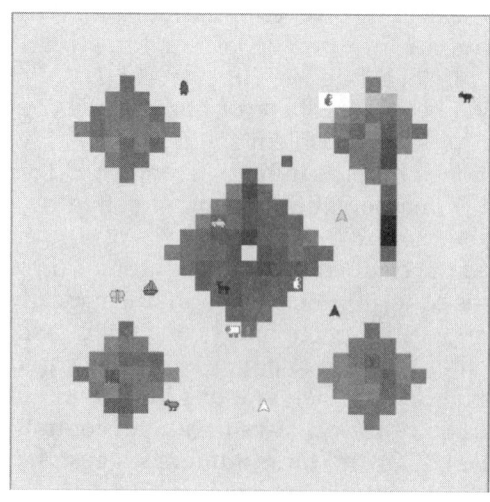

FIGURE 12–1. The resource distribution in which students foraged.

RESULTS

Figure 12–2 shows the typical relationship observed between average wait times for individual decisions and the cumulative resources found over the entire simulation. We use the term "wait times" here because "reaction time" often implies a minimum time to arrive at an answer ("How fast can one make a decision?"). Figure 12–2 shows that students who took longer to make decisions did better in the simulation.

Students who were interacting with the simulation at intervals less than one second were observed to be primarily interested in moving as quickly as possible through the resource space, paying little attention as to direction or distribution of the underlying resources. After the simulation, students reported that longer wait times were used to "think about where the resources might be." Using the students' terminology, longer periods of thinking before a move correlated with higher success in the problem environment.

To determine if students' behaviors were associated according to a power law distribution, we calculated the log of individual decision times and plotted them against the log of the rank of individual decision times—the rank is simply a rank

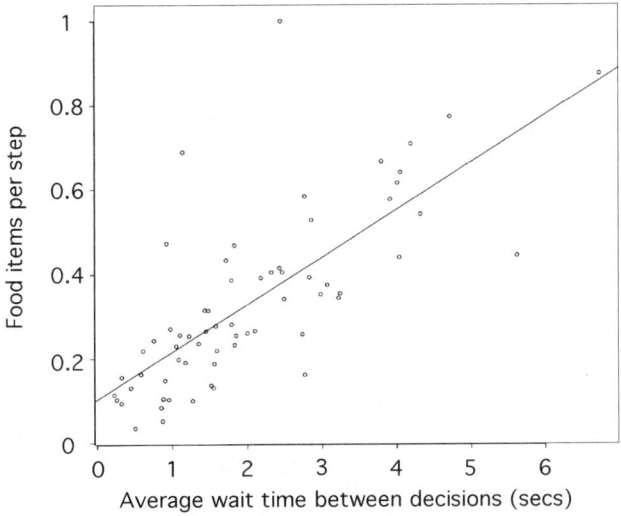

FIGURE 12–2. Students who took longer to make decisions found more resources per step in the simulation. Only one student took fewer than 20 steps (marked by the arrow). All other students took more than twenty steps and were observed to actively search for the duration of the simulation.

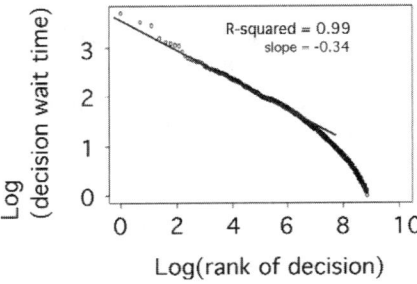

FIGURE 12–3. The Zipf distribution for all individual decisions ($n = 23,637$). Circles represent wait times between individual decisions. A linear regression fit to decisions longer than 3.3 seconds fits with an R-squared of 0.99.

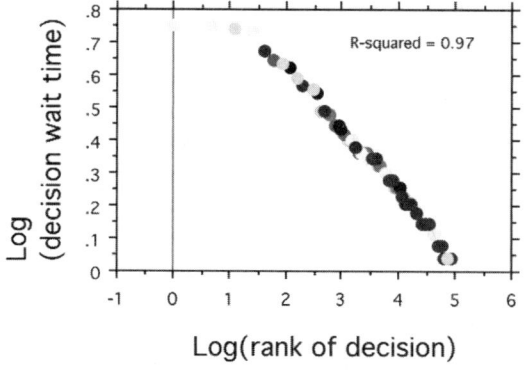

FIGURE 12–4. The Zipf distribution for average wait times per student in each simulation ($n = 297$). Circles represent the average wait times for individual students for a given resource distribution. A linear regression for wait times longer than 3.3 seconds fits with an R-squared of 0.96.

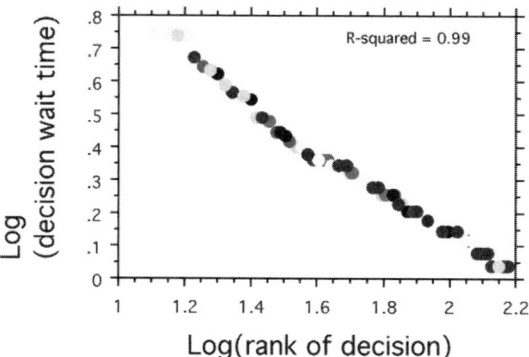

FIGURE 12–5. The Zipf plot for a single class during a single run (resource distribution). Individual decisions are represented by circles ($n = 139$). A regression for all decision times longer than 1 second fit with an R-squared or 0.97. A single shading represents all decisions made by one individual.

ordering of decision times by their duration. Figure 12–3 shows the Zipf plot for all decisions made by students in Figure 12–2.

This wait-time graph treats all individual behaviors as members of the group behavior and is highly consistent with a Zipf distribution—that is, the entire group's wait time behaviors are consistent with a scale-free power law distribution.

To get a sense of the robustness of the power law distribution, we also averaged decision times for individuals and plotted these in a similar fashion in Figure 12–4. This plot is less suggestive of a power law distribution, but does reveal a similar relationship above 3.3 seconds. It is also possible that these distributions fit a lognormal distribution better than they fit a log-log distribution (the distribution for a power law). We tested this possibility by calculating the R-squared for a log-normal distribution. For individual averages, the distribution fits a log normal with an R-squared of 0.97 over the entire distribution, but the log-log distribution fits with an R-squared of 0.90. This is not the case for the other log-log distributions shown in this paper, where log-normal distributions were a poorer fit. We further discuss the interpretations of the shapes of the distributions below.

If individual decisions over multiple simulations fit a power law distribution with accuracy such as that in Figure 12–1, we were also interested in understanding how individuals distributed their decisions within the class, and what an individual class looked like during a simulation. Figure 12–5 shows the distribution for an individual class and reveals that a student's decisions (marked with same colored circles) may show considerable variance within the class.

This suggests that most of the variance seen in the simulation is within individuals, not between them. In other words, individuals sample from a similar subset of

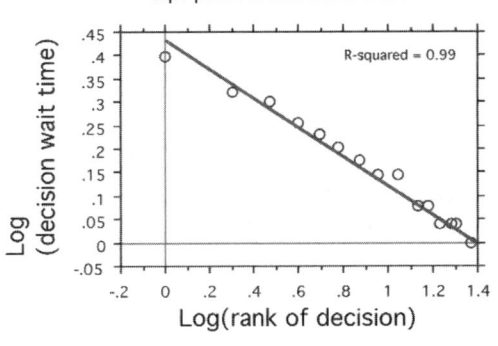

FIGURE 12-6. A Zipf plot for a single student during one simulation run. Circles represent decision times for individual decisions. A regression for all decisions longer than 1 second fit with an R-squared of 0.99.

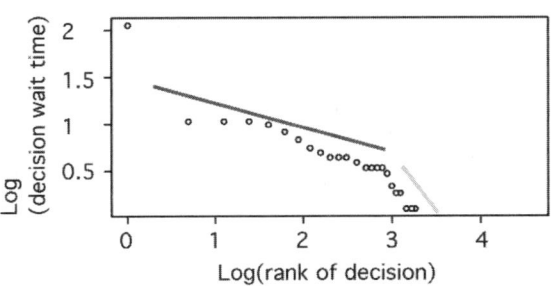

FIGURE 12-7. Zipf elbow for a single student.

decision times, and are more alike than individual averages would indicate. This is shown most dramatically in Figure 12-5 by observing the overlap of dark and light circles. There are many students who show considerable overlap in decision timing, despite the fact that their averages are likely to be different. Averaging, as a methodological approach, would misrepresent these results.

To further assess the scale invariance in decisions, we plotted the log-log distributions of individual students. Figure 12-6 depicts the Zipf distribution for a single student from Figure 12-5.

Like many students, this student's activity fit a power law distribution with an R-squared of 0.99. This is one of the better fits in the data set. On average students fit a log-log regression with an R-squared of 0.95. In some cases lognormal provided a better fit with the data, in other cases log-log distributions made the better fit.

Overall, distributions for all classes combined, individual classes, and individuals reveal a degree of similarity highly suggestive of scale invariance. As stated above this is one of the hallmarks of formal complex adaptive systems. Log-log distributions also reveal other characteristics about the nature of dynamic problem solving. Figure 12-7 shows a typical log-log plot for an individual student. For this student, the figure reveals a distinct elbow in the distribution at approximately 2 seconds. Numerous elbows at different times were observed in many of the data sets. Figures 4 and 5 both show evidence of similar elbows and it is possible that Figure 12-3 contains two elbows. In many cases the visual confirmation of the elbow is dramatic (as in Figure 12-7 and Figure 12-5) and it is not isolated to students or classrooms.

The elbows appear to be associated with different kinds of behavior in the simulation. For example, in Figure 12-7, the steeper slope on the right is associated with shorter wait times and may therefore be associated with straighter runs

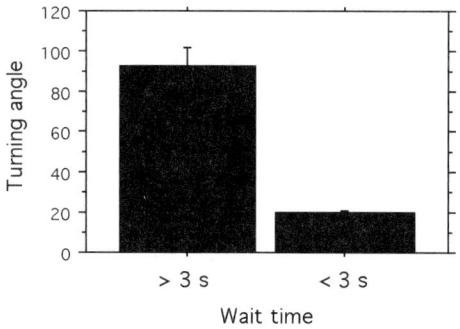

FIGURE 12–8. Comparison of turning angles taken after wait times of varying length. Decisions that took longer than 3 seconds to make are represented by '>3 s', whereas those taking shorter than 3 seconds are represented by '<3 s'.

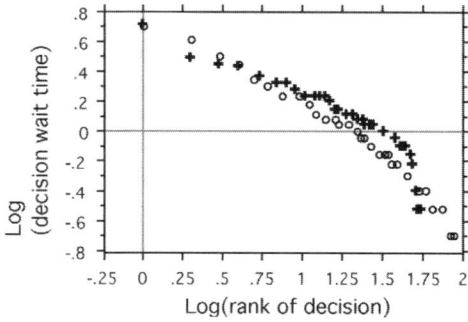

FIGURE 12–9. Data for an individual student during a single run of the simulation. Crosses represent decisions made during the first minute of the simulation, circles represent decisions made during the second minute of the simulation.

across the landscape, as opposed to high angled turns, which would be more characteristic of longer wait times on the shallower sloping distribution, to the left. As a preliminary test of this hypothesis, we compared the average turning angles after decisions lasting longer than 3 seconds and after decisions which took less than 3 seconds to make (Figure 12–8).

Decisions taking less than 3 seconds were correlated with significantly straighter movement than decisions taking longer than 3 seconds. Decisions times longer than 3 seconds typically ended in a right-angled turn.

It is also useful to note that individuals who performed most poorly in the simulations overall (see Figure 12–2), sampled from faster temporal distributions on average, which would therefore lie more on the vertical line to the right in Figure 12–3. Individuals who performed well in the simulations, sampled primarily from the slower temporal distributions, to the left in Figure 12–3.

The presence of a power law distribution was ubiquitous regardless of individual students' performances. Only five students failed to show a power law distribution for decision times lasting longer than 3 seconds. The presence or absence of a power law does not therefore appear to reveal much about the success of students in the simulation, however, the shape of the overall distribution is suggestive of other ways to characterize learning. For example, Figure 12–9 shows the log-log distribution for an individual student during a single run of the simulation. The presence of the elbow during the first minute of the simulation (represented by crosses) may point to a binary decision making process, in which the student is choosing between two strategies, one representing local foraging with high-angled turns and the other representing straight runs with no or very low turning angles (as shown in Figure 12–8). However, the second minute shows a much less distinctive elbow (represented by circles), and as a whole fits a power law distribution better than the first minute when all decision times are taken into consideration.

One possible explanation for this change in the shape of the distribution is that the student is adapting to the depletion of resources in the second minute by choosing a mixed search strategy that combines temporal features of both strategies used during the first minute. As resources are depleted, the patchy quality of the original distribution (Figure 12–1) begins to decay, such that the distribution of food is less spatially auto-correlated—finding one pixel is less informative about nearby pixels in the second minute than it is in the first minute. Therefore, the appropriate cognitive and behavioral response is to forage less locally around found resources, but also to rely less on running into large clumps of resources during straight runs. This is a hypothesis that would explain the difference in the early and late log-log distributions shown in Figure 12–9. This hypothesis is based on a very new approach to modeling cognition as a complex system, and much work remains to be done in formalizing that approach.

DISCUSSION

The evidence presented here is to our knowledge the first time human behaviors have been shown to be scale-invariant both at the level of the group and of the individual and, moreover, related to underlying cognitive processes that are predictive of success in a problem solving environment. That these behaviors follow a power law distribution implies that the distributions are self-similar and cannot be accurately described by a Gaussian distribution. Gaussian distributions have exponentially decaying tails and make behaviors much larger than the mean very unlikely. However, the power law distribution gives finite probability to individuals' having wait times much larger than the mean. This means that the *relationship* between the timing of decisions is preserved regardless of the time scale one chooses to observe the phenomenon—thus the self-similarity.

This work also suggests that behaviors are less alike than individuals, meaning that behaviors are constructed from a more strongly multiplicative organizational scheme than are the individuals themselves, who, though different, are more alike than any two randomly chosen behaviors might indicate. This evidence supports individualistic approaches to learning in the sense that studying single individuals can be very informative with respect to the general strategies used by all individuals. However, it also points out the limitations of treating individuals as averages. The data we present show that the cognitive processes involved in decision making reveal higher-order patterns that have not yet been investigated. These traits are inaccessible to static or individualistic approaches that fail to treat real-time thinking processes as dynamic representations of different 'kinds' of thinking. Averaging over thinking processes eliminates the ability to differentiate between different ways of thinking, and leads to the false implication that different ways of thinking are different primarily in quantity, not quality. The data we present here hardly support this claim. Figure 12–9 suggests this is not true even for one individual during a two-minute time span.

The Scope of the Complexity Claim

Not all dynamic spatiotemporal processes meet the requirements of formal complex systems (Table 12–1).

Power-laws are common, but not everywhere

• WHAT IS... • Corporate Firm Sizes • City Sizes • Species Extinction Rates • Avalanches • Movement times in insects • Distribution of tree sizes in forests • Links in world-wide web • Coauthorship networks	• WHAT ISN'T... • Traffic at airports • power grid for Southern California • Standardized test scores • Social networks in small communities (high school or church) • Connections between movie actors • Growth in your bank account

Systems That Do and Do Not Exhibit Power Law Relationships

Many processes, such as traffic flow at airports or bank account growth, are constructed of variables represented by Gaussian or other less common distributions. Future work investigating cognition as a complex system must naturally grapple with the divide between power law thinking and other kinds of thinking that are not formally complex.

In our analysis of the above data, we observed that for most individuals, successful performance in the foraging simulation was not correlated with the quality of the log-log fit—but appeared to be correlated with other characteristics of the distribution (see the discussion of elbows above). We checked this in a given classroom where individuals repeatedly foraged in the same resource distribution. While the amount of food per step increased two-fold on average, R-squared fits to a log-log distribution were unchanged (data not shown). The only evidence that R-squared fits are related to performance is seen in individuals that show very poor fits to the log-log distributions (R-squared < 0.85). These individuals performed far more poorly than their peers in food-per-step. Though they represent only a small fraction of the individuals in our data set (n = 5), we are inclined to believe that non-power law thinking holds ample information about the nature of cognition as a complex adaptive system. We are also inclined to believe that, to the extent that cognition is allowed to engage itself in tasks that nurture its formally complex features, these tasks will promote problem-solving abilities in general.

Designing activities to promote complex thinking may not be trivial, but our investigations do point to specific characteristics of design that may inhibit complex thinking. Foremost, problems with limited scope, that contain a single correct answer, are likely to prevent the kinds of distributions we witness here. This is apt to be especially true relative to the kinds of strategy switching associated with elbows. An aspect of problem solving that is seldom discussed is the ability of individuals to switch cognitive strategies on demand. This is different from giving up on failed attempts. Maintaining and sampling from several cognitive strategies may be more closely associated with attending to multiple aspects or features of a problem simultaneously. In the foraging simulation, the resource space consisted of patches of resources and empty spaces in between. Being able to construct and represent this space cognitively without having seen it, and to simultaneously switch between appropriate strategies while navigating that space, is clearly a feat of cognitive skill distinctly different from remembering the capital of the State of Indiana.

Structural Analysis of the Emergence of Power Laws

A number of theoretical explanations for power law distributions have been proposed. Two that may be relevant with respect to cognitive processes are the growth with preferential attachment theory proposed by Barabasi & Albert (1999), and the random stopping of exponential growth processes proposed by Reed & Hughes (2002). The Barabasi & Albert model (1999) suggests that scale-free networks emerge when they grow with preferential attachment to vertices that are already well attached. As new vertices arrive, they enrich previously well-connected centers, creating clusters with many nodes with few attachments, and few with many attachments. The Reed & Hughes model (2002) suggests that if processes grow with exponential expectations, but are randomly stopped, then scale-free distributions will be the result. The two models are probably not mutually exclusive, but may depend on whether relationships are spatial or temporal, respectively.

It is possible that cognitive processes behave similarly to many other phenomena described by power laws because they share similar underlying growth and organizational structures. For example, at the level of neural assemblies, if there are temporally weighted distributions associated with behavior, then these distributions, when active, may fatigue at exponential rates (temporal mechanisms of this type are reviewed in Hills, 2003). If other distributions are competing for activation, some of which may lead to terminal decisions and others of which may lead through distributed networks characterized by clustering, these processes taken together may generate power law distributed decision times, as in Figure 12–3. If time spent 'searching' the network is correlated with greater appropriateness of the final decision, this may in turn generate the relationship observed in Figure 12–2.

Growth with preferential attachment also has the characteristic ring of constructivist theories of learning, which suggest that the development of cognitive schemas is an organic growth process that depends on schemas that are already in the system, so to speak. Learning does not come from nowhere, but must marry itself to cognitive representations that are already available to the learner. For example, the ability to conceive of different kinds of horses is facilitated by the ability to recognize horses as a species. However, upon grasping the differences between an Irish Draught horse and a Welsh Mountain pony, one gains considerable cognitive leverage with respect to identifying horses in general and associating functions with their identity, which in turn opens up frames for characterizing other things and relationships. If horses are an important part of one's cognitive landscape, then new learning, and especially that kind of learning that happens outside of schools, is most likely to find itself in a cognitive relationship with horses.

The observation that individuals and classrooms share similar organizational properties with the forest fires and city sizes is promising in multiple respects.

For example, scale invariant phenomena allow one to make predictions about global behavior based on local observations. We can make predictions about the relationships between larger city sizes in the future based on observations of smaller city sizes now (Gabaix & Ioannides, 2004). In addition, as we begin to formalize our understanding of cognition as a complex system, the growing literature with respect to small world networks, power laws in nature, and self-organized criticality becomes an increasingly valuable resource in promoting understandings of cognition.

CONCLUSION

Making sense of what happens in classrooms is aided by thinking of the situation as a whole—in all its richness. That means considering the students and teacher as a contiguous set, mutually influencing each other and affecting and being affected by their combined contexts. Activity in the classroom never stops—the mutual constituency is a dynamic real-time process that is constantly evolving. When we think about classrooms in this inclusive sense the power of complex adaptive systems approaches to learning become clearer. Focusing at the level of individual and purposefully disconnected events in the classroom in the absence of their context and tensions we may be daunted by the "blooming buzzing confusion" that Ann Brown (1992, p. 141), following William James (1981, p. 462), speaks of. By using the tools of complexity theory, *patterns* of activity emerge from the confusion that provide us with potentially powerful frameworks and methods for understanding and promoting learning and for designing learning environments.

As our sophistication in applying complex systems approaches to learning matures we expect to see increased formalization and more mathematized language and methods develop for exploring complex learning. In recording, analyzing and reporting scientific observations of classroom activity at several levels from individuals to large groups the research reported here has provided an example of what we mean by increased formalization and has demonstrated what insights from a complex science approach might look like. The learning research reported in this chapter has identified remarkably ordered patterns in the apparently chaotic and disorderly behaviors of groups of learners and individuals. We have mathematically characterized relationships between decision wait times and success in foraging simulations and have demonstrated scale invariance in those relationships—the same log-log and log-normal functions appeared at levels of analysis ranging from individuals to classrooms to aggregations of classrooms. We provide evidence for at least two plausible types of strategies that the learners may be using, and suggest that the subjects may be applying multiple strategies flexibly, intermittently, and adaptively throughout the simulations. We point out connections to more established complexity sciences and use those connections to inform our thinking about learning.

Although it is very likely that not all learning can be categorized as complex, it is more likely that much of learning, especially learning in classrooms, can be shown to meet formal complexity criteria. If we accept the notion that learning is complex, then we can begin to apply some new and powerful tools for investigating into its nature and for designing effective instruction. Complexity science is helping to build knowledge and productive models in many natural and human phenomena and we see its application to human learning as being an important, natural, and viable direction for future education research.

ACKNOWLEDGMENTS

Some data and analyses included in this chapter were presented as part of a talk titled, "Scale-invariant power-laws in individual and group decision making" at the National Council Teachers of Mathematic Research Pre-Session held April 4–6, 2005 in Anaheim, California. Authors on that presentation were: Thomas Hills, Andy Hurford, Walter Stroup, and Uri Wilensky. Critical to the development of this line of research is ongoing funding from the National Science Foundation: Grant # 09093 entitled CAREER: Learning Entropy and Energy Project and grant # 126227 entitled Integrated Simulation and Modeling Environment. Texas Instruments has also generously supported this work. We gratefully acknowledge this support. The views expressed herein are those of the authors and do not necessarily reflect those of the funding institutions.

REFERENCES

Adamic, L. A., & Huberman, B. A. (2002). Zipf's law and the Internet. *Glottometrics 3*, 143–150.

American Association for the Advancement of Science. (1989). *Project 2061: Science for all Americans*. Washington, DC: Author.

American Association for the Advancement of Science. (1993). *Benchmarks for science literacy*. New York: Oxford.

Anderson, J. R. (1982). Acquisition of cognitive skill. *Psychological Review 89*, 369–406.

Anderson, J. R. (1983). *The architecture of cognition*. Cambridge, MA: Harvard University Press.

Anderson, J. R., Reder, L. M., & Simon, H. A. (2000). Applications and misapplications of cognitive psychology to mathematics instruction. *Texas Education Review 1*, 29–49.

Anderson, J. R., & Schooler, L. J. (1991). Reflections of the environment in memory. *Psychological Science 2*, 396–408.

Ares, N. (2005, April). *Culturally relevant design and analyses of network supported learning*. Paper presented at the annual meeting of the American Educational Research Association, Montreal, Canada.

Ares, N., Stroup, W. M., & Schademan, A. (in review). The power of mediating artifacts in group-level development of mathematical discourses.

Au, K. (1980). Participation structures in a reading lesson with Hawaiian children: Analysis of a culturally appropriate instructional event. *Anthropology and Education Quarterly, 11*(2), 91–116.

Axtell, R. L. (2001). Zipf distribution of U.S. firm sizes. *Science 293*, 1818–1820.

Bak, P., Tang, C., & Wiesenfield, K. (1988). Self-organized criticality. *Physica Review A 38*, 364–374.

Bak, P. (1994). Self-organized criticality: a holistic view of nature. In G. Cowan, D. Pines, and D. Meltzer, (Eds) *Complexity: Metaphors, Models, and Reality, SFI Studies in the Sciences of Complexity, Proceedings Vol. XIX*. Massachusets: Addison-Wesley.Barab, S., & Kirshner, D. (2001). Guest editors' introduction: Rethinking methodology in the learning sciences. *The Journal of the Learning Sciences. 10*(1&2), 5–15.

Barabasi, A., & Albert, R. (1999). Emergence of scaling in random networks. *Science 286*, 509–512.

Barron, B., Schwartz, D., Vye, N., Moore, A., Petrosino, A., Zech, L., Bransford, J., & The Cognition and Technology Group at Vanderbilt. (1998). Doing with understanding: Lessons from research on problem- and project-based learning. *The Journal of the Learning Sciences. 7* (3/4), 271–311.

Borovoy, R., Martin, F., Vemuri, S., Resnick, M., Silverman, B., & Hancock, C. (1998). Meme tags and community mirrors: Moving from conferences to collaboration. *Proceedings of the 1998 ACM Conference on Computer Supported Collaborative Work*.

Borovoy, R., McDonald, M., Martin, F., & Resnick, M. (1996). Things that blink: Computationally augmented name tags. *IBM Systems Journal, 35*(3), 488–495.

Bowers, J., Cobb, P., & McClain, K. (1999). The evolution of mathematical practices: A case study. *Cognition and Instruction 17*, 25–64.

Bransford, J., & Schwartz, D. (2001). Rethinking transfer: A simple proposal with multiple implications. In A. Iran-Nejad & P. D. Pearson (Eds.), *Review of Research in Education* (Ch 3., Vol. 24, pp. 61–100). Washington, DC: American Educational Research Association.

Callahan, P. (1999, April). *Generative content knowledge*. Paper presented at the annual meeting of the American Educational Research Association, Montreal, Canada.

Camazine, S., Denoubourg, J., Franks, N., Sneyd, J., Theraulaz, G., & Bonabeau, E. (2001). *Self-Organization in Biological Systems*. Princeton, NJ: Princeton University Press.

Campbell, J. R., Hombo, C. M., & Mazzeo, J. (2000). *NAEP 1999 trends in academic progress: Three decades of student performance*. NCES 2000–469. Washington, DC: 2000.

Carnegie Learning. (2003). *The cognitive tutor*. Retrieved July 3, 2005, from http://www.carnegielearning.com

Cancho, R. F., & Sole, R. V. (2003). Least effort and the origins of scaling in human language. *Proceedings of the National Academy of Sciences 100*, 788–791.

Casti, J. L. (1994). *Complexification: Explaining a paradoxical world through the science of surprise*. New York: HarperCollins Publishers, Inc.

Chapman, M. (1988). *Constructive evolution*. Cambridge: Cambridge University Press.

Chapman, M. (1988). *Constructive evolution: Origins and development of Piaget's thought*. Cambridge: Cambridge University Press.

Chi, M. T. H., Feltovich, P. J., & Glaser, R. (1981). Categorization and representation of physics problems by experts and novices. *Cognitive Science 5*, 121–151.

Clark, A. (1997). *Being there: Putting brain, body, and world together again*. Cambridge, MA: MIT Press.

Cobb, P. (1994). Constructivism in mathematics and science education. *Educational Researcher 23*, 13–20.

Cobb, P., Perlwitz, M., & Underwood, D. (1996). Constructivism and activity theory: A consideration of their similarities and differences as they relate to mathematics education. In H. Mansfield, N. A. Pateman & N. Bednarz (Eds.), *Mathematics for tomorrow's young children: International perspectives on curriculum*. Boston: Kluwer Academic Publishing.

Cole, B. J. (1995). Fractal time in animal behavior: The movement activity of Drosophila. *Animal Behavior 50*, 1317–1324.

Colella, V., Borovoy, R., & Resnick, M. (1998, April). *Participatory simulations: Using computational objects to learn about dynamic systems*. Paper presented at CHI '98, Los Angeles, CA.

Davis, S. M. (2002). *Research to industry: Four years of observations in classrooms using a network of handheld devices*. IEEE International Workshop on Mobile and Wireless Technologies in Education, Växjö, Sweden.

Davis, B., & Simmt, E. (2003). Understanding learning systems: Mathematics education and complexity science. *Journal for Research in Mathematics Education, 34*(2), 137–167.

Dewey, J. (1916). *Democracy and education*. New York: Macmillan.

Derry, S. J. (1996). Cognitive schema theory in the constructivist debate. *Educational Psychologist 3*, 163–174.

Diehl, E. (1990). Participatory simulation software for managers: The design philosophy behind microworlds creator. *European Journal of Operations Research, 59*(1), 203–209.

Diodati, P. F., Marchesoni, F., & Piazza, S. (1991). Acoustic emission from volcanic rocks: an example of self-organized criticality. *Physical Review Letters 67*, 2239–2242.

diSessa, A. (1993). Toward an epistemology of physics. *Cognition and Instruction 10*, 105–225.

Empson, S., (2003). Low-performing students and teaching fractions for understanding: An interactional analysis. *Journal for Research in Mathematics Education 34*, 305–343.

Ennis, C. D. (1992). Reconceptualizing learning as a dynamical system. *Journal of Curriculum and Supervision 7*, 115–130.

Ernest, P. (1996). Varieties of constructivism: A framework for comparison. In L.P. Steffe & P. Nesher (Eds.), *Theories of mathematical learning*, 335–350. Mahwah, NJ: Erlbaum.

Freire, P., Freire, A. M. A., & Macedo, P. (1998). *The Paulo Freire reader*. New York: Continuum.

Gabaix, X., & Ioannides, Y. (2004). Evolution of city size distribution. In V. Henderson and J. F. Thisse, (Eds.) *Handbook of Regional and Urban Economics, vol IV*. North Holland: Amsterdam.

Gardner, H., Kornhaber, M. L., Wake, W. K. (1996). *Intelligence: Multiple Perspectives.* Fort Worth, TX: Harcourt Brace.

Gleick, J. (1987). *Chaos: Making a new science.* New York: Penguin Books.

Gonzalez, N., Andrade, R., Civil, M., & Moll, L. (2001). Bridging funds of distributed knowledge: Creating zones of practices in mathematics. *Journal of Education of Students Placed at Risk, 6*(1&2), 115–132.

Goodman, N. (1976). *Languages of art: An approach to a theory of symbols.* Indianapolis, IN: Hackett.

Goodman, N. (1978). *Ways of worldmaking.* Indianapolis, IN: Hackett.

Hegedus, S., & Kaput, J. (2002). Exploring the phenomena of classroom connectivity. In D. Mewborn, et al. (Eds.), *Proceedings of the 24th Annual Meeting of the North American Chapter of the International Group for the Psychology of Mathematics Education* (Vol. 1, pp. 422–432). Columbus, OH: ERIC Clearinghouse.

Hernstein, R., & Murray, C. (1994). *The bell curve: Intelligence and class structure in American life.* New York: The Free Press.

Hiebert, J., Carpenter, T. P., Fennema, E., Fuson, K., Human, P., Murray, H., Olivier, A., & Wearne, D. (1996). Problem solving as a basis for reform in curriculum and instruction: The case of mathematics. *Educational Researcher 25,* 12–21.

Hills, T. (2003). Towards a unified theory of animal event timing. In W. Meck (Ed.) *Functional and Neural Mechanisms of Interval Timing.* New York: CRC Press.

Hills, T., & Stroup, W. (2004). *Cognitive exploration and search behavior in the development of endogenous representations.* Paper presented at the Annual Meeting for the American Educational Research Association, San Diego, CA.

Hmelo, C. E., Holton, D. L. & Kolodner, J. L. (2000). Designing to Learn about Complex Systems. *Journal of the Learning Sciences 9,* 247–298.

Holland, J. (1995). *Hidden order: How adaptation builds complexity.* Reading, MA: Addison-Wesley.

Holland, J. H. (1995). *Hidden order: How adaptation builds complexity.* New York: Addison-Wesley Publishing Company.

Holland, J. H. (1998). *Emergence: From chaos to order.* Reading, MA: Addison-Wesley Publishing Company, Inc.

Huberman, B. A., & Adamic, L. A. (1999). Growth dynamics of the world wide web. *Nature 401,* 131.

Hurford, A. (1998). *A dynamical-systems based model of conceptual change.* Paper presented at the Annual Meeting of the International Association for the Education of Teachers of Science, Minneapolis, MN.

Ioannides, Y. M. & Overman, H. G. (2003). Zipf's law for cities: an empirical examination. *Regional Science and Urban Economics 33,* 127–137.

Jacobsen, D. M. (1998). *Adoption Patterns and Characteristics of Faculty Who Integrate Computer Technology for Teaching and Learning in Higher Education.* Doctoral Dissertation, Educational Psychology, University of Calgary. [On-line]. Available: http://www.acs.ucalgary.ca/~dmjacobs/phd/diss/

James, W. (1890). *The principles of psychology* (2 vols.). New York: Henry Holt.

James, W. (1978). *Pragmatism and the meaning of truth.* Cambridge, MA: Harvard University Press.

James, W. (1981). *The Principles of Psychology.* Cambridge, MA: Harvard University Press, 1981. Originally published in 1890.

Johnston, A. C., & Nava, S. J. (1985). Recurrence rates and probability distribution estimates for the New Madrid seismic zone. *Journal of Geophysics Research B 90,* 6737–6751.

Kameenui, E., & Carnine, D. (1998). *Effective teaching strategies that accommodate diverse learners.* Upper Saddle River, NJ: Prentice Hall.

Kauffman, S. A. (1995). *The origins of order: Self-organization and selection in evolution.* New York: Oxford University Press.

Kelly, K. (1995). *Out of control: The new biology of machines, social systems and the economic world.* Reading, MA: Perseus Press.

Ladson-Billings, G. (1997). Toward a theory of culturally relevant pedagogy. *American Education Research Journal, 32*(3), 465–491.

Larsson, A., Törland, P., Mabogunje, A., & Milne, A. (2002). Distributed design teams: embedded one-on-one conversations in one-to-many. In D. Durling & J. Shackleton (Eds.), *Common ground: Design Research Society International Conference*.

Lave, J. (1988). *Cognition in practice: Mind, mathematics and culture in everyday life*. Cambridge: Cambridge University Press.

Lave, J. (1988). *Cognition in practice: Mind, mathematics, and culture in everyday practice*. New York: Cambridge University Press.

Lave, J., & Wenger, E. (1991). *Situated learning: Legitimate peripheral participation*. New York: Cambridge University Press.

Learning Technology Center. (1992). *Technology and the design of generative learning environments*. Hillsdale, NJ: Lawrence Erlbaum Associates.

Lesh, R., Carmona, G., & Post, T. (2002). Models and modeling. In D. Mewborn, P. Sztajn, D. White, H. Wiegel, R. Bryant, K. Nooney (Eds.), *Proceedings of the 24th Annual Meeting of the North American Chapter of the International Group for the Psychology of Mathematics Education* (Vol. 1, pp. 89–98). Columbus, OH: ERIC Clearinghouse.

Lesh, R., Hoover, M., Hole, B., Kelly, A., & Post, T. (2000). Principles for developing thought-revealing activities for students and teachers. In R. Lesh & A. Kelly (Eds.), *Handbook of research design in mathematics and science education* (pp. 591–645). Mahwah, NJ: Lawrence Erlbaum Associates.

Logan, G. D. (1988). Toward an instance theory of automization. *Psychological Review 95*, 492–527.

Malamud, B. D., Morein, G., & Turcotte, D. L. (1998). Forest fires: An example of self-organized critical behavior. *Science 281*, 1840–1842.

Mandelbrot, B. (1982). *The Fractal Geometry of Nature*. San Francisco: W. H. Freeman.

Mayer R. E, (1996). Learners as information processors: Legacies and limitations of educational psychology's second metaphor. *Educational Psychologist 31*, 153-55.

Moll, L. C. (1990). *Vygotsky and education: Instructional implications and applications of sociocultural psychology*. New York: Cambridge University Press.

Moll, L. C., & Greenberg, J. B. (1990). Creating zones of possibilities: Combining social contexts for instruction. In L. C. Moll (Ed.) *Vygotsky and education: Instructional implications and applications of sociocultural psychology* (pp. 319–348). New York: Cambridge University Press.

Montangero, J., & Maurice-Naville, D. (1997). *Piaget or the advance of knowledge* (A. Cornu-Wells, Trans.). Mahwah, NJ: Lawrence Erlbaum Associates.

Moss, B. J. (1994). Creating a community: Literacy events in African American churches. In B. J. Moss (Ed.), *Literacy across communities*. Cresskill, NJ: Hampton Press.

National Council of Teachers of Mathematics. (2000). *Principles and standards for school mathematics*. Reston, VA: Author.

National Research Council. (1996). *National science education standards*. Washington, DC: National Academy Press.

National Research Council, (1999). *How people learn: Brain, mind, experience, and school*. Committee on Developments in the Science of Learning. J. D. Bransford, A.L. Brown, and R.R. Cocking, (Eds.). Commission on Behavioral and Social Sciences and Education. Washington, DC: National Academy Press.

Neves, D. M., & Anderson, J. R. (1981). Knowledge compilation: mechanisms for the automization of cognitive skills. In J. R. Anderson (Ed.), *Cognitive Skills and Their Acquisition* (pp. 57–84). Hillsdale, NJ: Erlbaum.

Nicolopoulou, A. (1993). Play, cognitive development, and the social world: Piaget, Vygotsky, and beyond. *Human Development, 36*, 1–23.

Nuñes, T., Schliemann, A., & Carraher, D. (1993). *Street mathematics and school mathematics*. Cambridge: Cambridge University Press.

Ochs, E., Jacoby, S., & Gonzales, P. (1994). Interpretive journeys: How physicists talk and travel through graphic space. *Configurations, 2*(1), 151–171.

Pacey, A. (1983). *The culture of technology*. Cambridge, Mass: MIT Press.

Papert, S. (1990). A critique of technocentrism in thinking about the school of the future. *MIT media lab epistemology and learning memo no. 2*. Cambridge, MA: MIT Media Lab.

Peirce, C. S. (1909). A sketch of logical critics. *The Essential Peirce. Selected Philosophical Writings.* Vol. 2 (1893–1913), edited by the Peirce Edition Project, 1998 (pp. 460–461). Bloomington and Indianapolis: Indiana University Press.

Peirce, C. S. (1982) *The Writings of Charles S. Peirce.* 5 vols. to date. Edited by M. Fisch, C. Kloesel, et al. Bloomington, IN: Indiana University Press, 1982 to present.

Piaget, J. (1923/1959). *The language and thought of the child.* London: Routledge and Kegan Paul.

Piaget, J. (1924/1969). *Judgment and reasoning in the child.* Totowa, NJ: Littlefield Adams.

Piaget, J. (1929/1951). *The child's conception of the world.* London: Routledge and Kegan Paul

Piaget, J. (1968/1970). *Structuralism.* New York: Basic Books, Inc.

Prigogine, I., & Stengers, I. (1984). *Order out of chaos: Man's new dialogue with Nature.* New York: Bantam Books.

Prigogine, I., & Stengers, I. (1997). *The end of certainty: Time, chaos, and the new laws of nature.* New York: The Free Press.

Raup, M. D. (1986). Biological extinction in earth history. *Science 251,* 1530–1532.

Reed, W. J. & Hughes, B. D. (2002). From gene families and genera to incomes and internet file sizes: why power laws are so common in nature. *Physical Review E 66,* 067103.

Reiner, M., Slotta, J. D., Chi, M. T. H., & Resnick, L. B. (2000). Naive physics reasoning: A commitment to substance-based conceptions. *Cognition & Instruction 18,* 1-34.

Resnick, L. B. (1987). Learning in school and out. *Educational Researcher, 16*(9), 13–20.

Resnick, M., & Wilensky, U. (1998). Diving into complexity: Developing probabilistic decentralized thinking through role-playing activities. *Journal of the Learning Sciences 7,* 153–171.

Rogoff, B. (1995). Observing sociocultural activity on three planes: Participatory appropriation, guided participation, and apprenticeship. In J. V. Wertsch, P. del Rio, & A. Alvarez (Eds.), *Sociocultural studies of mind* (pp. 139–164). New York: Cambridge University Press.

Rubin, D. C., & Wenzel, A. E. (1996). One hundred years of forgetting: A quantitative description of retention. *Psychological Review 103,* 734-760.

Saxe, G. B. (1991). *Culture and cognitive development: Studies in mathematical understanding.* Hillsdale, NJ: Lawrence Erlbaum Associates.

Saxe, G. B. (1994). Studying cognitive development in sociocultural context: The development of a practice-based approach. *Mind, Culture, and Activity, 1*(3) 135–157.

Schoenfeld, A. H. (1988). When good teaching leads to bad results: The disasters of "well taught" mathematics classes. *Educational Psychologist, 23,* 145–166.

Schwartz, J., & Yerushalmy, M. (1985a). The Geometric Supposer: An intellectual prosthesis for making conjectures. *The College Mathematics Journal, 18,* 58–65.

Schwartz, J. L., & Yerushalmy, M. (1985b). *The geometric supposer.* Pleasantville, NY: Sunburst Communications.

Senge, P. M. (1994). *The fifth discipline: The art and practice of the learning organization.* New York: Doubleday.

Silbert, J., Stein, M., & Carnine, D. (1997). *Designing effective mathematics instruction: A direct instruction approach (3rd ed.).* Upper Saddle River, NJ: Prentice-Hall.

Silbert, J., Stein, M., & Carnine, D. (1997). *Designing effective mathematics instruction: A direct instruction approach (3rd ed.).* Upper Saddle River, NJ: Prentice-Hall Inc.

Skinner, B. F. (1954). The science of learning and the art of teaching. *Harvard Educational Review 24,* 86-97.

Slavin, R. E. (1990). *Cooperative learning theory, research, and practice.* Englewood Cliffs, NJ: Prentice-Hall.

Stein, M., Silbert, J., & Carnine, D. (1997). *Designing effective mathematics instruction: A direct instruction approach, 3rd Ed.* Upper Saddle River, NJ: Prentice-Hall, Inc.

Stor, M., & Briggs, W. L. (1998). Dice and disease in the classroom. *Mathematics Teacher, 91*(6), 464–468.

Stroup, W. (1997a) *Catalog of generative activities and what's a generative activity?* Retrieved July 3, 2005, from http://www.edb.utexas.edu/faculty/wstroup/gen_act_catalog.html

Stroup, W. (1997b). *Root beer game.* Unpublished software for TI–8x calculators based on *The Beer Game* by J. Forrester.

Stroup, W. (2002a, April). *The cognitive and affective affordances of new classroom network design for mathematics learning.* Paper presented at the annual meeting of the American Educational Research Association, New Orleans.

Stroup, W. M. (2002b, April). *The structure of generative learning in a classroom network.* Paper presented at the annual meeting of the American Educational Research Association, New Orleans.

Stroup, W., Kaput, J., Ares, N., Wilensky, U., Hegedus, S., Roschelle, J., Mack, A., Davis, S., & Hurford, A. (2002). The nature and future of classroom connectivity: The dialectics of mathematics in the social space. In D. Mewborn, P. Sztajn, D. White, H. Wiegel, R. Bryant, & K. Nooney (Eds.), *Proceedings of the 24th annual meeting of the North American Chapter of the International Group for the Psychology of Mathematics Education* (Vol. 1, pp. 195–203). Columbus, OH: ERIC Clearinghouse.

Stroup, W. M., & Wilensky, U. (2000). Assessing learning as emergent phenomena: Moving constructivist statistics beyond the bell curve. In A. E. Kelly & R. A. Lesh (Eds.), *Handbook of research in mathematics and science education.* Mahwah, NJ: Lawrence Erlbaum Associates.

Stroup, W., & Wilensky, U. (2003, April). *Mathematics structuring the social sphere (MS3): Rendering the interplay of utterance, gesture and artifact in participatory simulations.* National Council of Teachers of Mathematics, Research Presession. San Antonio, TX.

Stroup, W. M., Ares, N., & Hurford, A. (2005). A dialectic analysis of generativity: Issues of network supported design in mathematics and science. *Journal of Mathematical Thinking and Learning, 7*(3), 181–206.

Thelen, E., & Smith, L. B. (1996). *A dynamic systems approach to the development of cognition and action.* Cambridge, MA: The MIT Press.

Vygotsky, L. S. (1978). *Mind in society: The development of higher psychological processes.* Cambridge, MA: Harvard University Press.

Vygotsky, L., (1987). *The collected works of L. S. Vygotsky: Vol. 1. Problems of general psychology.* Including the volume "Thinking and speech" (N. Minick, Trans.). New York: Plenum.

Vygotsky, L. S., (1987). *The collected works of L. S. Vygotsky: vol. 1, Problems of general psychology.* Including the volume thinking and speech. (N. Minick, Ed. & Trans.) New York: Plenum.

Waldrop, M. (1992). *Complexity: The emerging science at the edge of order and chaos.* New York: Simon & Schuster.

Wertsch, J. V. (1985). *Vygotsky and the social formation of mind.* Cambridge, MA: Harvard University Press.

Wertsch, J. V., del Rio, P., & Alvarez, A. (Eds.), (1995). *Sociocultural studies of mind.* New York: Cambridge University Press.

Wilensky, U. (1999). NetLogo. *http://ccl.northwestern.edu/netlogo.* Center for Connected Learning and Computer-Based Modeling. Northwestern University, Evanston, IL.

Wilensky, U. & Stroup, W. (1999). Learning through Participatory Simulations: Network-based Design for Systems Learning in Classrooms. *Proceedings of the Computer Supported Collaborative Learning Conference, Stanford University, December.*

Wilensky, U., & Stroup, W. (1999). Participatory simulations: Network-based design for systems learning in classrooms. *Proceedings of the Conference on Computer-Supported Collaborative Learning,* CSCL '99, Stanford University.

Wilensky, U., & Stroup, W. (2005). *HubNet.* Available as part of the NetLogo download. Retrieved July 3, 2005 from http://ccl.northwestern.edu/netlogo

Wilensky, U., & Stroup, W. (in review). *Embodied science learning: Students enacting complex dynamic phenomena with the HubNet architecture.*

Wilensky, U., & Stroup, W. (2000). Networked Gridlock: Students Enacting Complex Dynamic Phenomena with the HubNet Architecture. *Proceedings of the Fourth Annual International Conference of the Learning Sciences, Ann Arbor, MI, June.*

Wittrock, M. C. (1991). Generative teaching of comprehension. *The Elementary School Journal, 92*(2), 169–184.

Wixted, J. T., & Ebberson, E. B. (1991). On the form of forgetting. *Psychological Science 2,* 409–415.

Yakubinskii, L. P. (1923). *O dialogicheskoi rechi* [On Dialogic speech]. Petrograd: Trudy Foneticheskogo Instituta Prakticheskogo Izucheniya Yazykov.

Zipf, G., (1949). *Human behavior and the principle of least effort: An introduction to human ecology.* Cambridge, MA: Addison-Wesley.

CHAPTER 13

Systemics of Learning for a Revised Pedagogical Agenda

Andrea A. diSessa
University of California, Berkeley

My primary aim in this essay is to expose some considerations that I feel have been vastly under-represented in thinking about innovation in education, including innovations based on technology. The fundamental observations are (a) that education (or, individually, math, science, or technology education) is in essence a very large-scale system, and (b) that workable and optimal configurations of such a system are strongly constrained by interactions among its parts. The intricacies of constraint in a large-scale system mean that, try as we might, innovation must be responsive to much more than what most of us regularly see and think about, even as innovators of technology or educational practice. Of course, as a community we certainly do not yet know all about such interactions; hence we cannot anticipate them optimally. Yet it seems only sensible to consider what interactions exist and hope to develop strategies for actively dealing with them.

My goal is to expose some important interactions, especially those that seem *most important*, but *least recognized*. For example, the need for "systemic reform" involves interactions in production and consumption of material, intellectual, and human resources in the professional practice of schooling. However, consideration of systemic reform is well recognized in the educational community, even if we are not yet coping optimally with the entailed interactions. So, I will not discuss it. Instead, my focus is primarily on *interactions among different stages and kinds of knowledge*.

In the background of these considerations I hope to respond in some degree to the charge given to us as chapter authors of this volume. In particular, I reframe (so as optimally to connect to my chapter goal) and simplify the given agenda in the following two main questions:

> **Main Question 1.** What math/science/technology knowledge and skills will be important in our high-tech and information-intensive future, and in what out-of-school contexts will we see their importance?
>
> **Main Question 2.** How do we understand competence in these areas, and how do we expect to support it and to see it develop?

When possible, I will also pick up one of the central themes of the conference that gave rise to this volume: the importance, meaning, and instruction of "modeling."

A word about terminology: In this chapter, I use the word "systemics" to denote the general consideration of order and interconnection in a system. I use the word systematicity to denote particular connections or orderliness.

VERTICAL AND HORIZONTAL SYSTEMICS

The Vertical Structure of Learning: New Considerations of Cumulativity

"Vertical systemics" refers to relations of knowledge across time: the way knowledge builds on other knowledge. A completely obvious fact about curriculum is that it has a sequence and an inescapable cumulativity. Students always learn on the basis of their existing base of skills and knowledge. The current educational system has a deeply embedded sense of progression, especially in mathematics. Emblematically, arithmetic comes first, then algebra and geometry, and then calculus.

The problem with the current vertical equilibrium is that it is ad hoc, unevaluated with respect to optimality, and based mainly on historical example and on untested intuitions of "simple," "prerequisite," and "more complex." In the following, I briefly treat four considerations that suggest the need for new and different vertical systemics, which would force reconsideration of the current vertical organization of learning.

Cognitive simplicity. The basic outlines of the progression of school-based mathematics and science instruction were established long before the cognitive revolution in the study of learning. But, we have learned a great deal since then about starting points and principles of accumulation. For example, the role of prior, domain-specific knowledge was all but completely ignored before the late 1970s and early 1980s. Then, "misconceptions studies" dramatized the role of intuitive knowledge on instruction. Unfortunately, the most publicly visible aspects of the influence of prior knowledge were negative. "Students have persistent 'misconceptions' that systematically interfere with learning." I believe, however, that the best studies of naïve ideas have shown fundamental productive roles in supporting learning via new starting points and new paths. See Smith, diSessa, & Roschelle (1993) for a discussion of these issues.

Two of the best examples of revised starting points and different principles of development (different principles of vertical systemics), to my mind, are the work on early arithmetic by Robbie Case and colleagues (1992) and the work on ratio, proportion, and fractions by Jere Confrey (1994). In my project's work, we found rich, direct, and useful intuitions about speed and movement that mean that one can start much earlier than currently expected in teaching such "advanced" concepts as acceleration and compound motion (diSessa, 1995a, 2000).

Re-mediation. The role of material symbol systems such as written language and algebra has mostly been invisible or implicit in considering educational scope and sequence. That is because we have had a stable infrastructure of core representations largely since calculus was invented, and so it has been easy to ignore representational issues or treat them as conceptual ones: "Calculus is too hard for elementary school students." But calculus may be difficult only because present versions of teaching the subject rely too much on a difficult symbol system, algebra.

Computer technology has brought with it a huge new set of possibilities for dynamic and interactive representations. Students can think and learn with these in substantially different ways, leading to new possibilities for curricular scope and sequence. Jim Kaput and his collaborators have chosen particular dynamic and interactive representations that they believe can radically change assumptions about who can understand what about calculus, and when they can understand it. In particular, important elements of calculus are truly and democratically accessible in elementary school (Stroup, 2002). In my project's work, we have chosen different fundamental representations, but with the same conclusion. As an emblem, we discovered that vectors, in dynamic and interactive computational form, become immediately intuitively accessible and even vividly useful to elementary school students (diSessa, 2000, chapter 2). Then, of course, the obvious question is when and how these observations of revised starting points and times will be capitalized on in the common curricular sequence, which now (for example) has vectors as an "advanced and difficult" concept for high school students?

Logic vs. sensible fabrics of activity. There are many ways to think about cumulativity in learning. Gagné (1985) provided and advocated one particular image: hierarchical decomposition of skills. Modern views of learning have shattered the plausibility of this and other logical and a priori methods of sequencing learning in at least two fundamental ways. First, intuitive or experiential knowledge can provide us "free" resources and excellent starting places for learning, independent of assumptions about where one must logically start, or assumptions about elementary units of skill. This issue was already broached under the topic of "cognitive simplicity." Second, modern approaches to instruction emphasize the "sensibility" of the activities in which students engage. Logical decompositions can lead to students' practicing skills that appear useless in themselves. In addition to affective issues, teaching a skill outside of a context in which it is evidently useful makes it difficult or impossible for students to judge the adequacy of their skill in terms of success in a sensible task. Student autonomy, engagement, and the effectiveness of learning through the feedback of experience all suffer.

More generally, the nature of the activities in which students are asked to engage is an important constraint on design. Activities (a) that motivate students, (b) that can be performed "nearly autonomously" by students (that is, without strong constraints provided by teachers and tasks), and (c) that have excellent community/collaborative properties are high-priority goals. Purely knowledge-structural characteristics, such as practicing sub-skills before skills, must sometimes, if not always, defer in some measure to these considerations. In several places (e.g., diSessa, 1992), I have conjectured that schooling should be concerned with fostering a progression of more refined activity types for students, in addition to the usual "progression of knowledge."

Goals for the future. Schooling is a preparation for future pursuits. Main Question 1 asks us to imagine future contexts and the skills they will demand. I defer some of my comments on this to later sections, but here I wish to make one overriding point. Flexibility and the ability to learn on-the-job (or between jobs) have come to assume a central role in the educational community's thinking about the future. I believe this has been forced by some basic issues concerning learning, general competence, and competence in particular contexts. But, in any case, our lack of knowledge about the particulars of future jobs, and rapid change in jobs and in requisite skills have pressed toward the same conclusion.

What is the nature of a curriculum that prepares students optimally to learn more later? Such a curriculum releases us somewhat from assumed constraints of directly teaching particular knowledge and job skills. So, perhaps we can worry a little less about what exactly students learn, and concentrate more on building on strength, teaching them more (guided by strong naïve knowledge), and concentrating on fostering a love for and skills at learning. At the core, this is a sort of "supply side"[1] view of learning and performance: Build the biggest, most robust pools of knowledge in students, and worry less about particulars of use. I feel there is a strong logic to this view. Can people who have learned an immense amount of mathematics and science, *independent of many details about what in particular they have learned*, really fail to be adaptable learners in the future?

On the other hand, it is sensible to make sure that we teach genuinely *extendible* skills, and knowledge that fits the imaginable future, at least in a general way. "Teaching for future learning" is a phrase John Bransford and colleagues have used (e.g., Bransford & Schwartz, 1999), and it is apt here. See also the discussion of several types and modes of transfer in diSessa and Wagner (2005).

The scale of redesign of mathematics and science curricula to cohere with these new considerations of vertical systemics—or even to bring scientific accountability to bear on current assumptions—is boggling. Many people are experimenting with new possibilities of starting points and progressions, at least in fragments of the curriculum. Almost none are dealing with revising the full-scale K–12 learning. The point of this chapter is to ask whether there should be some strategic consideration of how we can planfully and profitably engage the large-scale issues.

The Horizontal Structure of Learning: New Things to Learn and the Constraints Among Them

This section turns to "horizontal systemics," the question of what relations are important at particular instants of time, within the vertical trajectory of learning that I discussed above. The main work of this section is delineating different kinds of knowledge and thinking about relations among them. However, I begin with a brief elaboration of a central point.

I get nervous when I hear people talking about pure versions of teaching modeling or problem solving, independent of a program of deep conceptual development in particular scientific domains. One element of systemics is balance and complementarity. One should always prefer a balanced, diverse knowledge system, containing knowledge with complementary strengths and weaknesses in contrast to a narrow, highly tuned system, at least if future contexts of application are uncertain (which seems the main certainty concerning what future citizens will need to know). I believe high level (e.g., modeling) vs. lower levels (learning particular models) is one dimension along which we need to find a good balance. The well-known trade-off of weak methods vs. strong ones (breadth of application vs. focused, efficient power) reinforces the idea of maintaining a spread of levels. And one certainly should not forget that the intellectual heritage of civilization lays undoubtedly, in some substantial measure, in particular theories and particular methods, independent of recent fascination with higher-level knowledge.

---

[1]To explain the "supply side" metaphor, some economists have emphasized the importance of improving the conditions for the creation of products and services, in preference to worrying about the consumption side, such as demand (e.g., making workers richer). Supply-side learning similarly focuses on building the biggest supply of knowledge possible, and not on supposed "needs" or "constraints" on use. Ironically, supply side economics is generally associated with political conservatism, and supply side instruction is much more associated with advocates of "liberal education."

In addition to "balance and complementarity," *synergy* is an equally important issue in the systemics of knowledge and learning: What knowledge pools actually rely on each other for important effects? Although I believe a similar story could be told for modeling and domain-specific theories (or problem solving and specific methods), the synergies between levels that I have thought most about concern domain-particular knowledge and relevant *epistemological ideas*. Students' intuitive epistemologies have become much more visible in research concerning science and mathematics education in recent years. What do students know about knowledge and knowing, and how does that effect learning? Why do some students think school science is memorizing formulas, and others think it is about reasoning; and what are the consequences? For a recent synthetic review, see Hofer and Pintrich (2002).

I believe that there are two distinct and important synergies between epistemological knowledge and domain particular knowledge. Both synergies involve a degree of specificity of epistemological knowledge, even though most current research assumes a rather broad generality. First, I believe that the intuitive roots of epistemological knowledge (like the intuitive roots of knowledge about physics—e.g., diSessa, 1993) lie in observations and abstractions individuals make of their own experiences. Furthermore, the quality of observations and abstractions made out of experience depend on the quality of that experience. In the epistemological case, this means that one needs high-quality experiences concerning (first-order) knowledge and knowledge building in order to root excellent (second order) epistemological ideas. For example, I do not believe that it is plausible to teach someone what deep understanding is without having that person understand at least one thing deeply. To sum up, good epistemological knowledge depends on good specific learning experiences.

The second synergy of epistemological knowledge and subject knowledge goes in the other direction; good subject matter learning depends on a good epistemology. One of the essential powers of epistemological knowledge is in bootstrapping more learning. Students who think physics is a series of disconnected facts will not learn much deep physics. Indeed, we should probably measure the quality of epistemological knowledge precisely by the criterion of fostering good content learning, rather than by more common criteria, such as that it is a *correct* view of scientific knowledge, or even that *scientists have these views*.

To sum up, scientific knowledge per se and knowing about the nature of scientific knowledge are strongly synergistic. Epistemological ideas are rooted best in reflections about learning particular science, and they are valuable precisely with respect to their ability to foster productive further learning.

Balance, complementarity, and excellent synergies are particular systematicities to which we need pay attention, and about which, unfortunately, little research has been done. Two metaphors help me capture these considerations in a phrase. First, we must select "well-formed knowledge systems" as learning targets. What is the ideal composition of learning particular science and particular meta-knowledge? The connection some might instinctively make between "well-formed knowledge systems" and the idea of logical well-formedness may not be ideal, so I personally lean more toward a second, biological metaphor: "viable and generative cognitive ecologies."

The remainder of this section lists three classes of knowledge along with a few comments on horizontal systemics involving them.

Epistemology

I introduced personal epistemology to exemplify synergistic relations with domain-specific ideas. It goes into the list of new knowledge we need to cultivate, Main Question 1. For now, that is all that need be said about this class of knowledge.

Infrastructural Representations

I discussed the power of new, computational representations already in terms of vertical systemics: to wit, radical re-ordering of topics taught. Of course, horizontal systemics are an essential part of that story. If representations become much easier to learn in computational form (e.g., dynamic and interactive graphs; dynamic and interactive vectors), then certain conceptual knowledge follows along (a horizontal link) and may be learnable much earlier (e.g., linear and vector velocity and acceleration, including the relations of derivative and integral). Genuinely new representations are coming, too, not just pedagogically and conceptually better-adapted versions of old ones.

New representations, one at a time, can do remarkable things. But a new fabric of representation can shift the conceptual infrastructure of civilizations. Following the pioneering work of Seymour Papert and Alan Kay, my research group has been pursuing the possibility of true computational literacies. The advent of textual literacy was perhaps the most important intellectual event in the history of human civilization. We take the possibility of computational literacies *that* seriously, unlike what the phrase "computer literacy" is likely to evoke.

Algebra constitutes a benchmark example of a technical literacy that is now foundational to almost all professional mathematics and science. An important line in my group's investigations has been to see how simple programs can replace (or, better, prepare for and extend) algebra as a core, technical representation. For example, Figure 13–1 presents the one-dimensional *tick model* that shows the relations among position, velocity, and acceleration. The name "tick model" refers to the fact that the centerpiece of such models is what happens at each instant, at each "tick of the clock." Elementary school students are quite capable of learning and productively using models like this, even in vector form. See diSessa (2000, chapter 2) for an extended discussion of the pedagogical and conceptual advantages of the tick model over algebra. Using such models, we are confident that young students can generally learn a lot of scientific material that now is introduced only in high school, and they can learn it not in "toy" form, but with a high degree of mathematical generality and precision.

Synergies between representations and concepts are tight enough that computational kinematics, for example (a prime focus of our prior work), cannot be conceptually identical with the standard algebraic kinematics. Bruce Sherin has done the best recent work I know concerning the relations of representation to

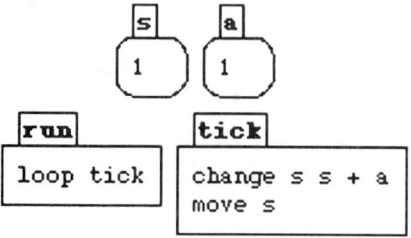

FIGURE 13–1 The tick model for accelerated motion: The main loop, called **run**, causes a procedure **tick** to be run over and over. **Tick** increments the speed **s** by the acceleration **a**, and causes an object to **move** a distance, **s**.

conceptualization, and he did his work precisely comparing algebra to computer programs (Sherin, 2001).

My group has used the tick model mostly as a given, instructed model. However, when students are even a little proficient with computational representations like this, a new range of student modeling is possible. *Modeling using generic modeling languages*, such as programming, has some very attractive properties. One of the most important is that, once learned, students are prepared immediately to model many things without needing to learn an idiosyncratic modeling language for new cases. Another advantage is cumulative expertise. Whenever you have a common representational form for many phenomena, true expertise with the representational form can grow gradually and facilitate all future uses of it. Algebra is an old example of a representational form that deserves extensive development of expertise, and programming is, we conjecture, a new one. As attractive as specialized modeling languages are, my feeling is that more generic ones will almost always have a favored place in the end, both conceptually and pedagogically.

Schwartz (this volume) discusses how effective pedagogical strategies almost always combine supplying models to students (such as the tick model) with having them construct their own models (such as using programming as a generic modeling language). This is a very sensible attitude. However, many educators have very limited expectations for student models. Student modeling is good for "process skills," they claim, but it is naïve to expect students to recreate any fundamental mathematics and science. After all, you just cannot expect to have many Newtons or Einsteins in the classroom. In general, I believe that expected limitations in student modeling need to be reconsidered, and we have at least three reasonably well-developed empirical cases to put forward in this regard.[2] The best researched case is the fact that students as young as sixth grade can "invent graphing" as a representational form, something that originally took the genius of Descartes to do. There are qualifications concerning students truly inventing graphing that I cannot go into here (see diSessa, 2004; diSessa, et al., 1991). However, in at least a half-dozen trials, we have validated this surprising result in essence.

Another case that has been replicated several times concerns students inventing Galileo's model of a dropped object, which is one of his central accomplishments concerning the theory of motion. Again, students as young as sixth grade reliably and spontaneously invent the two models that Galileo discusses— the correct, additive model (a constant additive difference in velocities; that is, constant acceleration), and an alternate, attractive but incorrect multiplicative model (increments in velocity involve a multiplicative constant, from one to the next). In this case, one caveat concerning students duplicating ancient works of genius is easy to express. Galileo dispatched the incorrect model of his pair, but children typically cannot manage this without substantial help. The third case of exceptional re-invention is only a case study, where a group of high school students "designed Newton's laws" (diSessa, 1995b).

The empirical facts concerning re-inventing substantial science are interesting, but the real value is theoretical. Clearly it is true that students are not generally geniuses of science. So how is it at all possible for them to redo "acts of scientific genius" in any respectable sense? I think the answer to this question is complicated, but here are some elements.

---

[2]There is a reasonably large literature on student-invented "standard" mathematics, including C. Kami (arithmetic procedures), R. Lesh (e.g., statistical measures, such as weighted average), & R. Nemirovsky et. al. (graphical representations). We do not review the literature here.

- People often dramatically underestimate the conceptual power of representational systems. Teaching students certain representational systems and associated concepts first makes other conceptual accomplishments much, much easier. This is an element in our general discussion of horizontal and vertical systemics of knowledge.
- People also dramatically underestimate common culture as a source of ideas and orientations that change the relationship of present-day students to conceptual innovations that required true genius the first time around. The idea that great effect can be had from mathematizing the physical world was a brilliant breakthrough by Galileo. But numbers and numerical views of nature are pervasive in common culture today. A second case in point is that contemporary students are exposed to a huge and powerful backlog of meta-representational ideas (see below), which facilitates inventing scientific representations, such as graphing.
- People underestimate the power of knowing that something is possible, which we can often simulate merely by asking students to do it. People also underestimate the support teachers can give subtly to students, without taking essential agency from them.

Meta-Representational Competence

I already fairly thoroughly anticipated the idea that students know, and can learn a lot more, concerning the nature of representations ("inventing graphing")"—what we call meta-representational competence (MRC). In particular, students can nearly autonomously design and evaluate cogent, new scientific representations. In retrospect, the idea of MRC should not be surprising. After all, the progress of science has been just as much a process of designing new, more apt and powerful representations as it has been a purely conceptual inquiry. If students can and should engage in conceptual inquiry, why should they not engage in representational design and critique? Furthermore, as the representational infrastructure of science becomes much more diversified in many controllable, computer-based representations, it seems obvious that the value and need of MRC has become much enhanced in the modern context. Here is a "new skill" to put on the list in answer to Main Question 1. MRC will be vividly entailed in most any future techno-scientific job, wherever computers represent or control the world, from nuclear power plants, to traffic flow in cities, to spatial displays of demographics for political or advertising purposes.

The most surprising thing about meta-representational competence is the excellent repository of naïve knowledge students have concerning it. This seems to contrast with incoming epistemologies, the meta level for scientific knowledge. Therefore, strategies of creating well-formed knowledge systems may differ between MRC and intuitive epistemologies.

Excellent base competence in MRC is mainly a recent discovery—at least specifically with respect to scientific and mathematical representations—and it complements the same sort of "misconceptions" literature, documenting incompetence, that began the study of intuitive ideas. That is, there have been many studies of representational misconceptions, but comparatively fewer studies of what we believe to be a strong pool of helpful ideas. See diSessa (2004) for a review. In terms introduced earlier, there are important *cognitive simplicities* concerning MRC that it behooves us to understand and use instructionally.

An important conjecture, which is far from empirical validation at present, is that MRC and learning particular representations are powerfully synergistic. Thus, for example, students who have difficulty learning particular representations may better be served by enhancing their MRC, compared to attacking, one by one, difficulties with particular representations.

I think an optimal (or even just *good*) future pedagogy will have to be much reorganized to take into account the fact of MRC and its relation to narrower versions of representational competence. That is, of course, just one more example of the main point of this chapter.

PATTERNS

General Sketch

In this section, I concretize some of the considerations above in terms of a new project my group has initiated. Although our results to date are tentative, many of this chapter's concerns are illustrated.

The best way into the project is via cognitive simplicity. In short, we intend to explore curriculum and learning in a particular conceptual area driven strongly by considerations of naïve competence; we have chosen a topic area primarily on the basis of our guess that strong intuitive conceptual resources exist and can be harnessed to yield high-quality scientific understanding. The topic is "patterns of change and control."

There is face validity to the idea that children have a great deal of knowledge about simple patterns of change and control. For example, toddlers build towers of blocks and discover an *increasing instability* as they build. They know they can push reasonably stable towers a little and they will recover. But they also know in some intuitive way that there is a *tipping point* beyond which a catastrophic new process takes over from perturbable stability. Older children know how to *pump* swings. It seems quite likely (and we have recent data on this) that they often think about social phenomena in terms of similar patterns. A friend can be teased to a point with a reliable return to "equilibrium." But, beyond some point, a flash of anger or even an unrecoverable split-up results.

The research base on this competence base is not barren, but it is fairly minimal. Some of my own work has uncovered certain relevant intuitions (e.g., diSessa, 1993), and some work by, Nemirovsky (1993) has been encouraging. But, by and large, an important first step in our program is to get a much better sense of the nature and content of intuitive knowledge about patterns of change and control.

The usual way to justify a teaching focus is to note that one is taking "an approach" to one of the traditional topics in existing school curriculum. But that is not the game here. Even many of the curricula that build self-consciously on intuitive ideas focus on traditional topics, and not on ones newly discovered to be feasible as instructional goals. On the other hand, just because something is easy to teach does not mean it is good to teach; we also need to defend the usefulness of the competence that we eventually intend to build.

To first approximation, we feel that the intuitive resources concerning patterns of change and control will be useful in coming to understand something like what is called dynamical systems theory. Like MRC, this is another (and more speculative) answer to Main Question 1. How can we justify this new topic, which is currently conceived as much too advanced for school science and mathematics? Let us briefly consider the systemics of this choice.

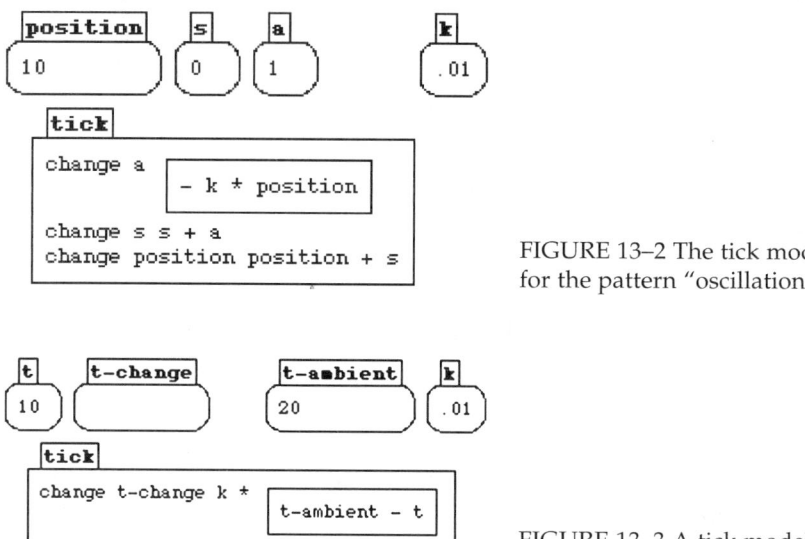

FIGURE 13-2 The tick model for the pattern "oscillation."

FIGURE 13-3 A tick model for the pattern "equilibration."

Cognitive simplicity. We aim to demonstrate in this project precisely that a rich and educationally useful pool of naïve knowledge exists for learning dynamical systems theory. More specifically we aim to explore teaching some form and slice of dynamical systems theory that is optimally supported by existing cognitive simplicities (and further systematicities, described below). This will certainly not be the form or slice written into current (advanced) textbooks on the topic.

Re-mediation. We would not think it plausible to approach dynamical systems theory well at early ages without computer mediation. This includes experiences mediated by simulations, but, more characteristic of our project, it includes developing computer-based models of essential dynamical phenomena. Following earlier discussion, we expect to present computer models as target representations that students will learn, but also to facilitate students' constructing their own models of basic patterns.

To illustrate, Figure 13-2 shows a tick model for the fundamentally important pattern of "oscillation." Readers will recognize the second two lines of **tick** simply as the tick model for accelerated motion, in which **move** has been explicitly written out as a change of position. The first line of **tick** expresses the fact that acceleration is proportional to position in such a way that, the farther an object is from 0 position, the more it is accelerated "backward," toward 0. Although from previous experience we believe this is a powerful instructible model, we do not know yet whether it is a student-inventable model.

Figure 13-3 is a tick model for the phenomenon of equilibration. The model is expressed here in terms of the domain of temperature equilibration, but, of course, it applies to many types and contexts of equilibration. In the model, the essential step is that the change of temperature at each instant (t-change) is simply proportional to the difference of the current temperature (t) and the "surroundings" (t-ambient). The constant k represents a complex confluence of geometry and matter parameters (including specific heat, volume of the sample, area of contact with the ambience, etc.). We do not propose to teach these details, but only the pattern, its meaning and its dynamic characteristics.

Sensible fabrics of activity. Activities are a central concern in our patterns work. There is quite a lot to say, but in this context I concentrate on one primary goal: We want students to become competent at discovering and elaborating the meaning and import of fundamental patterns. Of course, we will scaffold, as much as necessary, learning some of the most important and central patterns, like oscillation, equilibration, and pumping. But, we want students to become capable of working through complete cycles of discovery, description, explanation, and so on. Explication of fundamental patterns that we cannot leave to student discovery can serve as worked examples for students to emulate.

One of the especially interesting features of patterns inquiry is that discovery and elaboration of patterns is actually much more a theoretical inquiry than an empirical one. What does one mean, for example, by "stickiness"? Is any threshold phenomenon an example of stickiness, or should we, for example, separate off phenomena that involve sticky substances? Is "sticky tape," which wears out under repeated stick/unstick cycles, a prototype for a subclass of stickiness? What intellectual power do we get by studying the nature of stickiness? Developing and refining descriptions to maximize scientific power is the essence of theory development.

How should we (or students) know when we have developed an adequate understanding of a pattern? This is a principal aspect of Main Question 2 (what is the essence of competence?) concerning our patterns curriculum.[3] Below is a prototype rubric that we use to evaluate students' understanding of a pattern, and which we propose to students to evaluate their own understanding (meta-knowledge of patterns).

1. Definition: What is the essence and defining features of this pattern?
2. Extension: What are prototypical examples? What are some "boundary examples" that seem odd or different in some way, but are still examples? What are some "near misses" that appear similar to the pattern, but do not count in your definition?
3. Subclasses: Are there importantly different versions of the pattern? Why do they warrant separate mention, yet remain examples of the pattern?
4. "Internal Logic": What are important elements and relations among elements in the pattern? Can you provide a general explanation for the pattern? Can you make a computer model and explain how it results in the pattern?
5. Embeddings: Are there specialized principles that make the pattern work in different cases? For example, oscillation needs a "restoring force," but there are many different kinds of restoring forces: gravitation (a pendulum), elastic forces (springs, tree limbs), etc.

Epistemology and meta-knowledge. Internalizing and effectively using a rubric of evaluation, such as above, is one particular goal for our curriculum. Another very substantial goal has to do with the human process of developing scientific understanding of patterns. Prior work and work thus far in this project convinces us that students have some very particular difficulties in developing a science of patterns, which we briefly list below. These constitute epistemological goals for students (as well as just blocks to learning): understanding the processes of developing their own understanding.

---

[3]Note, however, that judging the adequacy of some "final" understanding is only a small part of the competence of developing the conceptualization of a pattern. Strategic moves in developing the pattern are another part.

We believe that students' intuitive resources concerning patterns have some very particular characteristics. Our "hyper-richness" hypothesis states that, at least in some areas, students have a huge repertoire of ideas that can be productively applied to patterns. But this seemingly very positive characteristic has some serious down sides:

1. Communicating with others may be difficult since each student may be seeing different things in the same situation.
2. Settling on a single meaning for a pattern may be difficult.

Note how the hyper-richness hypothesis contrasts to typical "misconceptions" hypotheses, where it is assumed that students have only false, unproductive knowledge; where it is assumed that students can articulate their ideas; and, where it is assumed student ideas are coherent to the point of constituting precisely one theoretical view. See discussion of fragmentation vs. coherence in diSessa, Gillespie, & Esterly (2004).

A different class of difficulties in students' intuitive patterns knowledge seems to have to do with level of abstractness. In our experience, students' preferred levels of explanation are often at a more specific and concrete level of description than patterns, as we define them. In one very formative study, we asked students to examine a "chaotic pendulum" that was driven by an array of magnets hidden in the base. A patterns-oriented inquiry would focus on modes such as stable oscillation, regimes such as small versus large oscillations, and things such as the limits of predictability ("stable" oscillation sometimes evolves into dramatic mode switches). Instead, students were preoccupied with whether there were, in fact, magnets in the base, or what alternatives existed. Rather than a patterns-level explanation of the behavior, they searched for a mechanism-level explanation (e.g., "magnets make it work").

Pursuing the wrong level of explanation and inquiry is a potentially critical difficulty, which perhaps emerges in the following way. Patterns-level understanding of nature, in fact, is quite rare. There are few powerful and explanatory patterns. So, naturally, people tend to spend more time closer to the concrete, mechanical structure of the world in searching for explanations. However, the rather mathematical level in which we are interested in the patterns project must become a sensible level of inquiry to students, at least if our patterns curriculum is to succeed in this fundamentally epistemic way.

Can students understand these difficulties and limits in their own beginning understanding of patterns, and develop strategies to overcome them?

Future learning. What value would students accrue from learning dynamical systems theory? First, dynamical systems is an excellent context in which to set other, more specific theories. For example, Newtonian dynamics is an important special case. Many other subjects, from ecology, to economics, to chaos, could draw on important ideas in the sort of intuitive dynamical systems theory we aim to teach. The ability of future citizens to control and understand systems critical to our survival—economic ("boom and bust" cycles), environmental (thermal "runaway"), ecological (diversity and stability), technological (controlling electrical grids or nuclear power plants), and social—creates face value for this topic beyond currently central topics, including currently sanctioned ones, such as Newtonian mechanics.

Some Details: The Case of (Thermal) Equilibration

This section illustrates in more concrete form many of the considerations of systemics described above. As well, it illustrates methods we intend to use to support learning our focal knowledge, which was Main Question 2.

In the summer of 2004, we ran a short experimental class on patterns, essentially to get our feet wet in the phenomenology of student learning about patterns. The class involved six students at the 10th and 11th grade level. Given a short time-scale (5 classes, 3 hours each), we abandoned the task of supporting students' own, extended inquiry into patterns, and instead focused on five important patterns-to-be-instructed. These were tipping-point, randomness, equilibration, oscillation, and pumping/resonance.

The equilibration class, unlike the others, focused on only one embedding of the relevant pattern, thermal equilibration. Early in the class we asked about what happens when one takes a glass of liquid out of a refrigerator (at just above freezing, a few degrees Celsius). In the ensuing discussion one student explained the graph he had drawn: "I basically thought that temperature will increase by...by, like, oh, will multiply itself by 2 every minute." Say, for example, it starts at 9:00 at 1 degree: It will go "2, 4, 16, and in between 9:03 and 9:04, it goes to 22 Celsius."

There are, of course, some serious problems with this model. First, exponential increase is inconsistent with intuitions about equilibrium that other students expressed. Indeed, it is inconsistent with intuitions that this same student expressed on other occasions. Such inconsistency would be typical of hyper-richness. Also, why the process stops abruptly when it gets to equilibrium is unclear.[4]

On the positive side, there are some striking properties of this early model. First, it is a quantitative model. Quantifying phenomena had a minimal role in earlier work in the class, so this seems to be a genuine student instinct to quantify. Earlier in this chapter we mentioned how common-sensical quantifying nature is in common culture and school, in comparison to Galileo's day. Beyond being a quantitative model, it involves a uniform process, constantly multiplying by a factor of two. That is a core element in a normative model (Figure 13–3), and it will turn out to mark a strong contrast from consensual intuitive models: Instincts in quantitative modeling are often more normative than qualitative conceptualizations, as we shall see. Although at this stage, little can hang on it, we note that this model is, in fact, an accurate mathematical model (!), provided that what gets multiplied is the *difference* of temperatures between current and equilibrium value, and provided that the factor is less than one.

The class did not pursue this quantitative model. Instead, a consensus arose gradually about the qualitative nature of equilibrium. Students agreed that equilibration was a three-phase affair. In the first phase, the temperature change will get "up to speed" gradually. Then, the middle phase is a constant, linear increase. Finally, temperature will more gradually close in on its eventual, ambient value. Altogether, temperature will equilibrate in an extended S curve (like a logistic curve), with increasingly and decreasingly sloped regimes sandwiching a linear, middle phase.

---

[4]At this early stage of discussion, the instructor did not press criticisms of points of view, so these issues were not raised in class.

There are a number of notable features of this model. First, it shows some limited, but expectable hyper-richness: Temperature equilibration is not one phenomenon, but it is three in succession. Second, this is an encouragingly abstract characterization. In particular, the assumption of homogeneity that probably drove the "multiply by two" model is here manifested most obviously as a linear middle phase. Not only linearity but also, more generally, homogeneity are properties of many of the most important scientific patterns.

Explanations of the first phase were provocative. One student made it clear that the first part of the graph would curve up gradually into the linear phase only if the liquid were taken directly from the refrigerator. If the liquid were already warming up, we would be seeing a straight line. Do students think temperature has momentum, so that it needs time to "get going from rest"? We do not believe this is the case, although there is minimal data in the video to dispute it. But consider that "change takes time to blossom" is a highly abstract pattern previously hypothesized in students' thinking.[5] This pattern is a simpler but more general one than momentum. Also, students explicitly rejected momentum-like behavior at other points in the temperature curve. For example, they summarily rejected the possibility that temperature would overshoot ambience.

It is worth noting that our temperature explorations with these students showed other manifestations of hyper-richness and of a possible preference for mechanistic explanations. First, some students voiced an expectation that differences in the initial temperature gap to ambience would make substantial changes in the shape of the equilibration curve. One student noted that equilibration of water in a test-tube immersed in a water bath would work differently than a glass of water in the air. Two other students argued that heating and cooling would be clearly distinct phenomena. After all, heating involves heat going from a huge (surrounding) space into a small, confined region (the glass of liquid), and cooling involves heat leaving the small region, spreading into the ambience. How could spreading out and concentrating work the same? Another argument was based on the fact that temperature is related to the speed of motion of molecules. Since it is much easier to slow things down than to speed them up, heating would be more difficult to accomplish than cooling.

The remainder of our temperature lesson had two major parts. First, we had students collect data in pairs using a computer and thermal probe setup. Second, we engaged in a scaffolded full-class attempt to build a computer model of temperature equilibration as empirically determined. The results of the data collection were described as "shocking" by one of the students. Their three-phase equilibration hypothesis was contradicted. Instead, equilibration followed a simpler, one-phase, "gradually decreasing slope" pattern, similar to what they had as the final of three phases.

For several reasons, we had limited expectations for the attempt to build a computer model. Students had very limited experience with programming; we did not feel we had time to build an infrastructural competence with programming as a generic modeling language. We also simply did not have the time to allow students multiple iterations in building a suitable model.

---

[5] See the "warming up" phenomenon discussed in diSessa (1993).

[6] At the time, the teacher was more focused on supporting the normative model, and he did not realize the mathematical sufficiency of the students' model. It is a good exercise for the reader to understand why the normative model, this model, and the model "multiply the difference between current temperature and ambience by a constant factor" all result in exponential equilibration.

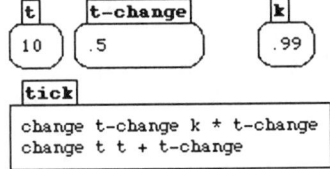

FIGURE 13-4 A student-constructed model that, in fact, produces exponential equilibration.

Still, all things considered, results were surprisingly good. Students constructed almost autonomously a model where the increment of temperature is constantly multiplied by a constant less than one (Figure 13-4). This is, again, a sufficient mathematical model in that it produces exponential equilibration, even if it is not the normative model, given in Figure 13-3.[6] With heavy scaffolding, the class also constructed the normative model (although under the given time pressure, this bordered on "showing the students" the model). Furthermore, students developed quite expert abilities to adjust parameters of the instructed model to match the data they collected. These abilities would count as "internal logic" in our rubric for pattern understanding.

While this first modeling experience falls short of demonstrating that students can almost autonomously construct the normative model of exponential equilibration, there are many positive features of what they did, summarized below. These findings will help us better negotiate future editions of the equilibration modeling activity.

1. Hyper-richness played out in a positive way, which, we conjecture, could be even better capitalized on. Students expected equilibration to have three phases, not one. Many students expected heating and cooling to behave differently, and they also expected that qualitative features of equilibration would depend on such things as temperature difference and phase of the ambience (liquid or air). The student shock at the simplicity of the result could be amplified if a wide diversity of experiments were performed and shown all to conform to the model. The fact of hyper-rich expectations, along with the ability of students to come up with and competently use good tick models, sets up an excellent drama to emphasize one of the main meta-points we want to make in our patterns instruction: Some patterns are powerful precisely for their striking simplicity and amazing range of application; our own human predilections are usually not to expect or believe in such powerful simplicities.[7]

2. Hyper-richness was manageable by focusing students' attention on mathematical formulations: graphs and then iterative calculations. However, we feel it would be important to return to reflect on mechanistic expectations in the context of discovering an amazingly simple and general model.

3. Students appear to have good instincts with respect to homogeneity of models. (Homogeneity is displayed in all of our to-be-instructed tick

---

[7] The drama of uncovering shocking simplicity and generality, we feel, is far preferable to trying to argue students out of prior conceptions with contradictory argument and data. "Love is a better teacher than duty."

models.) In particular, students nearly autonomously suggested two mathematically correct models of equilibration, even if they did not propose our normative target. In future versions of the modeling task, comparing and deciding the good and bad properties of these models would play into developing general criteria for better and worse models. For example, the normative model explicitly displays the role of ambient temperature, rather than the equilibration point's being a complex outcome of the model, as in the students' model in Figure 13–4. The normative model also shows the role of the gap between current and ambient temperature in "driving" temperature change, and that model transparently allows for varying ambient temperature during equilibration. In addition, the student model in Figure 13–4 has two free parameters, instead of one, and the meaning of initial t-change is opaque.

4. As we pointed out, students surprised us with their facility in coming to understand and control multiple parameters of these models so as to match their empirical findings.

CONCLUSION

This chapter aimed to expose a critical problem that we have to face in developing optimal learning environments for the future. We are just now gaining understanding of: (a) the existence of, and interconnections among, different kinds of knowledge (e.g., epistemological knowledge, meta-representational knowledge); (b) the representational basis for thinking and learning (and its revision, given the possibilities of computers); (c) "natural" trajectories of learning (given both the cognitive simplicities and intrinsic difficulties afforded by intuitive knowledge); and, finally, (d) the issue of instruction for future learning. These issues call into question nearly every aspect of the scope and sequence of scientific learning in schools at present. How will we come to grips with such a major, multiply-constrained job: designing a new, large-scale equilibrium? My chapter ends by urging a beginning: beginning a concerted, scientific and practical exploration of possible new equilibria, and how to reach them in the light of a better understanding of the systemics of learning. We intend that our patterns project, in searching out substantially different equilibia than currently exist, will be a part of that beginning.

ACKNOWLEDGMENTS

This work was supported, in part, by a grant from the Spencer Foundation to Andrea A. diSessa. The conclusions and interpretations drawn here are those of the author, and not necessarily those of the Foundation. The author has a financial interest in PyxiSystems LLC, which is the owner of the Boxer software in which simulations and models were produced for this work.

The patterns project work discussed in this chapter was done collaboratively by members of the Boxer Research Group. Bradford Hill and Zach Powers taught the experimental class, and the class was hosted by the Berkeley Graduate School of Education Academic Talent Development Program. Suggestions by members of the Boxer group on an earlier draft are gratefully acknowledged.

REFERENCES

Bransford, J. D., & Schwartz, D. L. (1999). Rethinking transfer: A simple proposal with multiple implications. In A. Iran-Nejad & P. D. Pearson (Eds.), *Review of Research in Education* (Vol. 24, pp. 61–100). Washington, DC: American Educational Research Association.

Case, R. (1992). *The mind's staircase: Exploring the conceptual underpinnings of children's thought and knowledge.* Mahwah, NJ: Lawrence Erlbaum Associates.

Confrey, J. (1994). Splitting, similarity, and the rate of change. In G. Harel & J. Confrey (Eds.), *The development of multiplicative reasoning in the learning of mathematics* (pp. 291–330). Albany, NY: SUNY Press.

diSessa, A. A. (1992). Images of learning. In E. De Corte, M. C. Linn, H. Mandl, and L. Verschaffel (Eds.), *Computer-based learning environments and problem solving* (pp. 19–40). Berlin: Springer.

diSessa, A. A. (1993). Toward an epistemology of physics. *Cognition and Instruction*, 10 (2–3), 105–225; Responses to commentary, 261–280.

diSessa, A. A., (1995a). The many faces of a computational medium. In A. diSessa, C. Hoyles, R. Noss, with L. Edwards (Eds.), *Computers and exploratory learning* (pp. 337–359). Berlin: Springer Verlag.

diSessa, A. A. (1995b). Designing Newton's laws: Patterns of social and representational feedback in a learning task. In R.-J. Beun, M. Baker, & M. Reiner (Eds.), *Dialogue and interaction: Modeling interaction in intelligent tutoring systems* (pp. 105–122). Berlin: Springer-Verlag.

diSessa, A. A. (2000). *Changing minds: Computers, learning, and literacy.* Cambridge, MA: MIT Press.

diSessa, A. A. (2004). Meta-representation: Native competence and targets for instruction. *Cognition and Instruction*, 22(3), 293–331.

diSessa, A. A., Hammer, D., Sherin, B. & Kolpakowski, T. (1991). Inventing graphing: Meta-representational expertise in children. *Journal of Mathematical Behavior*, 10(2), 117–160.

diSessa, A. A., & Wagner, J. F. (2005). What coordination has to say about transfer. In J. Mestre (Ed.), *Transfer of learning from a modern multi-disciplinary perspective* (pp. 121–154). Greenwich, CT: Information Age Publishing.

Gagne, R. (1985). The conditions of learning (4th ed.). New York: Holt, Rinehart & Winston.

Hofer, B., & Pintrich, P. (2002). *Personal epistemology.* Mahwah, NJ: Lawrence Erlbaum Associates.

Nemirovsky, R. (1993). *Motion, flow, and contours: The experience of continuous change.* Unpublished doctoral dissertation, Harvard University.

Smith, J. P., diSessa, A. A., & Roschelle, J. (1993). Misconceptions reconceived: A Constructivist analysis of knowledge in transition. *Journal of the Learning Sciences*, 3(2), 115–163.

Sherin, B. (2001). A comparison of programming languages and algebraic notation as expressive languages for physics. *International Journal of Computers for Mathematics Learning*, 6, 1–61.

Stroup, W. M. (2002). Understanding qualitative calculus: A structural synthesis of learning research. *International Journal of Computers for Mathematical Learning*, 7, 167–215.

CHAPTER 14

The *DNR* System as a Conceptual Framework for Curriculum Development and Instruction

Guershon Harel
University of California, San Diego

One of the main questions raised in the Foundations for the Future Conference was: What conceptual systems provide powerful foundations for success in mathematics or science classes? The chief goal of this chapter is to outline a conceptual framework—called *DNR-based instruction* (or *DNR*, for short)—claiming to be such a system. *DNR-based instruction* stipulates conditions for achieving the critical goals of provoking students' *intellectual need* to learn mathematics, helping them acquire[1] mathematical *ways of understanding* and *ways of thinking*, and assuring that they internalize and retain the mathematics they learn.

A critical element of *DNR-based instruction* is that mathematics teaching must not appeal to gimmicks, entertainment, or contingencies of reward and punishment, but focus solely on the learner's *intellectual need* by fully utilizing humans' remarkable capacity to be puzzled. Nor should mathematics curricula compromise the mathematical integrity of their contents. A subject matter is mathematical only if it adheres to and maintains the essential nature of the mathematics discipline. Thus, for example, a "geometry curriculum" is not geometry if deductive reasoning is not among its eventual objectives. Teaching correct mathematics, however, is not necessarily correct teaching. A teacher may maintain the mathematical integrity of the content he or she is presenting but neglect the *intellectual need* of the students or be mistaken as to what constitutes such a need for them. As a proof-free "geometry curriculum" does not teach geometry, an intellectually-purposeless "algebra curriculum," one in which students' actions are socially rather than intellectually driven, does not teach students. In *DNR-based instruction* the integrity of the content taught and the *intellectual need* of the

---

[1] In current mathematics education literature, verbs such as "to acquire" and "to attain" are often avoided because, some argue, they connote passive learning, and, instead, the verb "to construct" is commonly used. In this chapter, such verbs will be used freely and synonymously; their intended meaning is drawn from Piaget's theory of equilibration.

student are equally central. The mathematical integrity of a curricular content is determined by the *ways of understanding* and *ways of thinking* which have evolved in many centuries of mathematical practice and continue to be the ground for scientific advances.

The positions expressed in the previous paragraph reflect some of the elements constituting the worldview underlying *DNR-based instruction*. *DNR* can be thought of as a system consisting of: (a) premises (explicit assumptions underlying the *DNR* concepts and claims), (b) concepts (referred to as *DNR determinants*), and (c) *instructional principles* (claims about the potential effect of teaching actions on *student learning*). Not every *DNR instructional principle* is explicitly labeled as such. The system states only three foundational principles, the *duality principle*, the *necessity principle*, and the *repeated-reasoning principle*; hence, the acronym *DNR*. The other principles in the system are derivable from and organized around these three principles. Collectively, the three components that comprise the system—premises, determinants, and *instructional principles*—constitute a unified theoretical perspective about the learning and teaching of mathematics—a perspective that provides a language and tools to formulate and address critical curricular and instructional concerns. Some of these concerns were raised in the Foundations for the Future Conference; for example: What are some of the essential characters of problem-solving situations that lead to meaningful learning? What types of experiences facilitate or retard development?

The *DNR* system has been discerned from and, in turn, implemented in a long series of studies into the learning and teaching of mathematics in the elementary, secondary, and undergraduate levels, as well as studies with teachers of each of these levels. Earlier publications introduced a number of aspects of the *DNR* system—not always with this acronym and not necessarily in the version presented here (see, e.g., Harel, 1990, 1997, 1998, 2001). It goes beyond the scope of this chapter to do more than outlining the system and briefly pointing to its potential in guiding curriculum development and instruction. A more extensive publication is underway which will describe the system in its entirety, lay out its complete theoretical foundations, and demonstrate its capacity to constitute a conceptual framework for designing, developing, and implementing mathematics curricula.

The chapter is organized in four sections: The first section lays out several of the underlying premises and determinants of *DNR*. The second section outlines the three chief *DNR instructional principles—duality, necessity*, and *repeated reasoning*—along with examples of learning-teaching situations that fulfill or violate them, corresponding to experiences that facilitate or retard development. The third section discusses several instructional activities from an ongoing intervention with algebra teachers and points to their rationale in terms of the *DNR-based instruction* perspective. The fourth, and last, section recapitulates the essential elements discussed in the chapter.

UNDERLYING ASSUMPTIONS AND CONCEPTS

Premises

DNR instructional principles are not primary, in that they are based on certain premises and incorporate foundational concepts, called *DNR determinants*. The premises were not determined a priori, before *DNR* was formulated. Rather, they emerged in a process of reflection on and exploration of justifications for the *DNR* principles, and in defining the *DNR* determinants, as we will see. There are eight *DNR* premises;[2] for the purpose of this paper, only four are needed:

> *Subjectivity Premise:* Any observations humans claim to have made are due to the attribution of their mental structure to their environment.
>
> *Knowledge Development Premise:* The process of knowing is developmental in the sense that it proceeds through a continual tension between accommodation and assimilation.
>
> *Teaching Premise:* Construction of scientific knowledge is not spontaneous. There will always be a difference between what one can do under expert guidance or in collaboration with more capable peers and what one can do without guidance.
>
> *Mathematics Epistemology Premise:* Knowledge of mathematics consists of all the *ways of understanding* and *ways of thinking* that have evolved throughout the history of mathematics.

As the reader might have recognized, the first and second premises are inherited from known theories—the subjectivity premise from Piaget's constructivism and the knowledge development premise from the Piagetian theory of equilibration. Likewise, the teaching premise is Vygotsky's known concept of *ZPK* (Zone of Proximal Knowledge).

Determinants

As to the *DNR* determinants, it is beyond the scope of this paper to discuss each of them or provide the theoretical foundations and motivations for the ones presented here. I will focus only on the most essential for the presentation of the three *DNR instructional principles*.

A TRIAD OF DETERMINANTS: MENTAL ACT, WAY OF UNDERSTANDING, AND WAY OF THINKING

Humans' construction of knowledge involves numerous *mental acts* such as representing, interpreting, defining, computing, conjecturing, inferring, proving, structuring, symbolizing, transforming, generalizing, applying, modeling, connecting, predicting, reifying, classifying, formulating, searching, anticipating, and problem solving. *DNR-based instruction's* focus is not just on what mental acts students perform and how often they perform them but also on the *products* and *characters* of these mental acts. The distinction between *product* and *character* of a mental act is central, for it delineates the content of the cognitive objectives at which *DNR-based instruction* aims as well as the knowledge held by the learner:

Product is a particular outcome of a mental act carried out by an individual, whereas *character* is a particular feature of that mental act. Respectively, they are referred to as a *way of understanding* and a *way of thinking* associated with the mental act.

To illustrate the two categories of knowledge, ways of understanding and ways of thinking, consider the following example: Two first graders, Aaron and Betty, solve the problem "3 + 4 = ?". From conversations with the two, we may observe

---

[2] They are accepted as premises in the *DNR* theoretical perspective, but they might have been substantiated empirically or theoretically elsewhere.

that Aaron interprets the "=" sign merely as a command—add 3 and 4 and write the result in the place of the question mark—whereas Betty interprets the sign as equality between two quantities—the quantity that results from combining the two quantities 3 and 4 and an unknown quantity to be found. These different interpretations are products of Aaron's and Betty's interpreting act—they are their ways of understanding the "=" sign in the string of symbols "3 + 4 = ?". We may infer on the basis of multitude of observations the characters of, or the ways of thinking associated with, Aaron and Betty's interpreting acts. We may find, for example, that while Aaron's interpreting act is characteristically devoid of quantitative considerations, Betty's is quantitatively based.

Ways of thinking persist throughout grade levels from elementary school to college. Consider, for example, the string of symbols $y = \sqrt{6x - 5}$. Some of the *desirable* ways of understanding this string are:

$y = \sqrt{6x - 5}$ is an "equation"—a condition on the quantities x and y.

$y = \sqrt{6x - 5}$ is a "real-valued function"—for a real number x, there corresponds the value $\sqrt{6x - 5}$

$y = \sqrt{6x - 5}$ is a "proposition-valued function"—for an ordered pair of real numbers (x, y), there corresponds the proposition values "true" or "false."

These *desirable* ways of understanding are markedly different from many of high school and college students' ways of understanding equality (a string of symbols of the form $A = B$, where A and B are algebraic expressions, or functions). For many students, equality represents no quantitative reality except that symbols must be transformed according to some rules to get an answer judged correct or wrong by the teacher or textbook—consistent with Aaron's behavior. These students' behaviors manifest a way of thinking in which the mental act of symbolizing is characteristically free from meaningful quantitative referents. This behavior is referred to as the *non-referential symbolic* way of thinking.

Ways of thinking seem to be classifiable into three categories: *problem-solving approaches, proof schemes,* and *beliefs about mathematics.* Although these three categories may not constitute the entire universe of ways of thinking, they definitely represent an important portion of it.

Problem-solving Approaches

A problem-solving act is not of the same status as the other mental acts listed above. Any of these acts is, in essence, a problem-solving act. The acts of interpreting and generalizing, as well as those of inferring, structuring, symbolizing, proving and so on. are essentially acts of problem solving. Despite this, the distinction among the different mental acts is cognitively and pedagogically important, for it enables us to better understand the nature of mathematical practice by individuals and communities in the classroom and throughout history, and, accordingly, set explicit cognitive objectives for instruction.

The actual "solution" one provides—viewed as such by oneself—is a way of understanding because it is a particular product of one's problem-solving act. A problem-solving approach, on the other hand, is a way of thinking. For example, each of the approaches "look for a simpler problem," "consider alternative possibilities while attempting to solve a problem," and "look for a key word in the problem statement" characterizes, at least partially, one's problem-solving act; hence, they are instances of ways of thinking. Since *heuristics* are problem-solving approaches, they are also ways of thinking. In the literature, the term *heuristic* is often used for successful problem-solving approaches: "Heuristic strategies are rules of thumb for successful problem solving, general suggestions that help an individual to understand a problem better or to make progress toward its solution" (Schoenfeld, 1985, p. 23). I use the term "heuristic" in this sense, and further stress that, consistent with the subjectivity premise, the judgment as to whether a problem-solving approach is a heuristic is made from the viewpoint of the observer, not the observed. The observer, usually a mathematically experienced individual, is unlikely to judge the approach "look for a key word in the problem statement" as a heuristic. A student, on the other hand, may deem this approach successful if in her or his experience its application has resulted in a correct answer in a large percentage of cases—a not uncommon scenario in current school mathematics due to the type of problems students are assigned.

Proof Schemes

Like the problem-solving act, the mental act of *proving*, too, has a special status in that it is an act that is involved in one way or another in any mathematical activity. Indeed many ways of thinking are characters of the problem-solving and *proving* acts. While problem-solving approaches are instances of ways of thinking associated with the problem-solving act, *proof schemes* are the ways of thinking associated with the *proving* act. The mental act of *proving* has a special meaning in *DNR-based instruction*. As defined in Harel and Sowder (1998):

Proving is the act employed by a person to remove or instill doubts about the truth of an assertion.

Any observation one makes can be conceived either as a *conjecture* or as a *fact*:

A *conjecture* is an assertion made by a person who has doubts about its truth. A person's assertion ceases to be a conjecture and becomes a *fact* in her or his view once the person becomes certain of its truth.

There, in Harel and Sowder (1998), a distinction was made between two instances of the proving act: *ascertaining* and *persuading*:

Ascertaining is the act an individual employs to remove her or his own doubts about the truth of an assertion (or about the truth of its negation), whereas *persuading* is the act an individual employs to remove others' doubts about the truth of an assertion (or of its negation).

Conceptually and methodologically, it is difficult to separate these two acts, for in ascertaining for oneself, one considers how to persuade others, and vice versa. Pedagogically, however, the distinction is critically important, for students need to learn that it is necessary that they ascertain for themselves before attempting to convince others, and that what constitutes ascertainment for themselves may not convince others. The notions which have just been defined are the basis for the concept of *proof scheme*:

A *proof scheme* is a character of one's collective acts of ascertaining and persuading; hence, it is a way of thinking.

Note that while a proof scheme is a way of thinking, a proof—a particular chain of arguments one offers to ascertain for oneself or to convince others—is, by definition, a way of understanding. To illustrate, consider two individuals, say Aimee and Bob, who, through some experiences, made the conjecture that for any real numbers, a, b, c, and d, $(a + b)/(a - b) = (c + d)/(c - d)$ whenever $a/b = c/d$.

$$a/b = c/d \Rightarrow a/b+1 = c/d+1 \Rightarrow (a+b)/b = (c+d)/d \Rightarrow (a+b)/(c+d) = b/d$$
and
$$a/b = c/d \Rightarrow a/b-1 = c/d-1 \Rightarrow (a-b)/b = (c-d)/d \Rightarrow (a-b)/(c-d) = b/d$$
So:
$$(a+b)/(a-b) = (c+d)/(c-d).$$

Aimee evaluated this conjecture in several specific cases and concluded that it is true. Bob, on other hand, applied a different approach, such as:
Both Aimee and Bob have carried out the mental act of proving, but each produces a different outcome—a different proof. Respectively, these two proofs are Aimee's and Bob's ways of understanding the reason for why the conjecture is true. Clearly, the character of Aimee's proving act—her proof scheme—is different from the character of Bob's proving act: while the former is empirical the latter is deductive. In Harel and Sowder (1998), a broad taxonomy of students' proof schemes was identified by examining their proofs. Later, in Harel (in press), this taxonomy was refined and extended to capture more students' proof schemes as well as those of the mathematics community throughout history.

Beliefs About Mathematics

Problem-solving approaches and proof schemes are ways of thinking internal to mathematics; they characterize the mental acts one carries out "in doing" mathematics. It is necessary to distinguish them from *beliefs*: one's views "of" mathematics. More specifically, *beliefs* here are restricted to the character of one's interpretation of (a) what mathematics is, (b) how it is created, and (c) its intellectual or practical benefits. Examples of these beliefs include, respectively, (a) "Formal mathematics has little or nothing to do with real thinking or problem solving" (Schoenfeld, 1985), (b) "The solution of a problem should not take more than five minutes" (Schoenfeld, 1985), and (c) "It is advantageous to have multiple interpretations of a mathematical concept" (Harel, 1998).

WAYS OF UNDERSTANDING AND WAYS OF THINKING: DESIRABLE VERSUS INSTITUTIONALIZED

The terms way of understanding and way of thinking do not imply correct knowledge, as can be seen from the examples discussed this far. In referring to what students know, the terms only indicate the knowledge—correct or erroneous, useful or impractical—currently held by the students. It must be highlighted, however, that in DNR-based instruction the ultimate goal is for students to develop ways of understanding and ways of thinking compatible with those that have been institutionalized in the discipline of mathematics, those the mathematics community at large accepts as correct and useful in solving mathematical and scientific problems. For example, the goal is to gradually refine current students' proof schemes toward the proof scheme shared and practiced by the mathematicians of today. Such a scheme exists, by the epistemology premise, and is believed to be part of the ground for scientific advances in mathematics.

This goal is meaningless without considering the knowledge development premise, which implies that students' learning necessarily involves the construction of imperfect and even erroneous ways of understanding and deficient, or even faulty, ways of thinking. Teachers must be aware of this natural phenomenon when working toward a cognitive objective, and their teaching actions must be consonant with it. In particular, they must attempt to identify students' current ways of understanding and ways of thinking, regardless of their quality, and help students gradually refine them.

During this process of refinement, teachers may set their cognitive objectives in terms of *desirable*, rather than institutionalized, ways of understanding or ways of thinking. A way of understanding or way of thinking is *desirable* if a teacher sets it as an intermediate cognitive objective toward one held and practiced by the mathematics community at large. Clearly, any institutionalized way of understanding or way of thinking is also desirable, but the converse is not necessarily true. To illustrate this distinction between "desirable" and "institutionalized," consider the following example:

Several studies (e.g., Dubinsky, 1986, Ernest, 1984, Harel, 2001) have shown that students often have major difficulties with the concept of mathematical induction. There are many reasons for the difficulties students experience with this concept; the one that is relevant to the discussion here is that the standard instructional treatments introduce the formal principle of mathematical induction too abruptly and without ensuring that the students have a need—an *intellectual need*—for it. In a DNR-based teaching experiment, reported in Harel (2001), the concept was introduced in four instructional phases, each aiming at necessitating for the students a new way of understanding. Each phase is structured so that the repeated application of that way of understanding can lead to the construction of a particular way of thinking associated with the act of proving. The cognitive objectives in the first three phases were intermediate in the sense they consisted of desirable, not necessarily institutionalized, ways of understanding and ways of thinking. In Phase 1, for example, the students were engaged for a relatively long period of time in working on problems typified[3] by:

Find an upper bound to the sequence $\sqrt{2}, \sqrt{2+\sqrt{2}}, \sqrt{2+\sqrt{2}+\sqrt{2}}, \ldots$

Let n be a positive integer. Show that any $2^n \leftrightarrow 2^n$ chessboard with one square removed can be tiled using L-shaped pieces, where each of the pieces covers three squares.

The general solution approach that emerged from these problems was referred to in Harel (2001) as *quasi-induction*, which is exemplified by the following solution given by one of the students in the teaching experiment to first problem:

Since $\sqrt{2}$ is less than 2, $2+\sqrt{2}$ is less than 4, and so $\sqrt{2+\sqrt{2}}$ is less than 2.

Since $\sqrt{2+\sqrt{2}}$ is less than 2, $2+\sqrt{2+\sqrt{2}}$ is less than 4. Hence, $\sqrt{2+\sqrt{2+\sqrt{2}}}$ is less than 2. And so on.

Thus, the intermediate, desirable cognitive objective of this phase was for students to construct quasi-induction as a way of thinking—as a method of proof by which they ascertain for themselves or persuade others about the truth of certain kind of mathematical propositions. In this stage, the goal was "not" for students to prove such propositions by the institutionalized method, the formal principle of mathematical induction. That goal was left to fourth stage, the final stage of the instructional treatment.

---

[3]For the precise characterization of these problems, see Harel (2001).

The focus of DNR-based instruction on the two categories of knowledge, ways of understanding and ways of thinking, is entailed from the system of premises presented in the opening of Section 1. First, as was mentioned earlier, there is a need to identify students' ways of understanding and ways of thinking as they currently are—independent of their quality—for, by the knowledge development premise, students' construction of new knowledge is based on what they already know. Second, by the teaching premise, teachers' expert guidance is necessary to enable students develop new knowledge. Hence, it is essential that teachers build models for students' knowledge to inform their teaching actions; without such models teachers' planning and implementation of instruction is likely to be unguided and haphazard. Third, while necessary, such models are not sufficient: teachers' actions must be directed, in addition, by cognitive objectives formed in terms of desirable or institutionalized ways of understanding and ways of thinking, which, by the epistemology premise, comprise mathematical knowledge.

THE CONCEPT OF INSTRUCTIONAL PRINCIPLE AND ITS RELATED DETERMINANTS: TEACHING ACTION AND STUDENT LEARNING

What is an *instructional principle?* Consider the common conception "In sequencing mathematics instruction, start with what is easy." This conception might be interpreted as implying a cause-effect link between a *teaching action*—that of sequencing mathematics instruction—and *student learning*. The *teaching action* might be viewed as a "likelihood" condition: starting with what is easy for the students may help students learn. It might be viewed as a "necessary" condition: for students to learn teachers must start with what is easy for them. Or it might be viewed as a "sufficient" condition: starting with what is easy for the students will help students learn. Textbook authors and teachers seem to use this conception mostly in its necessary-condition form. This example can be abstracted to define the notion of *instructional principle:*

An *instructional principle* is a conception implying an effect of a teaching action on *student learning*. The teaching action may be conceived as a likelihood condition, necessary condition, or sufficient condition for the effect to take place.[4]

By teaching action is meant:

> A curricular or instructional measure or decision a teacher carries out for the purpose of achieving a cognitive objective, establishing a new didactical contract, or implementing an existing one.

Of critical importance is the definition of *student learning*:

> *Student learning* is a continuum of disequilibrium-equilibrium phases together with (a) the *ways of understanding* and *ways of thinking* that the learner utilizes or newly constructs during the various phases and (b) the cognitive, social, and affective stimuli that result from or instigate these phases.

Research in mathematics education has offered useful models for learning trajectories of various mathematical concepts and ideas. However, relative to the

---

[4]What is an instructional principle for one person might be a myth for another. See, for example, Harel and Sowder (2005) for a discussion on the pedagogical ineffectiveness of the conception "In sequencing mathematics instruction, start with what is easy."

broad scope of this definition these models are incomplete. This is expected due to the enormous empirical and theoretical difficulties in building models that incorporate phases of disequilibrium-equilibrium, their utilized or resultant *ways of understanding* and *ways of thinking*, and the cognitive, social, and affective stimuli that result from and instigate the various phases. Such comprehensive models, however, are desirable and the hope is that they will be constructible in the future. Ideally, the concept of "instructional principle"—a general conception about an effect of a teaching action on student learning—should be understood in terms of this encompassing definition of "student learning." However, given the obvious difficulties in fully incorporating all the elements of student learning, such an interpretation may not be practical. For this reason, *DNR* instructional principles will be limited in their scope, in that they are general conceptions about the effect of teaching actions on students' ways of understanding and ways of thinking and the intellectual stimuli that are needed for disequilibrium-equilibrium phases. While affective stimuli are important, they are not central to the theoretical perspective presented in this chapter.

DNR'S THREE FOUNDATIONAL INSTRUCTIONAL PRINCIPLES

The Duality Principle

The *duality principle* concerns the developmental interdependency between what the students produce and the character of their mental acts—between their ways of understanding and their ways of thinking. Before formulating the principle, let us consider one of the findings in our research on proof schemes. In all of the teaching experiments I conducted on this question, the most noticeable, prevalent, and persistent way of thinking among the participants—students as well as teachers—was the *empirical proof scheme*. That is, a scheme by which one proves (attempts to remove or instill doubts about the truth of a conjecture) either inductively—by relying on evidence from examples of direct measurements of quantities, substitutions of several numbers in algebraic expressions, and so on—or perceptually—by relying on evidence from visual or tactile perceptions. This way of thinking is especially robust among both students and teachers when they observe a pattern suggesting that one can generate as many examples as wants in support of the conjecture. This finding has been observed by other researchers as well (see Goetting, 1995; Chazan, 1993).

Based on the knowledge development premise, students do not come to school as blank slates, ready to acquire desirable knowledge independently of what they already know. Rather, what students know now constitutes a basis for what they will know in the future. This is true for all ways of understanding and ways of thinking associated with any mental act; the mental act of proving is of no exception. In daily life, the methods of justification available to humans are largely empirical. From childhood, when humans seek to justify or account for a particular phenomenon, they are likely to do so based on similar or related phenomena in their past. Given that the number of such phenomena in one's past is unavoidably finite, people's judgments are typically empirical. Through such repeated experience, humans' hypothesis evaluation evolves to be dominantly empirical, whereby the empirical proof scheme is established as a way of thinking characterizing their mental act of proving. A person's repeated experiences of justifying particular assertions and of accounting for particular phenomena in daily life thus shape the character of her or his mental act of proving summarizing:

1. The ways of understanding associated with one's mental act of proving impact the way of thinking associated with that mental act.

Of equal importance is the converse of this statement. Continuing with our discussion of the empirical proof scheme, this way of thinking does not disappear upon entering school, nor does it fade away effortlessly when students take mathematics classes. It continues to impact the particular justifications students produce for mathematical assertions—their way of understanding why particular assertions are true or false. It takes enormous instructional effort for students to recognize the limits of empirical evidence in mathematics and to construct alternative, deductively-based proof schemes. Even mathematically able students are not immune from the impact of the empirical proof scheme (see Fischbein and Kedem, 1982). Summarizing:

2. The ways of thinking associated with one's mental act of proving impact the way of understanding associated with that mental act.

This brief account points to a reciprocal cause-effect relation between ways of understanding and ways of thinking with respect to the mental act of proving: through everyday experience in justifying assertions and accounting for phenomena occurring in their physical and social environment, humans' mental act of hypothesis evaluation evolves to be empirically based. In turn, the way of thinking that most students bring to mathematics classes is dominantly the empirical proof scheme, which they readily apply in solving mathematical problems involving justification. Thus, together, Statements 1 and 2 above express a dual effect associated with the mental act of proving. The *duality principle* asserts that this dual effect is developmental and applies to all mental acts. Specifically,

The *Duality Principle*. Students develop ways of thinking only through the construction of ways of understanding, and the ways of understanding they produce are determined by the ways of thinking they possess.

There is reciprocity between ways of understanding and ways of thinking claimed in the duality principle: a change in ways of thinking brings about a change in ways of understanding, and vice versa. The claim intended is, in fact, stronger: not only do these two categories of knowledge affect each other but a change in one cannot occur without an appropriate change in the other. This assertion constitutes an instructional principle because it implies an effect of a teaching action on student learning. Specifically:

Students would be able to construct a way of thinking associated with a certain mental act or refine or modify an existing one "only" if they are helped to construct suitable ways of understanding associated with that mental act.

Conversely:

Students would be able to construct a way of understanding associated with a certain mental act or refine or modify an existing one "only" they are helped to construct suitable ways of thinking associated with that mental act in the form of problem-solving approaches, proof schemes, or beliefs about mathematics.

Thus, for example, the teaching action of preaching ways of thinking to students would have no effect on the quality of the ways of understanding they would produce. Similarly, talking to them about the nature of proof in mathematics, advising them to use particular heuristics, or telling them what beliefs about mathematics to adopt would have minimal or no effect on the quality of the proofs and solutions they will produce and the particular beliefs about mathematics they will hold. Only by producing desirable ways of understanding—by ways of carrying out mental acts of, for example, interpreting mathematical ideas through reading, writing, and oral communication, solving mathematical problems, and proving mathematical assertions—can students construct desirable ways of thinking.

Current textbooks and teaching practices do not pay enough attention to ways of thinking. For example, seldom do they consider questions such as: What specific ways of thinking do common instructional activities in algebra—such as simplifying, factoring, and so on.—promote? Or, what exactly are the ways of thinking that can or should be promoted by the use of computer technology in algebra? An immediate implication of the duality principle is that it is essential that teachers form instructional goals in terms of ways of thinking and devise and use appropriate instructional activities through which students can build ways of understanding that can potentially help them construct desirable ways of thinking. This implication leads to two critical questions: What constitute such appropriate instructional activities? What is the nature of instructional treatments that can help students construct desirable ways of understanding and ways of thinking? These questions are addressed by the other two *DNR* principles.

The Necessity Principle

Fundamental to *DNR-based instruction* is the knowledge development premise, which entails that problem solving should not be just a goal but also the means for learning mathematics (see also Brownell 1946; Davis, 1992; Hiebert, 1997). "Problem solving" is usually defined as "[engagement] in a task for which the solution method is not known in advance" (NCTM, 2000). Many of the situations students encounter in school satisfy this definition and yet they do not constitute problem solving because the situations are not *intrinsic*, but *alien*, to the students.

When we talk about problems in the context of mathematics curricula, we refer to learning-teaching events that involve two sets of interpretations (i.e., ways of understanding): the set that belongs to the poser of the problem and the set that belongs to the one to whom the problem is posed. An important consequence of this simple observation is that the two sets are not always identical and, in many cases, do not even intersect. Teachers are often unaware of this. They might present a problem to their class and incorrectly assume that their students share their interpretation(s) of the problem. Conversely, students might pose a question to their teacher, who either encounters difficulty making sense of what they are asking or interprets their question differently from what the students intended. Of particular interest are the scenarios where what is conceived as a problem by the teacher is unproblematic to the students, and vice versa. In general, a situation that is problematic to one person but is unproblematic to another is referred to as *intrinsic* (I) to the first and *alien* (A) to the second. Accordingly, mathematics problems in a learning-teaching setting are of four categories: (A, I), (I, A), (I, I), and (A, A), where the first component of these ordered pairs refers to the student, and the second to the teacher. (I, I)—the situation where a problem as stated is intrinsic to both—is the only desired category among the four. As long as the problem is intrinsic to both, significant learning is likely to occur, even when—or perhaps especially when—the teacher's interpretation is different from those of the students.

Unfortunately, none of the other three cases, (A, I), (I, A), and (A, A), is uncommon. The case (A, A), for example, occurs especially when the teacher lacks basic understanding of the mathematical content he or she is teaching—a not uncommon phenomenon that has been documented in the literature (Ball, 1990; Cohen, 1991; Ma, 1999; Post, T., Harel, G., Behr, M. & Lesh, R., 1991; Simon, 1993). To illustrate, consider the following episode: A ninth-grade teacher requires his class to accompany each assertion written on the left-hand side of two-column proofs by a "justification" on the right-hand side, both in algebra and in geometry. For example, a student in this class justified the assertions, "$AB \cong AB$" by the phrase "reflexive property;" the assertion "If $\angle ABC = 30°$ and $\angle CBD = 45°$, then $\angle ABD = 75°$," by "additive property;" and the assertion "$a + b = b + a$" by "commutative property." It turned out that neither the student nor the teacher understood the meaning of these phrases, and both viewed the three assertions as obvious—ones that require no justification—but each felt compelled to follow rules: the student had to follow those imposed by her teacher, and the teacher those imposed by the textbook. Thus, the task of justification was alien to both the teacher and his student—a clear (A, A) case. On the other hand, if this teacher possessed the *axiomatic proof scheme*—the way of thinking that mathematical structures are determined by a set of axioms—he would have understood the meaning and role of the reflexive and additive properties in Euclidean geometry, and of the commutative property in algebra, and in turn the situation would have been intrinsic to him. Cases of the (A, I) category are perhaps the most common in mathematics education in all levels, but especially at the college level, where the instructors have deep understanding of the problems they present to their students but usually are unaware that their appreciation of the problems is not always shared by their students. This supports the claim that while mathematical knowledge is indispensable for quality teaching, it is not sufficient. Teachers must also know how to address students as learners. An important application of the *necessity principle* is to assure that the mathematics students' learning grows out of problems intrinsic to them:

The *Necessity Principle*. For students to learn what we intend to teach them, they must have a need for it, where by 'need' is meant *intellectual need*, not social or economic need.

Most students, even those who are eager to succeed in school, feel intellectually aimless in mathematics classes because we—teachers—fail to help them realize an *intellectual need* for what we intend to teach them. The term *intellectual need* refers to a behavior that manifests itself internally with learners when they encounter an intrinsic problem—a problem they understand and appreciate. For example, students might encounter a situation that is incompatible with, or presents a problem that is unsolvable by, their existing knowledge. Such an encounter is intrinsic to the learners for it stimulates a desire within them to search for a resolution or a solution, whereby they might construct new knowledge. There is no guarantee that the learners construct the knowledge sought or any knowledge at all, but whatever knowledge they construct is meaningful to them since it is integrated within their existing cognitive schemes as a product of effort that stems from and is driven by their personal, intellectual need. Whereas one should not underestimate the importance of students' social need (e.g., mathematical knowledge can endow me with a respectable social status in my society) and economic need (e.g., mathematical knowledge can help me obtain comfortable means of living) as learning factors, teachers should not and cannot be expected to stimulate—let alone fulfill—these needs. Intellectual need, on the other hand, is prime responsibility of teachers and curriculum developers.

The Repeated Reasoning Principle

Even if ways of understanding and ways of thinking are necessitated through students' *intellectual need*, there remains the task of ensuring that students internalize, organize, and retain this knowledge. Namely, the question of concern is: What are the conditions on instructional treatments that are necessary for students to internalize, organize, and retain desirable ways of understanding and ways of thinking? Research has shown that repeated experience, or practice, is a critical factor in these cognitive processes (see, e.g., Cooper, 1991). The emphasis of *DNR-based instruction* is on repeated reasoning that reinforces desirable ways of understanding and ways of thinking. Repeated reasoning, not mere drill and practice of routine problems, is essential to the process of internalization, which is a conceptual state where one is able apply knowledge autonomously and spontaneously. The sequence of problems must continually call for reasoning through the situations and solutions, and must respond to the students' changing intellectual needs. This is the basis for the *repeated reasoning principle*.

The *Repeated Reasoning Principle*. Students must practice reasoning in order to internalize, organize, and retain ways of understanding and ways of thinking.

Consider the following scenario: Two elementary school children, Sam and Tami, were taught division of fractions. Sam was taught in a typical method, where he was presented with the rule $(a/b) \mid (c/d) = (a/b) \mid (d/c)$, and the rule was introduced to him in a meaningful context and with an adequate justification that he understood. Tami, on the other hand, was presented with no rule. Each time she encountered a division of fraction problem, she explained its meaning and the rationale of her solution. Sam and Tami were assigned homework problems on division of fractions. Sam solved all the problems correctly, and gained, as a result, a good mastery of the division rule. It took Tami a much longer time to do her homework. Here is what Tami—a real third-grader—said when she worked on $(4/5) \mid (2/3)$:

> How many 2/3 s in 4/5? I need to find what goes into both [meaning: a unit-fraction that divides 4/5 and 2/3 with no remainders]. 1/15 goes into both. It goes 3 times into 1/5 and 5 times into 1/3, so it would go 12 times into 4/5 and 10 times into 2/3 (She writes: $4/5 = 12/15$; $2/3 = 10/15$; $(4/5) \mid (2/3) = (12/15) \mid (10/15)$. How many times does 10/15 go into 12/15?... How many times do 10 things go into 12 things?... One time and 2/10 of a time...which is 1 and 1/5.

Tami had opportunities for reasoning of which Sam was deprived. Tami practiced reasoning and computation; Sam practiced only computation. Further, Tami eventually abstracted the division rule and learned an important lesson about mathematical efficiency—a way of thinking Sam had little chance to acquire. Further, though repeated solutions of this kind, Tami would likely develop the way of thinking of interpreting situations in terms of linear change—what is known in the literature as proportional reasoning.

The repeated reasoning principle is complementary to the other two principles in that its aim is for students to internalize the ways of understanding and ways of thinking acquired through the application of the other two principles. Through repeated reasoning in solving intrinsic problems the application of ways of understanding and ways of thinking become autonomous and spontaneous. This principle, as the other two principles in the *DNR* system, has been a keystone in our teaching experiments with students and teachers, and we attribute the internalization of important ways of thinking by our participants to the persistent application of this principle in our teaching. For example, in a linear algebra teaching experiment for mathematics and engineering majors, the aim was to eradicate students' faulty proof schemes by helping them develop alternative, deductive proof schemes. The students were engaged in numerous analyses of linear algebra questions in terms of systems of linear equations and, in turn, in terms of the meaning of row operations on the system's equations. The intellectual benefit of this activity is that students built images of the structure of the row echelon form of a matrix and the implication of row reduction to span and linear independence. Row reduction became for these students a conceptual tool—a way of thinking—to approach problems on the existence and uniqueness of solutions to linear systems and problems on span and linear independence in C^n.

DNR-BASED INSTRUCTION INTERVENTION: EXAMPLES FROM A PROFESSIONAL DEVELOPMENT CLASS FOR TEACHERS

A three-year NSF-funded project is underway to systematically examine the effect of DNR-based instruction on the teaching practice of algebra teachers and on the achievement of their students. I will conclude this paper with a discussion of instructional activities from this study and point to their rationale in terms of the underlying DNR-based instruction theoretical perspective.

It is clear from the formulation of the duality principle that the subjectivity premise is its most critical foundation. In our work with teachers a major effort is made to help them realize that the approach a student chooses to solve a problem depends on her or his way of understanding the problem; hence, not only are multiple solutions to a given problem possible but in many cases are inevitable. Adhering to the duality principle, the teachers come to this realization through their own multiple solutions to mathematical problems. In turn, they gradually acquire the ways of thinking, "a problem can have multiple solutions" and "it can be advantageous to solve a problem in different ways." Similarly, through their own interpretations of concepts, they learn that "a concept can have multiple interpretations" and "it can be advantageous to possess multiple interpretations of a concept." These ways of thinking, although essential in the learning and creation of mathematics, are often absent from teachers' and students' repertoires of reasoning. We engage teachers in activities aimed at promoting multiple ways of understanding concepts. For example, through their solutions to different problems the teachers learn that the concept of fraction, say 3/4, can be understood in different ways and it is advantageous to understand it in different ways. Such ways of understanding include: *unit fraction* (3/4 is the sum of 1/4 + 1/4 + 1/4); *partition* (3/4 is the quantity that results from dividing 3 units into 4 equal parts; *measurement* (3/4 is the measure of a 3 cm long segment with a 4-cm unit ruler); *solution to an equation* (3/4 is the solution to $4x = 3$); *part-whole* (3/4 is 3 out of 4 units). Similarly, the teachers solved problems through which they learned multiple ways of understanding algebraic expressions of the form $y = f(x)$ (e.g., $y = 3x^2 + x - 2$) and the significance of each interpretation. For example, they see how one can understand $y = f(x)$ in terms of a condition on the variables x and y, or in terms of a real-valued function, or in terms of proposition-valued function, as we have discussed earlier. These newly acquired desirable ways of understanding by the teachers are contrasted with the interpretations they or their students commonly possess. For example, through the actual work of students, the teachers see that for many students a string of symbols such as $y = 3x^2 + x - 2$ represents no quantitative reality (the non-referential proof scheme), except possibly that the "=" sign is understood as a "do something signal," where one side of the equation is reserved for the operation to be carried out and the other side for its outcome, as was documented by Behr, Erlwanger, & Nichols (1976).

In accordance with the necessity principle, the different interpretations of concepts are necessitated for the teachers through their own solutions to problems. For example, on one occasion the teachers, working in small groups, investigated the condition under which one fraction, say α, is smaller than another fraction, say β, where the difference between the numerators of α and β, which is given to be an integer, is equal to the ratio of their denominators. After 35 minutes, one of the teachers, Colleen, presented her group's solution. The group's solution process involved many quadratic inequalities. During Colleen's presentation there were numerous questions and suggestions from the class regarding the solution process. After about 25 minutes of classroom discussion, the final answer that the class derived from a complex system of inequalities was summarized by Colleen as follows:

We are to find when is $n/d < (n + x)/(d/x)$, where n and d are integers and x is a non-zero number. The solution is:

For $-4 < n$ or $n > 0$, $x > \left(-n + \sqrt{n^2 + 4n}\right)/2$ or $x < \left(-n - \sqrt{n^2 + 4n}\right)/2$;
For $-4 < n < 0$, all x values work
For $n = 0$ or $n = -4$, $x \cdot -n/2$.

Following this presentation, one of the teachers, Ken, suggested checking a few cases to see if they agree with the final answer. ("Evaluate your solution" is a way thinking strongly stressed throughout the intervention.) After about five minutes of group work, Ken indicated that he noticed a strange phenomenon: when different forms of the same fraction are used different results are obtained. He demonstrated his observation with the equivalent fractions $-2/5$ and $-6/15$. Indeed, for $-2/5$ any value of x works but for $-6/15$ only $x < 3 - \sqrt{3}$ or $x > 3 + \sqrt{3}$ work. This provided a great opportunity—intended and anticipated—for the instructor to point to the necessity to differentiate between "fraction" and "rational number." Although the ratio is the same, the fraction is not, and the solution with a different fraction, even if the ratio is the same, is also different. The solution for the fraction $-2/5$ is different from the solution for the fraction $-6/15$. This result was particularly amazing for the teachers. Their own solution to a problem that was intrinsic to them necessitated for them the distinction between "fraction" and "rational number."

On another occasion, the teachers worked on justifying the quadratic formula. Prior to this problem, the teachers had repeatedly worked with many quadratic functions, finding their roots by essentially completing the square. In doing so they repeatedly transformed a given equation $ax^2 + bx + c = 0$ into an equivalent equation of the form $(x + T)^2 = L$ for some terms T and L, in order for them to solve for x (as $-T + \sqrt{L}$ and $-T - \sqrt{L}$). They then abstracted this process to develop the quadratic formula. To get to the desired equivalent form, they understood the reason and need for dividing through by a, bringing c/a to the other side of the equation, and completing the square. For these teachers, the symbolic manipulation process was goal oriented and conditioned by quantitative considerations; namely, transformations are applied with the intention to achieve a predetermined intrinsic goal. In this case, the teachers practiced the way of thinking of transforming an algebraic expression into a desired form without altering its original quantitative value. This way of thinking—which is, in our view, one of the essential characteristics of algebraic reasoning—was known among the teachers as the changing-the-form-without-changing-the-value habit of mind. We see here the simultaneous implementation of the duality principle, the necessity principle, and the repeated reasoning principle. In particular, the repeated application of this habit of mind helped the participant teachers internalize it, whereby they become autonomous and spontaneous in applying it.

RECAPITULATION

Ways of understanding refer to products of mental acts, such as particular interpretations, generalization, solutions, justifications, and so on, which students produce as they are engaged in mathematical activities. Ways of thinking, on the other hand, are characters of one's mental acts; problem-solving approaches, proof schemes, and beliefs about mathematics are categories of ways of thinking. The non-referential symbolic proof scheme—a scheme by which one ascertains oneself or persuades others of the truth of a conjecture on the basis of the appearance of symbols alone, without attending to their quantitative or functional referents—is an example of undesirable way of thinking. Heuristics are desirable problem-solving approaches, judged so by an observer. Beliefs about mathematics are restricted to one's interpretation of what mathematics is, how it is created, and its intellectual or practical benefits. They might be considered as didactical contracts, in Brousseau's (1997) terms, which are linked to one's perceptions of mathematics.

Since reasoning deductively is the single most central way of thinking in mathematics, mathematical proof must be a central focus of mathematics instruction. There is little attempt in current mathematics curricula to alter the character of students' mental act of proving by helping them acquire deductively-based proof schemes. On the contrary, the empirical proof scheme that students acquire in daily life is reinforced throughout the school years, where a strong emphasis is put on justification by examples, physical manipulatives, and, in particular, reliance on generalizing from finite patterns. As a consequence, students remain in this vicious cycle where their empirical proof scheme affects the kind of justifications they produce, and the justifications they produce reinforce the empirical proof scheme they possess.

In sum, the theoretical perspective presented here entails four essential elements for DNR-based instruction. First, in designing, developing, and implementing mathematics curricula, ways of thinking and ways of understanding must be the ultimate cognitive objectives; they must be addressed simultaneously, for each affects the other. Second, meaningful concepts can only be elicited through solutions to problems that provoke intellectual needs for students. The concepts elicited and the ways they are elicited constitute and, at the same time, impact students' ways of thinking and ways of understanding. Third, ways of thinking and ways of understanding are organized, internalized, and retained through repeated solutions to intrinsic problems—problems the students understand and appreciate. Fourth, and lastly, proof and justification must be institutionalized as central means to create mathematics and as norms for mathematical discourse. As critical means of fostering desirable proof schemes, instruction must focus on carefully chosen problems—conceived as such by the students—through whose solution students gradually modify their ways of thinking as to what constitutes ascertainment and persuasion in mathematics. Such problems might be called "proof-eliciting problems" (Harel and Lesh, 2003; see also Lesh and Doerr, 2003).

REFERENCES

Ball, D. L. (1990). Prospective elementary and secondary teachers' understanding of division. *Journal for Research in Mathematics Education, 21*(2), 132–144.

Behr, M., Erlwanger, S. & Nichols, E. (1976). How children view equality sentence (PMDC Technical Report No. 3). Tallahassee: Florida State University. (ERIC Document Reproduction Service No. ED144802).

Brousseau, G. (1997). *Theory of didactical situations in mathematics.* Dordrecht, the Netherlands: Kluwer Academic.

Brownell, W. A. (1946). Introduction: Purpose and scope of the yearbook. In N. B. Henry (Ed.), *Forth-Fifth Yearbook of the National Society for the Study of Education: Part I. The measurement of Understanding,* University of Chicago, Chicago, 1–6.

Chazan, D. (1993). High school geometry students' justification for their views of empirical evidence and mathematical proof. *Educational Studies in Mathematics, 24,* 359–387.

Cohen, D. K. (1991). A revolution in one classroom: The case of Mrs. Oublier. *Educational Evaluation and Policy Analysis, 12,* 311–330.

Cooper, R. (1991). The role of mathematical transformations and practice in mathematical development. In L. Steffe (Ed.), *Epistemological Foundations of Mathematical Experience.* New York: Springer-Verlag.

Davis, R. B. (1992). Understanding "understanding". *Journal of Mathematical Behavior 11,* 225–241.

Dubinsky, E. (1986). On the teaching of mathematical induction. *Journal of Mathematical Behavior,* 5(3), 305–17.

Ernest, P. (1984). Mathematical induction: A pedagogical discussion, *Educational Studies in Mathematics, 15,* 173–189.

Fischbein, E., & Kedem, I. (1982). Proof and certitude in the development of mathematical thinking. In A. Vermandel (Ed.), *Proceedings of the Sixth International Conference of the Psychology of Mathematics Education* (pp. 128–131). Antwerp, Belgium: Universitaire Instelling Antwerpen.

Goetting, M. M. (1995). The college student's understanding of mathematical proof (Doctoral dissertation, University of Maryland, 1995). *Dissertations Abstracts International, 56-A,* 3016, Feb., 1996.

Harel, G. (1998). Two Dual Assertions: The First on Learning and the Second on Teaching (Or Vice Versa). *The American Mathematical Monthly, 105,* 497–507.

Harel, G. (1999). Students' understanding of proofs: a historical analysis and implications for the teaching of geometry and linear algebra. *Linear Algebra and Its Applications, 302–303,* 601–613.

Harel, G. (2000). Three principles of learning and teaching mathematics: Particular reference to linear algebra—Old and new observations. In Jean-Luc Dorier (Ed.), *On the Teaching of Linear Algebra* (pp. 177–190). Dordrecht, the Netherlands: Kluwer Academic.

Harel, G. (2001). The development of mathematical induction as a proof scheme: A model for *DNR*-based instruction. In S. Campbell & R. Zaskis (Eds.), *Learning and Teaching Number Theory.* Westport, CT: Ablex.

Harel, G. (in press). Students' proof schemes revisited: Historical and epistemological considerations. In P. Boero (Ed.), *Theorems in School.* Sense publishing.

Harel, G., & Lesh, R. (2003). Local conceptual development of proof schemes in a cooperative learning setting. In R. Lesh & H. Doerr (Eds.). Mahwah, NJ: Lawrence Erlbaum Associates.

Harel, G., & Sowder, L. (1998). Students' proof schemes. *Research on Collegiate Mathematics Education,* Vol. III. In E. Dubinsky, A. Schoenfeld, & J. Kaput (Eds.), AMS, 234–283.

Hiebert, J. (1997). Aiming research toward understanding. In A. Sierpinsda and J. Kilpatrick (Eds.), *Mathematics Education as a Research Domain: A Search for Identity* (pp. 141–152). Dordrecht, the Netherlands: Kluwer Academic.

Lesh, R., & Doerr, H. (2003). *Foundatioins of a models and modeling perspective on mathematics teaching, learning, and problem solving.* In R. Lesh & H. Doerr (Eds.). Mahwah, NJ: Lawrence Erlbaum Associates.

Ma, L. (1999). *Knowing and teaching elementary mathematics.* Mahwah, NJ: Lawrence Erlbaum Associates.

NCTM (National Council of Teachers of Mathematics). (2000). *Principles and standards for school mathematics.* Reston, VA: Author.

Post, T., Harel, G., Behr, M., & Lesh, R. (1991). Intermediate teachers' knowledge of rational number concepts. In E. Fennema, T. P. Carpenter, and S. J. Lamon (Eds.), *Integrating Research on Teaching and Learning Mathematics* (pp. 177–198). Albany, NY: SUNY Press.

Schoenfeld, A. (1985). *Mathematical problem solving.* Orlando, FL: Academic Press.

Simon, M. (1993). Prospective elementary teachers' knowledge of division. *Journal for Research in Mathematics Education, 24*(3), 233–254.

Sowder, L., & Harel, G. (1998). Types of students' justifications. *Mathematics Teacher, 91,* 670–675.

CHAPTER 15

Aspects of Affect and Mathematical Modeling Processes*

Gerald A. Goldin
Rutgers University

The premise of this chapter, based on research and practical experience, is that adequate foundations for studying and enhancing mathematical learning and development must incorporate the affective domain—not as an architectural add-on, but as a structurally essential building block.

I shall focus on some psychological aspects of affect in relation to mathematical thinking and modeling processes. After elaborating on several ideas that have come to the foreground in my earlier work on affect, I shall highlight briefly three affective/cognitive constructs that are not commonly associated with mathematics—*integrity, identity,* and *intimacy.* My thesis is that each of these, when well developed and functional in the individual in relation to mathematics, contributes essentially to power and effectiveness in mathematical learning, problem solving, and teaching.

Moreover, there is a natural connection with the processes that are here termed *modeling*. The development of a mathematical self-identity, the achievement of mathematical integrity, and the experience of intimacy with mathematics during engaged thinking and problem solving, as well as the awareness by teachers of these (affective) aspects of their students' inner mathematical activity, all involve or can involve modeling processes. Evidently the pertinent notion of modeling is not restricted to using mathematics to model non-mathematical situations, but is far more general. Furthermore, engagement in mathematical modeling itself fosters mathematical integrity and intimacy, as well as the formation of self-identity.

Such an overview, sketchy as it is, suggests that the functioning of affect in mathematical learning and teaching is far more complex than one might infer from just the surface consideration of emotional feelings. However, with some exceptions, the prevailing discussion of affect by mathematics educators has been of much more limited scope than proposed here. Traditionally this discussion includes attention to the roles of students' *attitudes* (positive or negative), and to the

---

*This work is based on research supported by the National Science Foundation under grant no. ESI-0333753. Any opinions findings, and conclusions or recommendations are those of the author and do not necessarily reflect the views of the NSF.

problems posed by *negative emotional feelings* such as anxiety or fear. It has also included the need to *motivate* students, by stimulating their interest in the topics under discussion. Sometimes the affective goals of educators have seemed to focus exclusively on making mathematics enjoyable and fun for children—while a countercurrent in the popular culture has been the idea that school mathematics should include plenty of drill-and-practice exercises; i.e., disciplined, repetitive activity that by its nature is not likely to be enjoyed, and perhaps *should* not be enjoyed.

My view is that these traditional issues do not do justice to the complexity of affect in interaction with cognition. The aspects explored in the present chapter form a part of that complexity.

Many of the ideas discussed here draw extensively on earlier, joint work with Valerie DeBellis. The reader is referred to this work for additional discussion and analysis. These ideas include the central notion of *meta-affect* (DeBellis, 1996; DeBellis & Goldin, 1997; Goldin, 2002), and the constructs of mathematical intimacy and mathematical integrity (DeBellis, 1998; DeBellis & Goldin, 1997, 1999, 2006).

SOME ENCOURAGING THEORETICAL TRENDS

Let us acknowledge at the outset the presence of tacit barriers to the serious study of affect in relation to mathematical cognition. One of these barriers is the perception of mathematics, both by mathematicians and non-mathematicians, as an essentially *logical* and *analytical* subject, with no role for the *irrationality* that people commonly associate with emotional reactions.

It is widely accepted that human beings experience and interpret situations by way of complex conceptual systems that form over extended periods of time. In discussing education we frequently focus on these developing conceptions of individuals (students or teachers). Sometimes we focus also on the shared, normative conceptual systems that influence individuals' development—those of social groups, societies, or cultures. In the domain of mathematics and mathematical reasoning, the expert culture places extremely high value on conceptual systems that are (to the maximum practical extent) logical and rational—i.e., conceptions that are based on precise definitions, on well-stated axioms abstracted and generalized from more specific examples, and on well-defined methods of reasoning for making and testing conjectures, reaching conclusions, and proving theorems.

As a mathematical scientist, I share this value. Indeed, I would support the view that the very *raison d'être* of mathematics derives from these attributes. Logicality and rationality are essential in accounting for the successes of mathematics—its power and consistency as the science of number and pattern, its value as a language describing the natural world quantitatively, and its flexibility as a way of understanding that is both abstract and applicable in modeling many concrete situations. The conventional systems that constitute so much of mathematics, including systems of symbolic representation (systems of numeration, of algebraic and functional notation, of graphical representation, and so forth) and accompanying definitional and axiomatic systems, all have rational and logical bases that can be, and of course should be, understood and characterized cognitively.

But to the extent that the popular culture and the culture of mathematics interpret logic and rationality as the opposites of emotional feelings, the latter are generally regarded as *inappropriate* to mathematical reasoning or discussion. With this I take issue. It is my perspective that affirming the importance of the affective domain in the understanding of mathematical cognition *does not contradict* the underlying rationality of mathematics—and I would not like this chapter interpreted as some kind of philosophical or post modern challenge to that rationality. Rather the focus on affect acknowledges that the cognitive conceptual systems of *people learning, doing, and teaching mathematics* (which, of course, include those conceptions we understand to be rational and logical) do not appear sufficient to describe or explain all or even most of people's interpretive and reasoning behavior. This observation pertains *even after we have taken into full account* the cognitive analyses of common misconceptions, alternate conceptions, beliefs, epistemological obstacles, and developmental patterns associated with children's cognitive growth.

When human beings set out to make sense of situations, including mathematical situations, they do not engage solely in logical processes, whether formal, heuristic, or intuitive. Rather there is persuasive evidence that human conceptual systems involve *in an essential way* emotional feelings, attitudes, emotionally-supported beliefs, values, aesthetics, and the *complex structures of meanings* that are connected with all of these. Whether in young children or in sophisticated researchers, doing mathematics seems to entail feelings that range from curiosity to avoidance, bafflement to elation, frustration and resentment to pleasure and satisfaction, accompanied by expressions pertaining to self-judgment, self-worth, and personal fulfillment or its absence.

It is this interplay between mathematical cognition, aesthetics, and what might be called *mathematical affect* that a growing number of mathematics education researchers now appear ready to consider seriously. Beyond the earlier work of McLeod and Adams (1989), recent work includes that of DeBellis (1996, 1998), Evans (2000), Gomez Chacon (2000a, b), Hannula (2002, 2004), Hannula, Evans, Philippou, and Zan (2004), Malmivuori (2001), Op't Eynde, De Corte, and Verschaffel (2001), Roth (in press), Sinclair (2004), and others.

The barriers to serious consideration of the affective domain in mathematical learning have not been situated exclusively on the mathematical, rationalistic side. For some decades, theoretical trends in mathematics education research have tended to follow one or another of the 'isms'—schools of thought ranging from behaviorism to radical and social constructivism and postmodernism—that highlight particular research methodologies, observational priorities, types or levels of analysis, and overarching philosophical ideas, while deemphasizing or dismissing those championed by rival perspectives. Elsewhere, I have argued that this phenomenon has had foreseeable, highly unfortunate consequences for mathematics education (Goldin, 2003). None of the prevailing 'isms' have placed serious emphasis on affect, resulting in its effective exclusion from the mainstream of research fashion.

But among recent, encouraging signs are the interesting wordings of two publication titles—'Beyond Constructivism' and 'Beyond Cognitivism'—which suggest at last a readiness to break the barriers imposed by such systems, and to embrace new perspectives that include the additional constructs associated with the affective domain.

The first title, *Beyond Constructivism: Models and Modeling Perspectives on Mathematics Problem Solving, Learning, and Teaching* (Lesh & Doerr, 2003a), is a direct precursor to the present volume. The book addresses the limitations of constructivism as a school of thought in its final chapter, concluding in part:

> Many constructivists have been leaders clarifying the nature of students' developing constructs in topic areas ranging from early number concepts...to geometry...to statistics...to early algebra concepts. So it is ironic that many of these same individuals have ended up supporting a philosophy that focuses on the verb *construct* rather than the noun *construct*. ... Mathematics educators have a long history of treating discovery (or

construction) as if it were an end-in-itself—rather than being a means-to-an-end (namely children's development of mathematical constructs). (Lesh & Doerr, 2003c, p. 534).

> Whereas social constructivists extend purely cognitive perspectives by viewing the development of individuals through the lens of social theories, [models and modeling] perspectives extend social constructivist perspectives by also viewing communities through the lens of cognitive theories. Also, whereas social constructivism recognizes that one way that ideas are validated depends on social norms and beliefs and approved procedures, [models and modeling] perspectives recognize that there are at least four distinct way[s]...(Lesh & Doerr, 2003c, p. 545)

An important chapter (Middleton, Lesh, & Heger, 2003) emphasizes strongly certain aspects of affect, mathematical self-concept, and mathematical self-efficacy. In effect, the book's intent seems to be to make possible an eclectic, unifying approach, that appreciates the complexity of mathematical thinking and learning, and that addresses affect seriously.

The second title is "Beyond Cognitivism: Toward an Integrated Understanding of Intellectual Functioning and Development" (Dai & Sternberg, 2004a). This is the lead chapter of a new book edited by the same researchers (Dai & Sternberg, 2004b). In the authors' words,

> The term cognitivism represents a broad movement in psychology in the second half of the 20th century known as the cognitive revolution . . . it manifests itself in many ways, and does not have a simple definition . . . Yet the main thrust of this movement was to treat the computer, a mechanical computational device, as a model of the human mind, and its main tenet is rule-based symbol manipulation . . . Cognitivism should not be confused with cognitive sciences, which represent interdisciplinary efforts to understand the mind, and cover all spectrum of cognitive, affective, and motivational issues, including the nature of consciousness, intentionality, intersubjectivity, and self . . .
>
> ...
>
> The first limitation of such cognitivism is its assumption of a pure cognitive system of perceiving and thinking, free of emotion and motivation (or treating them as peripheral or epiphenomenal).
>
> ...
>
> The second limitation of such cognitivism, related to the first one, is its exclusive focus on the constraints of what is called cognitive architecture on performance, independent of various supporting (and sometimes enabling) or debilitating emotions and motivations in functional contexts.
>
> ...
>
> The third limitation of cognitivism is its inability to include human phenomenological (i.e., subjective) experiences as a legitimate (and often essential) force for higher-order mental functions.... What is missed in a typical cognitivist approach is the role of consciousness, intentionality, and reflectivity.
>
> ...
>
> The failure to consider subjective experiences also creates blind spots such as how a thinker's values, attitudes, dispositions, self-understandings, and beliefs guide his or her thinking. (Dai & Sternberg, 2004a, pp. 5–7)

These excerpts suggest that the field of cognitive science generally, as well as the field of mathematical learning and development in particular, seem poised to move past the limits imposed by earlier fashions, reaching toward the inclusion of variables associated with the affective domain. Hopefully the discussion that follows will contribute to this endeavor.

THE AFFECTIVE DOMAIN: PERSPECTIVES AND CONSTRUCTS

The scientific study of affect and emotion, like the scientific study of cognition, involves various methodologies. We may rely on experiments in neuroscience and physiology in formulating a theoretical perspective, or we may draw inferences from systematic observations of mathematical behavior. We may construct quantitative measures such as attitude surveys, or we may carry out qualitative, structured interviews.

As an advocate of scientific methods of inquiry, it is important for me to acknowledge at the outset how difficult it is to arrive at an empirically-based, valid understanding of higher-level constructs in the affective domain. The ideas that follow are influenced (mainly) by qualitative behavioral observations, but they are to a considerable degree speculative and subjective. Judgment of their value toward a theoretical foundation for mathematics education research will have to depend on the outcomes of future investigations based on them.

Affect as Representational

As remarked above, affect is sometimes interpreted as a rather peripheral concomitant of cognition. In this view, the affective system consists of largely involuntary physiological-psychological responses to cognitive states, that feed back to influence cognition in broad, mostly context-independent ways. Possible examples of such feedback include the well-known aversion responses to feelings of disgust, or the "fight-or-flight" cognitions that occur in response to the feeling of fear. Thus a student who is emotionally upset finds it difficult to concentrate on a mathematical activity, while one who is having fun engages readily with the most complicated equations.

My perspective is that affect is far more complex, and far more pervasively intertwined with cognition, than this interpretation suggests. I adopt the view, developed in more detail elsewhere, that in human beings affect serves as a *system of internal representation*. As such it functions in interaction with other internal cognitive representational systems, such as verbal-syntactic, imagistic, formal notational, and planning/executive control systems (Goldin, 1992, 1998; Goldin & Kaput, 1996), in highly context-dependent ways. To express this idea in everyday language, we may say that *our emotional feelings carry meanings*.

In many ways, this perspective parallels the (much more commonly-held) view that other sorts of representational configurations termed *cognitive*—for example, words and symbols—carry meaning or signification for individuals. But the analogous interpretation of affect seems to be unorthodox among researchers of emotion. In my opinion, this may be so because the meanings for the individual associated with affective configurations are even more highly context-dependent than those associated with other modes of internal representation—and they are often preconscious or unconscious. Thus we sometimes don't know (consciously) the meanings of our feelings, what they signify to or about us, or the reasons why they occur.

This representational function of affect is, in my view, essential to cognitive functioning generally, and to mathematical cognition in particular. For example, the *feeling of frustration* as it may occur during mathematical problem solving can encode the apparent failure of a series of trials ("this is not working"), or a series of strategies ("I'm stuck, I don't know what to do"). It may evoke adoption of a different strategy ("let me try something else"), or a different heuristic approach ("let me start over with an easier problem"). But the feeling of frustration can also encode information about the individual's background knowledge state ("I don't know how to do this type of problem"), about the person's general likelihood of success ("this is what always happens to me"), about the approval of other people significant to the individual ("I am going to be in trouble at home if I don't do well"), or about the person's general sense of self in relation to mathematics ("I am someone who not very good at this, and I don't ever want to be."). Furthermore, while affect interacts intimately with the physiology of the individual (Damasio, 1999), emotional feelings are not wholly involuntary. Thus the individual's decision to pursue an alternate problem-solving strategy may, in itself, alleviate the feeling of frustration, or the person might (for example) readjust his or her expectations in the situation.

Regarding affect as a representational system renders plausible the hypotheses that it functions crucially during mathematical activity, and that ultimately it is an essential component of mathematical ability.

Affect as a System of Communication

Affect also serves as a *language of communication* among people—so that one normally is aware (to some degree) of the emotional states of other people, and sometimes shares them. Skilled teachers participate in and are responsive to the affect of their students.

Evidence of "emotion" in non-human mammals is well-known. But the complexity of human affective communication, like the complexity of natural language in human beings (with which affect interacts strongly), seems unique in the animal kingdom, having apparently evolved along with the evolution of the human species. It may be the affective system that made human kinship structures and human functioning at the tribal level possible, enabling culture to develop. It may thus be the affective system that allowed human children (unlike other mammals) to develop and learn over more than a ten-year span before becoming biological adults. These are speculations; but they help render plausible the idea that at least some aspects of a powerful internal affective representational system could be central to the growth of mathematical ability in the individual.

Affective messages may be sent by way of eye contact, facial expressions, hand gestures, posture, "body language," touch, tone of voice, tears, laughter, blushing, singing, and so on, as well as through ordinary language, expressive sounds, and interjections. To a considerable degree, such communication is tacit. Although most people are able to communicate affectively in this way, we normally find it very difficult to specify precisely what it is in another person's expression—e.g., eyebrow raised, lips just parted, eyes intent but looking occasionally up or to the side—that one might interpret (depending on the context) as indicating curiosity, amusement, serious engagement, or something else. It is probable that we often make highly inaccurate inferences of affect in others. Nevertheless, the system of communication seems to work by actually enabling two or more people to represent each other's emotional states, to "share" affect, and consequently to function effectively together.

When we encourage group problem solving in mathematics classrooms, the communicative aspects of affect in relation to mathematics become especially important—for example, one student's eager enthusiasm may inspire corresponding feelings in other students; but it may also evoke a sense of mathematical inadequacy in others, expressing itself as resentment or withdrawal. Such dynamics are likely to have important effects on learning, and possibly lasting effects on the development of mathematical self-identity.

Affect and Meta-affect

Meta-affect describes affect that is about (i.e., represents or encodes) other affect, affect that is about cognition that may in turn be about affect, and the monitoring of affect through both cognition and affect. Thus the concept of meta-affect captures the *self-referential* aspects of affect, which are manifold.

DeBellis and I have put forward the hypothesis that meta-affect is the most important aspect of affect (DeBellis & Goldin, 1997, 1999, 2006; Goldin, 2002). Meta-affect can transform affect—thus a negative emotion, like *fear*, may be experienced as pleasurable (as in taking an amusement park ride, or attending a horror movie).

Remarkably stable and powerful *towers of meta-affect* occur frequently. For example, a student may feel *love* for his father, but experience very keenly the *pain* of rejection in relation to that love, in that he thinks he is not meeting his father's high academic expectations (e.g., in mathematics). He may come to feel *anger* that protects him, to some extent, from the experience of pain. To this student, the anger represents the fact that his father has "no right" to have those expectations. But he also feels *guilt* about being angry with his father. The emotion that the student feels at a particular moment—for instance, the intense feeling of *anxiety* in beginning an important mathematics test—may thus simultaneously be *consciously* about the mathematics test, and perhaps *unconsciously* or *preconsciously* about the complex tower of feelings about feelings constructed in relation to his father. In this hypothetical example, the feeling at the base of the affective structure may be love. However, the negative meta-affect transforms it into a painful feeling, and the dysfunctional structure in relation to mathematics is stabilized by the further meta-affect of anger and guilt.

Of course, for a mathematics teacher to probe a student's affect to this depth is quite rare, and it is likely that the affective structures underlying "math anxiety" vary greatly from one individual to another. But consideration of meta-affect suggests two things. First, the most important goals of mathematics education in relation to the affective domain should be characterized as meta-affective. And second, these goals should *not* include eliminating negative feelings, or making all mathematical activity easy and fun. Rather, it should be our goal to develop stable, powerful affective structures in students. Then there may well be negative feelings associated with impasse or difficulty in mathematics, but the meta-affect *about* those feelings transforms them in a productive way.

Attention to the mathematics learner's beliefs and values are important to achieving this goal, as these influence the ecological function of particular emotions in the individual's personality. For instance, let us consider an ideal. The commonly occurring feeling of frustration in the course of work on a mathematical or scientific problem *could* and *should* be taken to indicate that the problem is challenging and interesting. The feeling should carry with it an expectation of positive feelings, such elation and satisfaction at understanding something new, or achieving something difficult. This response may presuppose a high value placed on achievement or on understanding, together with some degree of belief in one's likelihood of eventual success. Then the *frustration itself* can be experienced as a positive emotion, transformed by the meta-affect of anticipatory pleasure.

Domains of Affect

In earlier work, McLeod (1989, 1992) and others have suggested three components of the affective domain that are important in mathematical learning and problem solving—*emotions, attitudes,* and *beliefs*. These vary in duration, with emotions regarded as the most fleeting, and beliefs seen as the most stable. They vary also in the extent to which cognition is involved, with emotions being the least cognitive and beliefs being the most highly cognitive. DeBellis and I proposed adjoining a fourth component, *values* (that include ethics and morals as personally represented). Let us here comment briefly on each of these.

Emotions or emotional feelings refer to rapidly-changing states of feeling, of which the individual is consciously (or, possibly, pre-consciously) aware during mathematical (or other) activity. They are local, contextually-embedded, and may range from mild to very intense. They are often experienced as involuntary responses, and are sometimes accompanied by or partially encoded as overt changes in one's physical state (pulse rate, breathing, etc.).

Attitudes refer to predispositions or leanings toward having certain sets of emotional feelings in relation to particular contexts (e.g., mathematical contexts). They are moderately stable (more so than emotions), and they encompass an interaction between affect and cognition.

Beliefs refer to the attribution of truth or validity by the individual to propositions or other cognitive configurations. While beliefs (or systems of belief) are frequently highly stable, highly cognitive, and highly structured, affect is interwoven with them—beliefs are often such as to make one feel more comfortable or safer in (mathematical) situations that are likely to recur (Goldin, 2002).

Values (or systems of values) held by the individual, including ethics and morals, refer to the "personal truths" that express what a person cherishes. They provide a sense of personal meaning and help to motivate purposes. Values are stable, usually very highly affective as well as cognitive, and they may also be highly structured.

With these four components, we envision a tetrahedral model in which each vertex interacts dynamically with the other three as an individual engages in a mathematical task. Thus a student's belief in her own mathematical ability affects how she feels when she encounters a mathematical difficulty, and how she values the task on which she is working. The belief may also help her to sustain a positive attitude toward challenging mathematical problems. Furthermore, each vertex interacts with its counterpart in the affective domain of other people—a student's feelings may be immediately influenced by those of the teacher, or those she feels the teacher has, in the course of mathematical dialogue.

Affective Pathways, Competencies, and Structures

The term *local affect* refers to the changing states of emotional feeling experienced by individuals as they engage in mathematical (or other forms of) activity, together with the context-dependent meanings of these feelings and the cognitions with which they interact. Sequences of such states may tend to recur, and these we call *affective pathways*.

To summarize an idealized example introduced elsewhere (Goldin, 2000); a student might during mathematical problem feel *curiosity*, followed by *puzzlement* or *bewilderment* if the problem is nonroutine and unfamiliar. Successive attempts that bear no fruit may evoke *frustration*. Following a change of strategy, the student may feel *encouragement* at apparent progress, and then *elation* at a new understanding or insight—which may lead to a sense of *satisfaction* with what has been achieved. In an alternative, idealized affective pathway, the student's frustration may lead in succession to *anxiety, anger, fear,* and/or *despair*, bringing forth avoidance strategies or other defense mechanisms. When such sequences are repeated, they become well-worn affective pathways, with the meta-affect of comfortable familiarity and/or (possibly) foreboding and inevitability.

Now in studying mathematical cognition and learning in individuals, an important notion is that of *competencies*. This refers to capabilities of performing specific tasks, taking specific cognitive steps, or processing information in specified ways within a system of representation. Competencies do not occur in isolation; they connect with each other to form complex, cognitive structures. Analogously, we may consider *affective competencies* to refer to a person's capabilities of using affect to good purpose during mathematical activity—for instance, to act on curiosity to follow up a line of thought, or to interpret frustration as a signal that a strategy is not working and therefore to take a new approach.

Affective pathways such as those described may result in the construction of longer-term, *global affect*. This refers not merely to a generalized inclination toward positive or negative feeling in relation to mathematics, but to complicated *affective structures* in which emotional feelings are entwined with other emotional feelings, with beliefs and values, and with stories and memories about oneself and one's history. Some pathways may result in structures that facilitate enthusiasm, engagement, expectations of success, and a positive mathematical self-concept; some may facilitate distaste, avoidance, expectations of failure, and a negative self-concept.

It is well-known that many students and adults have affective structures that impede mathematical learning (e.g., resulting in the global affect known as "math anxiety"), but unfortunately, we do not know an easy way to transform these structures into more efficacious ones. It is important that future research be devoted to studying *mechanisms of change in global affective structures pertaining to mathematics*.

A cognitive analogy here might be with the way that construction of a proportionality schema can permanently transform the understandings associated with comparisons, from those that are exclusively additive to those that are multiplicative. An affective example, in a non-mathematical context, might be with the way that forgiveness or self-forgiveness can permanently transform or remove previously-developed structures of resentment, anger, and/or hatred.

ESSENTIAL AFFECTIVE STRUCTURES

Next I shall discuss briefly what I believe are three *essential* affective structures—mathematical integrity, mathematical self-identity, and the capacity for mathematical intimacy. In my view, each of these connects profoundly with mathematical development in students, and is fundamental to understanding the psychological structure of mathematical abilities in individuals. They deserve far more empirical research, and each in its own right could easily be the subject of an entire book.

Mathematical Integrity

The term *mathematical integrity* (including what DeBellis and I earlier called "mathematical self-acknowledgment") refers to the global affective structure and related substructures associated with *commitment to truth and understanding in mathematical activity* (DeBellis and Goldin, 1997, 1999, 2006). It includes the awareness (or lack of awareness) of the limitations of one's mathematical understanding at any given point. It includes the willingness to acknowledge those limitations, and to work on removing them.

Let me emphasize strongly the affective dimension of this structure. Mathematical integrity may include emotional feelings of pride and satisfaction associated not only with the solution of a problem, but also with the deepening of understanding of the mathematics behind it. On the other hand, when feelings of pain, embarrassment, or shame are associated with the absence of understanding, the individual's mathematical integrity may be sacrificed. Then, in place of acknowledging the insufficiency, the person may engage in bluffing, i.e., pretending an understanding to others that he or she knows is in fact not present, in order to alleviate the negative feelings. Alternatively, there may be self-deception, where the person who does not understand convinces himself or herself that there is nothing more to understand mathematically or nothing more that is worth understanding.

Neither of these responses is the same as the situation where an individual is authentically unaware of a gap in understanding, but (with mathematical integrity) is open to learning of it.

Often we observe students who focus mainly on performing the mathematical steps likely to satisfy the teacher, on doing what they are "supposed to do" in mathematical situations. Some students may identify their primary emotional satisfaction in mathematics with high scores on classroom tests, rather than focusing on achieving mathematical insight. Sometimes students (and mathematicians) endeavor to impress others with their mathematical speed or advanced knowledge base.

A highly developed affective structure of mathematical integrity leaves the individual uncomfortable with insight that is incomplete. The "why" behind mathematical steps is paramount—that is, the "meaning of the mathematics" in relation to previously developed cognitive representational structures. The emotion of discomfort or frustration is likely to encode the fact that this "meaning" is absent, as distinct from encoding the prospect of displeasing the teacher, or performing poorly on a test. And the meta-affect surrounding such *integritous discomfort* (to coin a phrase) is likely to be positive and encouraging, representing the path toward an anticipated deeper level of mathematical understanding.

Mathematical Self-Identity

The term *mathematical self-identity* refers to a complex, global affective structure encoding one's personal sense of self—"who I am"—in relation to mathematics, or encoding the part that mathematics plays in constituting the fabric of "who I am." In mathematical contexts, I sometimes prefer to use the term *self-identity* rather than *identity*, since "mathematical identity" has a quite unrelated meaning (pertaining to an algebraic or trigonometric equation that holds for all values of a variable).

Self-identity in this sense is constructed over time. Note that mathematical self-identity need not necessarily incorporate mathematics in a positive way—

many people seem to construct their very sense of self so as to exclude the mathematical, the analytical, or the abstract. Middleton, Lesh, and Heger (2003) explore "the fit between the activity and the self" (p. 410), stating, "Because mathematical self concept and self efficacy in mathematics tasks are learned knowledge (Middleton, 1999), an examination of students' experiences that contribute to the development of self-statements should shed light on the knowledge of, and about, the mathematics students are constructing." They discuss "structuring a mathematical self in a modeling environment" (p. 408), pointing out that "mathematical modeling and self-modeling are reflexive" (p. 411). In this spirit, Empson (2003) explores in detail how the "frequency of opportunities for identity-enhancing interactions" account for two students' mathematical success.

Let us try to characterize mathematical self-identity more specifically. To some extent, it may incorporate a feeling of *ownership* (or non-ownership) in relation to mathematics, a willingness or unwillingness to speak of working on "*my* math" or to think comfortably in mathematical ways. Such willingness, by its very existence, may confer a degree of mathematical power and flexibility. Self-identity may of course incorporate an emotionally important sense of or belief in one's own mathematical ability or lack thereof. One's self-identity may be consciously or unconsciously based on specific self-defining events from one's mathematical history. It may also involve the subtle emotional feelings associated with aesthetics—the sense that mathematical ideas, structures, or relations can be beautiful or even awe-inspiring.

Mathematical self-identity may involve feelings about other specific individuals, especially those who have been sources of inspiration or discouragement—e.g. teachers, parents, colleagues, friends, lovers, or historical figures. It may include reference to a set of people—a clique at school, an academic department, an employee group, or a professional community. This may be a group to which the person feels a sense of "belonging" (or to which others see the person as belonging), with all the accompanying role expectations, or it may be a group by which the person feels excluded or rejected. Thus self-identity as a structure involves some expectations of how one will be viewed by others.

At the most fundamental level, mathematical self-identity is not just a set of plausible stories about one's ability level, record of success, inspirational influences, or role played in the world of mathematics; it includes, but is not only, a "self concept". Rather it is also an affective structure that encodes (represents internally) a set of *current possibilities* for the individual. "Who I am" may thus be (or not be) someone who *investigates* mathematics deeply (*cf.* the discussion of mathematical integrity), who *explains* mathematics to others, who cares about and *appreciates* mathematics, who *inspires* others mathematically, who *takes responsibility* for solving mathematical problems; and (most importantly) to whom such possibilities, even when they are tacit, *matter emotionally* and *involve fundamental values*.

Mathematical Intimacy

The term *mathematical intimacy* refers to structures of emotional feelings, attitudes, beliefs, and values that are associated with deep, intense, highly engaged, and (most importantly) *vulnerable* interaction involving one's sense of self during mathematical activity. It characterizes the affect surrounding one's personal relationship with mathematics.

A child solving a problem in mathematics may lean over, cupping her hand about the paper as if to create a personal, protected space. When the teacher approaches to help, her behavior is to withdraw as if to say, "It's not ready yet." If she were to do this while working on a poem, or while drawing a picture, the prevailing expectation would most likely be for the teacher to respect the child's privacy (temporarily), until she became ready to share what she had produced. But in mathematics classes, the idea that this is an emotionally intimate moment for the child is relatively foreign. Many teachers, with the best of intentions, would interrupt to demonstrate the correct or most efficient way to solve the problem—with an attendant emotional cost.

Intimacy is a term often used specifically to characterize close personal relationships—those of love or friendship. DeBellis and I have adopted a more general interpretation, where intimacy characterizes an internal way of being and feeling in the individual (DeBellis & Goldin, 1999). The experience of intimacy *may* occur in relation to interaction with another person; but it may equally well occur in relation to interaction with a specific aspect of the individual's environment—art, literature, the natural world, music, or in the present case mathematics.

Yet the analogy with love or intimate friendship is worth pursuing. Personal integrity may be a prerequisite for the *trust* that allows a long-term, intimate relationship to grow, deepen, and progress. A developing personal identity—a sense of "who I am"—when brought to a personal relationship, enables connection and authentic closeness. The thesis here is that analogous affective structures come into play in one's long-term relationship with mathematics. Thus mathematical integrity and a developing mathematical self-identity make intimacy with the mathematics possible.

One feature of intimacy—engagement—may be directly associated with the experience of *flow* (Csiksentmihalyi, 1990; Lesh & Doerr, 2003c, p. 551). The high level of interest evidenced by one who experiences flow when doing mathematics is likely to be due, at least in part, to his sense of self-identity being involved—even as he seems to "lose himself" in the mathematics. There is a sense of trust associated with "letting go." There is intense concentration, a loss of the sense of time, and an overall experience that may range from pleasant to euphoric.

An accompanying aspect, however—vulnerability—means that as one engages deeply with mathematics, some emotional tenderness is involved. There is the potential, as in a personal relationship, for being disappointed or hurt in a profound way. Intimacy is not safe. One is not defended against experiencing "inner wounds" of a kind that do not heal easily, and early hurts may cause the person to avoid future intimacy with mathematics. It is perhaps surprising how many adults (including those who have continued with advanced study in mathematics) recount profoundly painful memories associated with school arithmetic. The level of pain suggests the vulnerability that is associated with early intimate experience.

In the next section, we shall briefly discuss possible relations between modeling and affect. A fundamental psychological component of modeling with mathematics is the assignment of personal meanings to mathematical symbols and configurations. These "semiotic acts" enable mathematical representations of non-mathematical situations to be constructed. Let us keep in mind that there is a sense in which internal, personal meanings are by their nature intimate. I think it is just this fact that makes mathematical intimacy an *essential* affective structure in learning and development.

THE AFFECTIVE DOMAIN FROM A MODELING PERSPECTIVE

Modeling refers to the construction of a meaningful structure within a symbol-system of some kind (possibly mathematical), that when appropriately interpreted, describes some aspects of a class of situations or phenomena. The book edited by Lesh and Doerr begins with the development of important aspects of models and modeling theory—including the following (Lesh & Doerr, 2003b; Lesh, Cramer, Doerr, Post, & Zawojewski, 2003):

> *Model-eliciting activities,* as distinct from traditional classroom mathematics problems;
>
> The notion of a model as both internal and external, *distributed* across a variety of representational media, and created through *model development sequences;* and
>
> Multiple *modeling cycles* that occur in problem solvers addressing model-eliciting activities, and that entail *local conceptual development.*

If we accept the idea of affect as a representational system (i.e., a kind of symbol system) within the individual, continually interacting with and exchanging information with other internal, cognitive representational systems, then these ideas from a modeling perspective have immediate application.

In particular, we may understand the affective structures developed by an individual over time as serving *to model himself or herself* in relation to mathematical activity. Such self-models may be *descriptive*, as with affective structures that encode a sense of one's own mathematical ability and interest, or lack thereof. But they may also be *prescriptive*, modeling an *ideal* of behavior, commitment, and/or ability level which the student seeks to emulate.

We would then expect the development of affective structures to occur through a sequence of modeling cycles, with local affective development entailed by each cycle as suggested in the description by Lesh et al. cited above. By the end of such a cycle, the individual has undergone global affective changes. Hopefully he or she brings to bear a more efficacious self-model in the next stage of mathematical activity, and maintains the willingness to enter a new cycle of affective development. In my view, this description pertains to the way mathematical integrity grows (or is suppressed) in the individual learner; to the way a mathematical self-identity develops or is reconstructed in the child, adolescent, or adult; and to the manner in which the individual experiences (or refrains from) intimacy with mathematics.

Mathematical modeling activity itself brings about not only cognitive development, but development of the affective system. Thus we should pay explicit attention to affective structures as desirable consequences of the mathematics curriculum.

Let us focus on an example from undergraduate mathematics. For me as a teacher, the best mathematical problem activities are not "one-shot" activities but investigations that can extend over several weeks or more. For instance, one may seek to model a dynamical process such as population change, exploring what can happen to competing fish populations in a medium-size lake, with a possibly limited food supply, as the rate increases at which fishing activity takes place. We make available increasingly complex and powerful methods—e.g., exponential growth, logistic models, coupled first-order differential equations describing predator-prey situations, computer simulations. We ask questions with practical consequences along the way; for instance, "Is there such a thing as 'overfishing'? If so, will there be warning that the lake is becoming overfished, or could a seemingly healthy population be on the verge of sudden extinction?" Numerous mathematical possibilities can be realized in the situation, such as the abstract phenomenon of bifurcation. And like many "real world" situations, the problem is not quite well-defined—for instance, there is the important choice between using a continuous model based on (parameter-dependent) differential equations, or a discrete model based on difference equations.

The cognitive goals in this example are themselves complex. At one level, the purpose of the modeling activity may be to have the students practice efficient techniques for solving various classes of differential equations, and learn some standard models for the dynamics of population change. These may be subsidiary goals, but they are sufficiently important so that such techniques and models are likely to be the primary focus of examination questions. At another level, we may want students to develop a *qualitative* understanding of the processes described by differential equations, and the ability to interpret classes of solutions. At the undergraduate level, this might include interpreting each term in an expression for the rate of change (the time derivative of one of the populations) as representing a partial contribution from a different component of the process. It may involve developing a "feel" for sources and sinks in the phase plane for a nonlinear differential equation, for what it means to linearize in a neighborhood of an equilibrium point, for what "stability" means, or for the ways in which the qualitative behavior of solutions can depend sensitively on initial conditions or parameter values. This has implications for the nature of *how* the mathematics is understood, and the ability to transfer those understandings to new mathematical techniques in new domains.

At a (still more general) "meta" level, a goal of the activity is to develop an understanding of what it means to create a mathematical model, and *what it means to have a "feel"* for a dynamical process—the self-awareness of knowing when this level of understanding is absent, and the doggedness to pursue it. In my view, the very characterization of this "feel" requires affective components. It is partially encoded cognitively, but partially in emotional feelings of comfort *vs.* discomfort, satisfaction *vs.* dissatisfaction, aesthetic pleasure *vs.* displeasure.

When such complex cognitive developments occur over a period of time, they are likely to happen in interaction with developing affective structures of the kind we have been discussing. Knowing what it means to have the deeper understanding, to "feel" the process being modeled mathematically, allows the possibility that proceeding without that understanding is unacceptable—an ingredient of mathematical integrity. Construction of an elaborate, successful mathematical model of a complex situation may contribute to a sense of mathematical self-identity—even as mathematics is used (meaningfully) to model the given situation, a model of the self as the efficacious user of the mathematics is also tacitly constructed. And inner emotional satisfaction may be associated with intimate experience during the successful modeling activity, contributing to a future sense of safety and self-confidence in intimate mathematical engagement.

The longer-term nature of a teacher-guided modeling activity lasting several weeks can thus have profound affective as well as cognitive consequences. Naturally it allows the teacher to attend to complex cognitive-structural goals. But perhaps still more importantly, it permits affective and meta-affective interactions to occur—opportunities for the teacher to encourage integritous inquiry, to confirm the individual student's developing self-identity, and to share in moments of mathematical intimacy. Thus the teacher can contribute importantly to students' affective-structural development.

REFERENCES

Csiksentmihalyi, M. (1990). *Flow: The psychology of optimal experience.* New York: Harper & Row.

Dai, D. Y., & Sternberg, R. J. (2004a). Beyond cognitivism: Toward an integrated understanding of intellectual functioning and development. In D. Y. Dai & R. J. Sternberg (Eds.), *Motivation, emotion, and cognition: Integrative perspectives on intellectual functioning and development.* Mahwah, NJ: Lawrence Erlbaum Associates. (pp. 3–38).

Dai, D. Y., & Sternberg, R. J. (Eds.) (2004b). *Motivation, emotion, and cognition: Integrative perspectives on intellectual functioning and development.* Mahwah, NJ: Lawrence Erlbaum Associates.

DeBellis, V. A. (1996). *Interactions between affect and cognition during mathematical problem solving: A two-year case study of four elementary school children.* Ann Arbor, MI: University Microfilms No. 96-30716.

DeBellis, V. A. (1998). Mathematical intimacy: Local affect in powerful problem solvers. In S. Berenson, K. Dawkins, M. Blanton, W. Coulombe, J. Kolb, K. Norwood, & L. Stiff (Eds.), *Proceedings of the 20th annual meeting of PME-NA, vol. II* (pp. 435–440). Columbus, OH: ERIC.

DeBellis, V. A., & Goldin, G. A. (1997). The affective domain in mathematical problem solving. In E. Pehkonen (Ed.), *Proceedings of the 21st annual conference of PME, Lahti, Finland, vol. 2* (pp. 209–216). Helsinki, Finland: University of Helsinki Department of Teacher Education.

DeBellis, V. A., & Goldin, G. A. (1999). Aspects of affect: Mathematical intimacy, mathematical integrity. In O. Zaslavsky (Ed.), *Proceedings of the 23rd Annual Conference of PME, Haifa, Israel, vol. 2* (pp. 249–256). Haifa, Israel: Technion, Department of Education in Technology and Science.

DeBellis, V. A., & Goldin, G. A. (2006). Affect and meta-affect in mathematical problem solving: A representational perspective. *Educational Studies in Mathematics, 60,* in press.

Empson, S. B. (2003). Low-performing students and teaching fractions for understanding: An interactional analysis. *Journal for Research in Mathematics Education, 34,* 305–343.

Evans, J. (2000). *Adults' mathematical thinking and emotions: A study of numerate practice.* London: Falmer Press.

Goldin, G. A. (1992). On developing a unified model for the psychology of mathematical learning and problem solving. In W. Geeslin & K. Graham (Eds.), *Proceedings of the 16th Annual Conference of PME, Durham, New Hampshire, USA, vol. 3* (pp. 235–261). Durham, NH: University of New Hampshire Department of Mathematics.

Goldin, G. A. (1998). Representational systems, learning, and problem solving in mathematics. *Journal of Mathematical Behavior, 17,* 137–165.

Goldin, G. A. (2000). Affective pathways and representation in mathematical problem solving. *Mathematical Thinking and Learning, 2,* 209–219.

Goldin, G. A. (2002). Affect, meta-affect, and mathematical belief structures. In G. C. Leder, E. Pehkonen, & G. Törner (Eds.), *Beliefs: A hidden variable in mathematics education?* Dordrecht, the Netherlands: Kluwer Academic. (pp. 59–72).

Goldin, G. A. (2003). Developing complex understandings: On the relation of mathematics education research to mathematics. *Educational Studies in Mathematics, 54,* 171–202.

Goldin, G. A., & Kaput, J. J. (1996). A joint perspective on the idea of representation in learning and doing mathematics. In L. Steffe, P. Nesher, P. Cobb, G. A. Goldin, & B. Greer (Eds.), *Theories of mathematical learning* (pp. 397–430). Mahwah, NJ: Lawrence Erlbaum Associates.

Gomez-Chacon, I. M. (2000a). *Matematica Emocional: Los Afectos en el Aprendizaje Matematico.* Madrid: Narcea, S. A. de Ediciones.

Gomez-Chacon, I. M. (2000b). Affective influences in the knowledge of mathematics. *Educational Studies in Mathematics, 43,* 149–168.

Hannula, M. S. (2002). Attitude towards mathematics: Emotions, expectations and values. *Educational Studies in Mathematics* 49(1), 25–46.

Hannula, M. S. (2004). *Affect in mathematical thinking and learning.* Doctoral dissertation, University of Turku, Finland. Annales Universitatis Turkuensis B 273.

Hannula, M. S., Evans, J., Philippou, G., & Zan, R. (2004). Affect in mathematics education: Exploring theoretical frameworks. In M. J. Høines & A. B. Fuglestad (Eds.), *Proceedings of the*

28th Annual Conference of PME, Bergen, Norway, vol. 1 (pp. 107–136). Bergen University College.

Leder, G. C., Pehkonen, E., & Törner, G. (Eds.) (2002). *Beliefs: A hidden variable in mathematics education?* Dordrecht, the Netherlands: Kluwer Academic.

Lesh, R., Cramer, K., Doerr, H. M., Post, T., & Zawojewski, J. S. (2003). Model development sequences. In R. Lesh & H. M. Doerr (Eds.), *Beyond constructivism: Models and modeling perspectives on mathematics problem solving, learning, and teaching.* Mahwah, NJ: Lawrence Erlbaum Associates. (pp. 35–58).

Lesh, R., & Doerr, H. M. (Eds.) (2003a). *Beyond constructivism: Models and modeling perspectives on mathematics problem solving, learning, and teaching.* Mahwah, NJ: Lawrence Erlbaum Associates.

Lesh, R., & Doerr, H. M. (2003b). Foundations of a models and modeling perspective on mathematics teaching, learning, and problem solving. In R. Lesh & H. M. Doerr (Eds.), *Beyond constructivism: Models and modeling perspectives on mathematics problem solving, learning, and teaching.* Mahwah, NJ: Lawrence Erlbaum Associates. (pp. 3–33).

Lesh, R., & Doerr, H. M. (2003c). In what ways does a models and modeling perspective move beyond constructivism? In R. Lesh & H. M. Doerr (Eds.), *Beyond constructivism: Models and modeling perspectives on mathematics problem solving, learning, and teaching.* Mahwah, NJ: Lawrence Erlbaum Associates. (pp. 519–556).

Malmivuori, M.-L. (2001). *The dynamics of affect, cognition, and social environment in the regulation of personal learning processes: The case of mathematics.* University of Helsinki Department of Education Research Report 172. Helsinki, Finland: Helsinki University Press.

McLeod, D. B. (1989). Beliefs, attitudes, and emotions: New views of affect in mathematics education. In D. B. McLeod & V. M. Adams (Eds.), *Affect and mathematical problem solving: A newperspective.* New York: Springer Verlag. (pp. 245–258).

McLeod, D. B. (1992). Research on affect in mathematics education: A reconceptualization. In D. Grouws (Ed.), *Handbook of research on mathematics teaching and learning* (pp. 575–596). New York: Macmillan.

McLeod, D. B., & Adams, V. M. (Eds.) (1989). *Affect and mathematical problem solving: A new perspective.* New York: Springer Verlag.

Middleton, J. A. (1999). Curricular influences on the motivational beliefs and practice of two middle school mathematics teachers: A follow-up study. *Journal for Research in Mathematics Education, 30*(3), 349–358.

Middleton, J. A., Lesh, R., & Heger, M. (2003). Interest, identity, and social functioning: Central features of modeling activity. In R. Lesh & H. M. Doerr (Eds.), *Beyond constructivism: Models and modeling perspectives on mathematics problem solving, learning, and teaching.* Mahwah, NJ: Lawrence Erlbaum Associates. (pp. 405–431).

Op't Eynde, P., De Corte, E., & Verschaffel, L. (2001). Problem solving in the mathematics classrooms: A socio-constructivist account of the role of students' emotions. In M. van den Heuvel-Panhuizen (Ed.), *Proceedings of the 25th Annual Conference of PME Education, Utrecht, the Netherlands* (vol. 4, pp. 25–32). Utrecht: Freudenthal Institute, Utrecht University.

Roth, W.-M. (2006). Mathematical modeling 'in the wild': A case of hot cognition. Chapter 4, present volume.

Sinclair, N. (2004). The role of the aesthetic in mathematical inquiry. *Mathematical Thinking and Learning, 6,* 261–284.

PART III

What Kind of Instructional Activities Are Needed to Support the Development of New Levels and Types of Understanding and Ability?

Richard Lesh
Indiana University

Jim Kaput
University of Massachusetts, Dartmouth

Chapters in Part I of this book focused on the question: What kind of mathematical thinking is involved in "real life" problem solving or decision making situations that occur outside of school classrooms? Part II shifted attention toward the question: What changes are occurring in the kinds of elementary-but-powerful mathematics concepts that might provide new foundations for the future? Now, Part III focuses on the questions: What does it mean to "understand" the preceding ideas and abilities? How can mathematics educators promote the development of new levels and types of understanding and ability—aimed at providing powerful "foundations for the future" for students who are preparing for productive lives where they will be functioning within learning organizations, within global societies, within knowledge economies, within a technology-based age of information?

When we investigate the nature of new levels and types of understanding and ability, Part III goes beyond observations of naturally occurring phenomena to also focus on designed and preplanned simulations of "real life" events. From the point of view of research, instruction, and assessment, these latter kinds of "designed experiences" provide many opportunities to witness patterns in behavior that are unlikely to be apparent in naturally occurring situations which often occur only once and without warning. For example, in designed simulations, it may be possible to observe comparable situations multiple times, from multiple perspectives, and with multiple groups of problem solvers. Consequently, expert observers often are able recognize the importance of things that are difficult or impossible to observe in naturally occurring phenomena.... On the other hand, every research methodology has weaknesses as well as strengths; and, any time contrived situations are used, questions arise about the authenticity of such tasks. So, the final summary section of this book will discuss a variety of these kinds of issues related to future directions for research design.

Part III begins with a chapter that describes a trend which is occurring in many countries, such as Singapore, where the students already score at the top of international testing programs that focus on basic skills. These countries are recognizing that, in a highly competitive global economy, basic skills are not enough! Other complex achievements also need to be cultivated; and, these include the development of creativity, and problem solving ability, as well as the ability to produce useful mathematical descriptions of complex systems, the ability to work productively within diverse teams of specialists, and the ability to adapt to rapidly evolving technical tools. All of these latter competencies are at the top of the list of attributes that are emphasized by employers in future-oriented professions and by admissions officers in highly selective universities. Yet, they tend to be ignored on easy-to-scorestandardized tests.

The second chapter in Part III emphasizes the fact that, when learning activities emphasize the development of complex conceptual systems, "what" is learned is strongly influenced by "how" it is learned. In particular, the understandings that students develop tend to be considerably different depending on whether (a) they are guided along narrow instructional paths toward cleaned-up versions of textbooks' portrayals of the meanings of relevant constructs, or (b) they are challenged to repeatedly express→ test→ revise their own ways of thinking about the relevant constructs. Nonetheless, even though the preceding points sound like things that constructivist philosophers emphasize, research on models and modeling suggests that, when learning activities emphasize the latter approach, very few of the processes that contribute to development fit the term "construction." For example, students sort out, revise, and refine ideas and abilities that already exist (at some intermediate level of development) far more than they construct ideas that are completely new; and, the understandings that evolve tend to be far more situated, far more representationally rich, and far more likely to integrate ways of thinking drawn from a variety of textbook topic areas, than traditional theories of learning suggest.

The third chapter in Part III emphasizes the fact that, when students interpret learning or problem-solving situations, the conceptual systems that they engage are not purely logical or mathematical in nature. They also include feelings, values, attitudes, beliefs, and dispositions (e.g., to adopt certain roles, or to engage in certain kinds of activities). In general, these attributes often are thought of as being associated with the student's identity as a mathematics learner and problem solver. When attempts are made to help students develop more productive problem solving personalities, the preceding feelings, values, attitudes, beliefs, and dispositions often are treated as if they were simply rules (or simple declarative propositions) to be learned. For example, it is easy to think of beliefs as "statements" of belief—and to ignore the "systems" of belief from which the subsumed statements (and other pieces of knowledge) derive most of their meanings. But, the meanings of most feelings, values, attitudes, beliefs, and dispositions have more to do with "seeing" than with "doing"; and, most things that learners or problem solvers do can lead to negative or positive outcomes—depending on contexts and purposes. In fact, no single, fixed, and inflexible profile of attributes is productive in all situations, or even across all stages of a multi-staged problem solving episode. This is why productive problem solvers tend to be people who are able to adopt a variety of personae (e.g., roles and dispositions) as they move from one problem solving situation to another, and as they move from one stage of problem solving to another within a single problem. They may be a brainstormer at one moment and a careful monitor and critic at another; or, they may be a leader in a group at one moment and a recorder or assistant at another. Yet, the principles that govern their behaviors at any given moment may function intuitively and subconsciously—rather than formally and analytically. Therefore, the goal of relevant instruction may be to use a variety of "reflection tools" to help students develop the ability to think "with" conceptual frameworks which would be far too complex and cumbersome to think "about" in a formal or analytic fashion.

The fourth and fifth chapters in Part III describe ways that case studies, design tasks, and other simulations of "real life" problem solving situations are used for both instruction and assessment in fields such as engineering—where successful professionals tend to be heavy users of mathematics, science, and technology. In general, design tasks appear to be specialized forms of the kind of modeling-eliciting activities that were describe in the second chapter in Part III. One reason this is true is because, in fields like engineering, the underlying designs (or models, or conceptual systems) tend to be extremely important components of properly designed artifacts.

Finally, the sixth chapter in Part III describes some of the most significant results from a large-scale curriculum reform project that emphasized the kind of models and modeling perspectives that have been described throughout this book. One reason why this study is remarkable is because participating students, teachers, schools, and school districts did not achieve higher performance on test scores by focusing exclusively on the kind of low-level "basic skills" that the tests emphasized. Instead, balanced attention was given to both (a) "basic skills" of the type emphasized in textbooks and on tests, and (b) deeper and more complex achievements that are needed for success beyond school.

CHAPTER 16

Beyond Efficiency: A Critical Perspective of Singapore's Educational Reforms

Mani Le Vasan
Universiti Brunei Darussalam, Brunei

Richard Lesh
Indiana University

Mardiana Abu Bakar
*National Institute of Education,
Nanyang Technological University, Singapore*

Around the world, governments, authorities and educators continue to grapple with the age old questions about the goal, design and implementation of institutionalized education. What is formal education for? What consequences does education have and should they have on individuals, communities and societies? Does education any longer have a future as a vocational training setting? Is it a seedbed for national or cultural cohesion? Can it be a site for developing important life-skills and knowledge? The fronts on which classroom, school, and system-policy reform are currently proceeding are many and varied (see, e.g., Darling Hammond, 1997; Jackson, & Davis, 2000; Lee, 2001; Newmann et al., 2001; Slavin & Fashola, 1998; Wiske, 1998). In many cases, however, the issues at stake come down to a familiar set of debates about:

> Classrooms as sites for cognitive learning and growth;
>
> Schools' capacities to maximise their students intellectual, technological, human and organizational resources; and
>
> Education systems as distribution sites for economic, social, cultural and intellectual resources.
>
> Education systems as development sites for a competitive labor in a globalized knowledge economy.

This essay is a critical perspective on the "success" of a highly state-mediated education system that has been held up as having much successes but which does not yet and may not have a happy ending with regards to implications for educational foundations for a future in a globalized knowledge economy. It suggests that inquiry-based learning built upon site-based, empirically sound evidence and warrants may be two perspectives that can be incorporated into current reform initiatives to achieve some of the goals of this nation-state.

Singapore's stellar performances in the Third International Mathematics and Science Study (TIMMS) (Kelly et al., 2000) and the Third International Mathematics and Science Study—Repeat (TIMMS-R) (Mullis et al., 2000) seem to signal to both mathematics and science educators that the nation's education system is on an enviable upward incline. However, a careful study of these two test data show that Singapore pupils have only performed well on items that were routine to them, that were formulaic in nature, and that tested what they had been taught in school (Kaur, 2003; Gopinathan, 2001). The performance of Singapore pupils on items that were non-routine or non-formulaic is well below the expectations of most teachers.

What this shows is that while Singapore has strong didactic, exam-driven pedagogic traditions that provide good scores, the system lacks sufficient focus on higher order and contextual application skills. Two of Singapore's most erudite policy watchers say this has quickly become impediments for post-industrialism (Sharpe & Gopinathan, 2002).

It is because of this concern that Singapore has now embarked on a slew of new reform initiatives to keep up with the needs of the nation state where the impact of globalization, and shake ups in the economic, political and technological environments, has rendered past initiatives inadequate for the kind of labour force needed to feed into the new knowledge based networked economy.

According to Ramsden (1992), the main aim of teaching is simple—to make learning possible. But, what kind of learning is most needed for citizens of a small nation like Singapore, where labour is its only capital? Gee (2000, 2000a) talks about "shape-shifting portfolio people" who can take on trajectories of learning that are aimed at filling up attributes, skills and achievements that can be rearranged as situations change and when the contexts require that they redefine themselves and this is what Singapore needs—students who are able to continually adapt and learn (Dewey, 1916). So, despite Singapore's "success" story in education, some challenges facing the country are how to move along the continuum from having obedient, compliant, marks-oriented, risk-averse students toward independent, risk-takers, creative problem solvers and autonomous life long seekers of knowledge.

Directions for progress have been clarified significantly by recent advances in our understandings about collaborative and constructivist learning (Jonassen, 1999), cooperative learning (Johnson & Johnson, 2003) and problem-based learning (Dochy et al., 2003). Many of these advances originated from a notion of 'situated cognition' (Vygotsky, 1962). Vygotsky's proposal that knowledge has a fundamentally contextualised character is often used to express dissatisfaction with models of formal knowledge and academic inquiry where most teaching-learning situations focus on the acquisition of inert concepts—for example, algorithms, routines and decontextualized definitions—which are of little or no possible relevance in everyday life because students cannot transfer or apply their learning to other situations.

PROBLEM-BASED LEARNING

Problem-based learning has been viewed as a general approach to remedy many of these ills. Proponents of PBL emphasize that its motivation comes partly from dissatisfied employers, who have traditionally complained that most graduates from secondary schools have poor written and oral communication, undeveloped abilities to solve complex problems, and difficulties workingcollaboratively with workmates.

Common descriptions of PBL emphasize the 'active' and 'situated' nature of learning. PBL approaches also typically emphasize:

> Experience-based learning;
>
> The development of inductive reasoning skills;
>
> The simultaneous use of and challenge of prior learning;
>
> Opportunities for both independent and collaborative activity;
>
> Context-specific knowledge use; and
>
> Complex problems in which students encounter ambiguity and multiple perspectives.

Thus, key arguments in favour of PBL are that:

> It enables students to learn new information—using 'deep learning strategies' through the activation of their prior knowledge;
>
> It prepares the students for the workplace through opportunities to learn social and communication skills; and
>
> It enhances the students' intrinsic interest in the subject matter.

There are also, however, major and persistent complaints about PBL. Notable amongst these are:

> It has not proven to reliably deliver the 'deep learning' that it promises (Glew, 2003); and
>
> It sometimes encourages 'ritualistic behaviour' in students leading to the failure of inquiry and team learning (Dolmans et al., 2001).

The trends of globalization and their impact on the need for human capital have shaped the changing educational imperatives across the world. It is obvious that there are no ready made or quick fix solutions to this continuing debate about the purpose of education and the appropriate ways of preparing the "future" of a nation. While PBL has advantages, its exact nature remains unclear; and, it does not address all of Singapore's needs for training a post-robotic worker with the following minimum characteristics:

> Has a repertoire or portfolio of skills
>
> Is an independent worker
>
> Is a creative and critical thinker
>
> Is a risk taker
>
> Is a problem solver
>
> Is hungry to learn

Before we discuss a research agenda appropriate for Singapore's proposed initiatives, it is important to be aware of the background and characteristics of the educational context in which these initiatives are to occur.

EDUCATION IN SINGAPORE

Dewey (1916) observed that there is something extraordinary, perhaps even vaguely miraculous, about education: "By various agencies, unintentional and designed, a society transforms uninitiated and seemingly alien beings into robust trustees of its own resources and ideals." Like many countries in Asia, Singapore regards education as "an engine of change" (Maclean, 2004) and seeks ways to effectively reform the education system to foster economic, social and political improvement. Consequently, in Singapore, in 1965, the impetus was given to policy formulation, planning, and implementation of an educational framework that would sustain the most valuable resource in the new nation state—its workforce.

Singapore's recent educational agendas have the underlying belief that human capital is best developed through well-resourced and innovative educational innovation. Consequently, Singapore has always been ready to apply methodological and technological innovations to all tiers of its educational system. To its credit, it has had no qualms about implementing additional fundamental changes at a time when its school system has gained international recognition for its high standards—as reflected in standardized tests such as TIMMS (Sharpe & Gopinathan, 2002). These ambitious plans have the objective of ensuring that Singapore retains its competitive edge.

With the impact of globalization and the needs of the knowledge based economy, some of the new initiatives partly stem from increased pressure to produce creative and innovative citizens with a wide range of competences, skills and experiences. Recent initiatives include *Thinking Schools, Learning Nation (TSLN)* (Goh, 1997), the *Master Plan for Information Technology in Education* (Ministry of Education, 1997). The *Thinking Schools Learning Nation* initiative focuses on developing all students into active learners with critical thinking skills and on developing a creative and critical thinking culture within schools. Its key strategies include the explicit teaching of critical and creative thinking skills, the reduction of subject syllabus contents, the revision of assessment modes and the greater emphasis on processes instead of on outcomes when appraising schools.

Innovation and Enterprise (I&E) is another initiative that was launched at the 2003 *Annual Ministry of Education Workplan Seminar* 2003. It is intended to continue to build on the foundations laid down by TSLN. I&E emphasizing the spirit of questioning, risk taking, and tenacity that the students need in order to respond to a fast-changing global landscape. This is facilitated by enhancing the flexibilities and choice in the education system and encouraging teachers and stakeholders to reflect and rethink the way things are done to better understand and realize the objectives of ability-driven education.

Similarly, the Ministry of Education Work Plan Seminar 2004 added another important dimension to the preceding ability-driven education drive. It announced the *Teach Less Learn More (TLLM) initiative* where schools are encouraged to cut back on the syllabus and develop local curriculum and pedagogical strategies; and teachers are encouraged to act as facilitators rather than authoritative dispenser of knowledge, with the Minister of Education praising and highlighting the work of teachers who have gone beyond the tried and tested ways of teaching to undertake teaching that involves process learning and situated hands-on work (see full speech at http://www.moe.gov.sg/speeches/2004/sp20040929.htm). More teachers also will be allocated to schools, and, autonomy will be given to schools as to how these teachers will be deployed. Furthermore, this package of initiative includes a component aimed at training teacher assistants to help with administration work in the classroom so as to free teachers to teach. Nonetheless, the mandate for *TLLM* is reverberating both optimism and worry among school leadership, teacher-training academics, and the teaching fraternity.

These latest initiatives open up more spaces and possibilities for the teacher to do less teacher-fronted content driven work and concentrate on facilitation work that will ensure more involved and problem-based learning. The question is: *How can Singapore schools, used for many years to a directed, top-down, tightly coupled educational system, now have the confidence and initiative to move into a more independent, loosely coupled educational system?*

Since the late 1970s, a series of significant changes have been undertaken that have moved Singapore's educational landscape toward a close linking to industrial needs and a clearer articulation of the need and rationale for a national system of education (Tan et al., 2002). The period from 1985 to 1990 however was also a time when the emphasis on excellence in education saw a move toward freeing schools from central control. The result was the establishment of a more variegated system of schools. For example, in 1988, the first independent schools were established that allowed high performing schools to hire staff directly, devise school curricula and choose textbooks while conforming to national education policies such as bilingualism and national examinations.

Similarly, in 1994, autonomous schools were established with lower echelon schools that are deemed to have "a good system in place to achieve the Desired Outcomes of Education. These initiatives have achieved consistently good academic and other results; and, they also have established parental support and public recognition" (Ministry of Education Web site, http://www.moe.gov.sg).

Since 1994, there also have been other strong initiatives to establish niche schools such as the *Singapore Sports School* (2004) and the *NUS High School for Mathematics and Science* (2005). Enrolment to both schools are done at Secondary one and three—students may join the schools direct after their six years of primary education and be at the school for the whole four years of their secondary education or they could apply for a place after their initial two years in the lower secondary school to enrol for their upper secondary education at these schools. The Ministry of Information, Communications and the Arts also intends to set up a *Specialized School for the Arts* for students aged between 13 and 18 years. The school will offer a six-year academic and arts program targeted to start in 2007.

Other new initiatives include a 6-year *Integrated Program (IP)* for students from Secondary one to their second year in junior college beginning 2004 IP students will take 'A' level examinations at the end of the junior college but will not have to do their "O" level, as the schools will conduct their own examinations. Currently, the IP is offered in four high-stream schools—the Raffles and Hwa Chong families of schools, National Junior College and Anglo-Chinese School (Independent). Temasek Junior College and Victoria Junior College will also offer the IP starting January 1, 2005.

A student in Singapore typically undergoes ten years of general education. This comprises six years of primary education and four years of secondary education. After sitting for the Primary School Leaving Examination at the end of Primary six, students will undergo four years of secondary education under the Special or Express course or five years of secondary education under the Normal course.

Upon completion of their GCE 'O' level examination at the end of secondary education, students will be able to proceed to a junior college for a two-year preuniversity course, or to a centralised institute for a three-year pre-university course. Or, students can subsequently pursue tertiary education in a local university.

Now, students who are talented in sports, mathematics and science can further develop their talents with the customised curriculum of specialised independent schools. These include the schools mentioned earlier: the *Singapore Sports School* and the *National University of Singapore (NUS) High School for Mathematics and Science* and the upcoming *Arts School*.

Freeing of the schools from central control and establishing of a more varied system, however, does not necessarily point to more enlightened pedagogy. Initial quantitative work done by the newly created *Centre for Research in Pedagogy and Practice* at the National Institute of Education and funded by the Ministry is designed to help with this difficulty. Tentative data show that, across the school system, pedagogy remains bogged down by what is deemed in the Singapore jargon as "worksheet culture"—an overdependence on assessment of the outcome of learning rather than focusing on the process of learning on a daily basis. Also, an overwhelming amount of time is spent on teacher-centered monologues and heavily guided questioning (see, e.g., Liu et al., 2004, and Mardiana & Abdul Rahim, 2005).

CHALLENGES THAT WAIT

Due to past education reform initiatives, curricula and school development have changed dramatically to accommodate the dissolution of ideas of the preceding years and to rise to the challenge of new frames of reference. Despite these noble and brave changes, however, what is of concern to this discussion is that the outcomes of the educational initiatives as originally perceived in the agendas are not what is done and felt in the classrooms. Challenges faced at the local school sites include variegated pick up rate and differentiated buy-in by schools.

What needs to be done is to develop a research agenda that will accompany and support Singapore's curriculum reform initiatives, and that will facilitate well-organized, site-specific, and user-friendly accumulate of expertise.

When redesigning pedagogies for Singapore's future, one key factor is having the confidence of relevant stake holders during the initial wait-out periods of the implementation. For any real change in the system, a period of disquiet and chaos is inevitable especially when the change is not just a cosmetic change (Tan, 2002; Mortimore et al., 2000; Tan & Tan, 2001). This unsettling time is something that must be appreciated by policy makers, teachers, researchers, and students -because it is a necessary and essential part of the coping and learning trajectories of these groups.

Past curriculum reform initiatives in Singapore also show that the following realities need to be kept in mind when implementing curriculum change: firstly that innovations have been too many or too rapid and secondly that performances at high stakes examination remain an important criteria for student selection and placement and school grants from the Ministry are decided by performance. In addition, a reward system for the teaching fraternity is predicated largely on performances of students while teacher needs and beliefs have not been fully addressed. Singapore also has only one teacher training institution and teacher training perspectives have thus far been slow to respond to the teachers' needs in the classroom (see Mardiana & Lim, 2005).

The discourse of the state is that of effort—everyone may succeed through the educational system if they only put in enough "effort". This has spawned a whole culture of parents and educators who invest their expectations in outcomes and not process. The people see the education structure as largely sound and rightfully competitive. Anyone who tries their best can make it and that the education system is an excellent structure—the elitist structure of the system is obfuscated by this macro discourse.

There have also been few questions raised about the transformative role of schools—it has just been assumed that the job is done. We argue that this outcome-driven situation continues to be perpetuated by the macro discourse which also negates the initiatives toward process-based learning.

While Singapore strives to focus on problem solving and creativity, the fact is nobody really knows what it means to "understand" these things; and on a large scale, nobody has ever succeeded in a curriculum reform movement that claims to focus on them. For example, in the United States, approximately every ten years the pendulum of curriculum reform swings back and forth between emphasizing "basic skills" and "problem solving." There has been much effort but the fact of the matter is that despite all the effort not much is known about the process; the how and the what, of the whole educational teaching and learning space.

RECOMMENDATIONS FOR A RESEARCH AGENDA

As educational researchers we would like to see various issues of the discourses surrounding education and the processes of the assumed transformative nature of education problematized. The authors of this chapter propose a research methodology now known as multi-tier design experiment research as one way to resolve this educational debate. Details about scientific principles governing *multi-tier design experiments* have been described in the *Handbook of Research Design in Mathematics, Science, and Technology Education* (Kelly & Lesh, 2000) and in a chapter in the *International Handbook of Research in Mathematics Education* (2002)—as well as in a special issue of the *Educational Researcher* (Kelly, 2003). Such research designs

> are based on approaches that are well established in design sciences, such as engineering, where: (i) the goals of projects typically involve the development of complex artifacts, but (ii) the underlying design is one of the most important components of the product that is produced. Thus, artifact development and knowledge development proceed in parallel and interactively; and, sequences of rigorous testing and revising cycles are emphasized throughout design processes. (Kaput, Lesh, & Hegedus, this volume)

Design researchers begin with the assumption that we (policy makers, teachers, etc.) need to grow in our understanding of what it really means to become more creative—and a better problem solver. It involves news ways of thinking about the nature of student development, of knowledge and ability and new ways about effective teaching, learning and problem solving. The goal is to investigate the ways of thinking of these groups of people by designing thought-revealing artifacts through a series of iterative cycles. By being actively involved in the process, the thinking becomes clearer and valuable trails exposing the process, that has been documented, become available for use and reuse as the development trail is documented and available for reflection.

Related to this are two things that have also been noticeably missing from past curriculum reform efforts aimed at "problem solving"(a) assessments to document deeper and higher-order achievements that students/teachers/schools are expected to emphasize, and (b) teacher development programs that will allow teachers to develop at the same time that their students are developing.

In design research, everybody is assumed to be in the model-development business. All parties are involved in descriptions and explanations about constructing tools and making models to make sense of complex systems. Students develop models (and accompanying conceptual tools) for making sense of mathematical problem solving situations. Teachers develop models (and accompanying conceptual tools) for making sense of students' modelling activities. And, researchers and policy makers develop models (and accompanying conceptual tools) for making sense of teachers' and students' model development activities. All of the preceding model developers develop new knowledge and abilities by expressing their current ways of thinking (e.g., about what it means to be a "creative" problem solver) in the form of artifacts and conceptual tools that need to be tested and revised multiple times so that there is existence of proof in the form of prototypes for use in schools.

If the preceding activities are designed wisely, then the express→ test→ revise cycles automatically generate auditable trails of documentation that provides hard-nosed data about what works and why. Tasks which serve this function should, first of all, involve the development of artifacts that embody important aspects of the relevant problem solvers' current ways of thinking. That is, they should be thought-revealing activities.

Second, the artifacts should also be sharable and re-useable. Why? Because conceptual systems are available to a given community of practice (students, teachers, policy makers or researchers) and can be thought of as an ecological system consisting of a teeming community of ways of thinking. So, if we want this ecological system to evolve, we need to ensure that basic Darwinian principles are satisfied. This would mean that the following elements need to be considered: Diversity. Selection. Propagation (i.e., sharability so that productive ideas spread throughout the community). Accumulation (i.e., Re-usability so that productive ideas are preserved).

To encourage diversity, what is a death knell is to pressure teachers to conform to a single way of doing things. To encourage selection, participants need timely feedback that so that they are encouraged to develop in productive directions. Students should not wait for end-of-year tests to give them feedback about how they are doing. Needless to say, what's being advocated here is exactly the opposite of what is typically done in the USA and Singapore. Not providing timely continuous feedback thorough the learning journey may be one reason why we have failed to achieve the educational objectives we set out to do.

In design experiment research, teachers express their current ways of thinking in the form of tasks developed for instruction or assessment. There are several different ways that teachers can test the ways of thinking that underlie these artifacts: (a) Trial by fire. That is, does the tool do what it was intended to do (i.e., get students to express→ test→ revise their ways of thinking in forms that produce self-documenting trails about what is being learned). (b) Trial by peer. That is, if the tools are intended to be sharable, then fellow teachers can provide valuable feedback about whether the tools that are being developed are useful. (c) Trial by authority. That is, "experts" can provide feedback about ideas that have worked at other sites.

The most effective kinds of activities (for students, teachers, or schools) are thought-revealing activities that provide self-fulfilling assessments of progress. That is, if we think of such activities as some form of progress tests, then they are not just neutral indicators of progress but carefully planned activities that encourage participants to develop in positive directions—while at the same time generating auditable documentation about the nature of developments that occur.

While the reform initiatives made by Singapore are praiseworthy, the decades of initiatives only say what is needed but they do not provide answers nor documented ways on how to find out what "creative, or problem solver, or independent thinker" really mean and how we go about finding out if this has been achieved. Singapore's top down educational initiatives seem too implementation simplistic- the assumption being that once an initiative is introduced, it will run the selected course regardless but we are dealing with humans with many permutations and we cannot assume anything.

What we have ended up with in Singapore is clockwork like student and teachers who have become experts in beating the system to achieve the measure of success that is valorised namely good grades at any cost. These tests measure the accumulation of the knowledge of facts and not the application of facts in real situations. This growing ingenuity at beating the system has brought short term gains that are visible but the more important long lasting intrinsic gains have been sidelined.

In addition, the Ministry of education, evaluation of reform initiatives, through the decades, are done through positioned measures of success that lack robustly documented trail or details on how the measurement was done, how successes were achieved, including the failures that were encountered.

In Singapore, top down policies initiate changes with expedience in mind or sometimes for popular—appeal effect. For example the TLLM was a speech made the day before and became a slogan that schools had to flesh out on their own starting almost the next day.

The reality is that this only ensures that changes that are seen may only be at the superficial level where schools make a song and dance of their responses without the substance and with very little buy-in at the level of the school. These top down policies do not take into account the site specific ecological complexities facing the interacting parties in the learning space: the researchers, teachers, and students.

The challenge to lasting reform is not the quantity or speed but thoroughness such that the end product is shareable and reusable with warrants from robust empirical evidence, through looking at the reform process as a learning journey of all concerned: policy makers, curriculum developers, teachers, researchers and students. The end objective is making a real difference in practice and in the process; it should or could also contribute to theory.

The multi-disciplinary, multi-tiered approach that we propose as part of our research agenda addresses many of these criteria. The purpose of design experimentation is to develop a class of theories (or proto-theories) about how learning occurs and how it can be supported and improved. This multi-tiered approach unpacks the learning trajectories of learners, teachers, researchers and the policy makers, thus providing as holistic a picture of the educational environment as possible to enable the reform agenda to take root and change to be brought about intrinsically. Theory of teaching and learning is therefore seen as co-constructed, multi-tiered, multi-levelled constantly discovering how variables interact in a given knowledge space. As stated earlier, such a process needs time and the confidence of policy makers and teachers to bear the unsettling period. This is because the account of what works and why is always under empirical contestation, and always modifiable. Design experiments are iterative and cyclic in their structure— theorisations are generated, refined or refuted, new conjectures are always developed and tested (Cobb, 2003; Kelly, in press).

Advocates of the design-based experimental approach argue that, in the end, it is the coherence, accessibility and strength of the ideas driving the innovations, and the convergence of research and development activities around those ideas, that can enhance productive application.

At present there are no empirically tested or documented evidence of what worked and what did not and why certain aspects of the curriculum have been retained while some have been left out. Some of the issues that we need to deal with before even the research agenda is set in motion is that:

1. Teachers need to learn to get into a different pedagogical mode—of non-fronted teaching and learn to lead from the front by example, become facilitators and mediators of learning instead of deliverers of a centrally prescribed curriculum with set texts that has seen very little input at local sites.
2. Students who have been in a learning environment that Gopinathan (2001) refers to as the bank deposit model must now learn to get into situated cognition mode of learning. The in-classroom processes of learning in Singapore has been called the black box of schooling (Tan et al., 1997) and it is an arena that is dominated by didactic pedagogy, reinforced by a common curriculum and emphasis on frequent assessment.

Our research agenda has the potential to put into place a system or process that will document these changes at the classroom level and track them both statistically and in micro details at the ethnographic level to provide reliable warrants for the next steps of what changes can take place, what went wrong, why and perhaps even why they should take place. We suggest that Singapore begins this process by first looking at the math and science curriculum at the primary school level. The first thing that needs to be done is to investigate and locate the key fundamental concepts that are crucial for understanding both these disciplines. Following the methodology of design experiment, unpack the underlying principles and concepts both for the teachers and the students. Some of the ways to prepare for the proposed processes of change are the preliminary steps are suggested below.

> Form a team of experts from these two fields to identify what key concepts are crucial at foundation level for these disciplines. This team, in addition to others, should also harness teachers currently teaching at these levels as part of the research team or invited to be part of the research process. The process of knowledge development is cyclical and interactive and teachers need to be an integral part of it. It is not like how a doctor–patient relationship can sometimes be—the patient asks the question and the authoritative and hopefully knowledgeable doctor provides the answers. All those whose beliefs and actions have impact on some way in the classroom and what goes on there, need to be represented in this team so a holistic picture can be constructed.
>
> Assessment and evaluation experts need to be called on for their expertise on alternative assessment that reflects the learning that has taken place and not merely the mastering of transfer of information.
>
> Teachers involved need to be trained in modelling so they know how to task activities and create artifacts that unpack students' cognitive and metacognitive maps.
>
> Teachers need to be trained on how to document their own change in the learning continuum in terms of creating the artifacts and changes in their belief systems.

Researchers need to know how students make their way in the new creative space given to them and to unpack their cognitive processes and articulate their own learning trajectories in a conducive non-threatening environment through the tools provided by the teachers.

The assessment modes that are needed in this agenda need to be researched and go through the same iterative cycles before they become a part of a pool of accepted modes of assessment that various teachers can draw from and more can be added to the pool as teachers stumble upon and discover new ways of measuring the students education journey.

With documented trails of the journey, and the rounds of revisions and retesting undertaken, there will be robust empirical evidence of what works in this cultural context. Policy makers can then make changes in the curriculum content and pedagogy according to what the documented trails of this journey reveal. The core materials of the research findings then become the reusable, shareable basics or foundations for other teachers to use but not be bound to so that there is some creative flexibility on the part of the teachers. If changes made are documented at each process of research until minimum change need to be made the tools task scaffolding becomes more stable and therefore ready for diffusion and scalability. It is common knowledge from literature that educational innovations work some of the time with some of the students and in some of the places but thus far very few studies have taken the time or done research to illustrate what works, when, why and with whom and what the short comings and why and what went wrong. Important aspects of implementation include cumulative knowledge that will inform and clarify all matters related to the innovation and implementation.

The changes made at this early stage of a child's formal educational journey then become true foundations for the future. While the suggestions made have been at the macro level, the principles underlying these ideas are crucial and necessary before the next stage of operationalizing the design and action plan.

REFERENCES

Bodilly, S. J., Keltner, B., Purnell, Reichardt, R., & Schuyler, G. (1998). *Lessons from new American schools' scale-up phase* (No. MR–942.0.NAS). Santa Monica, CA: The Rand Corp.

Clarke, D. J. (2003). International comparative studies in mathematics education. In A. J. Bishop, M. A. Clements, C. Keitel, J. Kilpatrick, and F. K. S. Leung (Eds.), *Second international handbook of mathematics education* (pp. 145–186). Dordrecht, the Netherlands: Kluwer Academic.

Clarke, D. J., Emanuelsson, J., Jablonka, E., & Mok, I. A. C. (Eds.). (2006). *Making connections: Comparing mathematics classrooms around the world*. Rotterdam: Sense Publishers.

Clarke, D. J., Keitel, C., & Shimizu, Y. (Eds.). (2006). *Mathematics classrooms in twelve countries: The insider's perspective*. Rotterdam: Sense Publishers.

Cobb, P., Confrey, J., de Sessa., A., Lehrer, R., & Schauble, L. (2003). Design experiments in educational research. *Educational Researcher, 31*(1), 9–13.

Conley, D. T. (1993). *Road map to restructuring: Policies, practices and the emerging visions of schooling* (No. ERIC Clearinghouse on Educational Managment). Eugene, OR.

Cooper, R. (1998). *Scio-cultural and within-school factors that affect the quality of implementation of school-wide programs* (No. ERIC Document Reproduction Service No. ED426173). Baltimore: Center for Research on the Education of Students Placed at Risk.

Corbett, H. D. (1990). *On the meaning of restructuring*. Philadelphia: Research for Better Schools.

Corbett, H. D., Dawson, J. A., Firestone, W. A. (1984). *School context and school change: Implications for effective planning*. New York: Teachers College Press.

CRPP, 2004 Initial Report. Unpublished internal report.

Darling Hammond, L. (1997). *The right to learn: A blueprint for creating schools that work*. San Francisco: Jossey Bass.

Desimone, L. (2002). How can comprehensive school reform models be successfully implemented. *Review of Educational Research, 2004*(3), 433.

Dewey, J. (1916). *Democracy and education*. New York: Simon & Schuster.

Dochy, T., Mien, M., Van den Bossche, P., Gijbels, D. (2003). Effects of problem-based learning: a metaanalysis. *Learning and Instruction, 13*, 533–568.

Dolmans, D., Ineke, H. J. M., Wolfgagen, C., Van der Vleuten, P. M. & Wijnen, H. F. (2001). Solving problems with group work in problem based learning: Hold on to the philosophy. *Medical Education, 35*, 884–889.

Fielding, M. (2000). Students as radical agents of change. *Journal of Educational Change, 2*, 123–141.

Fullan, M. G. (2001). *The new meaning of educational change (3rd edition)*. New York: Teacher's College Press.

Fullan, M. G., & Miles, M. B. (1992). Getting reform right: What works and what doesn't. *Phi Delta Kappan, 73*(10), 744–752.

Gee, J. (2000). Teenagers in New Times: A new literacies studies perspective. In J. Elkins & A. Luke (Eds.), *Re/Mediating Adolescents Literacy*. International Reading Association.

Gee, J. (2000a). The new literacy studies: From socially situated to the work of the social. In D. Barton, M. Hamilton, & R. Ivanic (Eds.), *Situated literacies: Reading and writing in context*. Simon and Routledge.

Glew, R. H. (2003). The problem with problem based learning: Promises not kept. *Biochemistry and Molecular Biology Education, 31*(1), 52–56.

Goh, C. T. (1997). Shaping our furture: "thinking schools" and a "learning nation." *Speeches* (pp. 12–20). Singapore: Ministry of Information and the Arts.

Gopinathan, S. (2001). Globalisation, the state and education policy in singapore. In J. Tan, S. Gopinathan, & W. K. Ho (Eds.), *Challenges facing the Singapore education system today*. Singapore: Prentice Hall.

Jackson, A. a. D. G. (2000). *Turning points: Educating adolescents in the 21st century*. New York: Teachers College Press.

Johnson, D., & Gunther, A. K. (2003). Globalisation, literacy and society: Redesigning pedagogy and assessment. *Assessment in Education, 10*(1), 2–11.

Jonassen, D. H. (1999). Designing constructivist learning environments. In C. M. Reigeluth (Ed.), *Instructional design theories and models, 2nd edition*. Mahwah, NJ: Lawrence Erlbaum Associates.

Jonassen, D. H. (2000). Toward a design theory of problem solving. *Education Technology Research and Development, 48*(4), 63–83.

Kaur, B. (2003). *An international comparative study on the teaching and learning of mathematics in primary schools*. University of Exeter Press, Exeter, UK.

Kelly, A. (2003). Research as design. *Educational Researcher, 31*(1), 3–4.

Kelly, A., & Lesh, R. (Eds.). (in press). *Handbook of design research in mathematics and science education*. Mahwah, NJ: Lawrence Erlbaum Associates.

Kelly, D. L., Mullis, I. V. S., and Martin, M. O. (2000). *Profiles of student achievement in mathematics at the TIMSS international benchmarks: U.S. performance and standards in an international context*. Chestnut Hill, MA: Boston College.

Lee, V. (2001). *Restructuring high schools for equity and excellence: What works?* New York: Teachers College Press.

Levin, B. (2000). Putting students at the centre in education reform. *Journal of Educational Change, 1*, 155–172.

Lim, T. M., & Mardiana, A. B. (2005) *Pedagogy & narratives of practice: Teaching normal technical students*. Paper presented at the Redesigning Pedagogy: Research, Policy, Practice Conference. June 2005.

Maclean, R. (2004). Educational change in Asia: An overview. *Journal of Educational Change, 2*, 189–192.

Mortimore, P., Gopinathan, S., Leo, E., Myers, K., Sharpe, L., Stoll, L., et al. (2000). *The culture of change: Case studies of improving schools in Singapore and London* (pp. 286–345). London: Institute of Education, University of London.

Newmann, F. E. (2001). School instructional program coherence: Benefits and challenges: Consortium on Chicago School Research.

Ramsdan, P. (1992). *Learning to teach in higher education*. London: Routledge.

Riley, K. (2000). Leadership, Learning and systemic reform. *Journal of Educational Change (1)*, 29–55.

Senge, P. (1990). *The fifth discipline: The art and practice of the learning organisation*. New York: Doubleday Currency.

Sharpe, T., & Gopinathan, M. (2002). After effectiveness: new directions in the Singapore school system? *Journal of Education Policy*, 2002, vol. 17, no. 2, pp. 151–16.

Simms, A., & Wilber, D. M. (1999). Innovative approaches to maximazing resources. In *Noteworthy perspectives on comprehensive schol reform* (pp. 49–55). Aurora: Mid-continent Regional Educational laborarory.

Slavin, R. E., & Fashola, O. S. (1998). *Show me the evidence: Proven and promising programs for America's schools*. Chicago: Corwin Press.

Tan, J. (2002). Education in the early 21st century: Challenges and dilemmas. In D. d. Cunha (Ed.), *Singapore in the new millennium* (pp. 154–186). Singapore: Institute of Southeast Asian Studies.

Tan, S., & Tan, H. (2001). Managing change within the physical education curriculum: Issues, opportunities and challenges. In J. Tan, S. Gopinathan, & W. K. Ho (Eds.), *Challenges facing the Singapore education system today* (pp. 116–189). Singapore: Prentice Hall.

Vygotsky. L. (1962). *Thought and language*. Cambridge, MA: MIT Press.

Wallace, M. (2004). Orchestrating complex educational change: Local reorganisation of schools in England. *Journal of Educational Change, 5*, 57–58.

Wiske, M. S. (1998). *Teaching for understanding: Linking research with practice*. Thousand Oaks, CA: Corwin Press.

CHAPTER 17

John Dewey Revisited—Making Mathematics Practical VERSUS Making Practice Mathematical

Richard Lesh and Caroline Yoon
Indiana University

Judi Zawojewski
Illinois Institute of Technology

> Why do students who get A's in school, and who score well on traditional standardized tests, often perform poorly in more complex "real life" situations where mathematical thinking is needed? Why do students who have poor records of performance in school often perform exceptionally well beyond school?

John Dewey emphasized that *making science practical* is significantly different than *making practice scientific*. One reason this is true is because "real life" problems often involve both too much and not enough information—as well as too little time, too few resources, and conflicting goals (such as low costs versus high quality, or getting things done fast versus getting them done right). Therefore, in fields such as engineering or other design sciences where solving real problems is just as important as trying to advance specific theories, experienced professionals generally consider it to be "common sense" that solutions to realistically complex problems usually must integrate ways of thinking drawn from more than a single textbook topic area or discipline. This observation about the "connected" nature of useful ways of thinking suggests one of the many ways that the knowledge and abilities needed beyond school often are significantly different than those emphasized in school. Others are described in this chapter.

Similar conclusions are emerging from current research that is based on *models & modeling perspectives* of mathematics problem solving, learning and teaching (Lesh & Doerr, 2003; Deifes-dux & Zawojewski, in press). For example, if students are engaged in problem solving activities that were designed explicitly to be authentic simulations of "real life" situations where important types of mathematical thinking should be useful, then the knowledge and the abilities that develop tend to be significantly different depending on whether the learning activities focus on:

- *making mathematics practical*—First, guiding students along (necessarily narrow) conceptual trajectories toward a textbook's (or teacher's) cleaned-up

version of the meaning of the relevant concepts or abilities. Second, applying what was taught in "realistic" situations.
- *making practice mathematical*—First, putting students in simulations of real life sense-making situations where they express>test>revise their own relevant ways of thinking. Second, analyzing, decontextualizing, systematizing, and formalizing student-generated conceptual tools to endow them with more elegance, power, sharablity, and reuseablity.

Of course, when we make such comparisons, these two teaching methodologies are being described as if they existed in "pure" forms which probably never occur in reality. In "real life" teaching and learning situations, sensible teachers tend to use strategic mixes of these and other approaches. In fact, the importance of mixed strategies and multi-disciplinary ways of thinking is a central point that was emphasized by Dewey and other *American Pragmatists* (William James, Charles Sanders Peirce, Oliver Wendel Holmes, George Herbert Mead). . . . *Pragmatists* rejected the notion that single "grand theories" should be expected to provide solutions to most "real life" problems—including those that arise for teachers or researchers who are trying to develop more useful ways of thinking about mathematics teaching, learning, and problem solving.

When we speak of comparisons between these two unrealistically "pure" teaching methodologies, our goal is not to advocate this or that strategy for teaching. Instead, our goal is to use such comparisons to help clarify the nature of concepts and abilities that can be (and perhaps should be) learned. Then, teachers, curriculum developers, and other practitioners can decide for themselves what mix of instructional strategies might be most useful for cultivating these understandings and abilities. . . . In fact, both of the instructional "treatments" that are compared in this chapter involved mixed strategies. Both were assembled using the same two components: (a) a *model-eliciting activity*—in which students express, test, and revise their own ways of thinking, and (b) a *model-exploration activity*—in which students are guided toward their textbooks' or teachers' ways of thinking about the relevant concepts and abilities. Differences between these two treatments result from the **sequencing** of components. As the table on page 315 shows, the treatments that focuses on *making mathematics practical* consist of a *model-exploration activity* followed by a *model-eliciting activity*; whereas, the treatment that focuses on *making practice mathematical* consist of a *model-eliciting activity* followed by a *model-exploration activity*.

WHAT IS THE NATURE OF "REAL LIFE" SITUATIONS WHERE MATHEMATICAL THINKING IS USEFUL?

In future-oriented fields that are heavy users of mathematics and science, powerful technologies for computation, collaboration, and communication are producing fundamental changes in the levels and types of mathematical understandings and abilities that are needed for success beyond school. Therefore, in fields that range from bioengineering to business administration, expert job interviewers consistently claim that future employees who are most sought-after should: (a) have histories of being able to make sense of complex systems, (b) work well and communicate meaningfully within diverse teams of specialists, (c) be skillful at planning, monitoring, and assessing progress within complex multi-stage projects, and (d) adapt rapidly to continually evolving conceptual technologies. . . . Again, similar conclusions are

A Comparison of Two Ends of an Instructional Continuum

Make Mathematics Practical—Teach first. Then, apply what was taught.

According to this approach, learning to solve "real life" problems is assumed to be more difficult than solving their decontextualized counterparts in textbooks and tests—because realistically "messy" problems require students to know context-specific information in addition to knowledge about relevant concepts and processes.

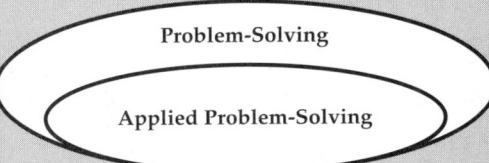

This teach-first-then-apply perspective considers "real life" problem-solving to be a special case of decontextualized forms of problem solving—where the messiness has been stripped away.

Make Practice Mathematical: Students express, test, and revise their own ways of thinking. Then, teachers help students "clean up" and empower their results.

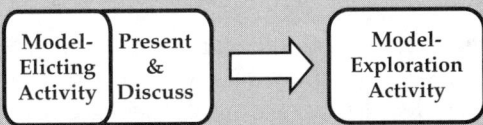

According to this approach, learning to solve meaningful (naturally occurring) "real life" problems is assumed to be easier than solving their decontextualized counterparts in textbooks or tests—because the latter requires students to make meaning out of symbolically described situations before sensible steps can be taken to generate solutions.

> Naturally Occurring Problems
>> Symbolically Described Surrogates for "Real Life" Problems

This *develop-first-then-harvest* perspective considers traditional conceptions of problem solving to be special cases of meaningful "real life" problems where meaningful details and purposes have not been stripped away.

Whereas, model development activities generally involve multiple cycles of the type shown below, traditional word problems tend to involve only getting from (pre-mathematized/computation-ready) givens to (mathematical) goals within a single cycle. Furthermore, even within this single cycle, students usually do not know who needs their mathematical result—or why. Therefore, because there seldom is any need to leave the world of mathematics, solution steps tend to emphasize derivation processes (computation or deduction). … Because such problems involve little mathematization, interpretation, or verification, they can be thought of as quarter-cycle (or half-cycle) modeling problems.

Solutions to Modeling Problems Involve Multiple Modeling Cycles

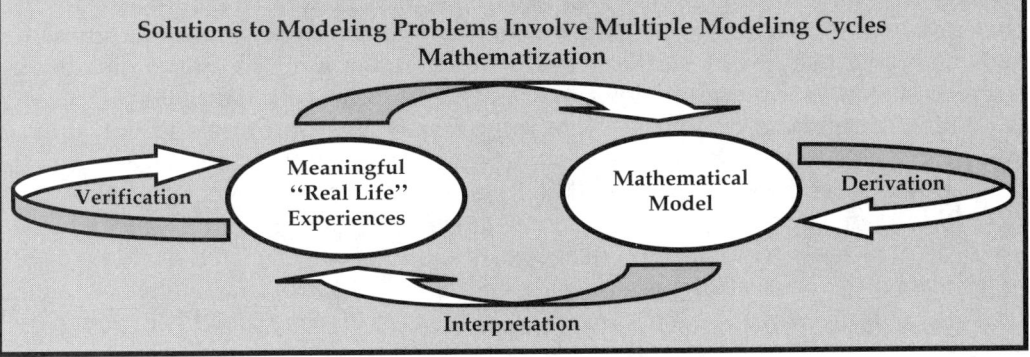

emerging when *models & modeling perspectives* are used to investigate students' thinking in problem solving situations that were designed explicitly to be authentic simulations of "real life" problem solving situations where some type of mathematical thinking is useful. For example:

- The mathematical products students produce generally involve more than easy-to-score answers to pre-mathematized questions. They often involve developing some complex artifact (conceptual tool, or conceptual model) which is designed for some specific decision maker and for some specific decision-making purpose—but which is seldom worthwhile to develop unless it is intended to be sharable (with others) and re-useable (beyond the immediate situation). Purposes of these tools range from optimization, to equalization, to ranking, to other types of decision-making.
- The "problem solver" often is not simply an isolated individual. Instead, the "problem solver" may consist of a team of diverse specialists who are using a variety of rapidly evolving technical tools, who represent a variety of different practical and theoretical perspectives, and who often use advanced technologies to communicate from remote sites.
- Because the solutions involve the development of complex artifacts (or conceptual tools), design processes usually involve a series of iterative development>testing>revising cycles in which a variety of different ways of thinking about givens, goals, and possible solution steps are iteratively expressed, tested, and revised—or rejected.
- Relevant ways of thinking seldom fall within the boundaries of a single, neat, and tidy disciplinary topic area. Instead, they usually draw on constructs and conceptual systems that come from a variety of disciplines and topic areas—and they usually are expressed using a variety of interacting representational media (each of which emphasizes and de-emphasizes somewhat different meanings of the underlying concepts and conceptual systems).

For people who are not familiar with the kind of mathematical thinking that is emphasized when groups of specialists use a variety of powerful technical tools to work on complex multi-stage projects, it might seem to be true that the need for mathematical thinking surely must decrease (or become easier) in such situations. *Don't tools and colleagues do some of the work for you? Don't complex or multi-stage projects enable you to focus on only a single component or stage of work that is needed? . . .* In reality, however, tools and colleagues tend to create as many conceptual challenges as they eliminate; and, complex multi-stage projects generally emphasize abilities that involve interpretation, description, explanation, communication, argumentation, and other higher-order capabilities. Whereas, computations and routine deductions tend to be carried out by low-level assistants (or technical tools).

This chapter will emphasize five important types of knowledge or abilities which are directly related to the preceding characteristics of problem solving in the 21st century. Each involves thinking that is based on some type of conceptual system: (a) *within-concept systems*, (b) *between-concept systems*, (c) interacting *representational systems*, (d) *basic skills whose meanings depend on systemic understandings*, and (e) *higher-order ideas or abilities whose meanings depend on systemic understandings*. . . . Each evolves naturally when students are engaged in *model-eliciting activities* in which they express>test>revise their own relevant ways of thinking. But, none are likely to be emphasized in activities that rely too heavily on *teaching first and then applying what was taught.*

WHAT ASSUMPTIONS ABOUT MATHEMATICS INFLUENCED THE CONCLUSIONS EMPHASIZED IN THIS CHAPTER?

Above all, the kinds of concepts and abilities that we emphasize in this chapter are influenced by the following assumptions about the nature of mathematics.

- *Thinking mathematically* is about expression and interpretation (description, explanation) at least as much as it is about computation and deduction. Unlike *physics*, which is the study of matter and energy, or *biology*, which is the study of living organisms, we believe that *mathematics* is the study of structure. In pure mathematical activities, these structures are developed and explored for their own sake; whereas, in applied mathematical activities, they are developed and used to create, or describe, or explain some other system. But, in the case of either pure or applied perspectives, we consider mathematics to be the study of "systemic" characteristics of the relevant structures.
- In mathematics and science, the conceptual systems that humans develop to make sense of their experiences generally are referred to as *models*; and, *mathematical modeling* includes quantifying, dimensionalizing, coordinatizing, systematizing, or (in general) mathematizing objects, relations, operations, patterns, and regularities which seldom occur in pre-mathematized forms in "real life" situations outside of school.

Because of these assumptions, when we investigate what kind of "mathematical thinking" goes on outside of school, we ask what kind of situations students can *describe* mathematically at least as much as we ask what kind of computations they can do in real settings. In other words, we consider the possibility that: (a) mathematics learners and problem solvers are model developers at least as much as they are information processors; (b) many of the most important types of situations where significant forms of mathematical thinking are useful beyond school might be *model-eliciting activities* where students express, test, and revise their current mathematical/conceptual systems; and (c) the development of powerful/sharable/re-useable models might be among the most important future-oriented goals of mathematics teaching and learning.

Another Significant Assumption About Students' Abilities To Develop Powerful Models

Limitations of instructional methodologies that *teach first and then apply* are not likely to seem significant to anybody who naively accepts the cliché that: *It took brilliant mathematicians hundreds of years to develop most of the powerful concepts in the elementary mathematics curriculum. So, it's not realistic to expect average ability children or adolescents to come up with such concepts in a few months or weeks—or during single problem-solving episodes.*

In contrast to the preceding point of view, our research that is based on *models & modeling perspectives of learning and problem solving* is filled with transcripts showing that average ability students routinely develop powerful, sharable, and re-useable constructs or conceptual systems (Lesh & Doerr, 2003). In fact, if the goal of instruction is to make significant changes in a student's underlying ways of thinking about important conceptual systems in mathematics, then perhaps the only way to induce significant conceptual change is to engage students in *model-eliciting activities* where they must express their current ways of thinking in forms that lead to testing and revision (or rejection). Therefore, we reject the notion that only a few exceptionally brilliant students are capable of developing significant mathematical concepts unless step-by-step guidance is provided by a teacher.

WHAT OTHER THEORETICAL FOUNDATIONS UNDERLIE *MODELS & MODELING PERSPECTIVES?*

The theoretical framework that we refer to as *models & modeling* traces its lineage to modern descendents of Piaget and Vygotsky - and also to *American Pragmatists* such as William James, Charles Sanders Peirce, Oliver Wendell Holmes, George Herbert Mead, and John Dewey. For example: (a) Dewey and Mead emphasized the fact that conceptual systems are human constructs and that such conceptual tools are fundamentally social in nature—partly because they seldom are worthwhile to develop unless they are intended to be sharable (with other people) and reuseable (in other situations beyond the one in which they were created). (b) Peirce emphasized that the meanings of these constructs tend to be distributed across a variety of representational media (ranging from spoken language, to written language, to diagrams and graphs, to concrete models, to experience-based metaphors)—each of which emphasizes and ignores somewhat different aspects of the constructs they are intended to express and/or different aspects of the "real life" experiences they are intended to describe. (c) Dewey emphasized that knowledge is organized around experience at least as much as it is organized around abstractions—and that the ways of thinking which are needed to make sense of realistically complex decision making situations nearly always must integrate ideas from more than a single discipline or textbook topic area. (d) James emphasized that the "worlds of experience" that humans need to understand and explain are not static. For example, humans are continually projecting their conceptual systems into the world—because the models (and underlying conceptual systems) that are developed to make sense of experiences also are used to mold and shape the "real life" situations in which these experiences occur. So, the world that needs to be understood is continually and rapidly changing; and, the understandings and abilities that are needed to make sense of such situations also are changing —and are shaped by human goals as well as by the existing state of things. (e) Dewey emphasized that, in a world that is filled with technological tools for expressing, communicating, storing, retrieving, and transforming ideas and procedures, it is naïve to suppose that all "thinking" goes on inside the minds of isolated individuals. For instance, at least since the age of written notation systems, mathematicians have been using conceptual tools to off-load formerly internal functions; and, outside of school mathematics classrooms, "problem solvers" often are diverse teams of specialists with different technical tools and areas of expertise.

WHAT CONCEPTIONS OF *MATHEMATICAL MODELS & MODELING* ARE EMPHASIZED IN THIS CHAPTER?

An entry-level conception of a *mathematical model* is that it is simply a (familiar) system which (for some obvious purpose) is being used to describe or explain some other (less familiar) system. For example, a single algebraic equation may be used to model some system of physical objects, forces, and motions. Or, a *Cartesian Coordinate System* may be referred to as a model of space—even though such a huge system seems to be more like a language for creating models rather than being a single model in itself. But, in any case, the main point is that a *mathematical model* has three parts: (a) a purpose (which Dewey called an *end-in-view*) which molds and shapes

Two Ways of Thinking About Problem Solving—or Modeling	
Traditional Perspectives	*Models and Modeling Perspectives*
Problem solving is defined to be *a process of getting from givens to goals when the steps are not immediately obvious.*	*Problematic Situations* are defined to be *goal directed activities in which adaptations need to be made in existing ways of thinking about givens, goals, and possible solution steps.*
Products that need to be produced tend to be thought of as being short answers to well defined questions about pre-mathematized situations.	*Products* that need to be produced often involve complex artifacts (e.g., conceptual tools) which are designed to be: a. powerful (for the specific purpose at hand), b. sharable (with others) and c. reusable (in other situations).
Conceptual tools are situated forms of knowledge. They are molded and shaped by the situations in which they are created or modified. Yet, they are not simply situation-specific knowledge that does not transfer. For example, if they are designed to be sharable and reuseable, then they represent generalizable achievements. Also, in realistically complex situations, useful ways of thinking usually must integrate concepts and conceptual systems that do not fit into a single textbook topic area—or even into a single discipline. One reason this is true is because it is usually the case that solutions to realistically complex problems not only need to be effective, but they also need to be cost-effective, durable, timely, politically acceptable, and so on. Therefore, when we say that knowledge is situated, part of what we mean is that the "chunks" of knowledge that develop tend to be organized around experience at least as much as around discipline-based abstractions.	
Problem solvers are thought of as being processors of information.	*Problem solvers* are thought of a being conceptualizers and creators of (complex) systems. Also, rather than being isolated individuals, "problem solvers" often are teams of diverse specialists with access to powerful tools for conceptualization, computation, and communication.
Relevant Knowledge is thought of as consisting of lists of condition-action rules. Note: Although cognitive scientists have largely abandoned the notion of using lists of LISP-based condition-action rules to simulate human thinking, rule-base conceptions of knowledge are alive and well in the behavioral objectives that underlie most standardized tests—and the skills that dominate "back to basics" curriculum standards documents.	*Relevant Knowledge* is considered to involve, above all, conceptual systems (or models) for describing or interpreting patterns, regularities, and other systemic properties of problematic situations. Because these conceptual systems are expressed in the form of tools, the tools sometimes are referred to as embodiments of the conceptual systems.
The development of underlying conceptual systems is not expected to be an all or nothing process. It is assumed that conceptual systems will develop along a variety of dimensions (e.g., concrete-abstract, simple-complex, unstable-stable, internal-external, situated-decontextualized, intuitive-formal). So, when they are needed to solve "real life" problems, most are expected to be at intermediate stages of development. Also, regardless whether the problem-solver is an isolated individual or a team of diverse specialists, their thinking usually involves a community of competing (yet perhaps undifferentiated) conceptual systems.	
Solutions Processes are thought of as linking together strategies (rules) along a linear path—or trajectory.	*Solution Processes* tend to involve iterative cycles in which existing ways of thinking are gradually expressed, tested, and revised (or rejected); and, development involves sorting out, integrating, refining or rejecting conceptual systems that you DO have—not trying to find missing ideas or abilities, and not trying to figure out what to do when you're stuck (i.e., when you're not aware of any relevant ways of thinking). So, progress seldom occurs along any single conceptual path. Instead, evolution tends to resemble genetic inheritance trees in which descendents inherit characteristics from a variety of ancestors.

Three Confusing Facts About Students' Models

1. Conceptual models that students develop to make sense of their experiences generally should not be thought of as being "toy worlds" that are internal copies of the external systems they are used to describe. Just as equations are not photograph-like copies of the systems they describe, every model has some properties that the system that it describes does not have; and, it also does not have some properties that the system that it describes does have. Yet, such models clearly must have some properties in common with the systems they are intended to explain—because the purpose is to provide powerful tools for describing, manipulating, and predicting the modeled systems.

2. Researchers' models of students' modeling behaviors are not the same thing as students' models. For example, suppose that a student is asked to develop a "conceptual tool" (or a "model") whose design "specs" demand that the underlying design principles must be included in the product that is produced. In such instances, acceptable products must explicitly reveal what kind of mathematical objects, relations, operations, and patterns are embodied in the design principles for the tool. Under these circumstances, the products that students produce are indeed the students' tools or models. They are not simply inferred by an observer based on indirect evidence.

3. When students develop models, whatever they explicitly reveal about their thinking is like the tip of an iceberg; a great deal cannot be seen; and, whatever is seen is likely to be in the process of change. For example:

- If the "problem solver" is a team of diverse individuals, then each individual's conception is likely to be somewhat different than the conceptions of other team members—as well as being different than the consensus that was developed by the team-as-a-whole.

- If the "problem solver" is an isolated individual, one of the purposes of externalizing current ways of thinking (e.g, using statements, or drawings, or experience-based metaphors) is to examine and modify these ways of thinking. So, the thinking of a person who draws a diagram cannot be assumed to be the same as the thinking of the same person who looks at the results a moment later.

A fundamental characteristic of conceptual models is that they are dynamic and continually adapting entities. When they are engaged, they change; and, when they function, they evolve. Like subatomic particles in quantum physics, causing conceptual models to appear in a form that is observable usually induces changes in them. So, researchers can never observe them as isolated entities; and, what can be observed tends to be only the residue from past trajectories of interactions and change.

When researchers develop models of students' models, there is a big difference between claims that are based on tasks in which (a) students were never asked to produce models and (b) students where given design specs which made it clear that students needed to explicitly express, test, and revise their own models, conceptual tools, or underlying ways of thinking.... In general, in this chapter, when we speak of students' models (or other conceptual tools), we are referring to research studies in which students were asked explicitly to produce models (or conceptual tools). In such research studies, the students are enlisted as co-researchers who interact with researchers while investigating their own ways of thinking. Therefore, what we report is what the students themselves reveal about the objects, relations, operations, and patterns that they intended to be embodied in the conceptual tools that they produce.

the kind of models that evolve by providing criteria for assessing the usefulness (power, sharability, re-useability) of alternative models, (b) an underlying conceptual system which includes mathematical objects+relations+operations+principles (or patterns which govern interactions among components of the systems), and (c) several interacting media in which interpretations, descriptions, and explanations are expressed.

> **Note:** Later, this initial conception of a model will need to be extended in a variety of ways. For example, because a model is used to interpret situations, and because interpretations involve more than simply engaging systems that are logical and mathematical in nature, interpretations also involve feelings, values, beliefs, and a variety of higher-order abilities and dispositions. However, according to *models & modeling perspectives,* the meanings of these latter aspects of knowledge and ability also involve systemic understandings; and, they are not simply learned separately and then added onto students' interpretations (or models). Instead, they are integral parts of the models (and underlying conceptual systems) that students' develop. In other words, the interpretations that students engage in any given situation determine which feelings, values, beliefs, and higher-order abilities and dispositions will function; and, vice versa. So, as interpretations change, feelings and other processes and dispositions also can be expected to change. . . . Finally, the meanings of feelings, values, beliefs, and higher-order abilities and dispositions continually develop in unison with the development of the conceptual systems in which they are embedded. Therefore, when conceptual systems are at primitive stages of development, initial stages in the development of associated feelings, values, and processes often are far more situation dependent and variable than traditional theories have suggested.

WHAT ARE *MODEL-ELICITING ACTIVITIES?*

As their name suggests, *model-eliciting activities* are activities which are designed to elicit powerful, sharable, and re-useable conceptual models. But, they also are designed to be simulations of meaningful "real life" problem solving situations. Furthermore, because they are intended to be used for research and assessment as much as for instruction, the goals (or ends-in-view) of *model-eliciting activities* are specified in such a way that the models (or other conceptual tools) that students produce are *thought-revealing*. That is, the products are not just short answers to pre-mathematized questions; they are complex artifacts (or conceptual tools) which reveal important aspects about the mathematical objects, relations, operations, and patterns that are embodied in students' underlying ways of thinking.

Several past publications have described and illustrated design principles for creating effective *model-eliciting activities* for students, teachers, or researchers (Lesh, et. al., 2001). For the purposes of this chapter, *model-eliciting activities* are significant because they are prototypes of learning activities that go beyond trying to *make mathematics practical* to also *make practice mathematical*. In particular, in well-designed *model-eliciting activities*, students are challenged to mathematize "real life" problem solving situations by going through modeling cycles in which they express, test, and revise their own current ways of thinking—rather than adopting and using teachers' or textbooks' ways of thinking.

The *Volleyball Problem,* given below, is an example of a *model-eliciting activity* that was designed to be a middle school version of a "case study" that we first saw being used in Northwestern University's *Kellogg School of Management*. The original problem was intended to be a simulation of a "real life" decision making situation that occurs quite frequently in management situations—as well as in many fields

The Volleyball Problem

Organizers of the volleyball camp need a way to divide the campers into fair teams. They have decided to get information from the girls' coaches—and to use information from try-out activities that will be given on the first day of the camp. The table below shows a sample of the kind of information that will be gathered from the try-out activities. Your task is to write a letter to the organizers where you: (1) describe a procedure for using information like the kind that is given below to divide more that 200 players into teams that will be fair, and (2) show how your procedures works by using it to divide these 18 girls into three fair teams.

Data From Volleyball Tryouts

Name	Height of Player	Vertical Leap in inches	40 Meter Dash in seconds	Number of Serves successfully completed out of 10	Spike Results (Out of 5 attempts) Note: D-R = Dink Returned, D-U = Dink Unreturned, O-B = Out of Bounds, I-N = In the Net				
Gena	6'1"	20	6.21	8	D-R	D-U	Kill	I-N	Returned
Beth	5'2"	25	5.98	7	Kill	Returned	O-B	D-R	Kill
Jill	5'10"	24	6.44	8	O-B	Returned	Returned	Kill	I-N
Amy	5'10"	27	6.01	9	Kill	Kill	D-U	Kill	Returned
Ana	5'6"	25	6.95	10	O-B	I-N	Returned	Returned	D-R
Kate	5'8"	17	7.12	6	Kill	D-U	Kill	Returned	Kill
Rhoda	5'3"	21	6.34	5	O-B	Kill	I-N	I-N	D-R
Christi	5'5"	23	7.34	8	I-N	Kill	Kill	Kill	D-U
Andrea	5'5"	24	6.32	9	I-N	O-B	I-N	O-B	Returned
Nikki	5'7"	19	8.18	10	D-U	Kill	Kill	O-B	Returned
Kim	5'9"	23	6.75	7	D-R	Kill	Returned	O-B	Kill
Robin	5'8"	15	5.87	8	Kill	Kill	Kill	D-U	I-N
Edna	5'4"	21	6.72	8	Kill	Returned	O-B	I-N	D-R
Lori	5'7"	19	6.88	9	O-B	I-N	I-N	Kill	Returned
Tina	5'1"	24	6.27	6	D-U	D-R	D-R	Kill	O-B
Angie	5'10"	23	6.54	8	O-B	Kill	O-B	O-B	D-R
Ruth	5'3"	26	7.01	9	D-U	I-N	Kill	Kill	Kill
Becca	5'9"	18	6.78	10	I-N	O-B	Kill	D-R	Kill

Volleyball Coach's Comments

Gena: Gertrude is tall but slow getting to the ball.
Beth: She is very agile on her feet.
Jill: Jill's height and jumping ability should prove to be an asset for any team.
Amy: She is an awesome leaper, but she needs to know when to use it.
Ana: She's not very aggressive partly because she's only played on unsuccessful teams.
Kate: Kate has great quickness to get to the ball after serves.
Rhoda: Rhonda has a tendency to loaf—playing best when the team is playing well.
Christi: She hides her abilities. Her family life has negatively impacted her ability to play well.
Andrea: She is exceptionally strong for her age.
Nikki: She does many things well. In particular she serves well.

Kim: Kim is a great blocker.
Robin: Robin is the hardest worker we've ever had at the high school. She makes everybody better she plays with.
Edna: Edna is a girl that others want to be with because whatever event she's in, she seems to always find a way to win.
Lori: Lori does not always get her serve over the net. But, when she does, it's a winner.
Tina: She is one of the most intense players we have ever seen. She works hard.
Angie: Her father coaches at a local school. So, she gets the most out of her native abilities.
Ruth: Her skills are good. She grew up watching her sister player at the Univ. of Alabama.
Becca: Rebecca is very coachable. She'll improve fast if she gets a chance.

where it is important to compare things or to rank things (e.g., products, people, places, or businesses) by aggregating a variety of different types of qualitative and quantitative information. For example, such aggregating and ranking occurs in consumer guidebooks that assess automobiles and other products; they occur in "places rated almanacs" that rank cities, states, or regions according to their "live-ability" or other factors; and, they also occur when teachers assign grades to students in their courses by merging information about performances on tests, quizzes, projects, and other assignments.

> **Note:** When middle school problem solvers are introduced to the *Volleyball Problem*, teachers usually begin with a homework assignment in which the students read (and respond to a few warm-up questions about) an article in a *math rich newspaper* that describes a summer sports camp specializing in girls' volleyball. This newspaper article provides background information about a situation in which, last summer, difficulties arose at a summer sports camp because it was difficult for the camp counselors to form fair teams that could remain together throughout two weeks of the camp.... Last summer, a random selection process was used to assign players to teams. But, using this process, some teams didn't get very many of the best players; or, other teams got four good spikers by no good servers. So, next summer, the camp councilors have decided to gather information from "try outs" that will be held on the first day of the camp. Then, data from these "try outs" will be used to form teams that will be as equivalent in ability as possible.... Because data for next summer are not yet available, the camp counselors have provided sample statistics for eighteen girls who attended the camp last summer. The goal of the problem solving session is to write a letter to the camp counselors describing a procedure that they can use next summer—and to show how to use this procedure using the sample statistics that are given from last summer.

How typical is this *Volleyball Problem* compared to other *model-eliciting activities*, or compared to other "real life" problems where mathematical thinking is needed beyond schools? ... On the one hand, there is perhaps no such thing as a typical *model-eliciting activity*; and, most *model-eliciting activities* do not involve tables of data or information that is already given in a pre-quantified form. So, in some ways, the *Volleyball Problem* is somewhat unusual. On the other hand, because modern newspapers are filled with such information—in sections that range from sports, to business, to local and national affairs—the task of making sense of tables of data is commonplace in a wide variety of "real life" situations. Furthermore, even though the *Volleyball Problem* presents an organized table full of pre-quantified information: (a) significant amounts of relevant information remain to be mathematized (e.g, some kinds of weights or values need to be assigned to different categories of information); (b) quantitative data that are given are not necessarily given in forms that are useful; and (c) many of the most important relationships that need to be considered are not apparent from the way the table is structured. Therefore, the *Volleyball Problem* involves a significant amount of structuring and mathematizing—even though some structure is imposed on the given information, and even though some of the relevant information is given in a form that is partly quantified.... Finally, one way to think about the goal of the problem is that it is a situation where students need to develop a useful "operational definition" that provides a way to deal quantitatively with the concepts of "fair teams" and "good volleyball players".... How common are such problems? Our experience suggests that finding a way to measure a quantity that cannot be seen or measured directly is one of the most common types of problems that occur in elementary science—as well as in everyday situations where mathematical thinking is needed (Lesh & Doerr, 2003).

How typical are the students' behaviors that will be described in this chapter? Over a period of more than twenty years, thousands of students from middle school children through adults have worked on this problem and other problems like it. Complete transcripts showing typical solutions to such problems have been given in a number of past publications (Lesh & Doerr, 2003); and, many also are given on the following web site: http://tcct.indiana.edu/foundations_for_the future/. Nonetheless, the most straightforward and persuasive response that we can give to this question is to point out that, at this moment, if readers of this chapter take a break from reading and instead work on the *Volleyball Problem*, then the chances are high that you too will find yourself using exactly the kind of concepts, skills, abilities, and processes that are emphasized in this chapter.

> **Note:** For readers who decide to work on the *Volleyball Problem*, or who decide to observe students working on this problem, it is recommended that: (a) problem solvers work in 3-person teams; (b) problem solvers use a calculator or a speadsheet whenever it is sensible to do so; (c) problem solvers should be allowed at least 60–90 minutes to finish the problem; and (d) problems solvers should be reminded that their goal is to produce the product that the problem statement asks them to produce. . . . *Who is the client? What do they need?* If problem solvers don't keep the client's needs in mind, then they'll discover that they have little basis for deciding among alternative ways of thinking—and no way of deciding "how good is good enough" for their final responses. . . . *Is a 5 minute response good enough? Is a 5 hour response needed?*

WHAT DIFFICULTIES AND INSIGHTS TYPICALLY EMERGE DURING SOLUTIONS TO THE *VOLLEYBALL PROBLEM?*

One factor that makes the *Volleyball Problem* interesting is that information from the "try outs" includes several different kinds of qualitative and quantitative information. Therefore, one "big idea" that can be developed around this activity involves aggregating several different kinds of information using linear equations of the form: $V = a*x + b*y + c*z + d*w$. An appropriate name for this "big idea" might be *weighted averages* or *weighted sums.*

- Teachers who emphasize *teach-first-then-apply methodologies* might begin a unit of instruction by using educational software such as *TinkerPlots* or *Fathom* to teach students about *weighted sums or averages.* Both *TinkerPlots* or *Fathom* are similar to graphing spreadsheets, such as Excel; but they also include easily transformable graphs and "sliders" which can be used to change the values of the "weights" in the preceding equation. Therefore, the results of using different values can be observed in dynamically changing graphs.
- Teachers who emphasize *developing-first-then-formalize methodologies* might begin a unit of instruction by having students express>test>revise their own relevant ways of thinking about the *Volleyball problem*. Then, after students present, discuss, analyze and assess similarities and differences in their work, the teacher may conduct a guided discovery discussion using *TinkerPlots* or *Fathom* in a manner similar to that described above for *teach-first-then-apply methodologies.* In this case, however, the discussion inevitably will be somewhat different because the students will be familiar with a meaningful purpose and context for the ideas that are introduced.

Recently, the preceding two methodologies or "treatments" were used with two equivalent classes of graduate students who were enrolled in an introductory course on *Statistics for Research in Education*. Thirty students were in each group; and, test scores from two midterm examinations indicated that the two groups were equivalent in ability. Results were not surprising. The *develop-first-then-formalize* group significantly outperformed the *teach-first-then-apply* group on an end of unit test. But, this was true mainly because the test not only emphasized the main "big ideas" of the unit, it also was designed to document the kind of understandings that evolved in *develop-first-then-formalize* group—but which did not occur in the *teach-first-then-apply* group. For example:

- When the *develop-first-then-formalize* group worked on the *Volleyball Problem*, each team of problem solvers tried out a variety of different way of thinking about the situation; each went through at least three or four modeling cycles that involved significantly different ways of thinking about the situation; and, each ended up emphasizing somewhat different ways of thinking. Therefore, when these groups presented and discussed their work, a wide variety of concepts, abilities, and skills came to the attention of all students in the class. So, students sometimes modified their own ways of thinking based on what they heard from others; or, when they saw alternatives, they often came to better understandings of their own own ways of thinking. . . . In general, differences focused on description rather than computation; and, relevant abilities also emphasized planning, monitoring, and assessing alternative ways of thinking.
- When the *teach-first-then-apply* group worked on the *Volleyball Problem*, each team assumed that they were expected to solve the problem using the software (*TinkerPlots* or *Fathom*) and equations of the form: $V = a*x + b*y + c*z + d*w$. So, little diversity in thinking occurred across these groups; and, the subproblems they encountered were very different than those that emerged for students in the *develop-first-then-formalize* group. Also, because the *teach-first-then-apply* group generally did not consider alternatives or go though multiple modeling cycles, they had no occasions to develop relevant planning, monitoring, and assessing abilities. Consequently, because of this lack of self-assessment, their final responses tended to be far less responsive to the needs of the client who was specified in the statement of the *Volleyball Problem*. Consequently, even though the results that these groups produced sometimes appeared to be more mathematically sophisticated than those that were developed by students in the *develop-first-then-formalize* group, clients generally would assess the work produced by the *develop-first-then-formalize* groups to be more useful (or "better").
- Another closely related result became apparent when questionnaires were sent to students who had been in similar *develop-first-then-formalize* groups during the preceding academic year. First, enthusiasm about such activities was reflected in the fact that more than 50% of these students responded to the questionnaire. Second, nearly all of these students were able to describe extraordinary details about their work on the *Volleyball Problem*. Third, instructors reported that, during other activities in their course, these students often referred to the *Volleyball Problem* (or similar *model-eliciting activities*) when they were trying to make sense of new learning or problem solving situations—and that these metaphorical references often provided powerful conceptual

frameworks (models) for making sense of new learning or problem solving situations. Therefore, when new ideas and abilities were developed, they often were integrated into the students' "volleyball thinking" model—more than they were organized around abstract concepts such as *weighted averages* or *operational definitions*.

The most interesting result from the preceding kinds of studies is not the fact that it is easy to produce quantitatively significant differences in test scores or responses to questionnaires. This is a trivial accomplishment once teachers and researchers clearly understand what kind of differences are likely to occur—and why. Therefore, the remainder of this section shifts attention toward describing distinctive characterististics of kind of concepts, abilities, and skills that tend to emerge when students go through multiple modeling cycles to develop solutions to the *Volleyball Problem*.

When *develop-first-then-formalize* methodologies are used, students' first-cycle interpretations of the *Volleyball Problem* tend to involve one of the following two general ways of thinking.

- *Rank-first-then-combine*: Problem solvers rank volleyball players within individual categories—such as jumping, running, serving, or spiking. Then, after these individual rankings have been completed, the problem solvers try to find a way to combine these rankings into something that could be thought of as a single *index of volleyball playing ability*.
- *Combine-first-then-rank*: Problem solvers try to produce a single "score" (which again could be thought of as some sort of *index of volleyball playing ability*) which combines different kinds of information that are given for each volleyball player. Then, these aggregated "scores" are used to rank volleyball players.

The *rank-first-then-combine approach* tends to focus on differences among volleyball players within each category of information (jumping, running, serving, spiking); whereas, the *combine-first-then-rank approach* tends to focus on different kinds of information that are given for each volleyball player. But, using either of these two general approaches, problem solvers' first-cycle interpretations of the *Volleyball Problem* tend to focus on only a small subset of the given information—and on whatever information happens to attract their attention. This information could be "serving ability" (while ignoring nearly everything else), or it could involve other categories of information such as running and jumping, or serving and spiking. In any case, first-cycle interpretations tend to be remarkably barren and distorted compared to later interpretations. They often focus on only a small subset of available information; comparisons among different types of information tend to be ignored; important distinctions are not noticed; and, some inappropriate or overly simplistic assumptions often are made. . . . Examples follow.

> **Note:** When we give the following descriptions of students thinking, it is perhaps important to emphasize that the thinking of both individuals and groups is by no means as monolithic as such simplified descriptions of thinking seem to suggest. That is, three individuals in a group may be thinking about the situation in quite different ways; and,

even the thinking of a specific individual may switch back and forth among different ways of thinking—without noticing.

Students who *combine-first-then-rank* tend to encounter the following kinds of issues quite early during their problem solving sessions.

- Quantities such as 5'6" generally are not recognized by calculators or spreadsheets.
- Information that is given about "spiking" (or "coaches comments") doesn't translate immediately into "scores" that can be operated upon.
- Not all of the information is equally important. For example, the problem solvers may consider "coaches comments" to be unimportant or too unreliable; or, in playing volleyball, running ability may be considered to be less important than the ability to jump, serve, or spike.
- Volleyball players are not equally "good" across all categories. For example, a player who is "good" in one category (serving) may be "poor" in other categories (jumping or spiking).
- Some categories of information should be combined. For example, it may be sensible to combine height measurements with jumping measurements to estimate how high each player can reach.
- Large numbers are "good" for some quantities (such as height, jumping, serving, or spiking), but small numbers are "good" for other quantities (such as running). *What does it mean to add such quantities? In fact, What does it mean to add inches and seconds?*
- Height measurements involve bigger numbers (e.g., 93, 87, 94); whereas, other measurements involve small numbers (6.21 seconds, 8 serves, 5 spikes). If these measurements are treated as if they were "raw scores" that can simply be added, then categories that involve bigger numbers are (implicitly) given more weight. . . . *What does it mean to add heights (which range from 80 to 100) to serving scores (which range from 5 to 10)?*

Table 17-1 shows one intermediate (partly flawed) way that problem solvers who *combine-first-then-rank* sometimes transform the "raw data" that are given for the volleyball try-outs. Notice that: (a) Each player's height and jumping data have been combined. (b) In the column that is labeled *running reversed*, the running scores have been converted so that high scores will be "good." (c) The raw data for "spiking ability" are converted to "points" that have no units associated with them. (d) Coaches comments have not been quantified in any way, and have been omitted from the table. (e) In the column labeled "totals", the information about jumping, running, serving, and spiking have been added without regard to the units of measure. That is, all of the measures are treated as if they were simply dimensionless "assigned points" (rather than being quantities associated with specific units of measure).

Table 17-2 is a graph that was generated using an *Excel* spreadsheed of the data in Table 17-1. However, in the first graph, the bars show the sums of raw scores; whereas, in the second graph, the bars show the sums of weighted scores. Together, these two graphs show that, when "raw scores" are used, the relative sizes of the numbers in different columns strongly influence rankings that are based on sums.

TABLE 17.1.
One Possible Way (Not Necessarily "Good") to Transform Data in the *Volleyball Problem*

Name Units	Height + Leap (inches)	Run (seconds)	Run Reversed (seconds)	Serves (counts)	Spikes (assigned points)	Totals (points)
Gena	93	6.21	1.97	8	4	106.97
Beth	87	5.98	2.2	7	5	101.2
Jill	94	6.44	1.74	8	3	106.74
Amy	97	6.01	2.17	9	7.5	115.67
Ana	91	6.95	1.23	10	1.5	103.73
Kate	85	7.12	1.06	6	5.5	97.56
Rhoda	84	6.34	1.84	5	2.5	93.34
Christi	88	7.34	0.84	8	7	103.84
Andrea	89	6.32	1.86	9	0.5	100.36
Nikki	86	8.18	0	10	5.5	101.5
Kim	92	6.75	1.43	7	5	105.43
Robin	83	5.87	2.31	8	7	100.31
Edna	85	6.72	1.46	8	3	97.46
Lori	86	6.88	1.3	9	2.5	98.8
Tina	85	6.27	1.91	6	4	96.91
Angie	93	6.54	1.64	8	2.5	105.14
Ruth	89	7.01	1.17	9	7	106.17
Becca	87	6.78	1.4	10	4.5	102.9
Sum	1594	119.71	27.53	145	77.5	

Table 17.2.
A Graph Showing Volleyball Players Ranked By (a) Sums of Raw Scores (b) Sums of Weighted Scores

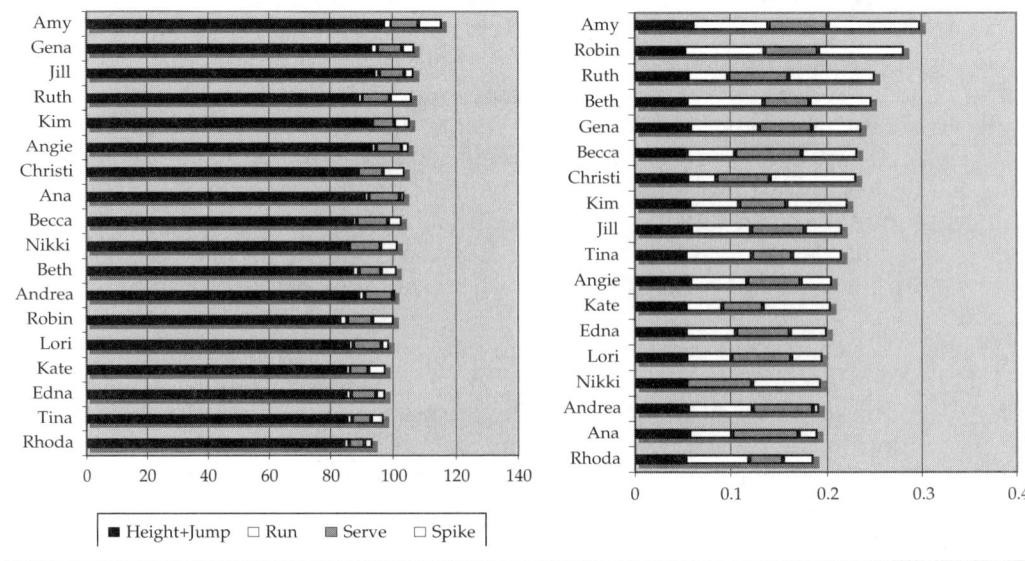

JOHN DEWEY REVISITED 331

In Table 17-2, notice that the two graphs result in significantly different rankings for the volleyball players. One way that problem solvers got the idea to create the second graph was by thinking of each category of information as being similar to scores in a mathematics class (for quizzes, projects, and larger examinations). So, they converted "raw data" to "test scores" on a scale that goes from 0 to 100. Then, later (see Table 17-5), if the problem solvers decided to give more weight to some categories than others, they simply assigned more points to this category. . . . For problem solvers who emphasize *combine-first-then-rank* approaches, the preceding sorts of issues (and ways of thinking) generally don't occur until third-cycle or fourth-cycle responses begin to emerge. But, not all problem solvers use this approach. For example, some use a the following rank-first-then-combine approach.

Students who *rank-first-then-combines* often are able to avoid many of the preceding difficulties that confront students who *combine-first-then-rank*. But, during second-cycle or third-cycle interpretations, the following kinds of difficulties often begin to emerge. . . . If *raw scores* are converted to *rank scores* (as shown in Table 17-3 for serving scores), then similarities and differences among campers tend to be distorted in ways that strongly influence comparisons that are made among volleyball players when ranks are combined across several categories of information.

To see how such distortions occur, consider the case of the information about serving ability that is shown in Table 17-1. Then, plot both columns of data on parallel number lines like the ones shown below. Finally, examine the way that the names of campers are distributed along the two number lines. Notice that the number line for "rank scores" assigns very different positions to volleyball players whose "raw scores" were identical! In particular, notice that volleyball players whose "raw score" were identical are assigned "rank scores" that make them seem to be quite different. Tables 17-5

TABLE 17.3.
Ranks and Raw Scores for Serving Ability

Name	Number of successful serves	Rank score
Gena	8	11
Beth	7	5
Jill	8	11
Amy	9	15
Ana	10	18
Kate	6	3
Rhoda	5	1
Christi	8	11
Andrea	9	15
Nikki	10	18
Kim	7	5
Robin	8	11
Edna	8	11
Lori	9	15
Tina	6	3
Angie	8	11
Ruth	9	15
Becca	10	18

TABLE 17.4.
Rank Scores VERSUS Raw Scores for Serving Ability

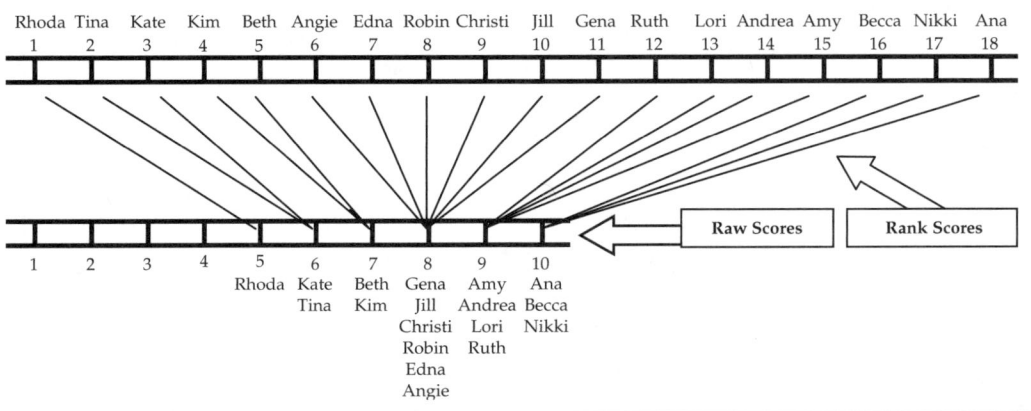

and 17-6 show some ways that software such as *Fathom* or *Tinkerplots* can be used to generate transformable graphs to investigate the preceding kinds of issues. In Table 17-5, the "sliders" that are shown are used to assign weights to each category of information; and, as the weights change, the figures in the graphs move back and forth across the computer screen to show how the total scores change. In this way, if the weight of a factor (such as jumping, running, serving, or spiking) is reduced to zero, then that factor is not taken into account in the sum. Or, if the weight of one factor is doubled relative to the other factors, then this factor contributes twice as much to the sum.... Such graphs demonstrate clearly that, when new "weights" are assigned to various quantities, the total scores for volleyball players often change significantly. For example, in Table 17-5, notice that the volleyball players who receive the top three scores are Amy, Ruth, and Gina; whereas, in Table 17-2a, the top three scores were given to Amy, Gina, and Jill; in Table 17-2b, they were given to Amy, Robin, and Ruth; and, in Table 17-6, where rank-first-then-combine methods are used, the top three scores are given to Amy, Gina, and Ruth.

TABLE 17.5.
Transformable *Fathom* Graphs Showing Weighed Sums Of Scores—When The Weights Are Equal For Jumping, Running, Serving & Spiking

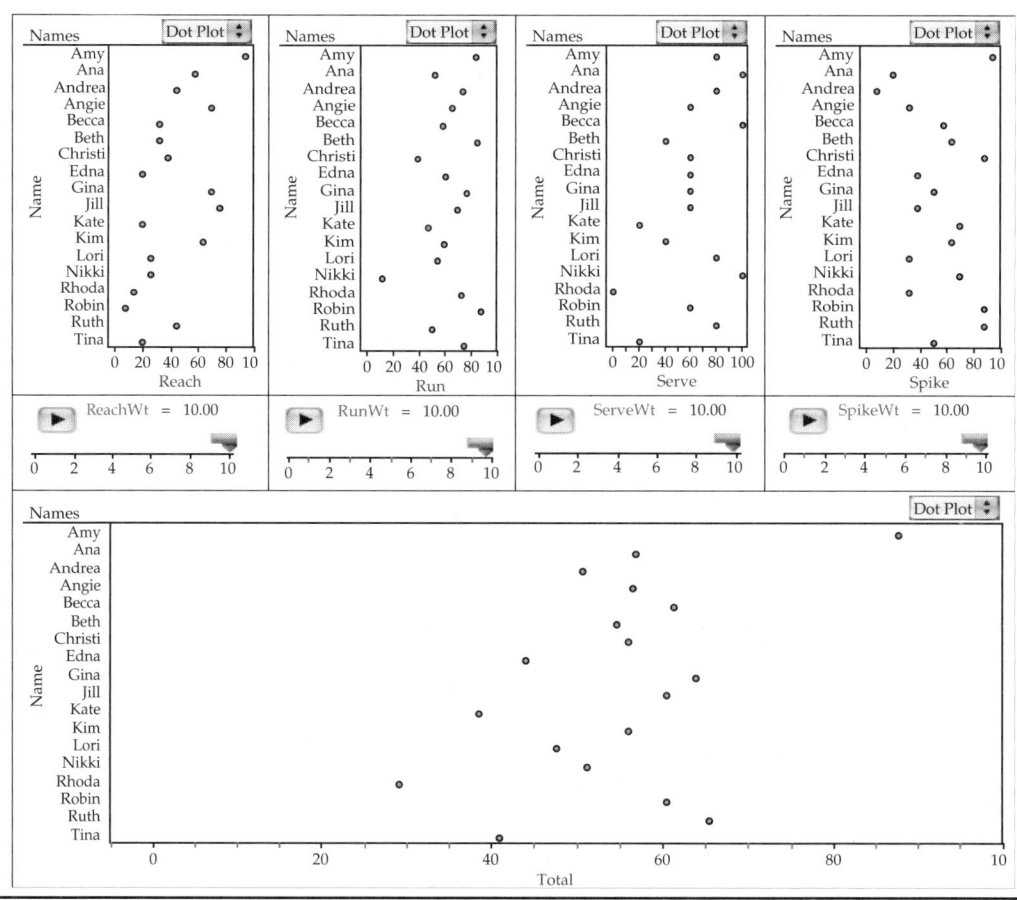

TABLE 17.6.
Transformable *TinkerPlot* Graphs Showing Weighed Sums of Ranked Scores—When the Weights are Equal for Jumping, Running, Serving, & Spiking

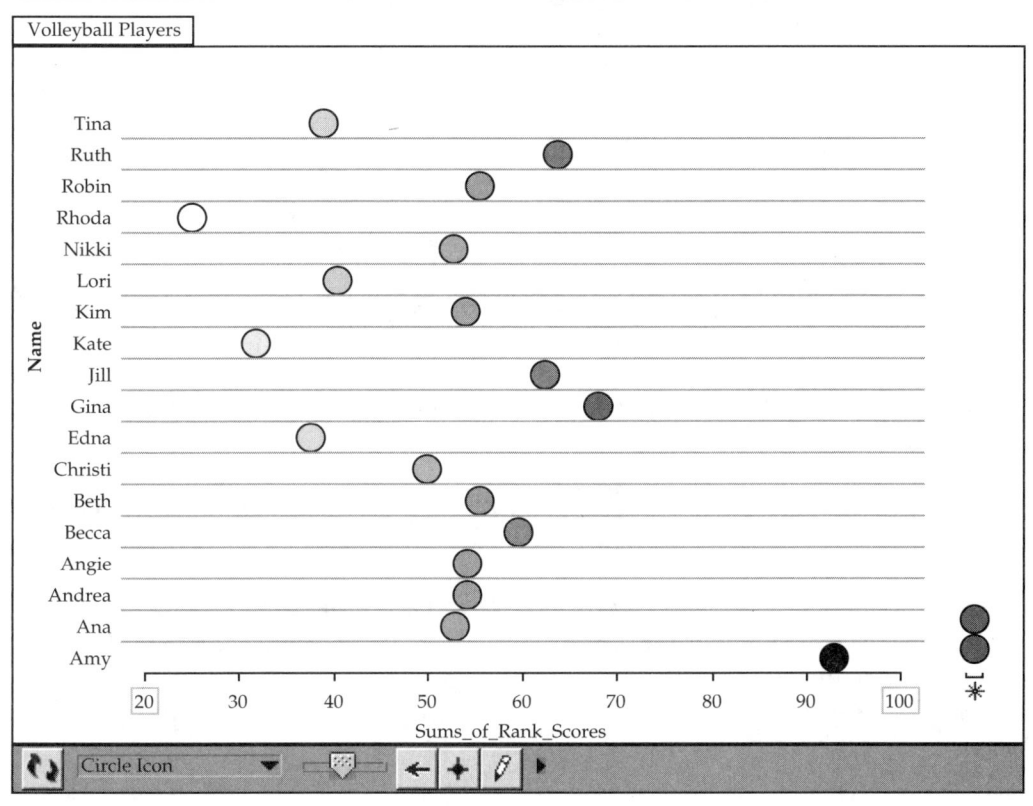

Transformable graphs, like those that are shown in Tables 17-5 and 17-6, can be marvelous tools for teachers and students to use during *model-exploration activities*—where the goal is to formalize conceptual systems that students have begun to develop during *model-eliciting activities*. But, using *develop-first-then-formalize* approaches to teaching and learning, it is only in very rare cases that students are likely to develop such graphs spontaneously. To do so would presuppose that the problem solvers already understand most of the organizational/relational/ operational systems that needs to be developed to generate useful ways of thinking about the *Volleyball Problem*. . . . For example, in Table 17-5: *How (and why) did all on the data end up being plotted on a number line that goes from 0 to 100. How (and why) did the data end up being distributed fairly evenly across the entire length of the preceding number lines? Why didn't the scores "bunch up" with long intervals where no scores appear?*

Note: One way to create comparable scales that have nicely distributed data is: (a) to subtract the difference between zero and the lowest score for each category of data, and (b) to multiply the preceding results by a factor which stretches the differences so that they are similar to those that exist for the other types of data. But, every technique injects some assumptions into the situation that clients (and problem solvers) may not want to make. Furthermore, these assumptions involve exactly the same organizational/ relational/operational systems that students are intended to develop when they work on the *Volleyball Problem*. So, it can be both deceiving and counterproductive to provide students with tools that embody conceptual systems that we want them to develop—if they have developed only primitive understandings of these systems. . . . We will return to this issue in the last sections of this chapter.

Regardless whether problem solvers emphasize *rank-first* or *combine-first* approaches, their final-cycle responses often begin to develop when the following issue emerges. If individual volleyball players are assigned to teams based on only an overall *index of volleyball ability* (which collapses a variety of different kinds of information into a single-dimensional score), then it is very likely that some teams will have no excellent servers, or no great spikers, or no outstanding defensive players. Therefore, if *develop-first-then-formalize* approaches are used, then students' last-cycle solutions to the *Volleyball Problem* often shift attention beyond simply quantifying what it means to be a "good individual player"—and they shift toward dealing quantitatively with what it means to be a "good team." . . . One way to accomplish this goal is to shift away from ranking players using only average scores (or total scores) and to shift toward some sort of iterative process—like the one described below.

Step 1: If you want to create N different teams, then deal out players to teams using a reversing sequence: 1, 2, 3, . . . , N, N-1, N-2, N-3, . . . , 3, 2, 1, and so on.

Step 2: Begin by dealing out the "best available spiker" to each team.

Step 3: Reverse the order, and deal out the "best available server" to each team.

Step 4: Reverse the order again, and deal out the "best remaining spiker" to each team.

Step 5: Reverse the order again, and deal out the "best remaining server" to each team.

Step 6: Reverse the order again, and deal out the "best available player" to each team—where this player is chosen based on an overall *index of volleyball playing ability*.

Step 7: Repeat step 6 until all of the players have been assigned to a team.

Step 8: As a cross-check on the preceding procedure, calculate each team's overall *index of volleyball playing ability* by finding the sum of the indices for individual players on the team. Then, if any team seems to have an especially low score, make a few "trades" among teams—based on coaches comments as well as individual players' *indices of volleyball playing ability*.

Note: If the preceding kind of iterative processes are used, then the ordering of steps essentially provides a way to assign different "weights" to different categories of information.

WHAT KIND OF CONCEPTS AND ABILITIES DEVELOP WHEN STUDENTS MATHEMATIZE "REAL LIFE" SITUATIONS?

In virtually every field where researchers have compared experts and novices in "real life" problem solving or decision making situations, results have shown that experts not only *do* things differently, but they also *see* (or interpret) things differently. Therefore, the following questions arise. *What is the nature of these "seeing" abilities? How do they develop? How can development be cultivated and assessed?*

Unfortunately, everyday usage of the term *seeing* suggests that information is being extracted from the things that are seen—rather than being projected onto them. So, when we focus on mathematical thinking, as opposed to other types of thinking, active terms such as *interpreting* or *conceptualizing* seem to be more appropriate than the passiveness suggested by *seeing*. This is because mathematics is not about surface-level pieces of information that can be seen. It is about patterns, regularities, and other structural or systemic characteristics of experiences which must be conceived—not simply perceived.

On the other hand, when researchers (or teachers) need to be precise about what it means to understand (or teach) *interpretation* or *conceptualization* abilities, then characterizing them as activities invites people to treat them as lists of things to do! Whereas, a main point that we want to stress in this chapter involves fundamental differences between abilities related to *seeing* and *doing*. . . . Each of the terms —*seeing*, *doing*, *interpreting*, and *conceptualizing*—is somewhat misleading in different ways.

With the preceding observations in mind, it is still appropriate to state that *model-eliciting activities* focus on abilities related to mathematical interpretation and conceptualization (description, explanation, communication) at least as much as computation or deduction. In particular, *model-eliciting activities* are designed to be contexts in which students develop more productive ways of *seeing* mathematics (and problem solving situations, and themselves as problem solvers). . . . Therefore, this section describes five types of mathematical knowledge or ability which all emphasize structural/systemic ways of thinking about experiences. They involve: (1) within-concept systems, (2) between-concept systems, (3) interacting representational systems, (4) basic skills whose meanings depend on systemic understandings, and (5) higher-order ideas or abilities whose meanings depend on systemic understandings.

1. **Understandings That Involve Within-Concept Systems:** *Within-concept systems* are perhaps the most fundamental among the five types of systems-based understandings that will be described in this section. If their nature is not understood, then other types of systems-based understanding also are not likely to be understood at more than superficial levels. On the other hand, beginning with *within-concept systems* is problematic because they have proven to be unusually difficult to explain. So, claims about them are exactly the type that evoke doubts and challenges from readers. . . . With these difficulties in mind, it is sensible to begin by pointing out that the claims we'll make are exceedingly modest forms of standard constructivist statements that *(all) mathematical knowledge must be constructed by learners*. . . . Our three main claims are the following.

- One of the most fundamental characteristics of most "big ideas" in elementary mathematics is that making judgments that involve them generally presupposes thinking that is based on some (holistically functioning) system of operations and relations. That is, the "big idea" is only meaningful if some organizational/conceptual system is used to describe the situation.
- When relevant organizational/relational/operational systems are insufficiently developed, the concepts and conceptual systems that are based on them tend to yield interpretations of learning and problem solving situations that are remarkably unstable, barren, and distorted compared with those that

develop later. That is, primitive conceptual systems focus on only a small number of superficial and disconnected *pieces* of information rather than emphasizing deeper *patterns* and *regularities*.

- If teachers attempt to teach systemically-based "big ideas" without first helping students develop relevant organizational/relational/operational systems, then the levels and types of understanding that students are able to achieve tends to be exceedingly limited; and, the teacher's ability to teach such concepts also is limited by the fact that the relevant operational/relational system cannot be presented to students in prefabricated forms.

The first claim is the main one that is not straightforward. For the other two, similar reasons explain why tennis serving ability, bicycle riding ability, and essay writing ability cannot be presented to children in prefabricated forms. Or, when a new visitor on a college campus asks for directions to get from point A to point B, the information that a direction-giver describes (or shows in pictures or diagrams) tends to be very different than the information that the visitor hears (or sees). This does not mean that it is useless to give maps to campus visitors. Nor does it mean that a child's understanding of mathematical conceptual systems is an all-or-nothing affair—so that no guidance or tools can be provided. But, it does mean that it is important to recognize that: (a) most mathematical concepts require thinking that is based on (holistically functioning) systems of relations and operations, and (b) students' levels of development of these systems determine the size of the mismatch between things that teachers say (or show) and things that students hear (or see).

Similar observations apply to teaching and learning experiences that teachers might create to surround *model-eliciting activities* like the *Volleyball Problem*. Claim #1 applies because *model-eliciting activities* are designed to involve developing some mathematically significant organizational/relational/operational system). Consequently, claims #2 and #3 also apply. Any demonstrations or explanations that teachers give prior to the activity are likely to be misinterpreted by students; and, whatever is learned is likely to be incorporated into existing conceptual/procedural systems rather than leading to the development of new conceptual/procedural systems.

One "big idea" that the volleyball problem could be use to emphasize has to do with aggregating different kinds of qualitative and quantitative information—using *weighted sums* or *weighted averages*. . . . When we unpack what it means to "understand" these concepts, and when we observe how these meanings develop during specific *model-eliciting activities*, what emerges is the importance of conceptual systems that must be imposed on experiences rather than simply being derived from them.

In the case of the *Volleyball Problem*, the relevant organizational/relational/operational systems involve: (a) comparisons among teams-as-a-whole, (b) comparisons among players based on multi-dimensional *profiles of competence* and/or single-number *indices of volleyball playing ability*, and (c) comparisons and/or combinations involving different kinds of qualitative and quantitative data—as well as transformations of the "given information" that involve: (d) quantifying qualitative information (e.g., spike results, coaches comments, weights for different categories of information), (e) combining, re-coding, or rescaling given data into forms that are more useful, or (f) observing patterns that describe different combinations of weights influence weighed sums or averages.

Notice that none of the preceding comparisons, transformations, or patterns are "given" in the original data. All must be projected onto the data; and, all presuppose thinking that is based on well organized conceptual/organizational systems. Therefore, in general, the process of developing relevant conceptual systems is similar to the process of developing well-coordinated action schemes—such as those that are involved in playing tennis or ballroom dancing. When students first begin to engage in such activities: (a) they are unlikely to notice more than a subset of the information that is relevant; (b) they are likely to impose inappropriate assumptions (and actions) due to prior expectations; and (c) they are likely to attend to surface-level information rather than to deeper patterns and regularities. So, in any demonstrations or explanations that teachers give, what is said (and shown) is quite different than what is heard (or seen).

In general, relevant conceptual systems cannot be provided to students before they engage in *model-eliciting activities* any more than coordinated ballroom dancing abilities are likely to be learned by watching and listening. But, could the relevant concepts and conceptual systems be learned through some type of guided discovery activities that emphasize applications of concepts being taught? . . . Discovery is a strange word to use when we are referring to the development of well-coordinated conceptual or procedural systems. *Can a child discover the ability to ride a bicycle?* In fact, even terms such as *guided construction* conjure up machine-based metaphors in which computers are being programmed, or automobiles and other machines are being constructed on assembly lines.

In fields ranging from athletics to the arts, where students are expected to develop complex abilities, successful coaches generally realize that two-pronged approaches to teaching and learning are sensible. First, don't neglect fundamentals because, to some extent, complex systems need to be built up. Second, don't forget to scrimmage because the systems that evolve in such complex activities tend to be quite different than those that develop in more highly guided activities. . . . Similarly, when mathematics teaching and learning is intended to focus on *within-concept systems* associated with important mathematical concepts, effective learning experiences probably should include early activities that emphasize *mathematizing reality*—by putting students in situations where they express>text>revise their own current ways of thinking (rather than being guided along efficient paths toward simplified versions of the conceptual systems that are described by their textbooks).

2. **Understandings That Involve Between-Concept Systems:** Whereas the preceding section drew heavily on constructivist philosophical perspectives, this section shifts attention toward pragmatist philosophical perspectives. In particular, two points are emphasized about the kind of mathematical understandings that evolve when learning activities focus on *mathematizing reality:* (a) The nature of the conceptual systems that evolve are influenced at least as much by the purposes that people (problem solvers and/or their clients) have in mind as by other contextual factors that exist in the problem solving situation. This *impact of purposes* on construct development is what pragmatists referred to as *abductive reasoning*. (b) The models (and underlying conceptual systems) that students develop to make sense of *model-eliciting activities* are "chunks" of knowledge that typically integrate ways of thinking drawn from a variety of textbook topic areas. . . . For both of these reasons, a great deal is lost if teach-first-then-apply techniques introduce ideas out-of-context—in situations where no purposes are apparent, and

where knowledge development is guided along narrow paths that do not stray outside of pre-specified textbook topic areas.

Pragmatists were strong advocates of the notion that solutions to realistically complex problems usually must draw on ways of thinking from more than a single textbook topic areas—and from more than a single practical or theoretical perspective. . . . In hundreds of transcripts from *model-eliciting activities*, students' solutions bear witness to this fact. For example, in solutions to the *Volleyball Problem*, a wide variety of ideas and procedures tends to be considered during the process of developing useful ways to:

- Compare several different kinds of quantitative and qualitative information,
- Formulate useful "operational definitions" of what it means to be a "good" volleyball player—or a "good" volleyball team.
- Transform "given" information into more useful forms.

So, relevant ideas and abilities often draw on topics ranging from graphing, to ranking, to alternative ways of scoring, to analyses of variation, to re-scaling, to weighted sums (or weighted averages), to iterative decision processes. Therefore, when three-person teams work on the *Volleyball Problem* within a classroom of thirty students, the reports and discussions of their work typically bring together a much wider variety of ideas and abilities than any teacher is likely to be able to introduce using *teach-first-then-apply methodologies*.

Teach-first-then-apply methodologies often predispose students toward lines of inquiry that restrict attention to a single, narrow conceptual path; and, in fact, even teachers' hints about "correct" ways of thinking often lock students into inadequate ways of thinking—and discourage them from going through a series of interpretation cycles in which several fundamentally different ways of thinking are investigated. Whereas, if students express, test, and revise their own ways of thinking, then the thinking of both individuals and three-person teams (as well as the class-as-a-whole) tends to resemble an ecological system in which communities of constructs and conceptual systems are struggling for survival (Lesh & Yoon, 2004). Consequently, any approach to teaching and learning that treats the class-as-a-whole as a single learner is not only completely unrealistic in terms of recognizing diversity in thinking; but, it also fails to draw on the enormous diversity of conceptual resources that exits within the class-as-a-whole, within smaller problem solving teams, and within the thinking of each individual student.

If teams of problem solvers go through multiple cycles in which they express, test, and revise (or reject) their own ways of thinking, then each team is likely to develop solutions that integrate ideas and procedures drawn from a variety of textbook topic areas; and, even in the case of ideas or procedures that are rejected, the meanings of both accepted and rejected ways of thinking tend to be enriched. . . . Sometimes, knowing why an idea doesn't work in a given situation is an important piece of knowledge; and, for both accepted and rejected ways of thinking, a large share of the meaning of any given concept or ability depends on connections to other concepts and abilities.

In general, students' solutions to *model-eliciting activities* bear witness to the pragmatist observation that the thinking of a given student is by no means as monolithic and one-dimensional as most current theories of learning suggest. Furthermore, when students work on the *Volleyball Problem*, or on other *model-eliciting activities*, most of the relevant conceptual systems exist at some intermediate level of development—not completely "mastered" but not completely unknown. Also, relatively well functioning conceptual systems can be expected to already exist for making sense of some parts of the problem. Consequently, the development of productive interpretations is not simply a process of constructing new conceptual systems. Instead, development generally involves sorting out, integrating, modifying, adapting, extending, or rejecting conceptual systems that already exist at intermediate stages of development—but which may be fuzzy, partly overlapping, and relatively undifferentiated. Therefore, just as in the case of the other types of complex and continually adapting ecological systems, the development of problem solvers' conceptual systems generally must involve diversity, adaptation, selection, preservation (of systems that are useful). For this reason, very few of the processes that contribute to development are suitably described using the mechanistic-sounding term *construction* (Lesh & Doerr, 2003). So, development tends to resemble a genetic inheritance tree (where grandchildren inherit characteristics from a variety of ancestors) — not a single linear path. (Lesh & Yoon, 2004).

Model-Eliciting Activities **Involve Local Conceptual Development.**
Yet, Development Doesn't Occur Along A Single Linear Path.

Model-Eliciting Activities Involve Local Conceptual Development. During *model-eliciting activities*, the development of useful conceptual systems generally involves going through a series of modeling cycles in which current mathematical interpretations (descriptions, explanations, or conceptualizations) are iteratively expressed, tested, and revised—or rejected. Consequently, in situations where the relevant conceptual systems correspond to concepts that have been investigated by Piaget-inspired researchers, the development cycles that problem-solvers go through during 60–90 minute *model-eliciting activities* often bear a striking resemblance to the kind of stages described by Piaget (Lesh & Harel, 2003). For example:

- Distinct "stages" can be identified because they involve somewhat different relations, operations, patterns, or mathematical "objects" (e.g., quantities, shapes, coordinates).
- Relatively primitive ways of thinking tend to be based on less refined, less complex, and less stable relational/organizational systems; and unstable systems tend to be relatively barren and distorted compared with later interpretations.

Development Doesn't Occur Along A Single Linear Path. In spite of the preceding similarities, there are significant differences between Piaget's developmental trajectories and the way conceptual systems develop during *model-eliciting activities*. For example:

- Because final conceptualizations of *model-eliciting activities* tend to integrate ideas from multiple-textbook topic areas, resulting chunks of knowledge tend to be organized around experience more than around abstract cognitive structures of the type emphasize by Piaget and his followers.
- During *model-eliciting activities*, construct development is far more situated than Piaget-inspired researchers have suggested (Lesh & Carmona, 2003). For example, when students' thinking evolves through several Piagetian stages during a single 60–90 minute problem solving episode, it is obvious that development is much more a matter of gradually increasing local competence rather than being a manifestation of the evolution of some general cognitive structure. Also, when two tasks are significantly different in difficulty, it is questionable whether students' thinking is organized around these abstractions—even though Piagetians would classify both of them as involving the same cognitive structure.

> ***Model-Eliciting Activities* Involve Local Conceptual Development. Yet, Development Doesn't Occur Along A Single Linear Path. (*continued*)**
>
> - During model-eliciting activities, model development (and the development of underlying conceptual systems) generally occurs simultaneously and interactively along a variety of dimensions. For example, concrete understandings sometimes evolve into abstractions. Intuitions sometimes evolve into formalizations. Simple conceptual systems sometimes evolve into more complex systems. External functions sometimes are internalized. Task-specific ideas sometimes evolve into generalizations. Contextualized knowledge sometimes becomes decontextualized. Unstable conceptual systems sometimes evolve into relatively stable conceptual systems.
> - In the next section of this chapter, we will describe how development also involves increasing representational fluency as meanings associated with a variety of representational media gradually are sorted out and coordinated—such as those that are expressed using spoken language, written symbols, diagrams, experience-based metaphors, or technical tools (Lesh & Doerr, 2002). Consequently, development involves much more than Piagetian transitions from pre-operational to concrete-operational to formal-operational thinking and more than Vygotskian transitions in which external functions are gradually internalized.
> - During intermediate stages in the solution of *model-eliciting activities*, the productivity of alternative ways of thinking is strongly determined by the purposes that are imposed on the situation by problem solvers or their clients; and, these purposes not only create the need for some kinds of comparisons and transformations (but not others), but they also provide criteria to assess the usefulness of alternative ways of thinking. In fact, if purposes are not apparent, then it is seldom clear whether one interpretation is better than another, or whether mathematical responses are better than responses that require very little mathematical thinking. It probably isn't even clear whether 5-hour responses are better than 5-minute responses or 5-second responses.

Is it possible to use direct instruction to teach students the kind of *between-concept* understandings that tend to develop naturally during *model-eliciting activities*? If instruction emphasizes guided inquiry, if it builds on students' prior knowledge and experiences, and if it emphasizes applications of the things that are being learned, then, in theory at least, the answer might be imagined to be: *Yes!* But, in practice, the answer is: *Not well! Only superficially!* This is because there are far too many distinctions and connections to fit within efficient teacher-led or textbook-led line of inquiry.

3. **Understandings That Involve Interacting Representational Systems:** William James, the father of modern semiotic perspectives, introduced the notion that many of the meanings of concepts derive from interacting representational systems. In the *Volleyball Problem*, just as in other *model-eliciting activities*, the development of useful conceptual systems usually involves using a variety of interacting representational media. These tools may range from diagrams, to spoken language, to written symbols, to concrete models, to gestures, to experience-based metaphors—each of which emphasizes and de-emphasizes somewhat different aspects of the system being described, or the conceptual systems being used. Such situations are similar to those that are common in physics—for example, where explanations of the behavior of light need to employ both wave models and particle models.

As stated earlier, every model has some properties that the system it describes does not have; and, it also does not have some properties that the modeled system does have. This is the nature of models. . . . In contrast, however, a distinguishing characteristic of polished theories, as well as elegant textbook descriptions of fundamental concepts, is that they generally try to embody as much meaning as possible within a single representational media—usually written symbols. Yet, in mathematics, just as in other disciplines, the following facts are well known.

- Many ideas and procedures are difficult to express in language—even though they are easy to describe using written symbols.
- Often, written symbolic statements seem incomplete unless they are ac-companied by spoken language descriptions. That is, spoken descriptions sometimes significantly enhance the meanings of written symbols. Yet, the language of mathematics often is poorly aligned with everyday language. So, confusions also may be introduced.
- A picture is sometimes worth a thousand words. Yet, many mathematical concepts and procedures are difficult or impossible to visualize.

Beyond the preceding observations, mathematics-in-the-making (of the type that occurs during *model-eliciting activities*) tends to be quite different than the formalizations that are embodied in textbooks and polished theories. For example, in the *Volleyball Problem*, students talk, write, draw, gesture, and use concrete materials to act out their thinking. Furthermore, they often use several media simultaneously—and weave back and forth among interacting media. So, representational fluency, or the ability to translate back and forth among alternative representation systems is an important part of what it means to understand many relevant concepts and conceptual systems.

Another point that emerges during *model-eliciting activities* is that some of the most powerful representations that students use tend to be embodied in experienced-based metaphors or in technology-based tools. For example, when problem solvers think about ways to aggregate data about different kinds of measurements in the *Volleyball Problem*, they sometimes refer to situations that are familiar to them — such as situations where grades are assigned to students based on scores from quizzes, projects, midterm and final examinations, and classroom participation in a mathematics course. Or, when they recode or rescale data, they may refer to "stretching a rubber band" or to some other familiar transformation of information.

Technology-based tools such as calculators or spreadsheets also facilitate some ways of thinking while making others difficult or impossible to use. But, they also often require "given information" to be transformed into technology-recognizable forms; and, they often produce results (such a graphs) whose assumptions and meanings are not clear. So, when such tools are used, a large part of problem solving efforts often ends up being spent trying to make the situation fit the tool. . . . Also, whereas self-generated representations often serve the function of helping problem solvers express their thinking in forms that can be examined and revised, adopted technical-based tools often serve exactly the opposite function. The often disguise assumptions and relevant reasoning processes in a "black box" which makes it difficult for them to be examined; and, they often lock problem solvers into a single way of thinking that leads them to assume they are not supposed to adapt or change to fit current circumstances.

Can *between-media systems* and *representational fluency* be learned through direct instruction and application? *Perhaps partly!* . . . One limitation to such approaches is that, for most mathematical concepts that it is desirable for students to develop, many of the media that contribute to understanding do not lend themselves to compact textbook presentations. Also, during problem solving sessions, many of the representations that are especially meaningful and useful to some students are too personal and idiosyncratic to be useful for teaching general audiences. Examples include the kind of experience-based metaphors that students use during *model-eliciting activities*.

During *model-eliciting activities*, when students are encouraged to use experience-based metaphors to express>test>revise their thinking, it usually becomes clear that most students already have developed relevant conceptual systems for making sense of many parts of the situation. So, to ignore these metaphors risks ignoring large and highly significant aspects of students' prior knowledge and abilities.

Other limitations of direct instruction can be seen by considering similarities between modeling abilities, writing abilities, and drawing abilities. Modeling, like writing and drawing, is largely about describing and explaining; and, it is obvious that written or drawn descriptions usually need to go through several drafts before they are sufficiently useful. Also, if descriptions are expected to be sharable (with others) and reuseable (in other situations), then they usually need to be modularized and in other ways put into forms that are relatively easy to unpack and reassemble for a variety of purposes.

Even though it is possible to learn a great deal about writing or drawing by listening to lectures—or by constructing things using a process of following carefully guided sequences of steps, a large share of what is needed in order to develop relevant writing or drawing abilities is not likely to occur without more open-ended explorations in which students actively engage in the development of complex artifacts. For instance, learning activities that carefully guide students (in a color-by-the-number fashion) to develop simplified versions of complex artifacts tend to lead to significantly different abilities and understanding than activities that require students to express>test>revise their own ways of thinking.

4. **Understandings That Involve Basic Skills Whose Meanings Depend On Holistic Conceptual Systems:** Three points are especially important to emphasize about the skills that emerge during *model-eliciting activities*. First, because *model-eliciting activities* are designed to be simulations of "real life" problem solving situations, the kinds of skills that emerge as being most important go beyond computation or deduction to also focus on conceptualization, mathematization (description), construction, explanation (manipulation, prediction), communication, representation, and interpretation. In fact, during 60–90 minute *model-eliciting activities*, relatively little time tends to be spent on computation. To see why this is true, consider the following problem characteristics—and notice that most of them create the need for skills related to communication, description, explanation, planning, monitoring, or assessing.

- The "problem solver" often is a team of specialists who are familiar with very different ways of thinking, language, tools, and practical or theoretical perspectives. So, explanation and communication skills tend to be important.
- The setting often allows access to a variety of powerful technical tools and resources. But, the information that is available is not necessarily given in a pre-mathematized form, nor is data necessarily in forms that fit the tools and resources that are available. So, mathematization and interpretation skills tend to be important.

- The product is a complex artifact (or conceptual tool) whose underlying design principles are important parts of the product that is needed. So, explanation and communication skills tend to be important.
- The client often needs a product that is powerful (for a specific purpose), and sharable (with others) and re-useable (in other situations). So, explanation and communication skills tend to be important.
- The development process often draws on resources from a variety of practical and theoretical perspectives, and it also often involves a series of iterative design cycles in which alternative ways of thinking are repeatedly expressed, tested, and revised. So, skills related to planning, monitoring and assessing tend to be important.

Second, the essence of "understanding" the preceding kinds of skills has as much to do with knowing when, where, and why to used them as it does with simply knowing how to do them. In fact, one of the most obvious characteristics of experts in most fields is that they are exceptionally capable at performing skills that are distinctive to that field. Yet, it also is true that many mediocre practitioners perform proficiently when relevant same skills are tested one-at-a-time and out-of-context. . . . Why do they continue to be mediocre? Because it is important to go beyond *doing things right* to also *do the right things*—by doing them at the right time, in the right situations, and for the right purposes. This is why, in fields that range from performing arts to athletics, or from carpentry to cooking, expertise includes not only the mastery of basic skills but also the development of more complex understandings and abilities that focus on *seeing* as much as *doing*. It also is why effective instructional programs in such fields generally include not only drills focused on basic skills but also scrimmages, recitals, and other complex activities.

Third, when students engage in *model-eliciting activities*, relevant skills tend to be at an intermediate stage of development. Consequently, when we observe students working on *model-eliciting activities* like the *Volleyball Problem*, we often get to see relevant skills when they are in-the-making—and at relatively primitive levels of development. That is, the skills that students use tend to be modifications or adaptations of those that they have used for other purposes in other situations; and, knowing how to perform such skills correctly tends to be closely tied to understandings of the contexts and purposes in which they are needed. . . . Consider the following examples that are needed in the *Volleyball Problem*.

- Quantify qualitative information. e.g., *Assign a score or a ranking to coaches comments — or to information about spiking ability (where, in five tries, one player might have one kill, one return, one dink return, and two in-the-net returns). Or, assign "weights" to reflect the relative importance of different types of information.*
- Aggregate a variety of different kinds of qualitative and quantitative information. e.g., *Combine different kinds of information using techniques such as weighted sums (or averages). Or, use iterative decision-making processes that do not depend on a single formula.*
- Generate and/or rescale graphs to eliminate misleading variability when comparing or combining different kinds of data for individual volleyball players or teams.

Students don't develop such skills by first learning them in decontextualized settings and then learning to use them for a specific purpose in a specific context. For example, consider the two *TinkerPlot* graphs that are shown in Table 17-7. If problem solvers use two such similar-looking graphs to compare volleyball players' "scores" from running and jumping, then, once the situation is conceptualized in a productive way, the nature of needed computations tends to be clear. In particular, when problem solvers notice details about differences between these two scales, they also begin to envision the operations (and skills) that can be used to put these data into forms so that it is sensible to combine (i.e., add) the results.

In general, once problem solvers recognize the need for such skills (that is, once they recognize the difficulties and distinctions that cause these skills to be needed), they often invent first-iteration ways to do them. For example, in the example shown here, once students recognize the need to equalize the variability in these two scales, they generally "see" ways to convert them into forms that can be added.

Skills, like other tools, acquire meanings that are molded and shaped to fit the contexts and purposes in which they are needed. This is why, in hardware stores or cooking stores, it doesn't make sense to think of first learning the tools and then learning how to use them in specific "real life" situations. Similarly, students do not first learn to communicate (or describe, design, model or mathematicize) in some decontextualized setting—and then learn to apply these capabilities for some specific purpose in some specific situation. Instead, purposes and contexts are important parts of what it means to understand them; and, in mathematics education, students should learn to communicate (or describe, or design, or model or mathematicize) in structurally significant situations....

TABLE 17.7.
Two TinkerPlot Graphs

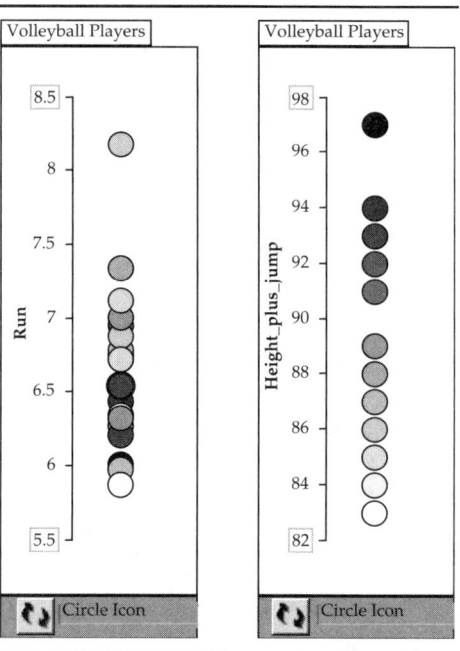

This is why a *model-eliciting activity* that is designed to encourage the development of a powerful conceptual system also serves as an ideal context for developing important skills. This is because, when important skills are developed in such contexts, then each time the relevant model is engaged to make sense of a new situations, the skills that accompany this model also tend to be engaged. For example, whenever a situation is seen as being similar to the *Volleyball Problem*, skills associated with rescaling also are likely to be engaged.

5. **Understandings That Involve Higher-Order Concepts or Abilities Whose Meanings Depend on Holistic Conceptual Systems:** In mathematics and science education, problem solving strategies—such as *draw a picture, work backwards, look for a similar problem,* or *identify the givens and goals*—have long histories of being advocated as important abilities for students to develop. But again, we ask: *What does it mean to "understand' them? How do these understandings develop? How can devel-*

opment be encouraged? . . . And again, our answers involve systems-based understandings and abilities.

To answer such questions about what it means to "understand" different kinds of problem solving strategies, we return to the fundamental question: *What is the nature of these constructs? (Are they content/context-independent strategies, or are their meanings closely tied to specific situations or concepts? Are they clearly defined skills (or rules), or are they just names for large categories of skills?)* . . . Clearly, the kinds of strategies that have been stressed in past math/science education research obviously have *descriptive* power. That is, experts often use such terms when they give after-the-fact explanations of their own problem solving behaviors, when they describe the behaviors of others they observe. But: *Do they have prescriptive power? Are they useful to provide guidance to novices?* Mathematics and science education researchers generally have agreed that the answer to this question is: *No!*—at least in the case of problem solving strategies that were specified in George Polya's famous book, *How to Solve It* (1945). Therefore, attempts have been made to convert Polya's *descriptive strategies* into longer lists of *prescriptive strategies* (Schoenfeld, 1985).

Results have not been impressive for this transition from *descriptive* to *prescriptive* functions (Lesh & Zawojewski, in press). In general, when lists of *prescriptive processes* are generated, they tend to become so long and situation-specific that knowing when to use them becomes the heart of what it means to understand them—and transfer of learning tends to be poor. Therefore, additional "higher order" (managerial) strategies and beliefs have been added; and, these are intended to specify when and why to use "lower order" prescriptive processes. . . . Unfortunately, when more higher-order and lower-order rules are added, these additions tend to exacerbate the following fundamental dilemma: (a) A small number of vaguely defined rules lack prescriptive power. (b) A large number of tightly specified rules make it difficult to know when to use them. Consequently, transfer of learning tends to be small.

For the preceding reasons, results have been unimpressive from attempts to teach prescriptive-processes+metacognitive-processes+beliefs. So again, transfer of learning is restricted; and, even in those few studies where significant achievements appear to have been occurred, the "treatment" usually involves an extraordinarily gifted instructor, complex and multi-faceted learning activities, and long periods of time. Therefore, it tends to be impossible to determine why learning occurs; and, even in these instances, transfer of learning tends to be small (Silver et al.). . . . Alternatives to this approach are emerging from research on *model-eliciting activities* (Lesh & Zawojewski, in press). The following table on page 344 describes some of the key elements of these *models & modeling perspectives*.

To briefly explain why the preceding *models & modeling perspectives* are plausible, consider Peyton Manning, the football quarterback for the *Indianapolis Colts*. Manning is well known to be one of the most capable decision makers to ever play his sport. Yet, his on-field decision making clearly is not characterized by concatenating sequences of prescriptive behavioral rules. In spite of the fact that "basic skills" is an important part of Manning's training regime, he simply does not have time during games for his thinking to be governed by analytic sequences of highly specific rules. So, the conceptual systems that he uses to make judgments are thought *with* (intuitively) rather than being thought about (analytically). Yet, they are continually being modified and adapted—because no two situations are ever exactly alike, because the "same thing" never happens twice, and because situations act back when they are acted upon. Circumstances are continually changing and so

Two Ways Of Thinking About Problem-Solving Strategies	
Traditional Perspectives	*Models & Modeling Perspectives*
Productive problem solving strategies are intended to provide answers to the question: *What should I do when I'm "stuck" (i.e., when I don't know what to do)?*	*Productive problem solving strategies* are expected to help problem solvers develop adaptations to conceptual systems that they *do* have—not to help them function better when none seem to be available (which almost never is the case in meaningful simulations of "real life" situations).
Higher-order (metacognitive) functions are those that operate on lower-order skills or processes by specifying which to use, when, and why.	*Higher-order conceptual systems* serve the function of helping problem solvers develop beyond current ways of thinking.
Increasing competence is characterized as becoming more proficient at generalizing prescriptive-level strategies and ways of thinking.	*Increasing competence* is characterized as becoming more proficient at particularizing descriptive-level strategies and ways of thinking. These proficiencies depend on developing useful models-of-modeling—which include productive conceptions of mathematics, problem solving, and one's own dynamic and malleable profile of dispositions, competencies, and attributes as a learner or problem solver.
Problem solving strategies and metacognitive functions (as well as dispositions, feelings, attitudes, values, and statements of belief about mathematics, problem solving, and one's own personal identity) are assumed to have positive effects in virtually all situations. So, the goal of productive problem solvers is to adopt a single invariant "productive profile."	*Problem solving strategies and metacognitive functions* (as well as dispositions, feelings, attitudes, values, and ways of thinking about mathematics, problem solving, and one's own personal identity) are expected to have effects that vary from one situation to another and from one stage of problem solving to another. So, the goal of productive problem solvers is to manipulate their own "profiles" to fit complex and continually changing circumstances.
Whether impacts are positive or negative depends on functions that need to be served. For example, drawing a picture can lock problem solvers into their current ways of thinking rather than helping them recognize alternatives—and rather than helping them develop beyond current ways of thinking. Similar statements apply to nearly any rule-based characteristic, capability, or action.	
Problem solving strategies and higher-order functions are assumed to be learned in two steps. First, learn them in simple situations. Then, learn to use them in a variety of "real life" situations where additional context-specific knowledge and information are required.	*Higher-order conceptual systems* are viewed as continually developing! And relevant dimensions of development are similar to those that apply to other conceptual systems—and concepts or abilities that depend on them. . . . Thus, higher-order conceptual systems are essentially models-of-modeling—or situated mini-theories of problem-solving. However, compared to the models-of-modeling that researchers develop, ordinary problem solvers think WITH the conceptual systems that they develop (intuitively and implicitly) rather than thinking ABOUT them (formally and analytically). The models-of-modeling that ordinary problem solvers create tend to be far more situated than the abstract and decontextualized models developed by researchers.

Two Ways Of Thinking About Problem Solving Strategies (*continued*)	
Traditional Perspectives	Models & Modeling Perspectives
Problem solving strategies and higher-order functions are expected to function best when they are engaged explicitly during problem solving processes.	*Problem solving strategies and higher-order functions* are expected to function implicitly in most cases during problem solving processes. Explicit attention focuses on current and continually changing conceptions of sub-goals within the task at hand.
Models & modeling perspectives anticipate that, when problem solvers interpret situations, they don't simply engage systems that are logical and mathematical in nature. Their interpretations also involve problem solving strategies, higher-order understandings and processes, dispositions, feelings, values, and beliefs. Furthermore, these latter aspects of knowledge and ability are not simply learned separately and then added onto students' interpretations (or models). Instead, they are integral parts of the models (and underlying conceptual systems) that students develop. So, as interpretations change, feelings and other processes and dispositions also change—and vice versa.	

are the conceptual systems that Manning uses to make sense of them. In fact, as his conceptions of the situations change, the situations themselves change.

Beyond the gradual and often painful process of learning through experience, how does Manning facilitate the development of powerful ways of thinking? One technique is to watch videotapes of past performances, to reflect on them, and to imagine them in future situations. In this way, descriptive levels of analysis provide frameworks so that, when prototypical performances are practiced, Manning is able to flexibly modify descriptive-level processes to fit a variety of continually changing needs and opportunities.

In this book, the chapter on *The Use of Reflection Tools in Building Personal Models of Problem-Solving* (Hamilton, Lesh, Lester & Yoon) describes how we use "reflection tools" to accomplish goals similar to those that Peyton Manninng accomplishes with videotapes and techniques to support reflection.

SUMMARY

Mathematizing reality is a term that we use to refer to the process of developing a purposeful mathematical description (or interpretation) of a problem-solving or decision-making situation. Such processes often involve quantifying, dimensionalizing, coordinatizing, or (in general) mathematizing objects, relations, operations, patterns, and regularities which do not occur in pre-mathematized forms. . . . One of the most important points of this chapter is that the development of powerful models is among the most important (yet neglected) goals of mathematics instruction. Furthermore, many concepts, skills, processes, abilities, attitudes, and beliefs that students develop quite often during model-eliciting activities tend to acquire very different meanings than when attempts are made to teach these same achievements out-of-context and unconnected to mathematical interpretations of specific situations.

Throughout this chapter, the examples and observations that have been given suggest that, regardless whether we focus on conceptual development or skill development, a large share of what it means to "understand" is likely to be neglected unless adequate attention is given to learning activities in which students *mathematize reality*—and express>test>revise their own ways of thinking, as opposed to being guided along artificially narrow paths toward idealized versions of their teacher's or textbook's ways of thinking. Nonetheless, this chapter is not about telling teachers how to teach. Our main goals have been to clarify the nature of important ideas and abilities whose meanings depend on some type of systems-based understandings, and to clarify how these understandings develop. In particular, we have focused on meanings that depend on: (a) within-concept systems, (b) between-concept systems, (c) interacting representational systems, (d) basic skills whose meanings depend on systemic understandings, and (e) higher-order ideas or abilities whose meanings depend on systemic understandings.

Certainly, we are not claiming that ALL mathematics instruction should consist of *model-eliciting activities*. For example, in chapter 2 in a recent book on models & modeling (Lesh & Doerr, 2002), we explicitly describe a variety of different kinds of instructional activities that should surround appropriate uses of *model-eliciting activities*. A variety of guided inquiry activities probably should play important roles in most forms of effective teaching and learning; and, there also are important roles for teaching strategies that involve direct instruction, demonstrations, and repetitive practice. But, if we are concerned about students' abilities to "think mathematically" beyond school, and if we look beyond students' performances on end-of-course tests to also consider what's left a few years after the course ends, then it seems sensible to focus on deep treatments of a small number of "big ideas" rather than being preoccupied with efficient "coverage" of a large number of relatively low-level understandings and abilities—of the type emphasized on easy-to-administer and easy-to-score standardized tests.

REFERENCES

Kelly, A., & Lesh, R. (Eds.). (2000). The Handbook of Research Design in Mathematics and Science Education. Mahwah, NJ: Lawrence Erlbaum Associates.

Lesh, R. (2003). Models & modeling in mathematics education. Monograph for *International journal for mathematical thinking & learning*. Mahwah, NJ: Lawrence Erlbaum Associates.

Lesh, R., & Doerr, H. (2003). *Beyond constructivist: A models & modeling perspective on mathematics teaching, learning, and problems solving*. In H. Doerr & R. Lesh (Eds.). Mahwah, NJ: Lawrence Erlbaum Associates.

Lesh, R., & Zawojewski, J. S. (in press). Problem solving and modeling. The second handbook of research on mathematics teaching and learning. Reston, VA: National Council of Teachers of Mathematics.

Lesh, R., & Johnson, H. (1976). Models and applications as advanced organizers. *Journal for Research in Mathematics Education, 7* (2), 75–81. Reston, VA: National Council of Teachers of Mathematics.

Piaget, J., & Beth, E. (1966). *Mathematical epistemology and psychology*. Dordrecht, The Netherlands: D. Reidel.

Pôlya, G. (1945). *How to solve it*. Princeton, NJ: Princeton University Press.

Schoenfeld, A. H. (1993). Learning to think mathematically: Problem solving, metacognition, and sense making in mathematics. In D. Grouws (Ed.) *Handbook of research on mathematics teaching and learning* (p. 334–370). New York: Macmillan.

Schoenfeld, A. H. (1985). *Mathematical problem solving*. New York: Academic Press.

Silver, E. A. (Ed.) (1985). *Teaching and learning mathematical problem solving: Multiple research perspectives*. Hillsdale, NJ: Lawrence Erlbaum Associates.

Vygotsky, L. S. (1978). *Mind in society: The development of higher psychological processes*. Cambridge, MA: Harvard University Press.

Zawojewski, J., Diefes-Dux, H., & Bowman, K. (Eds.) (under development). Models and modeling in engineering education: Designing experiences for all.

CHAPTER 18

The Use of Reflection Tools to Build Personal Models of Problem-Solving

Eric Hamilton
U.S. Air Force Academy

Richard Lesh, Frank Lester, and Caroline Yoon
Indiana University

This chapter describes a set of research and learning tools that are designed to benefit both investigators and participants who are working individually or in groups on *modeling-eliciting activities* (MEAs) or associated activities. We refer to them as Reflection Tools (or RTs) because, following a problem solving activity, RTs help students briefly record significant aspects about what they have done (e.g., changes that occurred in roles they played, strategies they used, feelings they had, or ways their group functioned). This allows them to later engage in reflections and discussions about the effectiveness of various roles, strategies, and levels and types of engagement.

  RTs serve as an observational device for research, but our paramount interest is in their potential to help students develop important conceptual frameworks for thinking about learning and problem-solving—and about themselves and others as problem solvers. Such frameworks may be considered personal models of problem-solving—or *models about modeling*. So, as researchers and participants collaborate in teams that design, modify and use RTs, they also iteratively express, test, and revise their current ways of thinking about: (a) the nature of solution processes MEAs, and (b) the ways that roles, strategies, and levels and types of engagement should change as problem solvers progress from modeling cycle to another—or from one type of MEA to another. Our underlying conjecture is that, as modelers become more sophisticated in the ways they think about problem-solving and themselves as problem-solvers, they also become more sophisticated in thinking about problems.

  In this chapter, following a brief elaboration of the notion of personal *models of modeling*, we give examples of the use of RTs. Then, we follow this section with a line of reasoning to suggest why RTs can be valuable both for the learner in developing modeling competences and for the researcher in understanding that development. The efforts reported here with RTs are promising but preliminary, involving assumptions and approaches that require far more testing and refinement. The intent of the chapter is to encourage MEA research and practice communities to consider the further development and systematic inclusion or adoption of RT-like activities. The chapter accordingly includes a section summarizing research directions that might help determine where and how RTs or similar mechanisms might be most useful.

PERSONAL MODELS OF MODELING

What constitutes problem-solving? What are distinctive aspects of my own personal characteristics as a mathematics learner or problem-solver? What kind of "mathematical thinking" is useful in my everyday life? Every individual brings to questions of this nature a unique set of experiences and meanings. Most students, and especially school-aged children, might not offer very sophisticated definitions or descriptions; and, their perspectives often are quite shifting and unstable—depending on circumstances. Nonetheless, the efforts we describe in this chapter seek to help them clarify, develop, and leverage those personalized understandings and identities, especially as they relate to simulations of "real life" situations in which some significant types of mathematical thinking are needed. Tools that help students to examine and critique their own ideas about modeling and about oneself as a modeler are expected to lead to important new understandings about both. For example, a few of these understandings might include nascent or evolving forms of the following:

> Group modeling entails a variety of activities such as organizing understandings, brainstorming, conjecturing, and testing, revising, and formalizing approaches; each of these activities is productive at some times but not at others.
>
> There are different roles that a person assumes during different MEAs, and during different stages in the solution of a single MEA and, each of these activities is productive at some times but not at others.
>
> Modeling is iterative. MEAs can produce interim solutions or ways of looking at a problem that might appear futile but actually can contribute to the formulation of a more adequate or encompassing model.
>
> The conceptual systems that I use to make sense of situations often change significantly from one modeling cycle to another, and they often operate without me being consciously aware of them—or thinking about them analytically.

Understandings of this nature also can lead to the following kinds of higher-level perspectives:

> *Models develop.* Reflections about modeling cycles, for example, may help problem solvers see: (a) how individual models develop for various MEAs; and (b) how changes occur in the roles, strategies, and levels and types of engagement that are productive during different stages of solution processes.
>
> *Modelers develop.* As discussed throughout this chapter, throughout this volume (e.g., Goldin, Roth), and elsewhere (e.g., Middleton, Lesh et al., 2002), how problem-solvers see themselves profoundly affects the ways they approach problem-solving activities. Constructs such as self-efficacy, persistence, cognitive flexibility, and tolerance of ambiguity all are enhanced by experiencing, seeing, and reflecting on personal development as modelers.

Developing a personalized awareness of these and other characteristics of modeling and modelers is a realistic precursor to leveraging or taking advantage of those elements. A *model of modeling* can be seen, in part, as one step beyond metacognitive *awareness* to a metacognitive *system* in which learners progressively form a personalized but organized way of thinking that blends awareness with competencies in altering the conditions about which one is aware. Within this system, we see an individual's repertoire of problem-solving strategies not as generic or decontextualized heuristics which are learned as explicit rules—and retrieved as needs arise. Instead, what it means to "draw a picture" or "assume a leadership role" or "engage in group brainstorming" tends to be molded and shaped by evolving interpretations of the situation—at least as much as interpretations are influenced by these strategies, roles, or group functions. Furthermore, all of these conceptualizations and functions may operate intuitively, or without formal awareness, rather than operating formally and analytically. Therefore, one of the goals of RTs is to help students develop these intuitively functioning abilities—which may operate without conscious analysis and without necessarily treating them as explicit rules to be learned.

As argued throughout this volume, the contexts in which strategies, roles, and personal attributes are generated originally are critical to their reuse and generalization to other problems.

EXAMPLES OF REFLECTION TOOLS

Our initial slate of four RTs is still in "early form," but it has evolved considerably through multiple iterations. The goals are: (a) to focus on a small number of RTs which emphasize diverse aspects of problem solving competence; (b) to create RTs which are not time consuming to use; and (c) to select RTs that provide information that is useful to stimulate productive reflections about relevant roles, strategies, or other aspects of a student's problem solving personae.

With the preceding goals in mind, the RTs that we have emphasized have consistently involved the following four aspects of *model-eliciting activities*: individual roles, group functioning, strategies and heuristics, and levels and types of engagement. Appendices A and B give RTs which, like all RTs, are considered to be at intermediate levels of development. That is, as long as students and teachers continue to use them, we expect users to continue to make modifications so that the RTs continue to fit their ever evolving conceptions of relevant roles, functions, strategies, attitudes, modeling cycles and other relevant constructs. This is because RTs are not considered to be "correct" ways of thinking. They are devices to stimulate productive reflections and discussions. Therefore, just as in the case of model-eliciting activities, even naïve or otherwise flawed tools sometimes lead to very productive advances to thinking—as long as they embody current ways of thinking about relevant constructs, and as long as they are submitted to testing and revision.

The RTs that are given in Appendices A and B were developed by students during a recent course on *modeling and problem solving*—where the goal was to engage students in activities that would challenge them to develop their own personal *models of modeling*. During this course, the preservice teachers worked on *modeling-eliciting activities* that were designed to be: (a) simulations of problem solving activities that might reasonably occur in their own everyday lives or those of their future students, and (b) similar to activities they might use in the mathematics classes that they eventually might teach. Then, these undergraduates also worked in teams (which included a graduate student and a faculty member who served as consultants) to design a variety of "tools" that they or their future students might find useful when they were engaged in model-eliciting activities. These tools included RTs; and, during the process of designing them, students expressed their current ways of thinking on the form of artifacts which needed to be tested and revised iteratively. In this way, these undergraduates devoted significant effort to helping to design, test, and refine both (a) their responses to model-eliciting activities, and (b) tools, such as RTs, that they used to enhance their experiences with model-eliciting activities.

In differentiating dynamic elements of group MEAs (such as individual roles, group functioning or strategies), an underlying goal is to find ways to encourage learners to approach modeling itself with some of the same approaches that they bring to bear on problems that they model. RTs not only are intended to help participants identify the structure and stages of MEAs, but they also are intended to set the stage to help them assess, independent of a teacher, their own evolving profiles as problem-solvers. Such assessments can become self-fulfilling as reflectors. As students express → test → revise their tools, they also express → test → revise their understandings of the underlying concepts that the tools are intended to embody—as well as their understandings of the modeling-cycles and modeling processes that are involved in *model-eliciting activities*. At the same time, RTs provide a powerful means for researchers to observe the formation of individual and group models (and models of modeling).

TECHNICAL CHALLENGES IN DESIGNING USEFUL RTs

There are numerous technical challenges that occur during the process of developing useful RTs. Several of these relate to timing—i.e., when to pose the reflections, how much time the tool should take to use, and how to structure the wording so that it is meaningful—but still "loose" enough to encourage students to grow their understandings of the relevant concepts (e.g., modeling cycles, and specific roles, problem solving strategies, or group functions). The intent is for understandings to develop through engagement in the activities; it is not simply to define-and-use concepts in the context of RTs.

Ideally, the RTs should not require more that five minutes to use at the end of MEAs. This is because RTs are intended to contribute to, not distract attention from, the development of substantive mathematical ideas and abilities—and because such tools are unlikely to be used in school classrooms if they require too much time. Also, after a modeling activity, participants generally are tired and wish to decompress, rather than reflect very substantively on their just-completed group time. Waiting for a subsequent class period, though, tends to result in a loss of important details. Finally, it is too distracting to try to capture such "meta-information" information during the 60–90 minute problem solving activities themselves.

In general, we have found that completing short debriefing forms, and preserving impressions immediately after a modeling activity, usually yields a rich set of sufficiently accurate observations to serve the purpose of facilitating productive reflections during subsequent sessions. We have also found that the more participants work together, the more they appear to form similar perspectives on the principle shifts that occur during an MEA. Thus, consensus building and testing-and-revising are the processes that lead to clarity of understanding—rather than relying too heavily of definitions (which are presented up-front) followed by applications.

During the process of designing RTs, there was a significant amount of discussion about whether four tools was the "right" number, about whether these particular tools (about individual roles, groups functioning, strategies, and engagement) focused on the most appropriate topics for reflection, and about whether other topics (such as skill use, habits of mind, or metacognitive beliefs) might be more interesting and productive. Also, even after a given general topic was selected, questions remained about what sub-concepts should be emphasized. For example:

> If the RT focused on problem solving strategies and heuristics, precisely which strategies and heuristics should be emphasized, and how should each be defined?
>
> If the RT focused on individual roles or group functioning, which roles and functions should be emphasized, and how should each be defined?
>
> If the RT focused on individual levels and types of engagement, should it emphasize feelings, values, beliefs, attitudes, or some other attribute?

The tools that our students have developed are in some cases quite directive in order to elicit focused reflections on modeling and avoid unproductive musings or "metacognitive failures" that have been found in other classroom discussions that report on group problem-solving (Goos, Galbraith et al., 2003). But, in our work, such issues are resolved through consensus building, and through iterative testing and revising processes, not though decrees or appeals to authorities.

INDIVIDUAL ROLES

The Individual Roles Reflection Tool is designed to attune problem-solvers both to the specific roles they take and to identifiable transitions in group modeling activities. Each of the different roles (facilitating/leading; recording; monitoring; strategizing; and other) in this RT is important in effective modeling. Appendix A includes early and late versions RT. Notice that in the later version more "teacher guidelines" are given. Yet, the preservice teachers who developed this tool agreed that their "final form" should not be the first one that they or others should use to introduce their future students to the relevant topic. This is because, as one student said: "*No matter how we define things at the start, the kids won't know what we mean. They need to learn by doing. I can't just tell the kids how to think. The point of the whole thing is for the kids to figure out what all of this stuff means. For starters, we just need to tell 'em enough so that they aren't completely confused—and can get started.*"

The RT on individual roles asks reflectors to map their roles against phases that they identify in their subsequent summary of the MEA. By coordinating roles with phases, the RT nurtures flexibility in recognizing and exercising useful roles depending on progress in the MEA. That is, one implicit message of the RT involves recognition that roles vary over the course of group problem-solving; such variation often reflects transitions in model development, and once understood, both transitions and roles can be manipulated in order to produce more effective models. This message is consistent across RTs. Productivity relates to function. Roles, group functions, engagement and strategies vary as a group model unfolds, and a given role or strategy may lead to either positive or negative effects. There is no invariant "correct role" or "correct strategy" but rather roles or strategies that may be useful or productive depending on the modeling context (Middleton, Lesh et al., 2002).

GROUP FUNCTIONING

In this instance, our students' RT on Group Functioning drew on themes of social Darwinianism through its focus on four prominent strands of group modeling.

The *diversity* strand involves sharing new ideas, interpretations, or hunches about a problem. This strand includes possible strategies, predictions or conjectures for selection and testing.

The *selection* strand shifts group members from the brainstorming of the diversity strand to scrutiny, clarification and testing of problem interpretations or potential strategies.

The *communication* strand focuses on activities that lead to the sharing of good ideas within the group; and, it's about coordinating the work so that people avoid duplicating one anothers' work.

The *preservation* strand focuses on activities that lead to the recording of work—so that good ideas are carried forward, and so that bad ideas aren't repeated—and it also deals with issues about whether a solution might be useful in novel contexts, or whether it is only useful for the situations at hand? That is, it asks participants to assess the generalizability of their selected approaches and ways of thinking.

Notice that all four strands involve functions that contribute in distinct ways to the development of new models. On the other hand, the development of new models also influences what it means to engage in each of these activities. Nobody ever masters communication (or selection, or diversification) in general; that is, nobody masters them for all purposes, and in all situations. Instead of being learned as prescriptive rules, these are descriptive terms whose meanings emerge in specific situations based on students' current understandings of the problems-at-hand and relevant problem solving processes.

This RT on group functioning is intended to nurture within an individual's model of modeling the notion of a group personality. How does the group express itself in modeling? What makes for a productive learning community? How can the group develop and maintain a productive problem solving personality? Does the group become *more* productive as it undertakes successive MEAs, and—using the self-fulfilling assessment ideas above—are there ways that emerge because of reflection to help the group develop more productively?

ENGAGEMENT

Elsewhere in this volume, several contributors (e.g., Roth, Goldin) emphasize affective, and motivational or experiential elements of model development that often are overlooked in problem-solving. Among these are conditions that contribute to a student's *engagement* in a task, the subject of the third RT. Engagement is sufficiently broad that it is useful to draw on several frameworks that have been used to analyze it.

Complex factors determine the depth of a student's engagement in a classroom activity. These factors include motivation and cognitive processing constructs such as self-regulation and strategy use (Miller, Greene et al., 1996; Wolters, 2004). They include more esoteric factors such as the subconscious mathematical processing leading to unexpected "Aha" experiences when one is ostensibly off-task (Hadamard, 1954) or the sudden insights recorded by Andrew Wiles in the solution of Fermat's Theorem.

While engagement is a multivariate construct, its apex—full and unbroken engagement in demanding activities—may be characterized as the state of *flow*.

Introduced as a psychological construct by Csikszentmihalyi (1975), this concept of *flow* has been widely researched—often in the context of examining intrinsic enjoyment or satisfaction while engaged in work or play, or the fully concentrated absorption in an activity. One of the most telling characteristics of an individual experiencing task flow is a loss of sense of time. This absorption is one of nine "flow characteristics" summarized by Csikszentmihaly (1996).

The summary that is shown in Table 18–1 is quite useful for suggesting indicators for engagement in an MEA. Additionally, because *flow* is associated with bursts of creativity and high performance, it is reasonable to ask both (a) whether it is possible to *design for flow* or to organize MEAs to help problem-solvers approach or attain high engagement and flow conditions, and (b) how model development phases coincide with such conditions. For example, does the group formulation of a stable or tractable model induce flow-like experience as the model is being tested or revised? In other words, does flow experience emerge more routinely once a stable model is available and the group is testing it? If so, an intriguing conjecture arises: Do more sophisticated modelers experience flow in the "pre-stable" formulation phase by dint of that phase being part of their own stable model of modeling? Such questions may be valuable in understanding why some individuals are not only tolerant of but relish ambiguity in problem-solving. This particular phenomenon, of relishing ambiguity in problem-solving, is a well-established characteristic of renowned scientists (Adelson, 2003). Is it available to so-called average learners? MEAs have already been shown repeatedly to elicit impressive competencies from such learners. It is realistic to ask whether metacognitive skills involved in developing a *model of modeling* might produce flow-like experiences of impressive foresight, perseverance and enjoyment in problem-solving more routinely for all learners.

The characteristics appearing in Table 18–1 fall into two categories: those that emerge from the learner (such as loss of a sense of time) and those for which at least some design or external considerations might realistically be involved. Classrooms, for example, are laden with distractions, and the notion of a student routinely experiencing flow in a classroom in traditional production-style school settings is almost oxymoronic. Yet, our experience suggests that MEAs can help induce the high concentration conditions that lead to learner flow by producing engagement conditions that help learners screen out distractions (Middleton, Lesh et al., 2002). Problems that have a meaningful and real-world application and that are negotiated and modeled in a small group setting lend themselves to such conditions.

TABLE 18–1.
Characteristics of Play and Work Flow Situations (Csikszentmihalyi 1996)

1. There are clear goals every step of the way.
2. There is immediate feedback to one's action.
3. There is a balance between challenges and skills.
4. Distractions are excluded from consciousness.
5. There is no worry of failure.
6. Action and awareness are merged.
7. Self-consciousness disappears.
8. The sense of time becomes distorted.
9. The activity becomes autotelic.

The engagement RT asks students to reflect on these external factors. In part, they help address the condition of designing for flow. But perhaps less ambitious than trying to attain flow is to make progress in conditions of productive engagement such as those enumerated in points 1–5 of Table 18–1. The Engagement RT is designed to help give insight into whether group members are experiencing these conditions and implicitly suggest to students ways to think about task engagement and behaviors (such as feedback to others) that might productively increase engagement.

Another principle that this tool helps to stress is that at certain points in a modeling activity, frustration or dead-ends are normal. If a dead-end is interpreted to mean that the problem is unsolvable, then disengagement is the likely outcome. On the other hand, if students can take frustration, dead-ends, or ambiguity as challenges-in-stride, developing a recognition and then tolerance for them, they are more likely to remain engaged and to seek new ways to interpret a problem rather than giving up and disengaging.

PROBLEM-SOLVING STRATEGIES

The first three RTs involve things that may appear to be *relatively content-free* (roles, group functioning, and engagement); and, in fact, even problem solving strategies often are treated by textbooks as a separate chapter or topic—relatively unrelated to substantive mathematical concepts. But, in our work, even early experience suggests that reflections about heuristics and strategies are mediated by and derive their character from content and context of an MEA. This is why the RTs stress phase changes and shifts while modeling. Reflections on shifts stress the variability of the topics reflected on and cultivate a nuanced understanding of roles and functions, for example that their usefulness depends on situation-specific factors.

The fourth RT involves *strategies* that the group used en route to model development. It is designed to help problem-solvers move away from thinking about heuristics in content-free isolation such as "draw a diagram" or "work backwards" but instead to reflect on them in terms of contexts from which they are situated. What comes first? Generation of a strategy or its formalization? Do we elicit strategies from students out of contexts where they are useful or impart formalisms that they can apply in contexts? This "elicit or tell dilemma" (Lobato, Clarke et al., 2005) phenomenon has multiple dimensions, but in general, there is strong reason to emphasize the learner, engaged in modeling and reflection activities, as a more potent source for *generating* a personal repertoire of problem-solving strategies than the teacher or textbook for *imparting* such strategies as decontextualized or formal heuristics. Although we do not propose an either/or approach to teaching heuristics and strategies, common practice, with its emphasis on pacing the introduction of strategies around the curriculum (NCTM 2000) or attempting to impart comprehensive pattern-matching schemes ("in case of x then do y") seems to bypass the student as the leading agent for developing a set of productive modeling tools.

WHY ARE REFLECTION TOOLS PROMISING?

The approach underlying the RTs in this chapter involves *structured* reflection that appropriately focuses attention on structural elements of modeling. We are attempting to stimulate systems-like—rather than rule-based—thinking about problem-solving and modeling. This design intent is challenging and merits continued refinement and experimentation. A closing section outlines research directions more fully, but forthcoming studies might yield a class of different RT approaches appropriate in different contexts or for different learners. For example, the group functioning tool uses a set of evolution constructs—diversity, selection, and preservation. But a somewhat different lens might entail constructs with slightly different emphases such as "orienting," "organizing," and "testing." The potential importance of well-designed RTs rests less in which specific tools are used when, and more in that they represent the fundamental idea that problem-solvers may develop skillful ways of explaining problem-solving that will make them more skillful in thinking about problems.

Some important ideas are developed below to help justify and clarify this assertion. They involve (a) the ubiquity of implicit knowledge in sense-making and problem-solving, and (b) the value of elaborating on and expanding such knowledge by making it explicit and public, especially in groups that share the same problem-solving experiences.

TACIT OR IMPLICIT KNOWLEDGE IN PROBLEM-SOLVING

A consistent theme of this volume is that problem-solving activities entail the coordination of multiple conceptual systems, and, for the purposes of this discussion, multiple forms of knowledge. While this theme is developed through the terminology of modeling and metaphors of complexity science, traditional classifications of knowledge also create a useful contrast for discussion. How such classifications are organized drives various theoretical frameworks in problem-solving research, and includes categories such as declarative, procedural, heuristic, semantic and strategic knowledge forms. Often these are treated in binary fashion—in other words, that one possesses a procedural skill or does not possess it. We suggest here, though, that conceptual systems that evolve in modeling are rarely of such a binary character. That is to say, they *develop*. Stressing how modeling competencies develop is a greater focus in this work than determining what competencies have been attained.

A different classification or cross-cut of knowledge forms—of explicit versus implicit knowledge—may shed light on such development and highlight the potential value of RTs in observing it and advancing it. Epistemologists have long stressed the importance of "tacit" or "implicit" knowledge, a vast reservoir of cognitive, affective and motivational competencies that are difficult to articulate but central to sensemaking and day-to-day living. It is especially in real-world and day-to-day problem-solving that tacit knowledge plays such a powerful role (Wagner and Sternberg, 1985). One of the earliest and most important modern-era philosophers to develop a philosophy of tacit knowledge, Michael Polanyi, introduced *The Tacit Dimension* (1967) with a simple existence proof. "This fact seems obvious enough; but it is not easy to say exactly what it means. Take an example. We know a person's face, and can recognize it among a thousand, indeed among a million. Yet, we usually cannot tell how we recognize a face we know. So most of this knowledge cannot be put into words" (p. 1). This particular example highlights a property of tacit knowledge, that it can underlie complex and parallel processing with essential connections that are difficult to articulate as simple lists of rules. Myers (2002) furnishes a summary of fields where intuitions and implicit understandings of connections and systems are widely recognized in every-day life—such as physicians who can develop spot and accurate diagnoses or athletes who make complex but instantaneous mental and kinesthetic calculations. He also cites less prominent occupations—for example, chicken sexers who can almost invariably determine the sex of baby chicks but not easily specify the criteria for doing so—where implicit understandings form complex recognition systems. Each of these examples entails an expertise that leverages tacit knowledge forms. We argue that implicit but deeply connected knowledge forms—ways of thinking about what to do when—can be cultivated through group modeling activities involving real-world or meaningful contexts and, importantly, through reflections that press group members to articulate and clarify structural elements of those modeling activities.

REPRESENTATIONAL TRANSLATION: EXTERNALIZING THE IMPLICIT

Mathematical communication competencies are prominently recognized as holding a central role in mathematical development in a variety of research approaches (e.g., Kieran, Forman et al., 2002; White, 2003; McCrone, 2005). But outside of the psychological literature on making implicit knowledge explicit in real-world problems (e.g., Sternberg, 1999) there has been little attention directed at communication fluency patterns that encompass a holistic articulation of factors such as context, intuition, mathematical interpretations and group processing in problem-solving.

Such communication is prized in MEAs and in subsequent RTs. Reflection Tools elicit elaboration of the mathematics embedded and manipulated in both social (group functioning) and situational (the underlying problem) contexts, and they attune the participants to their own respective levels of interest and engagement. This expansive view of what is communicated in reflections about mathematical problem-solving especially entails more than formalizations about mathematics but the full spectrum of meaning and experience that one brings to social, problem-solving or introspective tasks. It is because the full spectrum of meaning and experience, stored both tacitly and explicitly, cannot be separated from authentic group problem-solving that it is important to recognize the coordination of tacit and explicit knowledge in modeling.

During ambiguous or unclear parts of a modeling sequence trial-and-error pathways or predictions tend to entail both hunches and more easily articulated processes. Modeling activity is replete with uncertainty and intuitions that drive the use of more explicit strategies. Connections can be obscure, tentative or patchy. The process of eliciting from problem-solvers their accounts of how or why they interpreted problem-features in certain ways or followed certain pathways involves elaboration of both straightforward and more implicit and often deeper knowledge forms. Such an "externalization" process through a Reflection Tool helps to translate implicit to the explicit, and can lend clarity to those interpretations and pathways and why they were or were not useful to continue. This permits something of a re-internalization of clarified understandings by all of the members of a group, and helps a learner develop a personal empirical base of "what works when" and of recognition or "seeing" problems from a different vantage.

Initially at least, externalization may produce *descriptive* rather than *prescriptive* reflections. Yet, the instructional goal of a model of modeling is ostensibly prescriptive—to bring more systems-like thought and a more systems-like "field-of-view" to new problems. The combination of translating the internal to the external, though, and exposing it to public discussion, does in fact appear to cultivate more prescriptive understandings—the sort of "self-fulfilling assessment" phenomenon discussed earlier. For example, in the somewhat different context of group proof construction in college mathematics, Vidakovica and Martin (2004) referred to successive, verbalized translations back and forth between internal and external (group) conceptualizations of a problem as "parallel and successive internalization and externalization of ideas by individuals in a social context," (p. 465). Their finding of the significant role that such translations exert in building group and individual understandings is consistent with a general theme of the centrality of communication processes in mathematics development. In the contexts in which RTs stimulate internal-external-internal translations, they do indeed appear to help learners consolidate and clarify a personalized repertoire of prescriptive modeling strategies.

PRIOR KNOWLEDGE APPLIED TO NEW SETTINGS

One issue that reinforces the importance of elucidating implicit knowledge through reflection relates to the "prior knowledge principle"—one of the organizing ideas of applying learning sciences to instruction (Bransford, Brown et al., 1999). This principle suggests teaching new ideas as closely as possible to existing or prior knowledge structures. By its very nature as a weighted aggregate of prior sensemaking experiences, implicit knowledge coordinated with explicit knowledge in problem-solving is highly specific to each individual. The problem-solving strategies that implicit knowledge influences are nuanced and localized to their original and generalized contexts. "Prior knowledge" in problem-solving takes on new and more complex dimensions the more holistically it is understood—i.e., in this discussion, the more that it is recognized as the coordination of both easy-to-articulate formal competencies and a deeper reservoir of implicit knowledge. An individual learner arguably will not possess the most precise understandings of his or her knowledge or thinking patterns, but will possess the richest and most textured. The RT process allows exposure of those understandings and their clarification and amplification. Discussion or reflection on the microdynamics of a problem allows more intricate mapping of strategies to context. Reflectors articulate and then internalize both big ideas and understandings about problem-solving and micro-strategies about specific contexts that might be applied elsewhere. RTs may furnish a means to expose and represent such knowledge in a context that is especially conducive to "binding" new ideas or ways of thinking to existing conceptual models.

Other Domains of Reflection Involved in Making the Implicit Explicit

The practice of elaborating on tacit and implicit understandings has valuable analogs elsewhere. For example, structured reflective activity that follows real or simulated military engagements is increasingly adopted as a method for building strategic and leadership competencies for military commanders to prepare them to approach the inevitably different circumstances and complexities of subsequent engagements (Hughes et al., 2006). Such reflections are organized to expose and highlight connections between factors in a battlespace—in effect, to create understandings of a battlespace as a complex system of dynamic and interconnected variables and to build the capacity to rapidly test and convert intuitions about the system into action.

Elsewhere, many organizational theorists have recognized both the importance of tacit knowledge in spurring problem-solving and the role of explicating it to spur innovation and creativity (e.g., see Harvey, Novicevic et al., [2004] or Senker [1995]).

Nonaka and Takeuchi (1995) analyzed work groups within Japanese firms that successfully generate innovations on a continuous basis. They observed problem-solving and knowledge competencies within organizations forming and progressing as their members externalize and articulate tacit knowledge forms such as hunches or intuitions, summarizing with the suggestion that "human knowledge is created and expanded through social interaction between tacit knowledge and explicit knowledge." Takeuchi (1998) subsequently concluded:

> [T]he focus in the West has been on (1) explicit knowledge, (2) measuring and managing existing knowledge, and (3) the selected few carrying out knowledge management initiatives. *This bias reinforces the view of the organization simply as a machine for information processing.* What Western companies need to do is to "unlearn" their existing view of knowledge and pay more attention to (1) tacit knowledge, (2) creating new knowledge, and (3) having everyone in the organisation be involved. Only then can the organisation be viewed as a living organism capable of creating continuous innovation in a self-organising manner.

This "information-processing" and explicit knowledge orientation to knowledge management in professional organizations—or to problem-solving in mathematics curriculum—may make for more delineable and top-down strategies for building competencies, and it might make for cleaner or easier research methodologies, but it factors out the pervasive role that tacit knowledge plays in problem-solving.

Shifting back to formal education contexts, Japanese lesson study practice (Lewis 2002) gives insight into why reflection not only exposes but can improve problem-solving. As teachers use group settings to reflect on, analyze, and highlight flaws and strengths in actual classroom lessons, the lessons become progressively more refined tools for teaching for understanding.

RESEARCH DIRECTIONS ON REFLECTION TOOLS

Future studies involving reflections on modeling can benefit from an analysis of these and other domains of organized reflection. In the current MEA context, the "early form" RTs we have discussed are directed devices designed both (a) to engage problem-solvers in communication about mathematical models and about group modeling and (b) to give researchers tools for observing how models—and models about modeling—unfold.

Tracking the usefulness of RTs for helping *problem-solvers* build a model of problem-solving, and for helping *researchers* observe that process, requires at least two venues for observation. One is the reflections that RTs provide; the other involves observation of their subsequent modeling activities. In both RTs and the underlying MEAs about which reflections take place, several questions will help test the usefulness issue. The principal research question is whether group members express more informed or sophisticated understandings of modeling over time and whether that in turn produces more sophisticated modeling activities. This can be addressed in part with the following "subquestions":

> Do group members appear more fluid in adjusting or altering their roles or strategies during an MEA?
>
> Do hunches and intuitions seem to convert readily to testable directions?
>
> Do group members increasingly coincide in recognizing shifts or transitions in an MEA?
>
> Is it easier for group members to back out of unproductive strategies?
>
> Do dead-ends cue attempts to find new ways to think about relationships in the problem or do they induce frustration?

What are indicators of the development of a personal model of modeling? How do such models influence one's identity as a problem-solver?

If RTs help individuals construct a personal repertoire of problem-solving strategies—an internalized set of what to do when—how does that repertoire develop? What are the systems of thought that the repertoire nurtures and that nurture the repertoire?

Because Reflection Tools are still in the early stages of theory and development, a framework for exploring their value necessarily should also and in parallel to substantive questions attend to the technical issues discussed earlier. These include

What reflections are elicited;

Wording of the tools, including how directed they are;

When the reflections are elicited (for example, immediately after modeling or in a subsequent class session) and whether problem-solvers need to record events during modeling;

What portions of reflections involve group discussion versus individual recording?

All of these technical issues have a bearing on, and in some cases are driven by, the more substantive questions enumerated above. So, they must be addressed in conjunction with those questions as the tools mature and their usefulness is examined.

CONCLUSION

RTs appear promising as ways to help learners experience ways of thinking about modeling and themselves as modelers. Reflectors are able to embed their understandings of a problem structure into reflections about group processes, the roles that they exerted, and the nature of iterations in the modeling activity, in the context of their own experience.

In eliciting and discussing reflections, it is also clear that many intuitions converted to actions amount to the types of heuristics that are often associated with traditional problem-solving instruction. That is, we have found that learners themselves generate important heuristics or processes in modeling activities. But, rather than explicitly teaching condition-action rules, and then prescribing the myriad of contexts in which they might apply, we are finding that well-structured modeling activities help learners develop and internalize their *own* prescriptive strategies.

Every Latin student and most educators know that the word *educate* has its origin in the Latin verb *educare* to "draw out"—a stark contrast to modern instructional practice designed to "pack" procedures or skills into learners. No doubt education does impart explicit knowledge, including facts and skills to learners, and research and debate about "when to tell" and "when to elicit" will continue. But, we are finding that in "drawing out" of students their reflections about modeling, they not only clarify modeling behavior for the researcher/observer—but also for themselves. Pursuing approaches that help learners think about and more clearly see modeling behavior—that is, that help learners develop personal models of modeling—may position the learner more centrally in generating and developing strategic modeling competencies that fit their own respective understandings and abilities, rather than receiving and applying formalized strategies that do not originate in personally meaningful contexts.

ACKNOWLEDGMENT

This chapter was supported in part by National Science Foundation grant 04-33373. This support is gratefully acknowledged. The views in this chapter are those of the authors and do not reflect those of the National Science Foundation.

REFERENCES

Adelson, B. (2003). Issues in scientific creativity: insight, perseverance and personal technique—Profiles of the 2002 Franklin Institute Laureates. *Journal of the Franklin Institute* 340(3): 163–189.

Bransford, J., Brown, A., et al. (1999). *How people learn.* Washington, DC: National Academy Press.

Csikszentmihalyi, M. (1975). *Beyond boredom and anxiety.* San Francisco: Jossey-Bass.

Csikszentmihalyi, M. (1996). *Creativity.* New York: HarperPerennial.

Goos, M., Galbraith P., et al. (2003). Perspectives on technology mediated learning in secondary school mathematics classrooms. *Journal of Mathematical Behavior* 22(1): 73–89.

Hadamard, J. (1954). *The psychology of invention in the mathematical field.* New York: Dover Publications.

Harvey, M., Novicevic, M. M., et al. (2004). Challenges to staffing global virtual teams. *Human Resource Management Review* 14(3): 275–294.

Hughes, R., Ginnett, R., & Curphy, J. (2006). *Leadership: enhancing the lessons of leadership.* New York: McGraw-Hill.

Kieran, C., Forman, E., et al. (2002). *Learning discourse: Discursive approaches to research in mathematics education.* Dordrecht, the Netherlands: Kluwer Academic.

Lewis, C. (2002). *Lesson study: A Handbook of teacher-led instructional improvement.* Philadelphia: Research for Better Schools.

Lobato, J., Clarke, D., et al. (2005). Initiating and eliciting in teaching: A reformulation of telling. *Journal for Research in Mathematics Education* 36(2): 101–136.

McCrone, S. S. (2005). The development of mathematical discussions: An investigation in a fifth-grade classroom. *Mathematical Thinking and Learning* 7(2): 111–133.

Miller, R. B., Greene, B. A., et al. (1996). Engagement in academic work: The role of learning goals, future consequences, pleasing others, and perceived ability. *Contemporary Educational Psychology* 21(4): 388–422.

Myers, D. G. (2002). *Intuition: Its powers and perils.* New Haven, CT: Yale University Press.

NCTM. (2000). *Principles and standards of school mathematics.* Reston, VA: National Council of Teachers of Mathematics.

Nonaka, I., & Takeuchi, H. (1995). *The knowledge-creating company: How Japanese companies create the dynamics of innovation.* New York: Oxford University Press.

Polanyi, M. (1967). *The tacit dimension.* London: Routledge.

Senker, J. (1995). "Networks and tacit knowledge in innovation." *Economies et societes* 29(9): 99–118.

Sternberg, R. (1999). Epilogue—What do we know about tacit knowledge. *Making the tacit become explicit" in Tacit Knowledge in Professional Practice: Researcher and Practitioner Perspectives.* R. Sternberg & J. Horvath, Lawrence Erlbaum and Associates. Mahwah, NJ: 231–236.

Vidakovica, D., & Martin, W. (2004). "Small-group searches for mathematical proofs and individual reconstructions of mathematical concepts." *Journal of Mathematical Behavior* 23(4): 465–492.

Wagner, R., & Sternberg, R. (1985). "Practical intelligence in real—world pursuits: The role of tacit knowledge." *Journal of personality and social psychology*: 436–458.

White, D. Y. (2003). "Promoting productive mathematical classroom discourse with diverse students." *The Journal of Mathematical Behavior* 22(1): 37–53.

Wolters, C. (2004). "Advancing goal theory: Using goal structures and goal orientations to predict students' motivation, cognition, and achievement." *Journal of Educational Psychology* 96: 236–250.

APPENDIX A

Individual Roles Reflection Tool

Teacher Guide

The Individual Roles Reflection Tool is designed to (a) help students identify and assess the roles that they play during model-eliciting activities, and (b) to help students see how and why different roles may be productive during different stages of problem solving—and during different model-development cycles.

Directions:

1. Before beginning the problem-solving activity, show the tool on the overhead and briefly explain what the students are going to have to do when they have finished. Then, discuss the meanings of any terms that they do not understand. Such terms include the word 'transitions' as well as the various roles listed on the tool: facilitating/leading, recording, etc. This will ensure that the teacher and all of the students are on the same page when it comes to understanding what is being asked for on the tool.

   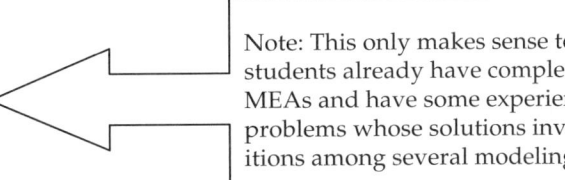

   Note: This only makes sense to do if the students already have completed several MEAs and have some experience with problems whose solutions involve transitions among several modeling cycles.

2. While students are working on the MEA, the teacher can post notes on the wall or bulletin board so that students will be reminded what decisions they made about the meanings of each of the preceding terms. This will ensure that meanings are not forgotten or mixed up.
3. Students also could be asked to jot down when they think a transition is occurring so they don't forget later, but this is optional so as to not take away from the actual problem-solving activity.
4. The first time you use this reflection tool, when the students have completed the activity, work through the reflection tool together as a class—focusing on one particular group. You may want to do this for the first few times until the students feel comfortable and confident about the tool and are ready to use it on their own.
5. After completing the reflection tool each time, ask students if they would like to change the tool, or modify the list of roles, or modify the definitions of any of the roles.
6. The students can use the information that is given on the reflection tool to discuss how well the students are working through the problem—and identify ways to improve the way the group functions.

Questions to Consider:

1. Are students playing multiple roles and contributing to the group?
2. Are students shifting roles in productive ways when they move from one stage of problem solving to another?

3. Are the roles students are playing helping the group-as-a-whole to function more productively.
4. Are students struggling through parts of the problem? What mini lessons could these struggles be turned into to develop student understanding?

INDIVIDUAL ROLES REFLECTION TOOL

Group Problem-solving Self-Reflection

Your name: _____ Name of activity/problem: _____

1. *Group Transition Points* Use the following timeline to show the various transitions your group made over the course of the activity. Please label each transition, explaining what happened during each phase.

   ```
   ----------------------------------------------------------------
   Start                                                     Finish
   ```

2. *Individual Roles* Recreate the lines from the timeline you created above to make as many columns on the chart as your group had transitions. Title each column using the same labels you gave in the timeline. Using the scale below, please rank the roles that you played during the problem-solving activity.

   4 – Mostly   3- Somewhat   2- A little   1- Hardly at all   0-None

Role	
Facilitating/ Leading	-----------------------------
Recording	-----------------------------
Monitoring	-----------------------------
Strategizing	-----------------------------
Other (specify)	-----------------------------

3. *Dominant Roles* What was the primary role each group member took during the problem-solving activity?

APPENDIX B

REFLECTION TOOL FOR GROUP FUNCTIONING DURING GROUP PROBLEM SOLVING

Your name: _____   Name of activity/problem: _____

EARLY PHASE
Brief description:

MIDDLE PHASE
Brief description:

LATE PHASE
Brief description:

The thinking that takes place in your group involves several processes: (1) **diversity**, (2) **selection**, (3) **communication**, and (4) **preservation of ideas**. Report using a scale of A to E, how you think your group participated in each of the four processes as your group worked on this activity.

DIVERSITY	EARLY	MIDDLE	LATE
A. Spent all of our time brainstorming ideas.	A	A	A
B. Spent most of our time brainstorming ideas.	B	B	B
C. Divided our time equally between brainstorming and working with current ideas.	C	C	C
D. Spent most of our time working with current ideas, did a little a brainstorming.	D	D	D
E. Spent no time brainstorming.	E	E	E

SELECTION			
A. Our attention was focused on one approach.	A	A	A
B. Our attention was focused mainly on one approach, with a backup idea.	B	B	B
C. Our attention was focused on two approaches equally.	C	C	C
D. Our attention was focused on several approaches.	D	D	D
E. Our approaches were all being considered equally.	E	E	E

COMMUNICATION			
A. There was a lot of communication – both written and verbal.	A	A	A
B. There was a lot of communication – either written and verbal.	B	B	B
C. There was some communication – written or verbal.	C	C	C
D. There was a little communication – written or verbal.	D	D	D
E. There was no communication.	E	E	E

PRESERVATION			
A. Our strategy was reusable.	A	A	A
B. Our strategy needed modification.	B	B	B
C. Our strategy was only applicable to this problem.	C	C	C
D. Our strategy was weak.	D	D	D
E. We did not have a working strategy.	E	E	E

CHAPTER 19

Diversity-by-Design: The *What*, *Why*, and *How* of Generativity in Next-Generation Classroom Networks

Walter M. Stroup
University of Texas at Austin

Nancy Ares
University of Rochester

Andrew C. Hurford
University of Texas at Austin

Richard Lesh
Indiana University

This chapter describes two interacting strands of 21st century learning and teaching: group-situated design and the use of highly interactive classroom networks. Relative to design, perhaps the most obvious and yet underutilized feature of most classrooms is that they involve *groups* of people. The potential exists for teaching to be much more than a parallel delivery system for individual students (e.g., tutoring); and "the class" can be much more than either a physically proximate collection of individuals or an undifferentiated monolith. Given this potential, we ask: How can we better design for group-situated learning and teaching interactions? How can a new generation of highly interactive classroom networks extend possibilities associated with group-situated design?

The goal of this chapter is to describe a practical approach for creating more fully participatory learning and teaching—in particular, one that makes full use of the capabilities of next-generation classroom networks in conjunction with what we call generative design. According to Stroup, Ares, & Hurford (2005, p. 188) generative design involves "orchestrating classroom activity in ways that occasion productive and creative engagement by participants, characterized by increased personal and collective agency." Such activities are generative in a sense similar to those described by Lesh, Yoon, and Zawojewski earlier in this book. That is, students generate knowledge by repeatedly expressing, testing, and revising their own ways of thinking—rather than being guided along narrow trajectories toward cleaned-up and oversimplified versions of their teachers' understandings.

To address the preceding goal, we describe the *what, why,* and *how* of an approach to formal education where diversity-by-design is emphasized both in terms of ideas and in ways of participating. Relevant forms of diversity may involve a variety of native languages, communication practices, and interaction patterns.

OVERVIEW

This chapter begins with an introductory account of *what* we mean by generative design and its relationship to issues of diversity—focusing specifically on group contexts. To do this, we extend and update several important theoretical foundations of prior research on generative teaching and learning. Next, we describe some ways that network technologies can effectively support generative classroom activities; we also describe a theoretical framework that has proven to be useful for designing for diversity—for moving beyond individualistic frameworks and pedagogies associated with traditional educational psychology. This framework integrates aspects of sociocultural perspectives on learning and teaching as well as pragmatist philosophy's perspectives on representational sense making. This theoretical framework then serves as the foundation for a pedagogically focused discussion of ways to advance diversity-by-design in actual classroom practice—with a particular emphasis on currently underserved students.

Using next-generation technology networks, this generative approach shifts classroom activities beyond monologues (teacher talk) toward more diverse and participatory forms of student-teacher and student-student communication; it encourages the development and use of shared communication patterns and the cultural and linguistic resources that students bring with them to class; in general, it helps students develop what Yakubinskii (1923) calls "apperceptual mass"—the tacit but powerful forms of shared, group-situated, communication patterns (p. 156, quoted in Wertsch, 1985, p. 87).

Overall, our goal is to create teaching and learning activities that promote the development of an emergent ecology where, as is the case in a healthy natural ecological system, diverse participation supports both the development of individuals in the class and the development of the community (or ecological system) as a whole.

UPDATING GENERATIVE TEACHING AND LEARNING FOR GROUP DESIGN

We focus on generativity partly because a significant research literature about generative approaches to teaching and learning already exists. Generative teaching, as discussed by Wittrock (1991), is "a model of the teaching of comprehension and the learning of the types of relations that learners must construct between stored knowledge, memories of experience, and new information for comprehension to occur" (p. 170). Generative learning in Wittrock's framework involves students' abilities to create artifacts that embody their constructed understandings in relation to prior knowledge.

In a similar fashion, researchers from the Learning Technology Center at Vanderbilt University emphasize creating "shared environments that permit sustained exploration by students and teachers" in a manner that mirrors the kinds of problems, opportunities, and tools engaged by experts (Learning Technology Center, 1992, p. 78). Here again, teaching involves "anchoring or situating instruction in meaningful, problem-solving contexts that allow one to simulate in the classroom some of the advantages of apprenticeship learning" (p. 78). While these and other approaches[1] are generative at the level of individual learners, or perhaps even at the level of a small group, they do not provide a clear picture of how to structure cross-individual or cross sub-group learning in classrooms.

Our current efforts are directed toward extending and, in some ways, reconceptualizing these earlier analyses of generativity to make them relevant to the development of group learning and classroom learning communities. The goal is to investigate how to use group-oriented technologies[2] to design instruction in a way that capitalizes on the multiplicity of learners' existing ideas and insights, and that supports the on-going development of powerful new forms of mathematical and scientific reasoning (Stroup, 1997a; Stroup, 2002a, 2002b; Stroup et al., 2002; Stroup, Ares & Hurford, 2005; Lesh et al., 2000).

WHAT ARE KEY CHARACTERISTICS OF GENERATIVE DESIGN?

Because generative design emphasizes student expressivity and inquiry, generative activities quite often are intended to be playful,[3] and, because (in our case) they are intended to encourage the development of important mathematical and scientific constructs, the underlying designs for the artifacts that students are challenged to produce should embody relevant concepts and conceptual systems. Furthermore, because the development of sufficiently useful ways of thinking generally requires several iterative cycles in which current conceptions are expressed, tested, and revised, it also is important for design specifications to be stated in such a way that they give rise to several rounds of generative exploration and/or detailed investigation. Therefore, they should be "thought revealing" in the sense described by Lesh, Hoover, Hole, Kelly, & Post (2000), or Lesh, Carmona, & Post (2002).

---

[1]Other examples include: Senge (1994) who claimed "for a learning organization, 'adaptive learning' must be joined by 'generative learning,' learning that enhances our capacity to create" (p. 14). Perhaps most significant in its connections to the sociocultural analyses taken up more directly later in this chapter is Paulo Freire et al.'s (1998) use of "generative words" in ways that explore the "creative play of combinations" (p. 87) to create new words as part of developing literacy with adult learners in Brazil (see also Callahan, 1999; Stroup, 1997a).

[2]Here we are pointing to the sympathetic and supportive interaction of technology, content, and pedagogy. We most certainly are *not* advancing any sense that the technology of next-generation networks, in themselves, serve to *determine* the nature of the learning and teaching experience (cf., Papert, 1990). Indeed, as we discuss herein and elsewhere, the technologies of classroom networks can be used to support distinctly ungenerative approaches to teaching and learning (cf., Stroup, et al., 2002; Stroup, Ares & Hurford, 2005; Ares, Stroup, & Schademan, 2005, in review; Ares, 2005). It is for this reason our inclusion of the phrase "next-generation" is to point to the possibility of moving beyond these previous uses.

[3]In considering how play actually works for children it is important to emphasize that play is not simply an "anything goes" state of affairs. Instead, play is an organized form of activity:

The premise that Durkheim, Vygotsky, and Piaget share "is that thinking and cognitive development involves participating in forms of social activity constituted by systems of shared rules that have to be grasped and voluntarily accepted.... The system of rules serves, in fact, to constitute the play situation itself. In turn, these rules derive their force from the child's enjoyment of, and commitment to, the shared activity of the play-world" (Nicolopoulou, 1993, p. 14).

Thinking and cognitive development related to play is structured by a system of shared rules that need to be understood and accepted. The power of this form of activity comes precisely from the participants' dedicated engagement related to being part of the "play-world." (see Stroup, Ares, & Hurford, 2005)

As communities of learners create collections of expressive artifacts and actions in relation to some shared task or set of rules (i.e., design specifications), the space of responses should be large enough so that they provide productive environments for group discussions about strengths and weaknesses of alternative ways of thinking. The kinds of behaviors and expressive artifacts that students create also should give teachers significant insights into the ways their students are thinking about the task.

By emphasizing *space-creating* play, we extend the ideas of co-operation[4] and emergent structure[5] to a range of classroom-based learning and teaching activities. The mathematical and scientific constructs that are created or embodied are not determined in advance but are developed though an iterative and recursive process of expressing→ testing→ revising in which groups of learners co-construct relevant artifacts—and the underlying constructs that they embody.

UPDATING THE ROLE OF NETWORK TECHNOLOGY

The kind of next-generation classroom networks that we employ are typically built with the classroom in mind; and, they also lend themselves to activities that emphasize generative design. First, using these networks, the tools that students have in hand enable them to engage in tasks in which they are asked to produce more than simply short-answers to teachers' questions. For example, their responses often involve the development of some complex artifact. Second, more than simply using the network to "aggregate" student responses, a diverse range of complex artifacts can be shared, compared, integrated, revised, and refined collaboratively. Rather than constraining the learning experience to be narrowly individualistic, these technologies can be used to support powerful kinds of social interaction and collaborative investigation.

Typically, using such systems, each student has a networked device that allows him or her to participate either synchronously or asynchronously in group-oriented activities and/or simulations. Often the devices have local input, display, and

---

[4]This is a reference to Piaget's tendency to talk about important forms of group-situated interactivity as "co-operation" (Montangero & Maurice-Naville, 1997, p. 140) in a way that extends equilibration, coordination and operational transformation well beyond the radical individualism that is sometimes projected onto Piaget's constructivism. The narrowly individualistic localization of knowing tends to be most prominent in some of the more recent theories of constructivism (especially as associated with various educational reform movements in the United States). It is a stance that is, perhaps, most clearly articulated in Ernst von Glasersfeld's influential account of "radical constructivism" where "knowledge, no matter how it is defined, is in the heads of persons" (1995, p. 1). By way of contrast, we would suggest in our approach to generative design that there are many co-constructed and interdependent "levels" of organization at which participatory agency, insight, and creative expressivity can be attended to (including at the level called "individual"). Ascribing all knowing to individual "heads," however, is an unnecessary and problematic restriction, especially in designing emergent learning activities meant to take place in group-situated contexts like school classrooms.

[5]What Piaget called *structures d'ensemble* (Chapman, 1988, pp. 343–347). We would also agree with Chapman's observation that "The kinship between Piaget's theory of equilibration and contemporary theories of self-organizing systems is particularly promising" (p. 340). Herein we are attending to emergence and self-organization in group learning contexts (see also Hills, Hurford, Stroup, & Lesh in this volume).

analysis capabilities (e.g., those of a graphing calculator); and students can interact with each other—and with other kinds of devices such as data-collection tools (e.g., motion detectors as used in the *People-Molecules* activity developed by Stroup and Wilensky [2004]). Then, the interactions and emergent results can be projected on a public display space—often in real-time—using a computer or calculator display system. Thus software, hardware, and learners work together to create this sort of emergent and "group-oriented" design. With this sort of infrastructure, both the processes and products are owned by the group itself as they achieve their own forms of what Piaget (following James, 1890, p. 627) called equilibration.

Because a significant number of the preceding kinds of networks are about to become widely available and are poised to become a major presence in classroom learning and teaching, the time is right for us to begin to develop approaches that optimize their generative potential. Although many of the kinds of activities that we will discuss in this chapter can be done without this new generation of network technology,[6] the highly interactive and group-focused capabilities of these systems provide ideal settings to investigate innovative ways to support generative design activities in school classrooms.

THEORETICAL FRAMEWORK: THE *WHY* OF OUR APPROACH TO DIVERSITY-BY-DESIGN

Moving from Individualism to Learning Communities

In traditional educational psychology, most theoretical and methodological commitments have centered on *individual* capacities and abilities (cf., Barab & Kirshner, 2001). For example, the study of individual intelligence (and/or intelligences) and the methodologies of standard psychometrics have been viewed as being among "the most compelling findings and contributions of psychology" (Gardner et al., 1996, p. 58). Accounts of success or ability assessed are individual tasks and group averages (e.g., Hernstein & Murray, 1994; Campbell et al., 2000, p. 40). Similarly, the existing instructional technologies of individual worksheets and textbooks—to say nothing of desks in rows and the absence of a medium of exchange that goes beyond raising hands and calling out individual answers—reinforce this commitment to individual-focused activities (cf. Slavin, 1990). This individualistic focus has had the effect that very few learning tasks in traditional classrooms are designed for groups.

In this chapter we move in a very different direction. We believe that participation in classrooms should be more like what gets enacted in the larger worlds of on-going, group-based practice and problem solving found outside of school (cf., Senge's [1994] analyses of businesses as potential "learning organizations," p. 14). As a result we ask: What would educators do if they actually designed problems or tasks intentionally to advance group-situated interactions where it would make sense to work as a team? How could the capabilities of next-generation group-oriented classroom learning environments support group situated design in classroom contexts? How is what is being proposed more than simply arranging desks in clusters and/or setting up "cooperative" groups to then work on relatively traditional tasks (cf., Slavin, 1990)?

A more robust and compelling notion of group design is called for—and it should be one that views a classroom as a place where a dynamic ecology of ideas

---

[6]Indeed, they have been used in pre-service elementary teacher education (cf. Stroup, 1997a).

and expressive artifacts can interact and develop. An explicit role for content also needs to be central to developing activities for group-situated learning and teaching. This is because the communities we have in mind are not simply communities of people, they are also communities of ideas, and our goal is to design activities that promote the development of each of these types of communities. This means developing activities that create the need for targeted concepts and conceptual systems and thus, ours is not a content-independent or content-neutral account of learning and teaching. Indeed, because of the kind of embodied knowledge that is emphasized in generative design, "the line between learning activity and content becomes so blurred and intermingled that the mathematics and science actually become the foundation for highly socially situated pedagogy" (Stroup, Ares, & Hurford, 2005, p. 187). Lesh, Hole, Hoover, Kelly, and Post (2000) also give some design principles for developing these activities. In such situations, complex mathematical and scientific artifacts, gestures, and dynamic discourse become the on-going medium of exchange and expression. Designing learning activities in this way, we believe, constitutes a paradigm shift from current practice (see also, Lave & Wenger, 1991; Moll, 1990; Rogoff, 1995).

To learn how group-based mathematical and scientific problem solving activities might function in schools, one promising approach is to investigate how such activities occur in learning communities outside of schools. For guidance in this, we look to the kind of studies reported in Part A of this book—and to well-established critiques of learning in school versus learning out of school (Lave, 1988; Nuñes et al., 1993; Resnick, 1987; Saxe, 1991, 1994; Schoenfeld, 1988).

Toward a Framework of Participation and Representational Truth-Making

Pragmatist philosophy describes why important links exist among our commitments to diversity, expressivity, and the central roles that social interactions and community play in participatory learning and teaching. C. S. Peirce, a brilliant scientist and mathematician as well as the founder of pragmatism, was forceful in drawing attention to the essential role that community plays as the origin of a conception of reality: "The very origin of the conception of reality shows that this conception essentially involves the notion of a COMMUNITY, without definite limits, and capable of a definite increase of knowledge" (Peirce, 1982, p. 311).

The writings of John Dewey (1916) also are emphasized to explain why participation in communities plays such a central role in pragmatist philosophy.

> There is more than a verbal tie between the words common, community, and communication. [People] live in a community in virtue of the things which they have in common; and communication is the way in which they come to possess things in common. What they have in common in order to form a community or society are aims, beliefs, aspirations, knowledge—a common understanding.... The communication which ensures participation in a common understanding is one which secures similar emotional and intellectual dispositions—like ways of responding to expectations and requirements. (pp. 4–5)

The preceding perspectives emphasize the fact that, for ecological systems to develop, diversity is important. As James (1978) notes, "Profusion, not economy, may after all be reality's key-note" (p. 93). So, "profusion" of ideas should be a key characteristic of generative classroom practices. This is because diversity (or "profusion of ideas") is what creates the space of possibilities that students and teachers

DIVERSITY-BY-DESIGN

need in participatory learning and teaching. At the same time, in order for development to occur in these complex ecological systems, not only diversity but also an affirming sense of selection is needed. In learning ecologies where students are intended to *make* mathematics and *do* science, pragmatists would emphasize the proactive sense that *truth is what truth does*. Representational truth must, as William James (1978) suggested, have "cash value" (p. 97).

TAXONOMY OF GENERATIVE ACTIVITIES

This section presents a compact taxonomy of several kinds of classroom-situated generative activities that has proven to be especially useful in our work with teachers and students. The taxonomy is organized in terms of pathways and endpoints. Pathways are intellectual and/or behavioral routes for arriving at given endpoints. Endpoints are outcomes or artifacts created by learners that represent some form of completion of the generative task (see Figure 19–1). As with all taxonomies, such an outline is merely a *heuristic*. In this case it is a heuristic for supporting generative design and should be evaluated in terms of its utility (or "cash value") in supporting participatory learning and teaching across a range of important learning tasks. We begin our discussion with "nominally" generative learning tasks that are most like those used in traditional classroom practice. We then move on to learning tasks that are more fully generative in ways that support participatory and group-oriented learning and teaching.

Nominally Generative Activities

Using a pathways-and-endpoints representation, the form of generative activity shown in Figure 19–2 is only barely, or nominally, generative. This is because the structure of the activity has an agreed upon endpoint and a single pathway to this endpoint. An example might involve asking students to simplify the expression

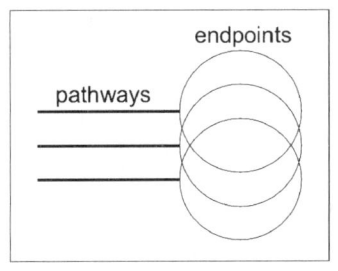

FIGURE 19–1. The taxonomy centers on a pathways and endpoints analysis of generativity.

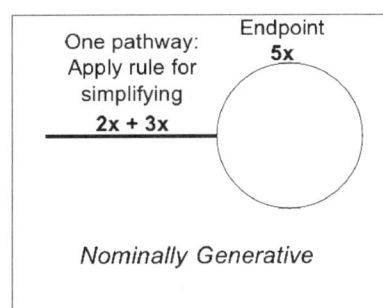

FIGURE 19–2. Nominally generative activities have one pathway to get to a single endpoint.

$2x + 3x$ or to balance a given chemical equation using well-rehearsed procedures. For the example shown in Figure 19–2, the endpoint would be $5x$ and the pathway for getting to this endpoint would be the application of a specific rule for simplifying (e.g., "combining like terms").

This form of activity is nominally generative because the "space" of participation and analysis is only about whether a given response is "right" or "wrong" in applying a given rule or procedure. Nominally generative activity accounts for much of traditional classroom practice. Group behavior is treated as the simple aggregate of individual responses and/or as the average score on a "right/wrong" type assessment. As a design approach, such activities are also relatively ineffective in putting students' ideas at the center of classroom discourse and learning.

Although nominally generative instruction is aligned with many types of "direct instruction" (Silbert et al., 1997; Kameenui & Carnine, 1998), in our work with teachers, we have found it helpful to talk about nominally generative questioning as being associated with one-on-one tutoring. Unfortunately, these nominally generative approaches tend to "break" as they are pushed to scale beyond one-on-one situations. They break primarily because, during the time period when teachers are asking one student such a question, other students are left with little to do. Furthermore, the right-wrong nature of answers provides little opportunity to compare alternative student responses and ways of thinking.

Activities Involving Multiple Pathways and an Agreed Upon Endpoint

Nominally generative tasks sometimes can be made significantly more generative simply by reconsidering the form of the question. For example, rather than asking students to simplify 2/4, they can be encouraged to create ten fractions that are the same as 1/2. Rather than asking students to simplify the expression $1x + 3x$, students can be asked to find five functions that are the same as $f(x) = 4x$ (see Figure 19–3). For a middle school science activity about Hooke's Law (applied force = spring constant times deformation), rather than ask students to find the spring constant for a given spring, ask them to create, using a box of various springs and some pulleys, a spring-based system that has a spring constant in a target range. Space-creating *play* (Stroup et al., 2002; Stroup, Ares & Hurford, 2005), not simple item-response convergence, is a central feature of our approach to generative instructional designs.

These sorts of multiple-pathway-agreed-upon-endpoint tasks have the potential to significantly advance the group's engagement with mathematical and/or scientific structures. Creativity, flair, and novel insights are more likely to be acknowledged and celebrated; and, some important issues of cognitive transfer (cf. Bransford & Schwartz, 2001) can be addressed as each student can see his or her thinking as a

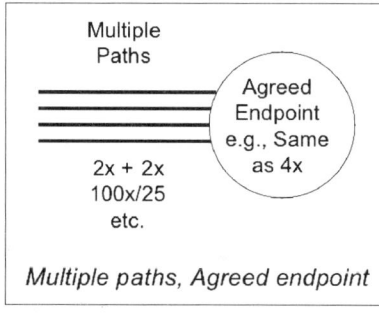

FIGURE 19–3. Visual representation of multiple path, agreed upon endpoint task.

kind of instance situated in relation to potentially unifying forms of scientific and mathematical participation.

When collaborating with pre-service and in-service elementary and secondary teachers, we began working on this kind of multiple-pathway-agreed-upon-endpoint design well before the latest generation of highly interactive classroom networks was developed (cf., Stroup, 1997a). In learning to create these kinds of activities, teachers found it useful to think about turning the *answer* to a tutoring-type question ("$4x$" as in "$1x + 3x = 4x$") into the *question* (Can you find expressions that are the same as $4x$?).

Highly interactive networks support this kind of generative design by allowing the complex expressive artifacts to be readily shared, recorded and displayed together. Moreover, students can use the network to submit responses anonymously, and then decide later if they want to take ownership of a particular solution in whole-group discussions (Davis, 2002).

Activities that involve multiple pathways and an agreed upon endpoint move us down the road of supporting participatory approaches to learning and teaching by illustrating how relatively traditional tasks can be modified to serve more generative and engaging activities. Expressivity, creativity, and inventive representation become more central than simple denotation (i.e., comparing with a single correct answer or simply practicing a given procedure). Furthermore, students develop the sense that truth is being made, not just pointed to. Truth becomes what truth does.

Using such approaches, our experience suggests that *there are no significant topics in any standard curriculum that cannot be engaged in a generative and participatory way*. Moreover, through the display and exchange of complex artifacts (e.g., $1,000,004x - 1,000,000x$) next generation classroom networking can serve to broadly support and enhance this paradigmatic shift to group-situated learning and teaching practices.

Modeling—Multiple Pathways and Endpoints Where Fit with Data Is Central

Modeling has the potential to be an especially effective type of multiple-pathway-and-multiple-endpoint activity. The artifacts tend to be sufficiently complex so that many levels and types of responses have some merit, yet comparative strengths and weaknesses also can be discussed. Because learners can create different models that yield distinct outcomes, a central feature of the subsequent conversations is how well the outcomes of the models fit with data (real or anticipated) or experience (broadly conceived).

Unfortunately, modeling in classrooms is often pursued as if it was nominally generative. For example, much of current laboratory work in science classrooms has this collapsed structure. Students are asked to use a prescribed model (a single pathway) to create computed outcomes that are then mechanically compared to the actual data collected using tightly scripted laboratory procedures. A similarly collapsed notion of modeling is also what gets discussed when the application of a particular mathematical idea is presented in textbooks or as part of classroom presentations.

Modeling at its best, however, is closer to what is suggested in Figure 19–4, where learners *create* a range of models and use them to *generate* model outcomes (implications or predictions). These outcomes can then be discussed in terms of "goodness of fit" to real or anticipated data. Structural conversations about the ways in which various models might be similar or distinct, their range of utility

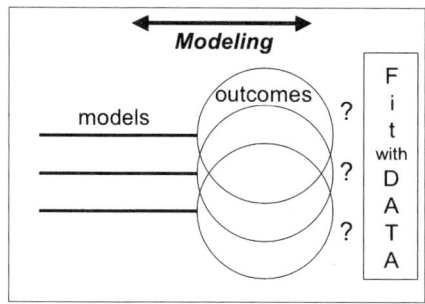

FIGURE 19-4. Modeling characterized as series of pathways, endpoints, and comparisons with data (experience).

(Would your idea work in space? Under water?), their usability to others, and so forth, can and should occur. In addition, issues related to what it means to fit data can be discussed, thereby provoking significant conversations about representativeness, the need for multiple trials, and so on. In Figure 19-4, the double arrow over modeling indicates that models and outcomes interact, iteratively, in the evolving sense-making of learners.

The pragmatists' notion of "truth" is especially applicable to modeling—"models consist of purposeful conceptual systems that are expressed using a variety of interacting media (concrete materials, written symbols, spoken language)" (Lesh, et al., 2000, p. 214) and are used to organize experiences and actions in the world. While a generative sense of modeling certainly can be carried out without network technology, next generation network capabilities can allow students and teachers to make visible and act on the "interacting media" used to express a given set of models. Whether it is a drawing, a sketch in a network-enabled geometry environment, a finite-difference equation, text, voice, or a computer program (e.g., the HubNet system supports aspects of this latter capability [Wilensky & Stroup, 2005]), next generation networks offer the potential for making the machine-based interacting media associated with mathematical and scientific ideas the coin of the realm in pursuing generative approaches to modeling. Models can be made more visible to the group and can be acted upon directly in a network space. Moreover, network mediated role-playing (including using participatory simulations as discussed below) can underscore for students that useful and informative modeling is often the result of negotiated participation in *groups* of interacting modelers.

Situating modeling activity in this way helps to scaffold students' understanding of how modeling can and does work in professional communities outside of school. The practices of many communities of science and mathematics center on the ability to create and compare models. For example, debates about global warming, monetary policy and approaches to providing health care are just some of the contexts of where communities of practice, and citizens in a democracy, are called upon to compare the quality of models, outcomes, feasibility, and fitness. Examples of activities that move in this direction, and that can be mediated using network-based capabilities, can be as simple as asking students to sketch and submit, using their individual devices, renderings of what the position-time graph would look like for a softball player sliding into home plate. The graphs could be displayed publicly and all the graphs sent back to the students. Using low-cost motion detectors, data can be collected on the school playground and then sent out to the students. The question might then become which of the student-generated graphs best "fit" the actual data?

More ambitious examples include topics or issues the students would choose as meaningful for them to model. Socially significant questions about what should be modeled can be engaged. In short, a true model-making community of practice would be enabled, and the group oriented design of next-generation network technologies stand to play an important mediating role in these efforts.

Designing modeling experiences with a multiple-pathway-and-multiple-endpoints design, where outcomes and fit with data play a significant role, has proven helpful in our efforts toward invigorating more generative approaches to modeling. On the other hand, nominally generative approaches to modeling, especially as practiced in many current science curricula, can actually subvert the effort to make authentic modeling central to formal learning and teaching.

Design Activities—Multiple Pathways and Endpoints Where Satisfaction of Goal I Central

Design activities are similar to modeling discussed in the preceding section. Both should be multiple-pathway and multiple-outcome tasks; and, in order to be significant from the point of view of education, both should involve models or designs whose underlying constructs are mathematically or scientifically significant. The primary difference between these two types of activities is that, for design tasks, fit with a goal or a set of design specifications replaces the analysis of fit with data (see Figure 19–5).

Of course, design tasks, just like modeling tasks, can be carried out in a nominally generative way where the approaches are so severely constrained, or overtly suggested in advance, as to make the pathways very few, the outcomes relatively uninspiring, and the conversations about the quality of fit with the goal so limited as to be all but perfunctory. Alternatively, as powerful as a good design activity can be, there is still the danger that—as is true with modeling—the opportunities to engage larger "structural" issues surrounding design can be ignored.

More than just creating specific solutions, students should be invited to compare designs. Are there patterns in the design approaches? Would a successful (or unsuccessful) design in one context be optimal in other contexts? What role, if any, should aesthetics like elegance play (e.g., both a BMW and an economy car can move passengers from point A to point B, so why would anyone buy a more

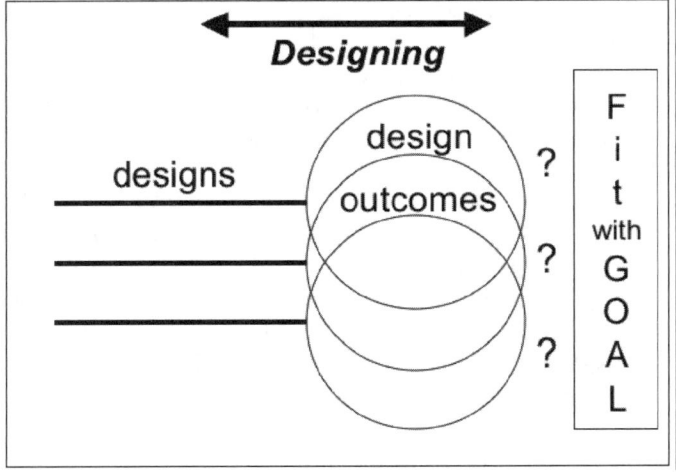

FIGURE 19–5. Generative design tasks are like modeling tasks except fit with a goal is central.

expensive BMW)? Absent these larger structural conversations about the always-present (if often only tacit) social context of design and absent an analysis of how design-based learning can be about producing more than a collection of solutions, the full potential of a design task is cut off. In other words, as "fun" as design projects can be, simply having students design singular instances of solutions to a design task without having this larger conversation misses the opportunity for the group's learning to become more than the sum of these singular solutions.

In the world outside of school, the aesthetics and other social norms surrounding a solution can matter as much as the functionality (cf., Pacey, 1983). Solutions that are more economical or more in line with current practices and/or institutional norms can win out over adequate, even inspired, solutions that also fit the goal. Iterative rounds of design (see double-ended arrow in Figure 19–5) as carried out in the in the world of work are often highly situated in ways that students should learn to engage as part of making school-based design practices more like authentic design practices outside of school.

When these types of lessons are approached in a more generative way, socially situated and cross-case structural issues can be explicitly addressed. This approach is, perhaps, closest to what researchers at Vanderbilt refer to as generative teaching and learning. For these researchers, generative learning and teaching are seen as being about mirroring practices found in apprenticeships, where instruction is embedded in professional communities' problem-solving processes and traditions (Learning Technology Center, 1992, p. 78). The apprenticeship now can grow up and around the unique and compelling diversity of design expertise found in the classroom. In this way design comes to be situated in a group in ways more closely resembling the interaction of abilities and generative insights that are the hallmark of successful design teams outside of school contexts.

As with modeling, network supported design tasks can make the complex design artifacts (e.g., a CAD-like drawing) and representations public and even interactive (see, for example the fly-through capabilities associated with modern architectural design). Cross design analyses related to what makes a "good" design are more readily supported and apprenticeship begins to be seen as a more fully shared, group-based activity.

ROLE-PLAYING ACTIVITIES

Mathematical and scientific role-playing activities represent another important kind of emergent group activity that is well supported by next generation networks. In role-playing simulations, learners often assume *iconic* (like-the-thing)[7] roles in a system, and through their interactions they create emergent behaviors of that system. Then, in a classroom network with projection capabilities, learners can use their individual devices to, as examples, relocate their individual points in a Cartesian coordinate system according to a rule like "move to a place where your y-value is two-times your x-value" (TI Navigator 2.0™), turn a light green or red in a simulated traffic grid (Wilensky & Stroup, 1999; Stroup, Ares, & Hurford, 2005), or control the motions of simulated elevators projected simultaneously for the whole class (Hegedus & Kaput, 2002).

---

[7] This characterization of "iconic" closely follows that of the pragmatist C. S. Peirce (1909), "... I had observed that the most frequently useful division of signs is by trichotomy into firstly Likenesses, or, as I prefer to say, *Icons*, which serve to represent their objects only in so far as they resemble them in themselves..." (pp. 460–461).

While role-playing activities have a long history of use in education, our interest for this chapter is in their implementation in highly interactive classroom networks. A huge space of possible role-playing activities can be supported in these new systems, and much of what we have to say is intended to apply to most forms of network-supported role-playing activity. We are, however, particularly interested in a special kind of emergent role-playing activity called "participatory simulations."

Participatory simulations are role-playing activities that are often more overtly about ideas related to dynamic systems and/or complexity theory (e.g., feedback, the butterfly effect, etc; see Wilensky & Stroup, 1999).[8] Students assume roles and the emergent behavior created in the network then becomes the object of attention and analysis. Roles can be varied or refined by students and the resultant emergent behaviors can be iteratively explored.

As is suggested by the two upward-turned arrows in Figure 19–6, the classroom-based analyses of the emergent role-playing behavior can be framed as either a modeling (Figure 19–4) or a design task (Figure 19–5) or as some combination of these. For participatory simulations there is the added requirement that the modeling- or design-related analyses include overt attention to systems-related constructs.

For example, when introduced to the Gridlock participatory simulation (Wilensky & Stroup, 2005), students are presented with the following scenario: The Mayor of the City of Gridlock is unhappy with the traffic congestion in town and she has commissioned the class to improve the situation (Wilensky & Stroup, in review). The goal of the activity is for the students to find ways of optimizing traffic flow for the simulated city (see Figure 19–7). With a focus on the goal of improving traffic flow (or getting the Mayor re-elected) the Gridlock activity can be considered a kind of design task. Optimization, feedback, and emergence are among the systems-related ideas that are expected to be engaged.

To start the conversation, students are asked what they know about traffic flow. In response to the teacher's asking, "What are some of the things that you guys listed that would be indications that traffic would be good or bad to you?" the learners articulate a wide range of factors that may impact complex phenomena like traffic. Some of these responses involve behaviors of individual drivers (slow drivers), and others are related to the structure of the roadways or the context for the behavior (e.g., barrier walls too close, lights too long). Collectively, these lists suggest that students do have an initial appreciation of how individual agent behavior (drivers) in an environment can have consequences for the emergent features of a complex system (traffic).

---

[8] As is noted in this publication and others (e.g., Wilensky & Stroup, in review; Stroup, Ares, & Hurford, 2005), possibly the first major instance of where participatory role-playing was used in the context of systems dynamics and systems learning was *The Beer Game* as developed by Jay Forrester and his systems dynamics group at MIT in the early 1960s. In a way that alluded to the military use of simulator environments in WWII, these participatory simulations were called "flight simulators." There is a significant literature related to *The Beer Game* and interest in this participatory simulation has been recently revitalized as a result of its appearance in Senge's widely read *The Fifth Discipline* (1994). Diehl (1990) appears to have been the first to use the phrase "participatory simulations" to describe these activities. Over the years different technologies have been used to implement participatory simulations. These implementation range from the use of simple paper and pencil (e.g., Senge, 1994; Stor & Briggs, 1998), to the use of electronic badges (so-called "Thinking Tags," see Colella, Borovoy, & Resnick, 1998; Borovoy et al., 1996; 1998), handheld technologies (Stroup, 1997b; Soloway et al., 2001 [using a Palm OS device]); and a new, network-based HubNet architecture (Wilensky & Stroup, 1999, 2005, in review). Our focus herein is on participatory simulations as implemented in next generation classroom networks.

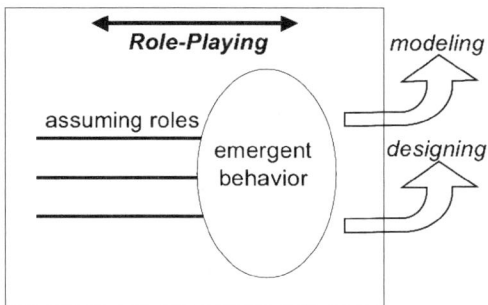

FIGURE 19–6. A range of role-playing activities, including participatory simulations, are seen as generative.

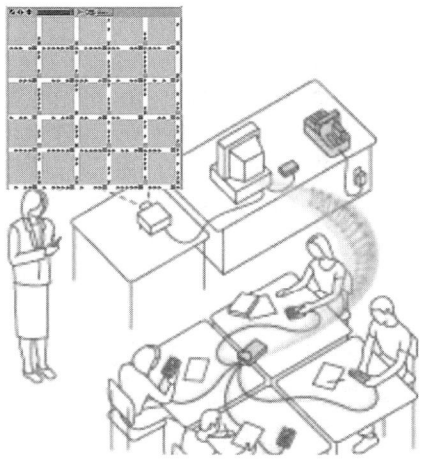

FIGURE 19–7. Students use interactive classroom network to control lights in a simulated traffic grid.

During the simulation, students control individual traffic lights and also call out directions and strategies to each other aimed at optimizing flow (see Figure 19–7). The electronically mediated gesture of pushing a button and physical gestures like pointing assume *social* significance in ways more thoroughly discussed later in this chapter. Students call out, "Change your light"…"I got it" and similar exclamations as part of the highly interactive and animated engagement with the activity. These socially situated gestures and utterances then become available for "inner speech" (Vygotsky, 1978, p. 57) in ways that we discuss toward the end of this chapter.

After participation in the role-playing activity, whole-class analyses of the traffic system can center on students' exploring the dynamic relationship between individual or sub-group behavior, and the resulting, emergent, behavior of this complex dynamic system. In the Gridlock simulation, these insights come both from using the students' personal experiences outside school as well as from their participation in creating and analyzing a dynamic system. This participatory engagement with complex, dynamic activity is intended to support the development of more robust, incisive, and powerful understandings of the kinds of complex phenomena students and researchers tend to find most interesting about the social and natural systems outside schools.

While to date this Gridlock role-playing activity has been used primarily as a design-focused generative activity, other role-playing simulations tend to be more modeling oriented (e.g., a disease simulation where students iteratively enact the spread of a disease through a population and then develop versions of something like the logistic curve to model their experiences). Design-based uses of the Gridlock simulation notwithstanding, all such role-playing simulations are themselves models and the Gridlock simulation is no exception. Models are evaluated in relation to their fit with, and ability to organize, experience. While much about the Gridlock participatory simulation is not exactly like real traffic (e.g., the traffic only flows from left to right or from top to bottom, the cars cannot change their direction, and the traffic grid wraps back on itself [is toroidal]), there are also significant ways in which it fits "enough" with aspects of real traffic that the conversations about reaching the goal of optimizing traffic flow can be engaged in subtle and truthful ways.

For example, the students begin to understand—as a result of trying to compare their different strategies—that they can't just report to the Mayor of Gridlock that they "kinda, sorta think this strategy might be best." They realize that they need to work to create and evaluate the utility of distinct metrics (e.g., average wait time, number of stopped cars, etc.) for traffic flow. These discussions can then implicate broader, highly socially situated, issues related to what it means to optimize traffic flow in a way that might better help the Mayor win re-election. This could mean the students might optimize traffic flow in terms of a metric of average speed but then find out, through creating and distributing a questionnaire, that drivers actually value more highly continuous motion (fewer and shorter stops) over higher average speed and that optimizing for continuous flow (over higher average speed) might better satisfy the Mayor's re-election goal.

The social contexts for modeling and design are made salient and accessible through the use of this kind of role-playing simulation. Significant sense-making and representational invention (e.g., exploring different metrics for traffic) emerge and develop in ways that are not limited to being only about how this system is literally, or exactly, like something else; in this case, like real traffic. Expressive and coherent truth-making can and does occur without requiring the activity be judged only in terms of being an exact copy of something else. Indeed, there may be no formulaic way of deciding in advance what forms of like-ness, how much like-ness (or even divergence) or even whether possibly jumping into a completely "abstract" or self-contained system of interactions (e.g., moving a point around a coordinate system) makes the most sense for learning and teaching particular "big ideas" in mathematics and science. These options have to be explored in terms of the work they do for real learning and teaching.

With participatory simulations—as an important subset of network-supported, role-playing activities—"big ideas" related to systems dynamics and complexity theory are to be explicitly embodied and discussed. These topics may include feedback, characteristics of emergence, relations between micro- and macro-states, the butterfly effect, and optimization. With the publication of widely read books about research in systems dynamics and complexity theories, interest related to these ideas has spread from the scientific community to popular culture (cf., Gleick, 1987; Senge, 1994, Kauffman, 1995; Kelly, 1994; Holland, 1995; Waldrop, 1992). Elsewhere we have discussed in greater depth advancing systems-related learning with participatory simulations (cf. Wilensky & Stroup, 1999), and we believe an engagement with these topics, like the use of emergent generative design, should be seen as one of the hallmarks of future-oriented learning and teaching. Participatory simulations represent an important instance where innovative content and innovative classroom practices intersect in substantial ways.

As with modeling and design tasks, if the sole outcome of participating in a role-playing activity is only that a certain behavior emerges—and there is no subsequent analysis of how the behavior might have developed, how it might be different in another iteration or under different conditions, or how it might be like or unlike other systems—then little learning of a fully generative or structural nature is likely to occur. However, if thoughtfully utilized, next-generation networking can assume a particularly significant standing relative to using role-playing activities to advance the teaching and learning of a wide range of "big ideas" in mathematics and science, including systems-related ideas.

Regarding the latter, the complexity of classroom-based, network-supported, activity maps well onto the complexity of dynamic systems like traffic or population ecology. There is a good fit between aspects of classroom-based interactivity supported by next-generation networks and aspects of a range of kinds of complex phenomena (i.e., students, behaving as agents using local strategies, can create emergent behaviors in a simulated complex system). As a result, the group and its interactions stand to become a good "manipulative" for exploring the behavior of complex systems. Highly interactive networks (e.g., HubNet) capable of implementing the necessary level of functionality are, we believe, important to the successful realization of the potential of these and other role-playing activities for 21st century classroom learning and teaching. Both the *what* and the *how* of 21st century learning and teaching are to be changed by the focus on learning system dynamics using next-generation classroom networks.

EXPLORATIONS OF KIND AND QUALITY OF PATHWAYS

These are possibly the most challenging of the generative forms to describe. Relative to the pathways and endpoints depiction, the focus here is not so much on getting to an endpoint as it is on exploring the "quality" and different kinds of possible pathways (Figure 19–8). Not only might this type of exploration involve many ways, for example, to prove the Pythagorean Theorem, but it also would include a long list of broad issues related to how reasoning moves forward, what the nature of development or improvement is, establishing cross-context similarity in the structure of a given system, structural reasoning proper, generalization, reflective abstraction (as it appears in Piaget's writings), or how situated-ness helps to determine the nature of reasoning and belief. Even notions about general systems and complexity theories can be seen to exist as kinds of explorations into the nature of structural reasoning.

This attention to forms (kinds) of reasoning and the quality of pathways can be engaged mechanically or by rote, as is often the case with many students' experience of geometric proof or with students' experience of studying algorithmic design in computer science (e.g., the different approaches to sorting). But to the extent that these ideas can be engaged generatively, the potential of students' attending to the *forms of reasoning* can assume greater significance in group-based learning settings.

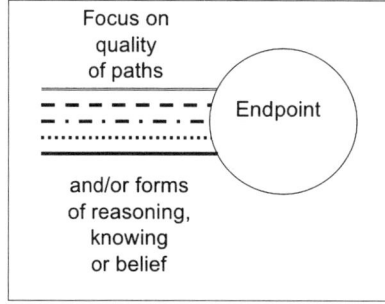

FIGURE 19–8. Exploring the kind and quality of pathways is generative.

The particular power of network-supported capabilities relative to this type of activity comes from making visible multiple instances of a particular kind of reasoning. Because this form of generativity is focused on rhetoric and logic, broadly conceived, students and teachers are supported in engaging the senses in which the particular forms of reasoning they use are inter-related. What *kind* of reasoning, for example, allows us to speak of this expressive artifact "$3x + 1x$" as being the "same" as "$2x + 2x$"? Even more provocatively, how is this form of reasoning like or different from the forms of reasoning that allow us to say that two pieces of music are both jazz? Conversations about the ways in which forms of expressivity and exemplification might be seen as inter-related (or distinct) can be engaged. Network supported interactivity allows the pathways learners use to become visible (e.g., a collection of proofs in a networked version of something like the *Geometric Supposer* [Schwartz & Yerushalmy, 1985a, 1985b], or a collection of sort algorithms students submit) and thus supports this sort of analysis of quality or kind of pathway.

ELABORATION OF THE SOCIAL SIGNIFICANCE OF GENERATIVE DESIGN

Up to this point, our attention has focused primarily on the emergence of structural patterns and the senses in which content can serve to organize the social activity of the classroom. The significance for seeing such learning and teaching as socially mediated is also vital to our belief in the value of generative design, especially as it is supported by next generation networking. We now shift attention toward an account of *how* generativity, as an approach to diversity-by-design, works in group-situated learning and teaching.

In relation to the social aspects of learning and teaching, our previous work has emphasized the central roles assumed by participation—playing a part rather than playing along—and agency, where willing and doing are more fully unified (Stroup, Ares, & Hurford, 2005). We add to this previous work by drawing attention to how generative design can both minimize barriers and serve to heighten participation for all students, and particularly for students whose cultural and linguistic backgrounds often are under-valued or excluded.

By supporting dialogic interactivity, generative design encourages participation and the use of a discursive "apperceptual mass" (discussed in a subsequent section) that emerges during class or that students bring with them to class. We describe a variety of ways of participating, as supported by next-generation technologies and generative design, that encourages diverse engagement with mathematics and science.

In Support of Linguistic and Cultural Diversity

We make extensive use of the sociocultural analyses Wertsch (1985) presents for understanding how Vygotsky (following Yakubinskii, 1923) distinguishes between monologic and dialogic forms of speech. Within this sociocultural framework, what distinguishes monologic speech from dialogic speech is "not the number of individuals involved...[r]ather what distinguishes the two speech forms is the degree to which both parties participate in a *concrete speech setting* to *create* [italics added] a text" (Wertsch, 1985, p. 86). Generative activities of the sort discussed earlier in this chapter provide just such settings where students work to create mathematical and scientific text through shared discourse and network-mediated interactivity.

Contrasting with the monologic practices of "teacher talk" found in most classrooms, dialogic communication provides a more fully participatory approach that is especially useful for engaging a diverse array of students—and for drawing on the varied cultural resources that they bring to classroom activities. By expanding the range of ways that students can participate, generative activities encourage the development of a critical apperceptual mass—tacit, mutually understood and co-constructed forms of communication and acts of signification—which allows teaching and learning activities to be more natural and efficient in supporting the development of students from a variety of cultures and backgrounds.

Monologue Versus Dialogue

In discussing how Yakubinskii's analyses of monologic and dialogic communication influenced Vygotsky, Wertsch (1985) notes that "dialogue is a genetically prior and 'natural' form of verbal interaction, whereas monologue is a later and 'artificial' form" (p. 86). Moreover, meaningful communication can be achieved with less effort and fewer words due to a reduced need to create shared understandings. Wertsch (1985) quotes Yakubinskii:

> If everything that we wished to express were enclosed in the formal meanings of the words we used, we would have to use many more words than we in reality use in order to express any thought. We speak only through the use of necessary hints. (p. 87)

The more apperceptual mass we have in common, the fewer words we need to use.

> The more our apperceptual mass has in common with the apperceptual mass of our interlocutor, the easier we comprehend and perceive his speech in conversation. In this connection our interlocutor's speech may be incomplete; it may abound in hints. (Yakubinskii as quoted by Wertsch, 1985, p. 87)

These insights are central to our analyses of the importance of generative design in supporting participatory approaches to learning and teaching.

Fostering Shared Meaning

In our classroom observations, students engaging in generative practices supported by next-generation network technologies are able to communicate much more readily in terms of "hints," shorthand, gestures, and abbreviated utterances that emerge in relation to shared forms of expression (cf., Ares, 2005). Consequently, this reduction in the overhead of elaborated and formal language enables a corresponding reduction in the burden of many words having to do all the work of communication. (Wertsch, 1985, p. 88).

The development of an apperceptual mass also is common in communities of practice outside of school. For example, engineers and physicists rely heavily on the presence of shared forms of communication to support their abbreviated and highly productive forms of dialogic communication (see Ochs et al., 1994; Larsson et al., 2002). What we suggest here is that just such shared forms of expressivity and invented representation should also be emphasized in school

---

[9]Ironically, these productive abbreviated forms of communication serve to make even more challenging the already significant methodological issues associated with attempting to code and analyze the complex classroom discourse situated relative to the use of generative design and next generation networks.

classrooms. So, generative design is intended to encourage the development of just these kinds of hints, efficient and tacit kinds of participation, and abbreviated forms of utterance[9]—opening up new trajectories of participation even as the threshold of participation is lowered for all students, especially those in linguistically and culturally diverse populations (Ares, 2005).

Diversity of Resources

In other work (Stroup, Ares, & Hurford, 2005) we described how—during the use of a particular generative activity involving integers, rate, and slope—two Latinas collaborated in Spanish throughout the activity, sitting next to two European American boys who did their work individually and then compared their results in English. Both pairs' interactions were seen as important and appropriate, expanding the ways of participating seen in more conventional teaching.

In addition to Spanish and English languages serving as cultural resources, choosing to collaborate versus choosing to work independently also probably had gender-related and/or culture-related roots. The dynamic structuring of their activity occurred not only through the mathematics involved, but also through the students' use of individual and collective social, cultural, and academic resources—vitally important aspects of an apperceptual mass they brought with them to the classroom. Thus, opportunities to participate in varied ways made good use of important resources that each of the students brought to an engagement with novel content.

In another example (Ares, 2005) where co-construction was an explicit *requirement* of the activity, varied cultural and linguistic resources were used to encourage the development of a community of learners—rather than encouraging "competitive individual achievement" (Ladson-Billings, 1997, p. 480). Similar examples of coordination and co-construction have involved:

- the collective construction of stories in native Hawaiian communities Au, 1980),
- *confianza* or mutual trust networks of relations in Mexican communities in Tucson, Arizona (Moll & Greenberg, 1990; Gonzalez, Andrade, Civil, & Moll, 2001), and
- call and response traditions in African American churches in Chicago (Moss, 1994).

Much more work is required to fully understand the implications and affordances of students' existing apperceptual masses in generative and network-supported learning, however these findings point to potentials that should be explored.

Socially Situating Electronic and Physical Gesture

Electronic and physical gesturing are among the communicative "shorthands" that are prominent and well supported by generative group design and the associated capabilities of next-generation classroom networks. We now turn to sociocultural analyses to help us understand how gesture requires a social context in order for physical or electronic actions to become "true gestures."

Vygotsky (1978) gives the example of a young child's pointing. He notes that "[i]nitially this gesture is nothing more than an unsuccessful attempt to grasp something, a movement aimed at a certain object which designates forthcoming activity" (p. 56). The shift from physical movement to true gesture happens when "[p]ointing becomes a gesture for others" and the child's "unsuccessful [grasping] attempt engenders a reaction not from the object he seeks but *from another person*" (p. 56). So,

> [a]t this juncture there occurs a change in that movement's function: from an object-oriented movement it becomes a movement aimed at another person, a means of establishing relations. *The grasping movement changes to the act of pointing.* (italics in original, p. 56)

Then,

> [a]s a result of this change, the movement itself is then physically simplified, and what results is the form of pointing that we may call a true gesture.... It becomes a true gesture only after it objectively manifests all the functions of pointing for others and is *understood by others as such a gesture* [italics added]. Its meaning and functions are created at first by an objective situation and then by people who surround the child. (p. 56)

Here, the use of gesture as external sign is radically reconceived as it comes to be understood as a communicative act. In turn, this communicative reconstruction connects with larger features of Vygotsky's (1978) sociocultural theories where "[t]he developmental changes in sign operations are akin to those that occur in language. Aspects of external or communicative speech as well as egocentric speech turn "inward" to become the basis of inner speech" (p. 57). Socially mediated sign use becomes available as a resource for inner speech and thought and we see just such development in physically and electronically mediated gestures in network-supported inter-activity.

Although examples can be provided from any of the kinds of generative activities discussed in this chapter, we illustrate this radical reconstruction and its developmental significance relative to group design with the Gridlock traffic participatory simulation discussed earlier. In the simulation, students' physical and electronic movements (e.g., pressing a button to change a light) start off not having appreciable social significance. Often students' movements center on physically locating themselves on the grid and making sure their lights change color when they press the appropriate button. These actions, however, are radically reconstructed when students begin pointing to an intersection in the grid that needs to change its light. At this point a physical or electronic action (changing the color of a simulated traffic light) can become a true gesture and is coordinated with other expressive acts (e.g., prompting others to change their lights). It becomes a true gesture by this action now being situated in relation to the socially mediated significance of needing to coordinate traffic flow across the entire grid. In turn, the external sign of, say, pointing to a problematic intersection can impact this particular individual's understanding of what he or she might need to attend to in improving traffic flow.

Aspects of these gestures as external communication then have the potential to turn inward and become for this student, *and* for the class as a whole, the basis of inner speech to be used for future thought and action. The *individual's and the group's* inner speech is thereby enriched and extended. As discussed earlier, once the gestures and other expressive forms are more fully socially situated, they can add to the shared apperceptual mass, and then become increasingly simplified (hint-like) as part of the ongoing development of dialogic speech.

Of course, our extension of Vygotsky's ideas related to the development of *inner* speech to *group-level* analyses provokes questions concerning whether such inward-directed speech can be actually observed in the discourse patterns and forms of

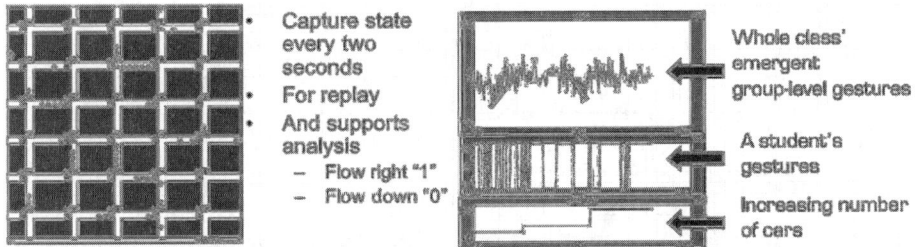

FIGURE 19-9. Emergent whole-group and individual gestures can be observed in the traffic grid (left) and in network-supported mathematical analyses of behaviors (Stroup & Wilensky, 2003).

interaction found in the group. We argue that versions of this group-level, but still internal (to the group) speech, are what mediate the emergence of the various self-referential, group-level strategies (which then also become available for individual appropriation). Instances like a "traffic cop strategy" (i.e., observe which direction the most cars are backed up and change to green in that direction), an "anticipation strategy" (i.e., observe which direction the biggest group of cars is coming from, and change to green in that direction), an "everyone-change at fixed time intervals strategy," and a strategy to synchronously cascade the lights along a row or column of lights are all products of, and then serve to iteratively produce, internal group-level speech. Physical action at the group and individual levels assumes social significance and becomes true gesture whereupon it becomes available for inner speech related to the socially negotiated goals of the activity.

In addition to such speech being observable in the discourse of the classroom (as exemplified in the overt articulation of group-level strategies), further evidence can be found in the emergence of group-level gestures. These can be seen to emerge over time in the running of the Gridlock simulation. Instead of just an individual changing his or her light, the class as a whole begins to implement, initially in tacit ways, a "turn all the lights green in one direction [allow flow to the right], wait... turn all the lights green in the other direction [allow flow downward]" strategy (in moving toward articulating the "everyone change" strategy mentioned above). This physical coordination becomes true group level gesture when it assumes social significance. When such patterns, or ways of gesturing, become understood by others or recognized as a sign operation the gesture then becomes available to the group as a kind of signifying action mediated by the network. Once recognized as a true gesture at the *group* level of agency, it then becomes available to the *group* for inner speech to support thought. Whole-group strategies emerge in relation to the group's inward appropriation and use of group-level gesture.

Not only can this group-level gesturing be recognized by the participants' overt observations of the patterns in the lights, the advanced features of next-generation networks can help make these sorts of emergent patterns of behavior more clearly visible and thus more available to the group. Using the real-time, data-capture capabilities of the HubNet system we can create alternative representations of the emergence of group level interaction as, in the case shown in Figure 19-9, more cars are added to the traffic system (see the "Increasing number of cars" graph on the lower right of Figure 19-9).

The evolving state of the networked activity is captured and rendered, in this case, in two-second intervals and can be used either for replaying the activity or for supporting additional analyses. By coding "1" for each intersection allowing traffic flow to the right and "0" for each intersection allowing flow downward in the grid, we can track an individual student's gestures (see middle graph on the right of Figure 19–9) and also sum these values to track the emergent "waves" of allowing flows in one direction and then the other (see the lines overlaid on the top-most graph of Figure 19–9).

At the group level, this increasingly coordinated signifying behavior (seeing all the lights turning green in one direction as highlighted with the lines in top-right graph in Figure 19–9) is an instance of making visible a group-level gesture and, when socially situated in ways supported by network interactivity, "becomes a movement aimed at...establishing relations" (Vygotsky, 1978, p. 56). Iteratively, this external sign or gesture can then be used internally to foster the production of additional signs or speech for individuals *and the group*. In so far as we can now attend to a range of forms of group expressivity, this can serve as further evidence for treating the group itself as having *Subject*-ivity. The signifying agency associated with a signifying Subject—the group in this case—can be made visible and assume social significance in analyzing both individual and group level actions.

These forms of signification center on creating expressions and invented forms of shared representation. Prior to their assuming social significance, the patterns are simply patterns in physical action. When seen by the group as having group-level significance as a natural part of collectively *making* sense together, these actions become fully realized as true gestures. The significance of the gestures is *made* in relation to the network-supported activity. Representational truth is made in relation to individuals and in relation to the groups' being able to successfully inter-act in improving traffic.

Overall, then, symbolization and signifying acts, including gesture and verbal utterance, become more fully integrated and increasingly judged as meaningful speech acts in relation to expressivity and successful symbolic invention and not as the simple consequence of referring to some presumed and/or prior "true" referent. The students come to *make* the significance their gestures are seen to have. Then, these gestures become available for inner speech and sense making. So, truthful representation emerges and develops in relation to the intra-activity of the group.

In emphasizing the emergence of expressive significance, the practical and scientific arts (including mathematics and science proper, but also engineering and medical arts as two more examples of socially situated and community based problem solving) can join with the "fine" arts, as discussed by Goodman (1976, cf. p. xi), to come to be seen as increasingly freed from the expectation that meaningful truths should, or even could, be the result of a simple comparison to, or pointing at, some singular, self-contained or fixed state of the world. As supported by generative design and the capabilities of next-generation networks, developing and managing the *expressive significance itself* can be emphasized in the socially situated, and fully pragmatic, meaning-making processes and analyses. Gesture and other expressive forms can emerge and become available for inner speech—and thus the advancement of thought—at both the individual and group levels of insight and inventive agency.

SUMMARY—*WHAT, WHY,* AND *HOW* OF DIVERSITY-BY-DESIGN

The pathways and endpoints taxonomy presented in this chapter is intended to be useful in clarifying the relations among different kinds of generative activities and in exemplifying internal aspects of generative design. These approaches to generative learning and teaching can be integrated with next-generation classroom networks to produce significant improvements in mathematics and science teaching and learning. A playful, emergent, and evolving space of possibility and participatory interaction typify generative design and the classroom-oriented practices discussed in the taxonomy can be seen to represent the *what* of diversity-by-design.

This *what* can be integrated with the *why* and *how* of diversity-by-design if we see this kind of classroom activity as more fully coordinated with the forms of truth- and design-making activities found in the wider worlds of work and life outside of school. Diversity-by-design is supported by generative activity in classrooms, but diversity-by-design also asks us to see classroom activity as part of a trajectory toward meaningful participation in life beyond formal education. Toward this end, attending to the socially situated emergence of structure is intended to address the needs of all learners and is to be particularly attentive to the capabilities and richness of resources found in linguistically and culturally diverse classrooms. One purpose of valuing diversity as a vehicle of productive engagement in classrooms is to scaffold similar productive engagement beyond the classroom. In particular, new forms of compact, shared, and highly productive classroom communication emerge in a way that is likely to parallel the use of such forms of communication in productive activity outside of schools.

We view this evolving ecology of classroom-based truth making as distinct from traditional accounts of ecosystems in that the resources we are most interested in are not principally of the finite or limited varieties. In traditional discussions of biological evolution, resources like water or food are limited and thus must be competed for. If, however, the resources of most interest in our accounts of learning and teaching are not finite in this way, but can and are expected to grow, then the dynamics of the ecology change. Some of the resources emphasized in education are understanding, participation, and agency—understanding algebra, for example, is a resource that can and should grow in relation to educational activity. This growth should be found at many levels of educational activity and attention, including the individual, the class, and the larger society (e.g., the number of people who are expected to attain "proficiency" in aspects of introductory algebra in the United States has increased dramatically in the last century). For this kind of growth-capable resource, the rules of the ecosystem are expected to be distinct from those of finite-resource ecologies like those typically studied in, for example, environmental biology.

Instead of conceiving of a classroom as the site of zero-sum competitiveness between individuals, diversity and co-operation would be even more valued as powerful vehicles for broadening and deepening the kinds of experiences that lead to growth in resources like understanding, insight, and agency at all levels of the system (including, but not limited to, individuals). Diversity-by-design, then, would have us attend to the ways that we can nurture these non-finite ecologies in the confines of classrooms and, in so doing, provide a framework for advancing understanding and active participation beyond schooling. The *why* and the *how* of diversity-by-design are powerfully linked back to the *what* of generative design, including the use of next-generation network technologies, in group-situated learning and teaching.

ACKNOWLEDGMENTS

In addition to the Web site from 1997 (referenced below) versions of the taxonomy contained herein were presented in a paper titled *Technological Resonance and*

Generative Activity: Design Tools for Listening to and Extending Students' Constructive Voice at the Annual meeting of the American Educational Research Association in Montreal, Canada (1999) and more recently in a paper titled *A Taxonomy of Generative Activity Design Supported by Next-Generation Classroom Networks* published in the 2004 Proceedings of Psychology of Mathematics Education-North America. (Ontario, Canada. [pp. 837–846]). The authors on the latter paper were Walter Stroup, Nancy Ares and Andrew Hurford (2004).

Also critical to the development of this line of research is funding from the National Science Foundation: Grant # 09093 entitled CAREER: Learning Entropy and Energy Project (W. Stroup, Principal Investigator) and grant # 126227 entitled Integrated Simulation and Modeling Environment (with U. Wilensky and W. Stroup serving as Principal Investigators). Texas Instruments has also generously supported this work. The kind assistance and copious patience of editor Thea Freygang was invaluable. We gratefully acknowledge this funding and support. The views expressed herein are those of the authors and do not necessarily reflect those of the funding institutions.

REFERENCES

American Association for the Advancement of Science. (1989). *Project 2061: Science for all Americans*. Washington, DC: Author.

American Association for the Advancement of Science. (1993). *Benchmarks for science literacy*. New York: Oxford.

Ares, N. (2005, April). *Culturally relevant design and analyses of network supported learning*. Paper presented at the annual meeting of the American Educational Research Association, Montreal, Canada.

Ares, N., Stroup, W.M., & Schademan, A. (in review*). The power of mediating artifacts in group-level development of mathematical discourses.*

Au, K., (1980). Participation structures in a reading lesson with Hawaiian children: Analysis of a culturally appropriate instructional event. *Anthropology and Education Quarterly, 11*(2), 91–116.

Barab, S., & Kirshner, D. (2001). Guest editors' introduction: Rethinking methodology in the learning sciences. *The Journal of the Learning Sciences. 10*(1&2), 5–15.

Borovoy, R., Martin, F., Vemuri, S., Resnick, M., Silverman, B., & Hancock, C. (1998). Meme tags and community mirrors: Moving from conferences to collaboration. *Proceedings of the 1998 ACM Conference on Computer Supported Collaborative Work.*

Borovoy, R., McDonald, M., Martin, F., & Resnick, M. (1996). Things that blink: computationally augmented name tags. *IBM Systems Journal, 35*(3), 488–495.

Barron, B., Schwartz, D., Vye, N., Moore, A., Petrosino, A., Zech, L., Bransford, J., & The Cognition and Technology Group at Vanderbilt, (1998). Doing with understanding: Lessons from research on problem- and project-based learning. *The Journal of the Learning Sciences. 7* (3/4), 271–311.

Bransford, J., & Schwartz, D. (2001). Rethinking transfer: A simple proposal with multiple implications. In A. Iran-Nejad & P. D. Pearson (Eds.), *Review of Research in Education (Ch 3., Vol.24, pp. 61–100)*. Washington, DC: American Educational Research Association.

Callahan, P. (1999, April*). Generative content knowledge*. Paper presented at the annual meeting of the American Educational Research Association, Montreal, Canada.

Campbell, J. R., Hombo, C. M., & Mazzeo, J. (2000). *NAEP 1999 trends in academic progress: Three decades of student performance*. NCES 2000–469. Washington, DC: 2000.

Carnegie Learning (2003). *The Cognitive tutor*. Retrieved July 3, 2005, from http://www.carnegielearning.com/

Chapman, M. (1988). *Constructive evolution*. Cambridge: Cambridge University Press.

Colella, V., Borovoy, R., & Resnick, M. (1998, April). *Participatory Simulations: Using computational objects to learn about dynamic systems*. Paper presented at CHI '98, Los Angeles, CA.

Davis, S. M. (2002). *Research to industry: Four years of observations in classrooms using a network of handheld devices*. IEEE International Workshop on Mobile and Wireless Technologies in Education, Växjö, Sweden.

Dewey, J. (1916). *Democracy and education*. New York: Macmillan.

Diehl, E. (1990). Participatory simulation software for managers: The design philosophy behind microworlds creator. *European Journal of Operations Research, 59*(1), 203–209.

Freire, P., Freire, A.M.A., & Macedo, P. (1998). *The Paulo Freire reader*. New York: Continuum.

Gardner, H., Kornhaber, M. L., Wake, W. K. (1996). *Intelligence: Multiple Perspectives*. Fort Worth, TX: Harcourt Brace.

Gleick, J. (1987). *Chaos: Making a new science*. New York: Penguin Books.

Gonzalez, N., Andrade, R., Civil, M., & Moll, L., (2001). Bridging funds of distributed knowledge: Creating zones of practices in mathematics. *Journal of Education of Students Placed at Risk, 6*(1&2), 115–132.

Goodman, N. (1976). *Languages of art: An approach to a theory of symbols*. Indianapolis, IN: Hackett Publishing Company.

Goodman, N. (1978). *Ways of worldmaking*. Indianapolis, IN: Hackett Publishing Company.

Hegedus, S. & Kaput, J. (2002). Exploring the phenomena of classroom connectivity. In D. Mewborn, et al (Eds.), *Proceedings of the 24th Annual Meeting of the North American Chapter of the International Group for the Psychology of Mathematics Education* (Vol. 1, pp. 422–432). Columbus, OH: ERIC Clearinghouse.

Hernstein, R. & Murray, C. (1994). *The bell curve: Intelligence and class structure in American life*. New York: The Free Press.

Holland, J. (1995). *Hidden order: How adaptation builds complexity*. Reading, MA: Addison-Wesley.

James, W. (1890). *The principles of psychology* (2 vols.). New York: Henry Holt.

James, W. (1978). *Pragmatism and the meaning of truth*. Cambridge, MA: Harvard University Press.

Kameenui, E., & Carnine, D, (1998). *Effective teaching strategies that accommodate diverse learners*. Upper Saddle River, NJ: Prentice Hall.

Kauffman, S.A. (1995). *The origins of order: Self-organization and selection in evolution*. New York: Oxford University Press.

Kelly, K. (1995). *Out of control: The new biology of machines, social systems and the economic world*. Reading, MA: Perseus Press.

Ladson-Billings, G. (1997). Toward a theory of culturally relevant pedagogy. *American Education Research Journal, 32*(3), 465–491.

Larsson, A., Törland, P., Mabogunje, A., & Milne, A. (2002). Distributed design teams: embedded one-on-one conversations in one-to-many. In D. Durling, & J. Shackleton, J. (Eds.), *Common ground: Design Research Society International Conference*.

Lave, J. (1988). *Cognition in practice: Mind, mathematics and culture in everyday life*. Cambridge: Cambridge University Press.

Lave, J., & Wenger, E. (1991). *Situated learning: Legitimate peripheral participation*. New York: Cambridge University Press.

Learning Technology Center, (1992). *Technology and the design of generative learning environments*. Hillsdale, NJ: Erlbaum Associates.

Lesh, R., Carmona, G., & Post, T. (2002). Models and modeling. In D. Mewborn, P. Sztajn, D. White, H. Wiegel, R. Bryant, K. Nooney, (Eds.), *Proceedings of the 24th Annual Meeting of the North American Chapter of the International Group for the Psychology of Mathematics Education* (Vol. 1, pp. 89–98). Columbus, OH: ERIC Clearinghouse.

Lesh, R., Hoover, M., Hole, B., Kelly, A., & Post, T. (2000). Principles for developing thought-revealing activities for students and teachers. In R. Lesh & A. Kelly (Eds.), *Handbook of research design in mathematics and science education* (pp. 591–645). Hillsdale, NJ: Erlbaum.

Montangero, J., & Maurice-Naville, D. (1997). *Piaget or the advance of knowledge* (A. Cornu-Wells, Trans.). Mahwah, NJ: Lawrence Erlbaum Associates, Inc.

Moll, L. C. (1990). *Vygotsky and education: Instructional implications and applications of sociocultural psychology*. New York: Cambridge University Press.

Moll, L. C., & Greenberg, J. B. (1990). Creating zones of possibilities: Combining social contexts for instruction. In L. C. Moll (Ed.) *Vygotsky and education: Instructional implications and applications of sociocultural psychology* (pp. 319–348). New York: Cambridge University Press.

Moss, B. J., (1994). Creating a community: Literacy events in African American churches. In B. J. Moss (Ed.), *Literacy across communities*. Cresskill, NJ: Hampton Press.

Ochs, E., Jacoby, S., & Gonzales, P. (1994). Interpretive journeys: How physicists talk and travel through graphic space. *Configurations*, 2(1), 151–171.

National Council of Teachers of Mathematics, (2000). *Principles and standards for school mathematics*. Reston, VA: Author.

National Research Council, (1996). *National science education standards*. Washington, DC: National Academy Press.

Nicolopoulou, A. (1993). Play, cognitive development, and the social world: Piaget, Vygotsky, and beyond. *Human Development*, 36, 1–23.

Nuñes, T., Schliemann, A. & Carraher, D. (1993). *Street mathematics and school mathematics*. Cambridge: Cambridge University Press.

Pacey, A. (1983). *The culture of technology*. Cambridge, Mass: MIT Press.

Papert, S. (1990). A critique of technocentrism in thinking about the school of the future. *MIT media lab epistemology and learning memo no. 2*. Cambridge, MA: MlT Media Lab.

Peirce, C. S. (1982). *The Writings of Charles S. Peirce*. 5 vols. to date. Edited by M. Fisch, C. Kloesel, et al. Bloomington, Indiana: Indiana University Press, 1982 to present.

Peirce, C. S. (1909). A sketch of logical critics. *The Essential Peirce. Selected Philosophical Writings*. Vol. 2 (1893-1913), edited by the Peirce Edition Project, 1998. Bloomington and Indianapolis: Indiana University Press. pp. 460–461.

Resnick, L. B. (1987). Learning in school and out. *Educational Researcher*, 16(9), 13–20.

Rogoff, B. (1995). Observing sociocultural activity on three planes: Participatory appropriation, guided participation, and apprenticeship. In J. V. Wertsch, P. del Rio, & A. Alvarez (Eds.), *Sociocultural studies of mind* (pp. 139–164). New York: Cambridge University Press.

Saxe, G. B. (1991). *Culture and cognitive development: Studies in mathematical understanding*. Hillsdale, NJ: L. Erlbaum Associates.

Saxe, G. B. (1994). Studying cognitive development in sociocultural context: The development of a practice-based approach. *Mind, Culture, and Activity*, 1(3) 135–157.

Schwartz, J., & Yerushalmy, M. (1985a). The Geometric Supposer: An intellectual prosthesis for making conjectures. *The College Mathematics Journal*, 18, 58–65.

Schwartz, J. L. & M. Yerushalmy. (1985b). *The Geometric Supposer*. Pleasantville, N.Y.: Sunburst Communications.

Schoenfeld, A. H. (1988). When good teaching leads to bad results: The disasters of "well taught" mathematics classes. *Educational Psychologist*, 23, 145–166.

Senge, P. M. (1994). *The fifth discipline: The art and practice of the learning organization*. New York: Doubleday.

Silbert, J., Stein, M., & Carnine, D. (1997). *Designing effective mathematics instruction: A direct instruction approach (3rd ed.)*. Upper Saddle River, NJ: Prentice-Hall Inc.

Slavin, R. E. (1990). *Cooperative learning theory, research, and practice*. Englewood Cliffs, NJ: Prentice-Hall, Inc.

Stor, M., & Briggs, W. L. (1998). Dice and disease in the classroom. *Mathematics Teacher*, 91(6), 464–468.

Stroup, W. (1997a). *Catalog of generative activities and what's a generative activity?* Retrieved July 3, 2005, from http://www.edb.utexas.edu/faculty/wstroup/gen_act_catalog.html

Stroup, W. (1997b). *Root Beer Game*. Unpublished software for TI-8x calculators based on *The Beer Game* by J. Forrester.

Stroup, W. (2002a, April). *The cognitive and affective affordances of new classroom network design for mathematics learning*. Paper presented at the annual meeting of the American Educational Research Association, New Orleans.

Stroup, W. M. (2002b, April). *The structure of generative learning in a classroom network*. Paper presented at the annual meeting of the American Educational Research Association, New Orleans.

Stroup, W., Kaput, J., Ares, N., Wilensky, U., Hegedus, S., Roschelle, J.,Mack, A., Davis, S., & A. Hurford (2002). The nature and future of classroom connectivity: The dialectics of mathematics in the social space. In D. Mewborn, P. Sztajn, D. White, H.Wiegel, R. Bryant, & K.

Nooney (Eds.), *Proceedings of the 24th annual meeting of the North American Chapter of the International Group for the Psychology of Mathematics Education* (Vol. 1, pp. 195–203). Columbus, OH: ERIC Clearinghouse.

Stroup, W., & Wilensky, U. (2003, April). *Mathematics structuring the social sphere (MS3): Rendering the interplay of utterance, gesture and artifact in participatory simulations.* National Council of Teachers of Mathematics, Research Presession. San Antonio, TX.

Stroup, W. M. & Wilensky, W. (2004). *A Guide to Calculator-HubNet based Participatory Simulations: Network-based Design for Systems Learning in Classrooms.* Version 1.0, January 13, 2004. http://ccl.northwestern.edu/ps/guide/part-sims-guide.html . Center for Connected Learning and Computer-Based Modeling, Northwestern University, Evanston, IL.

Stroup, W. M., Ares, N., & Hurford, A. (2005). A Dialectic Analysis of Generativity: Issues of Network Supported Design in Mathematics and Science. *Journal of Mathematical Thinking and Learning, 7*(3), 181–206.

Vygotsky, L. S. (1978). *Mind in society: The development of higher psychological processes.* Cambridge, MA: Harvard University Press.

Vygotsky, L., (1987). The collected works of L.S. Vygotsky: Vol.1, Problems of general psychology. Including the volume "Thinking and speech" (N. Minick, Trans.). New York: Plenum.

Waldrop, M. (1992). *Complexity: The emerging science at the edge of order and chaos.* New York, NY: Simon and Schuster.

Wertsch, J. V. (1985). *Vygotsky and the social formation of mind.* Cambridge, MA: Harvard University Press.

Wilensky, U., & Stroup, W. M. (1999). Participatory simulations: Network-based design for systems learning in classrooms. *Proceedings of the Conference on Computer-Supported Collaborative Learning*, CSCL '99, Stanford University.

Wilensky, U., & Stroup, W. M. (2000). Networked Gridlock: Students Enacting Complex Dynamic Phenomena with the HubNet Architecture. In B. Fishman & S. O'Connor-Divelbiss (Eds.), *Fourth International Conference of the Learning Sciences* (pp. 282–289). Mahwah, NJ: Erlbaum.

Wilensky, U., & Stroup, W. M. (2005). *HubNet.* Available as part of the NetLogo download. Retrieved July 3, 2005 from http://ccl.northwestern.edu/netlogo/

Wilensky, U., & Stroup, W. M. (in review). *Embodied science learning: Students enacting complex dynamic phenomena with the HubNet architecture.*

Wittrock, M. C. (1991). Generative teaching of comprehension. *The Elementary School Journal,* 92(2), 169–184.

Yakubinskii, L. P. (1923). *O dialogicheskoi rechi* [On Dialogic speech]. Petrograd: Trudy Foneticheskogo Instituta Prakticheskogo Izucheniya Yazykov.

CHAPTER 20

When the Model Is a Program

Fred G. Martin
University of Massachusetts Lowell

Margret A. Hjalmarson
George Mason University

Phillip C. Wankat
Purdue University

Learning the principles of computer programming and computer simulation are common goals in engineering, computer science, and other technical fields that are part of the foundations for the future. In such work, the student is responsible for designing, evaluating and analyzing both programs and their output. This process integrates knowledge from multiple science and mathematics disciplines.

In addition, each program has a specific, intended purpose or goal to fulfill. The program is supposed to do something for someone. In one example presented here, the students are designing a robot that is supposed to behave in certain ways. In the second example, the students are using a simulation to design chemical separation processes. In both cases, there is a stated purpose that enabled students to test whether or not the program was indeed successful.

Students' programs represent their interpretation of the problem situation and the assumptions made about the situation. By examining their code, instructors can understand the knowledge that students bring to the situation. In other words, the program is both a model of student thinking about the situation *and* a model that represents how the students have integrated both interdisciplinary knowledge and the constraints and affordances of the problem context.

All models (e.g., mathematical models) are representations of understanding. This chapter explores a special case of modeling activities where computer programming is the modeling activity. We answer two questions. First, what characteristics of models and modeling are shared with computer programming? Second, what are the unique characteristics of modeling activity when the goal is to produce a computer program or develop a design with a simulation? For simplicity, we refer to both of these modeling activities as "programming"—either writing code or using drop-and-drag procedures.

We explore two example domains: one from computer science (robot design) and one from chemical engineering (process design using simulators). Both examples are used to demonstrate the characteristics of the modeling processes that are part of programming. Before the description of the programming activities, we explain aspects of the models and modeling theory of conceptual development that connect to the programming.

MODELS AND MODELING FOR PROGRAMMING

This section describes how models and a modeling perspective on teaching, learning and problem solving is well-suited to thinking about the design of computer programming tasks for engineering and computer science students. At the foundation of the models and modeling perspective are the models designed and developed throughout the tasks. According to Lesh and Doerr (2003),

> Models are conceptual systems (consisting of elements, relations, operations, and rules governing interactions) that are expressed using external notation systems, and that are used to construct, describe, or explain the behaviors of other system(s)—perhaps so that the other system can be manipulated or predicted intelligently. A mathematical model focuses on structural characteristics (rather than, e.g., physical or musical characteristics) of the relevant system. (p. 10)

In the examples described in this chapter, the representational system is the computer program. As with any representational system, programming languages have limits and implicit assumptions. There are structures that cannot be represented in the language and assumptions have to be made about the real-world scenario in order to accomplish a representation in the programming language (particularly in the case of the chemical engineering simulations). There are also conceptual systems behind the design and use of the programs. In the programming examples discussed, the conceptual systems include parts of mathematics, science, engineering, computer science, technology and the integration of multiple disciplines.

Modeling activities have long emphasized the design of procedures to solve problems (e.g., Lesh, Hoover, Hole, Kelly & Post, 2000). In addition, a fundamental principle for modeling activity design is self-assessment. Namely, the students should be able to determine whether or not their procedure is performing the intended function. In the case of the robot design and computer simulation, the students develop programs and simulations and can test them directly. For instance, the students can test how well their robot follows a line on the ground. The procedure should be generalizable (e.g., the robot should be able to follow any line set on the ground). The generalizability of the model also means it should provide students a way of thinking about similar situations. For instance, students should take knowledge about programming the robot to follow a line to other robot programming situations with increasing complexity.

Another aspect of modeling activities is that the design process should provide metaphors and ways of thinking about future situations. So, designing one robot for one purpose should help students design future robots and future programs. In the case of the chemical engineering simulations, each experience with the simulator should provide ways of thinking about other simulation situations—either with the same simulation program or not. For instance, some aspects of the simulation design process have a trial-and-error quality. Students should take information from the trial-and-error process from one simulation problem to refine the trial-and-error process in future situations. So, generalizability applies both to the final program or simulation the students design as well as the process of designing the computerprogram. Throughout the semester, the design and simulation problems should become more complex and require the integration of knowledge from previous tasks.

In both examples, students follow an express-test-revise cycle for the design and use of the computer programs. First, they write a program which expresses their understanding of the task and the goal for the task. Then, they test it according to the conditions given in the task and revise accordingly. The focus on an iterative design process is consistent with the workplace students can expect as future engineers and computer scientists. In the robot example, the students can adjust certain values that determine how the robot moves in response to certain conditions (e.g., running into a wall). For the computer simulation example, the students have to adjust some values by comparing results with data and using their professional judgment as they are testing their procedure.

Process is used in two different ways, and the programming task involves two different processes. The first process is the program itself which is a process of accomplishing a goal. The second process is the design process the students use to design the program. As discussed previously, aspects of both types of processes should be generalizable to future programming tasks. Considering programs as models also highlights a need for tasks whose end-product is a process. The quality of the process the students produce is determined by how well the program accomplishes the stated goals of the task. The design process used to create the program includes the technical content as well as other professional skills like teamwork, communication, and problem-solving emphasized by accreditation boards such as ABET (Accreditation Board for Engineering and Technology) as desirable characteristics of successful graduates in engineering and technology. The process of designing the program means students have to sort out the assumptions in the task, the limitations of the program, and the conceptual systems that will apply to the program.

Writing and using computer programs is an essential aspect of work and education in technical fields. The examples given here are examples of how a view of that programming as a modeling process can move the tasks beyond programming-for-programming's sake toward programming for a purpose, with an end-in-view. So, rather than learning lists of commands, language, or procedures in decontextualized ways, students are using the programming language to solve a real problem and generate a usable solution generalizable to other situations. Both examples use real-world programming problems to situate the tasks. This approach also emphasizes that learning about the process of programming is as important as the final product.

MODELING BY WRITING PROGRAMS

Here, we explore how students come to understand the behavior of small mobile robots as they develop the robots' physical characteristics, select and mount sensors on them, and finally write code to control the robots' movements.

The work discussed here was performed by undergraduate computer science majors in a junior/senior level course at the University of Massachusetts Lowell. The computer science department offers a two-semester undergraduate sequence in mobile robotics; the work described here was done in the first course, entitled "Robotics I." In the course, students build small robots from LEGO components, mount sensors on them, and then write code to control the robots' movements. Figure 20–1 shows a typical student robot built during the course of the semester.

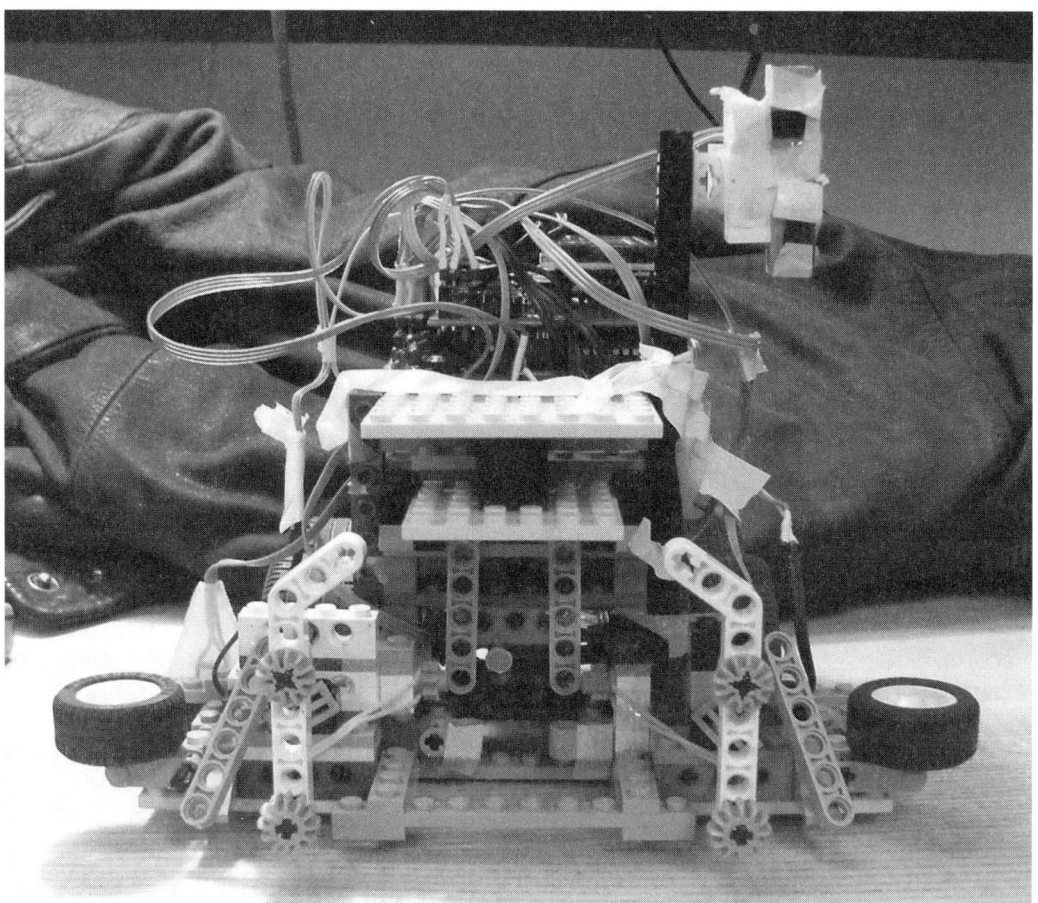

FIGURE 20–1. Typical student robot.

The Robotics I course is organized in three major sections.

- In the first section, students proceed through a series of 6 weekly labs, which introduce them to mechanical design, sensors, programming, and the implementation of typical mobile-robot behaviors, such as obstacle-avoidance (that is, backing up and turning away when an obstacle is hit), line-following, and wall-following.
- In the second section, students are given a task description for a competitive robot game. In a recent semester, a simplified version of the Robot Egg Hunt game was assigned (Beer et al., 1999). Students then individually design and implement robots to perform the task; their robots play against one another in a classroom competition.
- The final section of the course is an open-ended design project. Based upon the ideas introduced in the course, students design and implement a robotic system of their own conception.

This chapter presents results from the first two sections of the course (which are substantially more structured than the last). We focus on the code that students developed to control their robots, and what this code reveals in students' thinking and understanding.

Literally, the code is the control program for a given robot. It therefore controls the behavior of the robot; precisely, it is procedural specification for what the robot should do.

But it is not all this simple. Even though the code seems precise and simple, there is a collection of implicit assumptions embodied in any robot program. Any example of student code is code *for a particular robot*, which itself is designed *for a particular environment*. Various changes, even perceptually small ones, to the robot or its environment may cause the robot to fail in its nominal task.

Let us start by examining one of the early labs. Students have already built a simple two-motor vehicle, and equipped it with a left- and right-side touch sensor (see Figure 20–2). Their task is to program it to "avoid obstacles"—that is, to back up and turn away when it runs into something. Students are asked to compare their robot's behavior to that of a "Weazel ball," a battery-powered children's toy that uses a weighted mechanism to roll around the floor. The Weazel ball almost never gets stuck in a corner; its mechanical randomness causes it to eventually roll away from almost any situation. Do students' robots do as well?

Figure 20–3 shows typical student code for the obstacle avoidance task. Many details of their robot design can be discerned simply by reading through their code

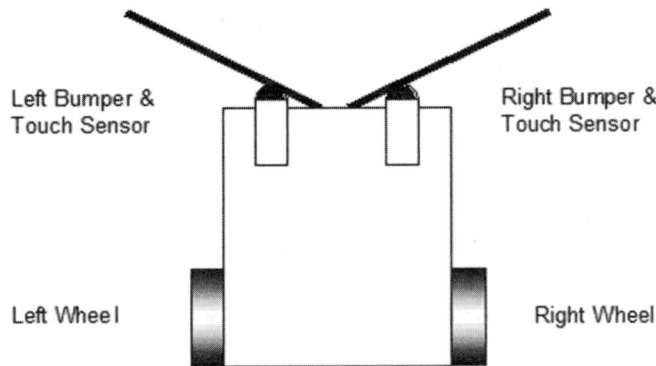

FIGURE 20–2. Diagram of basic student robot.

```
void main(){
    // start the car
    while(1){
        fd(2); fd(3);
        // if right bumper hit,
back up and turn to left
        if(digital(10)){
            bk(2); bk(3);
            sleep(1.5);
            fd(3); bk(2);
            sleep(0.5);
        }
        else{
            // if left bumper
hit, back up and turn to right
            if(digital(11)){
                bk(2); bk(3);
                sleep(1.5);
                fd(2); bk(3);
                sleep(0.5);
            }
            else{
                //if both bumper
hit at the same time, back up and
turn to
                // either left or
right for 3 seconds
                if(digital(10) &
digital(11)){
                    bk(2); bk(3);
                    sleep(1.5);
                    fd(2); bk(3);
                    sleep(1.0);
                }
            }
        }
    }
}
```

FIGURE 20–3. Typical student code for obstacle-avoidance task.

with their embedded comments. Their robot uses two motors, which are plugged into output ports 2 and 3. The students make their robot move forward with the command sequence "fd(2); fd(3);" and they make it go backward with the sequence "bk(2); bk(3);".

The other important characteristic that we see represented in this program is time. The students have explicitly expressed *how long things take to happen.* For example, in the code associated with the action "if right bumper hit, back up and turn to left," the students have decided that the robot should go backward for 1.5 seconds "sleep(1.5)" and then turn for 0.5 seconds ("sleep(0.5)") before resuming with forward movement.

These particular values are typical for the class of robots that would be built by the materials the students are provided, but they are hardly universal constants. Rather, the students obtained these values through experimentation—by trying different numbers to see what worked best or simply well enough.[1]

In this simple back-up-and-turn example, the exact values chosen typically are not critical. But these "timing constants" are typically littered throughout student control code, and there are other, more complex situations where the range of values to get the robot to "work" is quite narrow. Even in this simple case, one can imagine a case in which the backup-time constant is too small, and the robot fails to pull itself away from the obstacle before trying to turn, causing the turn to fail. Further, these timing constants are often more brittle that students expect, because robot motor performance can vary significantly with changing battery levels.

Line Following

After the obstacle-avoidance task, students are asked to design a robot that can follow a line. The line is a curved stripe of black electrical tape, about 1/2" wide, stuck down to the linoleum tile lab floor.

Students are provided with reflective light sensors that are well-suited for detecting the line. As part of solving this problem, though, students must decide *what algorithm they will use for following the line,* including, critically, *placement of the sensor(s) on their robot.*

A typical solution is shown in Figure 20–4. The diagram shows a two-wheeled robot positioned atop a line. The line sensors are affixed to the front of the robot, on either side of the line.

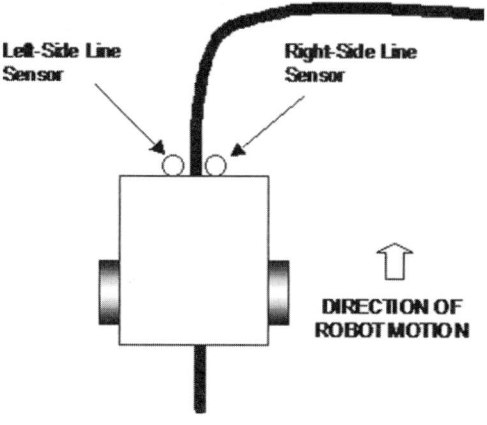

FIGURE 20–4. Typical line following robot.

---

[1]This is a well-known engineering heuristic!

A sample of student code for controlling the two-sensor robot design is shown in Figure 20–5. The code repeatedly tests the sensor state and then advances the robot. Each time through the loop, any of three possible moves may be executed:

- If neither sensor sees the line ("both sensors see white"), then the robot moves straight ahead.
- If the left-mounted sensor sees the line, then the robot moves in an arc toward the left.
- If the right-mounted sensor sees the line, then the robot moves in an arc toward the right.

In practice, what happens? First, note a key assumption: the robot must start with its sensors straddling the line (as is shown in the diagram). From this initial state, neither sensor sees the line, so the robot drives straight ahead. The robot will continue driving straight until it veers off one side of the line, or, as in the diagram, the line itself curves. Then, one of the line sensors will see the line, and the robot will turn. If driving over a line like that in the diagram, the robot's right-side sensor would be likely to "see" the line, and then the robot would make a turn to the right.

There are additional hidden assumptions that must be satisfied in order for the algorithm to work as intended:

- The robot must be able to turn in a sharper arc than the curve of the line.
- The robot must move slowly enough that the sensors can detect when the robot moves over the line, and cause the robot to change its direction before it has crossed completely over the line.
- Both line sensors should not "see" the line at the same time. In the sample code shown, this case is not explicitly tested, and if it occurs, the robot will continue moving in whatever fashion it had chosen from a previous (successful) reading.

```
void line(){
    //start the motors up
    fd(0);
    fd(2);
    while(1){//this loop will
keep going effectively it means
that the robot should be able to
keep following a line till its
bats die
        //as long as both sensors
see "white" move foward
        if(analog(5) < 150 &&
analog(6) < 150){
            fd(0);
            fd(2);
            msleep(100L);
        }
        //if the left sensor is
reading 'black' turn left
        if(analog(5) > 225){
            off(0);
            fd(2);
            msleep(100L);
        }
        //if the right sensor
reads 'black' turn right
        if(analog(6) > 225){
            off(2);
            fd(0);
            msleep(100L);
        }
    }
}
```

FIGURE 20–5. Code for two-sensor line-following robot.

The "line-between-two-sensors" design is the most popular, but other sensor configurations are possible. In the 2-sensor space, the sensors can be put right next to each other, with the intention of both traveling atop the line. Also, the problem can be solved with just one sensor, having the robot take a "drunkard's walk" back and forth across the line. This solution has the elegance of requiring just the one sensor, but students tend to not like it because the robot never actually drives straight ahead.

Students tend to implement the first algorithm that occurs to them, and tweak it until it works, but not systematically analyze its strengths and weaknesses, nor the hidden assumptions in their solution.

Seeking Open Space

In one of the more complex labs in the course, students are given two specialized components and asked to build a robot that seeks out and drives toward open space. These components are:

1. An ultrasonic sonar sensor. This device emits an ultrasonic "ping" and measures how much time elapses for the ping to hit something and the echo to return. Since sound travels relatively slowly (about 1 foot per millisecond), simple microprocessor electronics can measure this time delay and yield decent readings for obstacles from a few inches to about 10 feet.

2. A servo motor. This small motor can rotate to any commanded position within an 180° angular sweep. The sonar sensor is mounted on the shaft of the servo motor, so it can be rotated through multiple positions, from which distance readings can be taken.

In a recent version of the class, a suggested algorithm was provided to the students:

> One way to do this is have the servo motor turn the sonar to 5 different positions, spaced evenly through the servo's range, and take readings at each one. Then move your robot in the direction that had the farthest readings; this will lead your robot into open space.

This problem is given to students in week 6 of the course. At this point of the semester, students have come to realize a crucial insight: *sensors don't always work*. The sonar sensor exhibits this problem particularly well; while in principle it *should* work reliably, the environment in a typical lab or office building includes many objects that do not reflect sound echoes well, such as office corridors. Students discover that large flat surfaces act like sonar "mirrors" that reflect sound away from its source. The sonar ping is never returned, and the hallway wall may actually become invisible to the sonar sensor.

The understanding that sensors are fallible is then represented in students' code. Figure 20–6 shows sample code that illustrates this. The student's comment at the top of the function states it clearly: "get and store readings, ignore any that could be wrong." Then within the body of the function, the student checks for out-of-bounds readings and discards them.

The function includes only rudimentary error-checking; only readings that are completely illegitimate are rejected. Nevertheless, it is instructive because it does show that its author has formalized his understanding of at least one way in which the sensor fails.

```
//get and store readings, ignore any that could be wrong
void sense()
{
    for(i=0; i<6; ++i){
        servo0=angles[i];
        msleep(400L);
        readings[i]=sonar();
        if(readings[i]==32767 || readings[i]<0)
            readings[i]=0;

        printf("%d\n",readings[i]);
    }
}
```

FIGURE 20-6. Student's sonar code includes error-check.

Summarizing, we conclude that:

- Students' programs have lots of implicit assumptions about the how their robot behaves and the characteristics of its world.
- While the programs are "formal" in the sense that they are objective and machine-executable, they are not formal in the mathematical/proof sense.
- Tweaking and other such development strategies are common.
- Some students more than others become aware of the brittleness of these strategies.
- It would be good to have robot design challenges that require more mathematical analysis.

Modeling Perspective on the Robot Example

The robot example highlights the balance between what should theoretically happen in a problem situation and what actually happens. For instance, sensors might not work as planned. So, students have to account for that in their program. The program represents student understandings of mathematical constructs such as distance, rate, time, and the relationships between them. Students also need to analyze and weigh the relative strengths and weaknesses of competing models. As the students express, test, and revise their programs for the robot and as the robot problem scenarios increase in complexity, the students have more information to sort and may bring more conceptual systems to bear on the task.

In the robot example, there could be an argument for two types of representational systems. The first is the program that controls the robot. The program represents the students' interpretation of the best possible means for controlling the robot. It also represents their understanding of the conceptual systems required for the task and the constraints and assumptions of the problem situation. The second representational system is the robot itself. The robot, while a concrete object, represents the motion of living things. The students have to consider how the robot is like living things and how the robot is an approximation of live movement in order to design their program. Coordinating the representations available in the program with the physical nature of the robot is the focus of the modeling activity in this scenario. Students, then, have to coordinate two representational systems and the accompanying conceptual systems.

USING COMMERCIAL SIMULATORS TO DEVELOP MODELS

Practicing engineers develop and use models routinely. These models vary in a continuum from simple conceptual models to very complete computer programs that include all known aspects in the mathematical description. The most useful approach is often to use a combination of the simplest, "back-of-the-envelope" models, and the most complex, sophisticated computer programs. The simple conceptual models are useful as thinking tools, but they do not include enough details for accurate design. The complex computer programs have the details necessary to design, but are too complex to use to think.

Despite their constant use of models that are computer programs, most practicing engineers do not code using computer languages such as C, C++, FORTRAN, or Java. For relatively simple calculations they use computer tools such as spreadsheets or mathematical calculation packages such as MATLAB, Mathematica, or Mathcad, and may do some "coding" within these packages. For detailed design calculations they normally use a commercial simulator. There are a number of companies that create, test and sell very detailed simulations. To be specific, we will discuss the use of AspenPlus, which was created by Aspen Technology, Inc. for chemical engineers; however, the application of other simulators in other engineering disciplines will be similar.

Commercial simulators contain the detailed mathematics for designing parts or processes, but the creative development of the model is still under the control of the user. For example, when creating a process to separate a chemical mixture using AspenPlus, the engineer must first decide on the basic flowsheet model. Back-of-the-envelope ideas are useful for this creative stage. Then the engineer draws the flowsheet in the simulator using drop and drag procedures. To create the mathematical model of the flowsheet the engineer must choose the appropriate mathematical models for each unit in the flowsheet and for the physical properties. This step requires professional judgment. Next, appropriate specifications and initial guesses for equipment sizes and operating conditions are input into the simulator. One then crosses one's fingers, presses the start button, and hopes for the best. Except for rather simple routine problems, the simulator never works the first time. The simulator may inform the user that there is an error or that convergence was not obtained, or the user may realize the results are not reasonable. The engineer must debug the simulation by checking input variables, refining the solver methods chosen, simplifying the flowsheet by cutting recycle loops to obtain initial answers and so forth. Once a preliminary solution has been obtained, the engineer will vary conditions to study the system behavior and obtain an optimized solution. Although the coding, the benchmarking, and debugging have been done in commercial simulators; their use still requires both professional engineering judgment and artistry. The main advantage of using a *good* simulator is the time for learning how to use it plus time for debugging the simulation is significantly less than the time to develop a program from scratch. This time saved is then available to explore many alternatives or to go have a beer.

Since the modern practice of engineering is heavily based on commercial simulation programs, engineering colleges want their students to have experience with this approach. If there is limited competition in a given area, commercial simulators can be very expensive. When there are a number of alternative simulation programs on the market, which is the case with AspenPlus, the costs charged companies will be significantly less although more than universities can afford to pay.

Fortunately, most of the companies producing simulators (or in some cases very large users of the simulators such as GM or Ford) realize that it is in their best interest to have new engineers trained in school on their simulators. Thus, they either donate programs to universities or license them at very reasonable prices. These licenses will be restricted to educational use, will not include training and often provide very limited troubleshooting assistance.

There is really nothing in the commercial simulation programs that a team of graduate students in engineering and computer science could not develop if they had a lot of time. However, the commercial programs tend to be more user-friendly, more robust, more capable of solving complex problems, more sophisticated in the use of mathematical solution techniques, and more likely to undergo significant beta testing than student written programs. In addition, they typically have a lot more options since student programs are often written for a single use. As an example of the enormous power of the current commercial simulators, I (PCW) currently give students non-isothermal absorption problems to do individually in two hour lab periods with AspenPlus. I first encountered similar but simpler problems 35 years ago as a graduate student. At that time these problems were assigned as a month-long project for a team of four chemical engineering graduate students. The difference is that the team of graduate students had to write the code.

Most engineering colleges in the United States use commercial simulators in their senior capstone design course. Most Chemical Engineering departments in the United States use AspenPlus or a similar steady state simulation program such as HySys, ChemCad, or Pro/II in their design courses. In design classes, the students are assumed to know the fundamentals required to understand all of the models in the simulator. If the simulator is unfamiliar to the students, they will need some help in getting started. (Very few commercial simulators have good simple instruction manuals. Instead, the manuals tend to be encyclopedic, which is useful for experts but not novices.) This help can be provided by marching the students through a simple design problem step-by-step. Then the students are essentially turned loose to use the simulator for the course project. As noted previously, because of the options, engineering judgment is required to achieve a good design.

Students will be much better prepared for their design course if they first use the appropriate simulation package in required "lecture" courses. Since commercial simulators represent "real" engineering, their use can be very motivating. Many schools start students on the simulator they will use for design in either their sophomore or junior engineering courses. Usually, only a small part of the simulator is used in each lecture course. When simulators are first used in a lecture course, the professor invariably tries to use it as an add-on homework assignment. Just as invariably, this attempt fails. Commercial simulators, unlike courseware, were not designed for use by students who are learning the material at the same time. Unfortunately, the use of simulators will probably be demotivating if the students become stuck and are not successful.

The Chemical Engineering department at Purdue University has found that integration of computer laboratory sessions into the junior "lecture" course on "Equilibrium-Staged Separations" is an effective way of introducing simulators (Wankat, 2002). Very little lecture time is devoted to discussing the simulator since lecturing on how-to-simulate tends to be ineffective and boring. Instead, the students are assigned to two-hour computer laboratory sections that typically have twenty students plus a teaching assistant (TA). There are currently eleven laboratory sessions including a computer laboratory test interspersed throughout the fifteen week semester. At the first session the students are given a cookbook recipe of how to start and run the simulator. Invariably students will become stuck when first learning how to use these complex computer programs. If there is no help available, being stuck often leads to intense frustration and a desire to smash the computer. In the computer laboratory environment other team members, the TA, or the professor can usually get the student moving forward again in a few minutes.

Students are given relatively simple trial-and-error problems for flash distillation during the second laboratory session. These problems had previously been assigned as homework to do by "hand" (actually with a calculator or spreadsheet). If the problems are chosen with a little care, the students see that with the simulator they can solve even relatively simple problems much faster than with other tools. The difficulty of the problems rapidly escalates so that students are soon solving realistic problems such as multicomponent distillation or non-isothermal absorption that would be very difficult if not impossible to solve by hand.

What do the students learn when they use a simulation package? 1. How to do design using commercial simulators. Since this is the way most modern engineering design is done, we want our students to be good at it. This experience also helps the students be competitive in the job market. 2. How to solve open-ended problems and optimize. Since the vast majority of the coding has been done, the students can be asked open-ended problems that require a large number of simulations. The junior year is really the first time they get a significant amount of open-ended problem solving in chemical engineering at Purdue University. In design class (senior year) the students have time to focus on optimization. 3. The importance of prerequisite material such as thermodynamics. One of the choices the designer must make in AspenPlus is the thermodynamic model to use for the equilibrium behavior. Having students use both good and bad choices of vapor-liquid equilibrium packages brings home the importance of thermodynamics more than anything else (they don't believe their sophomore thermodynamics professor). To know what is a good or bad choice the students compare their predications with experimental data that they found by searching the literature or the web. The choice of equilibrium data is critical since millions of dollars have been lost in industry from wrong choices. Thus, the correct model clearly depends upon the context—the chemicals to be separated. In other areas of engineering, the topics students realize are important may be different, but there will be at least one. 4. They discover the behavior of real systems. [Actually, they are discovering the behavior of a complex model of a real system. But in some cases (e.g., distillation, which is the major topic of the junior course) the models are so good that chemical engineers build distillation columns that cost in excess of a million dollars straight from the model without pilot testing.] The students are forced to ask *what if* questions about the behavior of complex equipment. For example, one of the computer labs asks, what happens in the design of a distillation column if the pressure is increased? The students explore this question by running the simulator and determining for themselves how condenser and reboiler temperatures, column height and diameter, relative volatility, and the concentration of azeotropes change as pressure increases. The answers they discover for themselves tend to stick.

Ideally, simulators are not a black box—the students hopefully understand the models from study in class and from "hand" calculations. By teaching the simulator as part of the lecture course, the professor can require students to make connections between theory and the models used in the simulator. For example, in the second lab the students simulate the flash distillation of a mixture of benzene and water. Since benzene and water are immiscible, the liquid separates into two layers and the vapor composition remains constant as the fraction of feed that vaporized increases. This result mystifies the students until they apply a simple theory that immediately clarifies the situation. In reality, despite these efforts, parts of the simulator will be a black box. For example, the programs often use some numerical methods that students, and often the professor, are not familiar with; however, the students should be familiar with the models for all of the pieces of equipment or parts of the object being designed.

The Chemical Engineering program requires a number of laboratory courses with real equipment to develop engineering judgment and to prevent students from developing an uncritical over-reliance on simulation predictions. Experimental data remains essential in areas of chemical engineering such as adsorption or catalysis where small amounts of an impurity can, over time, poison the adsorbent or catalyst. The commercial simulators can be too good and hide the details of trial-and-error calculations, which makes the simulator more of a black box. To prevent this, we neglect to tell the students about some of the features of AspenPlus. The black box issue remains a real one—a minority of engineering professors refuses to use simulators because of this issue.

Developing models that are computer programs and operating detailedcommercial simulators are both worthwhile learning exercises. Unfortunately, crowded engineering curricula do not have time for extensive amounts of both. Most engineering departments that used to require writing code currently believe that what students learn running a commercial simulator is more important than what they would have time to learn if they had to do the planning, coding, debugging, testing and running of a number of detailed programs. One would expect that Computer Science departments would conclude that developing the program is more important.

Modeling Activity in the Commercial Simulator Example

The commercial simulator is distinct from the robot scenario in that the students are not writing the code for the simulator (although this would be another form of modeling activity). In this situation, they are operating the simulator and generating models for chemical phenomena using information gathered from the multiple simulation trials. The students are testing procedures for chemical engineering in a simulated environment. Because it is a simulation, the program carries assumptions and limitations connected to the real environment that students have to consider. The modeling activity in the simulation scenario occurs as students apply conceptual systems from engineering, mathematics, and chemistry to the simulation representation of the chemical phenomena under investigation. As with the robots, the simulator is providing a representation of real-world phenomena so the students are applying conceptual systems about real-world phenomena to a representation of those phenomena. In addition, the students have to generate a process for running the desired simulation and producing the intended results. Thus, there are two levels of models. One is the model the students generate to run the simulation. The other model is the simulation program itself which is a model of real-world phenomena. In effect, the students are generating a model to analyze and interpret a simulation which is a model itself. As in the robot example, learning the process of designing the simulation is as important as learning the technical content to run the simulation. The students are learning about the simulation design process as well as the particular simulation for chemical engineering. Both are generalizable to future simulation and programming tasks.

CONCLUSION

Both examples represent modeling tasks requiring the production of a computer program that is a process for accomplishing a goal. Both examples require the integration of content knowledge from multiple disciplines. Both examples are also as much an experience in learning about design process as learning about the technical content required for the course. When the model is a program, the students are naturally expressing their interpretations, assumptions, and understandings of the situation in a meaningful representation (i.e., it is sensible to produce the representation in the context).

Both tasks have practical application as part of the programming. In the computer science/robots example, programs are tested in a physical environment. In the chemical engineering/process design example, programs are tested in a simulated environment. But in both cases, there is fidelity to the technical work and ways of working that graduates will be required to do in their careers. For instance, the computer simulators are representative of simulators used by corporations. Both situations also allow for the complexity of the tasks to increase and for students to bring knowledge from prior experiences to bear on their current work.

REFERENCES

Accreditation Board for Engineering and Technology, Inc., "Criteria for Accrediting Engineering Programs: Effective for Evaluations During the 2003–2004 Accreditation Cycle," Baltimore, Nov. 2002. Retrieved from www.abet.org

Beer, R., Chiel, H., & Drushel, R. (1999). Using autonomous robotics to teach science and engineering. *Communications of the ACM, 42*(6), 85–92.

Lesh, R., & Doerr, H. M. (2003). Foundations of a models and modeling perspective on mathematics teaching, learning and problem solving. In R. Lesh & H. M. Doerr (Eds.), *Beyond constructivism: Models and modeling perspectives on mathematics problem solving, learning and teaching* (pp. 3–34). Mahwah, NJ: Lawrence Erlbaum Associates.

Wankat, P. C. (2002) Integrating the use of commercial simulators into lecture courses. *Journal of Engineering Education, 91*, 19–23.

CHAPTER 21

Uncertainty and Iteration in Design Tasks for Engineering Students

Margret A. Hjalmarson
George Mason University

Monica Cardella
University of Washington

Robin Adams
Purdue University

"Engineering is a profoundly creative process. A most elegant description is that engineering is about design under constraint." (National Academies, 2004, p. 7) This statement which begins *Engineering 2020: Visions of Engineering in the New Century* encapsulates two aspects of engineering design we focus on here. Engineering is defined as a process and, more specifically, a "creative process." However, alongside the creativity are constraints to the design. One interpretation is that creativity enters engineering because of the constraints to a design. The constraints around an engineering task may be economic, social, scientific, or technological. Ambiguity may be present because of the number of variables and the amount of information related to any context. The ambiguity relates to the process of engineering and the iterations engineering designs go through over time because the ambiguity can motivate re-design. This chapter will focus on two aspects of the ambiguity of the process (uncertainty and iteration) that we have included in design tasks used in undergraduate engineering.

By uncertainty, we mean aspects of a context that may not be clear or may lack clear information to guide decision-making. By iteration, we mean a process of revision of a product through multiple cycles. Students start with an initial idea, gather more information, analyze the data available, and revise their ideas over time. Iteration is used as a noun to describe the versions of a solution students develop over time. Iteration is also used a verb in order to describe the process of revision. We emphasize two aspects, the product and the process, in order to describe the interrelationship between uncertainty, iterations of solutions, and the iterative process of engineering design.

In many of the tasks students have completed in mathematics courses prior to university, the available information is clear, the goal is clear, and the student's task is to find the correct procedure (Hiebert et al., 2003; Stigler & Hiebert, 1999). This is in contrast to engineering design tasks where the end-goal may not be clearly delineated. Designers have to sort and gather information, and designers might have to make decisions with a certain level of uncertainty (e.g., Gainsburg, 2003; Hall, 1999). Since engineering is very rarely a part of the pre-college curriculum, students often have limited understanding of what engineering is and what engineers do when they enter an engineering program. This chapter will describe three studies at three different levels of undergraduate engineering education that served as opportunities for students to work with uncertainty and iteration in a design task.

The goal of the chapter is to describe two characteristics of the design and problem-solving process in engineering. We limit our discussion to the role of iteration and uncertainty as they are common themes in engineering tasks (particularly in our own research) and because they are particularly challenging for engineering students whose conceptions of problem solving and design may be limited in scope and depth. In this chapter we will examine what happens when engineering students are placed in a problematic situation and asked to design something for a client. Whether that process is design or problem solving may be in the eyes of the beholder. In any case, although we examine design in an engineering context, there are aspects of the processes which could apply to design in other contexts.

We employ three engineering design contexts in order to illustrate how uncertainty and iteration are related to each other. The first context is an Industrial Engineering capstone project to design a distribution center and redesign a supply chain for an industry partner. Second, design tasks (in particular, a playground design task) posed to freshmen and seniors illustrating the role of iteration and students' understanding of iteration in the design process. Finally, the use of modeling tasks in freshmen engineering which illustrate interactions between uncertainty and iteration. Table 21–1 gives a summary description of the task. The three studies represent investigations into learning from the introductory to the advanced undergraduate level. The size of the projects ranges from a one-hour laboratory activity to a 2-quarter capstone project. Two projects are team projects. In all three, there is a clear final product or end-in-view (Dewey, 1938; English & Lesh, 2003) which the students are moving toward and we will refer to as the design for our purposes here. At all stages of each project, students find ways to account for

TABLE 21–1.
Task Descriptions

Task Feature	Capstone Project	Playground Design	Nanoroughness
Nature of problem	Design distribution center and redesign parts of supply chain	Design a community playground	Design a procedure for measuring surface roughness
Work Structure	Team of 4 students	Individual	Team of 4 students
Level of students	Juniors and Seniors	Freshmen and Seniors	Freshmen
Time to complete	2 academic quarters	3 hours	2 hour laboratory session

uncertainty and revise their solutions. In all three projects, iteration of the design is both related to uncertainty and a component of improving responses.

Iteration, as we have defined it here, occurs when students revise their current way of thinking about a task and shift to a new or modified interpretation. We anticipate that the students' first interpretation or design will not be the best one. As students (working individually or in a team) gather more information, encounter other perspectives, or solve sub-tasks embedded in the larger task, the design goes through iterations. Hopefully, the iterations result in higher quality final designs. Iterative cycles can be used to analyze how designs change over time and what impacts shifts from iteration to iteration.

Uncertainty across the projects comes in a variety of forms. The first and most obvious form is a lack of information that might be helpful to completing the design. For the students, part of the design task could be finding ways to get information they need in a useful format. A question at this point is why information would be withheld from students working on a design task. The time available to introduce all the information for a task is limited by the time in a laboratory class session or a semester to work on a project. From an instructional perspective, a lack of information is a realistic part of design tasks in the engineering workplace where deadlines have to be met and not all the information is immediately available in a user-friendly format. Additionally, in the workplace, information is not simply given to engineers; instead they must determine what information they need and then devise a method for getting that information.

As shown in Table 21-1, the three contexts are a capstone design project for seniors, a playground design tasks for seniors and freshmen, and a nanotechnology task for first-year students. The next three sections will describe each of the contexts and the uncertainty and iteration encountered by the students in each context. Even though the tasks encompass a range of ability levels and different levels of detail, all three tasks include uncertainty and iteration. The description of the capstone project includes a narrative description of the group's work in order to illustrate in detail the design process and the interactions of the team that connect to uncertainty and iteration. The playground task section describes the analysis of responses across multiple students in order to generate means of aggregating characteristics of the students' design process and iteration in particular. The nanotechnology task section will describe interactions between different types of uncertainty and iteration in a cutting-edge area of engineering.

UNCERTAINTY ENCOUNTERED IN A CAPSTONE DESIGN PROJECT

Let us begin to investigate uncertainty and iteration by stepping into the Industrial Engineering (IE) undergraduate computer lab. Ten senior IE students and 4 juniors are working here tonight at 10 p.m. The team that we are observing is designing a new distribution center and redesigning parts of the rest of the supply chain to account for the new distribution center for SCS, their industry partner. SCS is an internationally known company with a local office that distributes products but does not itself engage in manufacturing. The five seniors on our team (Ben, Diego, Fil, John and Mei[1]) are meeting to work on their capstone design project. For these students, this is one of the first open-ended, "real-world" design projects that they

---

[1] We have replaced the name of the Industry Partner as well as each of the students involved in this research project with pseudonyms.

encounter. Throughout winter and spring quarters, these students will be meeting with each other, their instructor, and their industry partners as they apply everything they have learned in their undergraduate education to an actual need provided by the industry partner.

Planning: Meeting Two[2]

A few weeks earlier, halfway through winter quarter, the students met on-site: in a break room at SCS. Observing the team at this point in time, we see that the students are talking about the information that they need in order to proceed with their project. They make a list of questions that they can ask their representative at SCS, and also develop a plan to observe some of the SCS employees as they perform their job functions. They discuss the feasibility of these observations and talk about the type of information that they hope to gain. Using a mathematical language, they talk about finding the average time on task for various job activities. This meeting is still early on in the project, and as the students have explained to us, they are trying to make plans for next quarter and also define what the problem is that they are trying to solve. They are also preparing to submit a project proposal. They have spent the first half of winter quarter reviewing what is involved in engineering design on a general level, and are now beginning to focus on their specific problem. They must come up with a plan for how they will proceed during spring quarter. So far, they have used estimates as placeholders in their report. They are now, however, expressing some anxiety about the estimates, and hoping to be able to replace the estimates with more accurate information before submitting the proposal.

In addition to considering the types of information that they will need to finish their project, they are also anticipating the types of skills and methods that they will be using during the second quarter. They tell us that they will be doing more modeling and using more math during the second quarter. They decide to list in the proposal all of the methods they may use, and they will decide later which ones will actually be beneficial and which methods will actually be used.

As they review the data that they do already have, Fil remarks that he would like to start the data analysis already. Mei hesitates because the team has not yet decided what they are going to do with the data. She sees there are a multitude of analyses that can be done with the data, and the team must decide the best way to proceed.

Developing the Design: Meeting Ten

We return to observe the team again two and a half months later. The team now has three and a half weeks left to complete their project. They meet in the IE undergrad lab. Diego and Fil are modeling the process of unloading volume from a truck and then sorting the volume. Using an Excel counting function, they are trying to determine the queue length for the volume that has been unloaded but is waiting to be sorted. Mei and John are working on cost analyses to compare their alternative solutions for the new distribution center and the revised supply chain. Right now they are looking at the costs of two different sizes of trucks that might

---

[2]The researcher observed a total of 22 of the team's meetings over the course of five months. Meeting Two is the second meeting that the researcher observed. During these meetings, the researcher audio-recorded the team's conversations and took field notes using a Tablet PC.

transport volume to and from the distribution center. They combine the two and find the average to represent the cost for an average truck. Mei wonders if the volume level that needs to be transported will be smaller on a Saturday; John suggests that they make an assumption about this and move forward with the cost analysis. Mei next adds employee wages into the cost analysis and remarks, "this is a real rough estimate."

The team moves to a group meeting, and after commiserating over the midterms they have been taking for their other classes, begin to discuss their need for error distribution to see the variance in volume levels for different days. They have already analyzed the data for the volume that was transported in March and September, but would like to know the "spread on base service levels" for other months. They decided to use Excel's filter and sort functions, and John and Fil take turns directing their teammates in which Excel functions they should use to create the scatter plots showing the distribution for their data (different volume levels on different days).[3]

After considering many other costs and savings (distances traveled, storage of vehicles, and other transportation costs) the team begins to think they should "quantify other things." They consider what kind of future business they can anticipate for the region served by the distribution center they are designing. John asks his teammates how they can go about making these forecasts, and Diego suggests that they talk to the people in SCS's marketing department.

Fil then directs the team to move from the cost analysis on to the next update: the models they are creating to simulate the scoring process within the distribution center. Diego explains "we can't do more until we get more realistic rates." He has been using an assumed volume level for his preliminary model, but at this point he wants more exact numbers for both the sorting rate and the volume levels. The team talks about the sorting rates at other distribution center locations, and Diego suggests that they "need to see one in action" to get a better sense of what the rate would be for the new distribution center that they are designing. Diego continues to talk about the fine sort model and updates his team on the queue lengths he has found. A little later, after further exploring the possibility of visiting another distribution center, Diego decides that the team needs to "find out all the details, like to develop other alternatives" and figure out how to model the sorting rates.

The team discusses the unload rates a little more, shifts to a discussion of the structure of their proposed distribution center, and then considers making a list of questions to email to their representative at SCS. As Mei makes a list, Ben comments that he would like "to know how the other groups are doing" on their projects. Fil reminds him of the group that worked with SCS for the previous year's capstone project, and Mei notices that the previous group's final poster is still in their computer lab. They realize that the previous group focused on a much smaller problem—only the sorting model piece of the problem. Diego realizes that the poster gives them some of the information that they need to determine the fine sort rate for their project: "based on this unload rate...assume a complete random distribution...but it would be different for a different volume level." He adds the information he has gleaned from the poster to his Excel model of the fine sort process and considers using Arena[4] instead because of the variance in the volume levels.

---

[3]Excel is a software program trademarked by Microsoft.
[4]The software program Arena allows the user to create simulations to model manufacturing systems and document, animate, test and analyze new processes before actually implementing them.

When the team redirects their attention to a group discussion again, they consider the scoring rubric that they are creating to rank their alternatives. Fil suggests, "cost is everything," Diego suggests that time is also very important, and John reminds them that their professor has instructed them to include other considerations. After the team suggests customer relations and expansion as other categories, Diego realizes that they can "weight [each category] so much, but we need to ask SCS what their weights are." Before convening for the evening the team discusses other questions that they still have about the project, different car sizes, the amount of space they will need for their proposed distribution center and the previous SCS team's poster again.

Finalizing the Design: Meeting Seventeen

It is the night before the big deadline: each group will be giving an in-class presentation on their project and turning in a final paper tomorrow. The team meets in the lab, along with 12 other classmates (and one student's dog). At 3 a.m. they begin to describe what they see as the major problems with the estimates they have used—inaccurate forecasts (they have often taken into account the solution's ability to respond to increased volume due to future population growth or annual peak or high-volume seasons) and the annual impact from a daily difference of 50 cents.

The team is still struggling with their "12-hour constraint": ensuring that the main driver does not work more than 12 hours in a day. They decide to meet this constraint by eliminating the 20 minutes of prep time at the beginning and end of his day, but still continue to wrestle with the tight constraint. They finally justify one of their decisions about the design of the distribution center by saying that they can choose it because the alternative is only hypothetical and not based on actual data. They reason, "If we're scheduling up to the minute, it's hard to do that with a theoretical model."

UNCERTAINTY

Through the course of twenty-two observations of the team's meetings and six interviews with the team members, it became quite apparent that as part of the capstone project, engineering students have the opportunity to encounter a great deal of uncertainty. The uncertainty that students encounter can come from a number of sources, and the students can respond to the uncertainty in a number of ways. Unlike mathematicians (Schoenfeld, 1992) and practicing engineers, engineering students are very uncomfortable with uncertainty. When engineering students encounter uncertainty it can become a stumbling block to their progress in their design solution—either because they redirect all their resources towards eliminating the uncertainty or because they decide that they are stuck until the uncertainty is resolved.

The uncertainty that engineering students encounter in capstone design projects can come from a number of sources: "missing" information they has not yet been gathered, questionable information or data that must be verified, information that does not exist (e.g., the exact sorting rates for a hypothetical model when it is actually put into practice) or multiple possible approaches to solving each problem. Just as there are many pathways for students to encounter uncertainty, there are many options for how students can respond to uncertainty, with varying levels of effectiveness and appropriateness.

In the capstone project study, students responded to uncertainty in a variety of ways: they expressed frustration that information was missing, devised strategies for collecting the information, used estimation to fill in the missing information (similar to the strategy of conservatism that Gainsburg (2003) observed in her study of structural engineering practitioners), decomposed the problem into something more manageable, or used a strategy Schoenfeld (1992) calls "guess and verify" (Cardella & Atman, 2005). Additionally, when they encountered uncertainty, due to the many possible approaches to solving the problem, they monitored their progress to ensure they chose the best approach.

Generally, the students did not choose to use mathematical modeling to deal with the uncertainty. Diego and Fil did use Excel and Arena to make a model of the sorting process, but they ensured that the numbers that they used for the models were based in data and expressed anxiety until they verified and re-verified with SCS that the numbers they were using were accurate. In an interview after the project was completed, one student commented that when the group was working on the cost analysis to compare different solutions and they were missing some of the costs they

> ...could have put in variables you know, and then get it the next day and have it to you. So it was a lot of useless time, I thought, to be sitting around discussing things...how much it would cost. We didn't have the cost for those things, but we could easily add it into the big picture later on instead of 'here's all our data we need to completely understand these, these tiers of the data, and stuff like that.' (Interview with Ben)

ITERATION IN DESIGN

In the previous section we described ways that uncertainty emerges in an authentic engineering team design task and ways that the team of designers dealt with uncertainty. In this section, we delve into how uncertainty in design evokes iteration. Iteration is inherent in complex design problems because the starting point is ambiguous and open to interpretation and the stopping point is neither definitive nor systematic but based on an evaluative process against an elusive target goal. As Archer (1979) noted, "Design activity is commutative, the designer's attention oscillating between the emergent requirements ideas and the developing provisions ideas, as he illuminates obscurity on both sides and reduces misfit between them." (p. 18)

Iteration indicates cycles of converting an ill-structured problem into a well-structured solution, where the designer revisits and resolves ambiguity on both ends. As such, aspects of iterative behavior include gathering and filtering information, monitoring progress and understanding, and creating and synthesizing new knowledge to improve upon potential solutions. Iterations may result in incremental corrections or transformative shifts in understanding.

In the sections below we highlight findings from a lab-based task in which individual freshman and senior engineering students were told that they were an engineer in a mid-sized city and that their community asked for their help in designing a playground (Adams, 2001; Turns & Atman, 2001). Participants were given 3 hours to design a fictitious playground and were told that they could ask the administrator for additional information such as safety guidelines, results from a neighborhood opinion survey, and the kinds and costs of materials available to build a playground. Participants were also told that their design needed to address a set of functional requirements and that they had to work within a set of constraints. Requirements and constraints were presented in subjective language for which little information was provided to the participant for evaluating when and how these were met in their solution. As an example the final design needed to have at least three different playground activities, be safe for children ages 1–10, and not cost too much. Similarly, participants were told to use common materials, create a design that could be ready in 2 months, and that someone should be able to build the playground from the solution they develop without any additional information about the design. Participants were asked to talk aloud as they engaged in the task and were videotaped; these sessions were later transcribed into verbal protocols.

A cognitive model of iteration in design was developed and used to empirically describe when iterations occur (frequency and duration over the design task), why they occur (what triggers an iteration and what is the result), and where iterations occur (in relation to a design process model) (Adams, 2001; Adams & Atman, 1999). An exploratory comparative analysis of 4 freshmen and 4 seniors' iterative behaviors identified emergent hypotheses across differences in design performance and levels of experience. This was followed up with a confirmatory analysis across 24 participants (12 freshmen and 12 seniors) of which key findings around issues of uncertainty are described in the paragraphs below.

Iteration Is Prevalent in Design

Before delving into the ways and reasons iteration comes into play in design, Figure 21–1 illustrates the prevalence of iteration in design. As shown here, regardless of experience level or the quality score given to each design,[5] participants iterated for a considerable portion of their total design time. Averaging across the whole group, participants spent 35.6 percent of their total design time revisiting and modifying earlier conceptions. Overall, time spent iterating was 1) significantly higher for seniors than freshmen ($p < .01$), and 2) significantly and positively correlated with final quality scores ($p < .05$) and the number of requests for additional information ($p < .01$).

Looking across design task timelines, iteration was observed to occur throughout the process (Adams, 2002). Although a noticeable level of iterative cycles were evident at such points in the process as evaluating whether or not the design solution was at a "stopping point", participants iterated an average of once every 1.6 minutes (range of once every 0.6 to 2.9 minutes) over an average total design time of 94 minutes. Whether or not iterations were planned or occurred spontaneously, the average duration of an iterative cycle (40.8 seconds, range of 21 to 91.8 seconds) suggests that iterations were either quickly resolved or dropped.

If one view of iteration is an effort to address uncertainty, the level and frequency of iteration observed in the design processes of these 24 participants supports a perspective that uncertainty is prevalent throughout design. Similarly, the average duration of an iterative cycle suggests that efforts to manage uncertainty in an *individual design task* are not long in duration or uniquely associated with a particular stage of the design process. But what is going on during these

---

[5]Evaluation of quality was based on a combination of the following: how the overall design met the stated requirements and constraints, supplementary scores on particular aspects of the design based on existing playground design guidelines (e.g., safety guidelines for a swing set), and ranking on a set of subjective scores (e.g., creativity, aesthetics). Scores were normalized to 1.0 (or 100 percent).

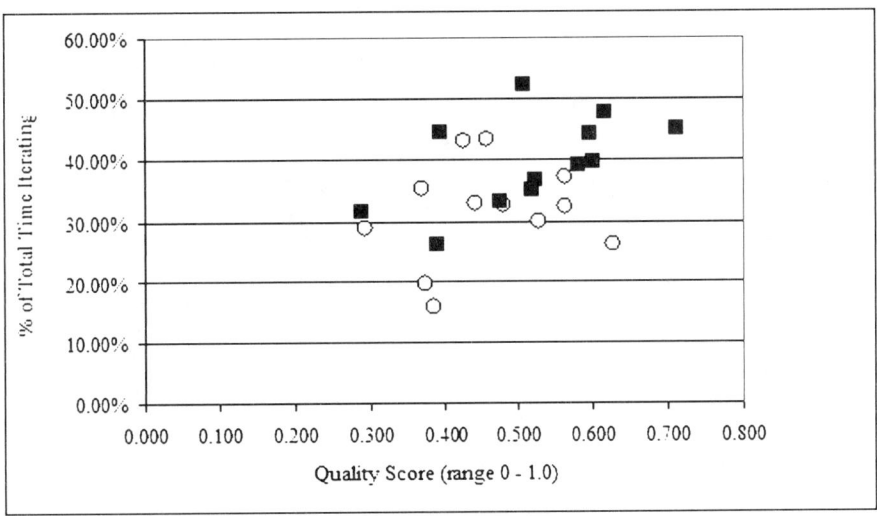

FIGURE 21–1. Scatter plots illustrating relationship between % of total time iterating and final quality score (○ = Freshmen, ■ = Seniors). For the freshmen, the average time spent iterating was 32.7%; for seniors, 42.9%.

iterative cycles? What triggers them to occur and what kinds of uncertainty are associated with revisiting earlier ideas? In the next two sections we provide insights into these questions.

Iteration Is Most Often Directed Towards Understanding the Problem

The figures below, *iterative web diagrams*, illustrate where iterative sequences occur within a design process model (see Figure 21–2a). The ellipses in the diagrams represent common steps within the design process (i.e., Atman and Bursic, 1998) and include (moving in a clock-wise sequence) defining the problem, gathering information, generating solutions, modeling solutions, analyzing feasibility of solutions, evaluating solutions, deciding which solution (among a set) to implement, and communicating the final solution so it may be implemented. The arrows in the diagram represent iterative sequences of moving from one design activity to a previous design activity. The beginning of the arrow illustrates where in the process iterations were triggered and the direction of the arrow signifies the goal of an iterative sequence (e.g., Feasibility to Modeling, Modeling to Gather Information). The percentages in the diagrams refer to the amount of total time in that iterative sequence. For the case of iterating within a design activity (e.g., Modeling), percentages are located within the associated ellipse; for iterating across design activities, percentages are located on the arrow associated with that iterative sequence.

The iterative web diagrams provided in Figures 21–2a and 21–2b illustrate important differences across experience and performance observed across the full dataset. For example, the web diagram in Figure 21–2a is representative of those participants (senior and freshmen) who received high quality scores for their final design solutions[6] and spent a considerable time iterating. As shown in the figures participants engaged in a variety of iterative sequences (e.g., the number of arrows

---

[6]The quality score was normalized to 1.0.

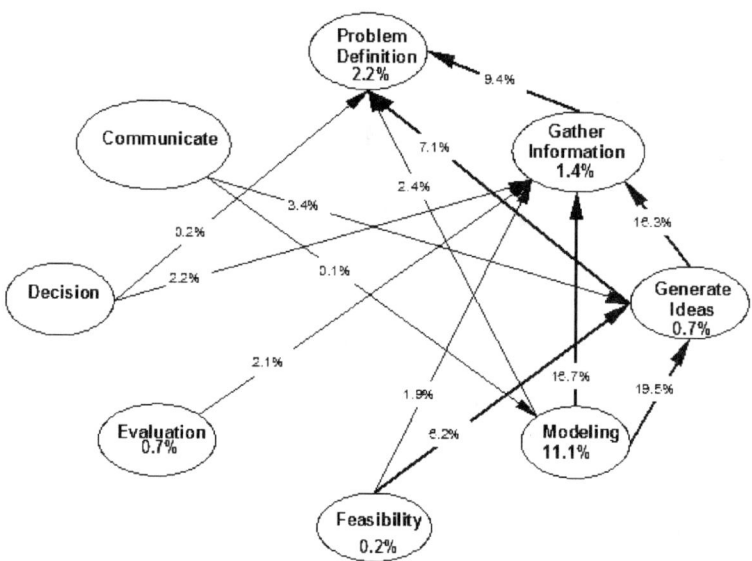

FIGURE 21-2a. Iterative web diagram for a senior who was above average for time spent iterating (39.8 percent) and had a high quality score (.597).

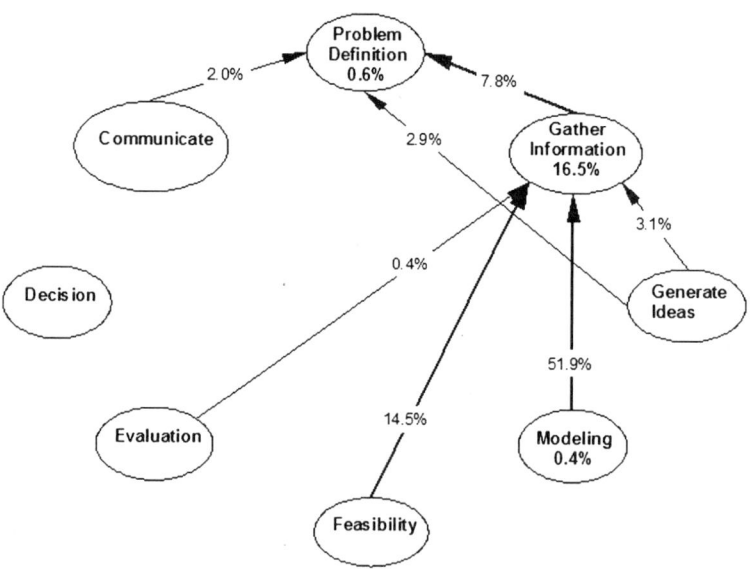

FIGURE 21-2b. Iterative web diagram for a freshman who was below average for time spent iterating (19.5 percent) and had a low quality score (.375).

in the web diagrams), and a greater number of sequences correlated with more experience, more time iterating, and higher quality scores. One interpretation is that the number of iterative sequences may be indicative of a willingness to monitor, seek out, and address uncertainty.

From another perspective Figures 21-2a and 21-2b illustrate that a predominant goal of iterative sequences are activities associated with understanding the nature of the problem (particularly revisiting, clarifying, and gathering information). This suggests that information about the problem is assimilated throughout the design process and often occurs in the context of specifying and developing a particular solution. From a landscape view, patterns evident in the web diagrams suggest that iteration is a conversation across representational spaces: between conceptualizing a solution (e.g., generating ideas, modeling, determining feasibility, evaluation) and defining the nature of the problem (e.g., problem definition, gathering information). Such dialectical processes are evident in other complex problem-solving activities. For example, Schön (1993, p. 68) describes the process of reflection-in-action as being triggered by an unexpected event that causes a conceptual shift that stimulates a transformational conversation with a situation's "back talk" (see also Adams et al., 2003). Similarly, in the context of writing Bryson et al. (1991) describe the process as a living conversation moving between content and rhetorical goals where the goals recursively evolve. In the next section we explore these ideas with specific examples from the verbal protocols.

Iteration Is Coupling Uncertainty at Both Ends of the Process (Problem and Solution)

The web diagrams in Figure 21–2 are evocative of a conversation across problem and solution spaces where the predominant iterative activity is managing abstraction at both ends of the process. In the coding scheme used to characterize iterative behaviors, these types of iterative activities were defined as "coupled" iterations. The examples below are drawn from the verbal protocol and provide greater detail on the causes and processes of iteration in design.

Consider the following passage for a senior engineering student who was above average for time spent iterating (45.9 percent) and final quality score (.515). The passage occurred fifteen minutes into the design task when the senior was already working out solution ideas, and illustrates an effort to clarify a particular problem requirement (comments from the task administrator are included):

OK, as far as like a full scale baseball diamond, first, second, third base...ok so it says 12 children should be kept busy at any one time so how many children are expected on this lot? I mean is that up to me?
Administrator: *yeah.*
If 12 children should be kept busy what about their parents?
Administrator: *ok uh that's you're decision.*
Uh am I assuming the parents are there or not?
Administrator: *you can do whatever you want.*
OK, uh it would be easier if I said their parents weren't there. But how many one year olds go out without parents? OK, we'll assume the parents are there we'll make a bench for them to sit on.

Here, while generating ideas and mapping out preliminary solutions the participant revisited the problem requirement "twelve children should be kept busy". General questions were asked to the administrator but no details were provided. In the process of trying to interpret this requirement the participant proactively appended the problem statement to include the needs of the parents who would be taking their young children to the playground. At some level this also became an issue of safety and as such the preliminary design was modified to include benches for the parents to observe their children. At about halfway through the process the same participant continued refining a conception of the role of the parent in the playground design:

The seesaw should fit in the middle and go this way north to south. The swings um should face this way towards back and forth towards the parents so at least they can keep an eye on them. And I'm designing this so the parents from their position should they choose will be able to observe their kids unobstructed in every direction within the compounds of the playground. And the monkey bars in this quadrant I'll have going diagonal for the very same reason.

Here, including a previously unarticulated requirement from the perspective of the parent provided justification for a decision on where to place the equipment in the playground lot. This became a concrete requirement that impacted the placement of the remaining equipment.

Sometimes participants were explicit in their efforts to resolve ambiguity. For example, the passage below is from a senior who actively sought out potential failures early in the process in relation to preliminary solutions. This provided a mechanism to define safety in terms of specific characteristics associated with the equipment being open to the elements. As with the previous example, this senior spent a considerable amount of time iterating (39.9 percent of the total design time) and received one of the highest quality scores (.585).

O.K. what else do we need? We need uhh let's see. Children use the playground 1 to 10...12 children any one time...3 different types of activities. Any equipment...be safe...remain outside all year long.
What can remain outside? Sandbox can remain outside...get wet but it can remain outside Umm ahh we want some...on the slide. On the steps we want some gripping stuff. Uhh anti-skid stuff_ what do we call it ahh...skid-resistant...skid-resistant...umm tape I guess they call it on the...on the ahh the stairs or the steps. Alrighty...now a rail. That'll be a good idea_ a rail for them to hold on to.

Here, assumptions about the climate became integrated into the task—illustrating a strategy of using a particular solution to map out features of the problem. For example, as the participant moved between particular solution contexts and abstract problem requirements ("be safe" and "be outside") the playground was redesigned to include features that would limit children slipping and falling. As another example the following passage is from a freshman (iterated 55.7 percent of the total design time and received a .577 quality score). This passage illustrates an effort to clarify the impact of the general climate on the playground which began with asking for information from the administrator.

um, what is the, what is the general climate of this area.
Administrator: *the climate*
Yes, that's always good (inaudible) unless it's raining, damp or anything, thank you. [Receives information from administrator] I just called the um, weather station, and they said the climate, the yearly climate of this city is consistent with an, a Mid-Atlantic state, like Pennsylvania, the keystone state, average temperature ranges as, are as follows, in the winter it is ten, it ranges from about ten to thirty five degrees Fahrenheit, spring is about thirty five to sixty five degrees Fahrenheit, summer is around sixty five to seventy...., ninety five degrees Fahrenheit, fall is around thirty to sixty degrees Fahrenheit...so there is a some what type of pre frosting in there between the winter and spring, and the fall/winter, that

means that this material could be a metal, a, ah, metal or wood, but it has to last a long time, at least remain outside all year long, that means, that I should not have some type of flat metal surface cause that will freeze up in the winter, too hot in the summer, so usually a gravel, if, if this is to be a walked surface not a playing surface, gravel type of surface, would..., work best because it drains well, doesn't freeze up, it's pretty easy to get to snow off, if (inaudible) packed ground, loose gravel, for bike riding it's also good.

With this new information, this participant interpreted "remain outside" and incorporated this immediately into the solution, in the process expanding this requirement to include drainage issues.

Across the full study cohort, features of "coupled" iterations include: gathering information on a just-in-time basis, justifying design decisions by qualifying or quantifying problem requirements, synthesizing information into solution elements as generated, describing behavior or function of a solution in terms of problem needs, evaluating solutions while clarifying evaluation commitments, and clarifying problem requirements while selecting preliminary solutions. Common activities that triggered these iterations include self-monitoring and reviewing progress, searching for potential failures, clarifying and interpreting the meaning of the problem, and examining solution behavior to identify potential failures. Overall, coupled iterations were transformative processes that involved redefining problem elements through levels of abstraction and integrating this new information into the evolving design solutions. Finally, seniors were significantly more likely to engage in coupled iterations and these kinds of iterations correlated positively with higher quality scores. This suggests that simultaneously managing uncertainty at both ends of the process (problem and solution) and using emerging solutions as a concrete context to interpret problem requirements appear to be an effective iterative design behavior.

Iteration and Managing Uncertainty

The iteration lens illustrates that efforts to deal with uncertainty are frequent and throughout the process as well as quickly resolved or dropped. This gives a sense of scale and frequency to the results presented in the previous section. Also, iteration in design suggests that managing uncertainty is directed towards understanding the problem and involves revising problem and solution elements simultaneously (coupled iterations). These coupled iterations lead to conceptual shifts in understanding and may be markers of a conversation across problem and solution spaces. Characteristics of coupled iterations also provide insights into forms uncertainty takes in design and the strategies designers use to deal with uncertainty, including an awareness and willingness to seek out and address uncertainty. Finally, the iterative behaviors described in this section suggest that dealing with uncertainty is related to greater experience and better performance.

MATHEMATICAL MODELING[7]

In this section we describe mathematical modeling tasks that were used to provide design experiences in a first-year course for engineering students. In the previous two sections, the students were required to design a distribution system and a playground. In both cases, the students encountered uncertainty and iterative

---

[7]This material is based upon work supported by the National Science Foundation under Grant No. 0120794. Any opinions, findings, and conclusions or recommendations expressed in this material are those of the author(s) and do not necessarily reflect the views of the National Science Foundation.

cycles of solution occurred as a result of the conditions of the design context. In this example, the context is an emerging field of engineering (nanotechnology) with uncertainty even for experts in the field. The students produced a range of solutions which represent possible iterations in a design cycle. By organizing a sample of responses, we could observe a range of solutions and develop hypothesis about possible iterations of solutions over time.

A 4-year curriculum reform and development project for a first-year engineering course in problem solving and computer tools (UNIX, Excel, MATLAB[7]) focused on the design of modeling tasks set in engineering contexts. In designing the tasks, we followed six principles for the development of modeling activities which fit with the goals of the course (Lesh et al., 2000). The goals of the course included introducing students to the application of the computer tools to engineering problem-solving situations as well as particular content in statistics, economics, and mathematical modeling. The introduction of the modeling activities allowed for the integration of various course topics within a particular context. In addition, the modeling tasks asked students to design models and procedures using a process and criteria similar to authentic engineering work. The tasks presented a client with a problem, a data set for the context, and background information about the context. For example, one task, Nanotechnology, asked students to design a procedure for measuring roughness of materials at a nanoscale in order to aid in the design of materials for hip replacement (Moore & Diefes-Dux, 2004). The students were given three digital images of the surface of samples of the material at different scales showing the peaks, valleys, and bumps only visible at a molecular level and asked to generate a procedure for ranking the samples in order of roughness. The grey-scale value of each pixel in the image corresponded to the height of the surface at the point. The students drew on their knowledge of statistics and statistical sampling in order to generate a data set of points from the image and analyze the data points with statistical measures. In engineering work, there are a number of procedures available to measure surface roughness used depending on the context. For example, drawing random lines on the image and developing criteria for picking points along any line (e.g., the ten highest points, the ten lowest points). Some of the student responses were primitive versions of engineering methods. For the purposes of this chapter, we will focus the discussion on uncertainty and iteration related to the student responses to the task. Hjalmarson (2005) gives a more detailed analysis of the evaluation and assessment of the Nanotechnology task responses.

Role of Iteration in Modeling

There were two types of iteration as part of the Nanotechnology task. There is iteration within the implementation and in the range of student responses to the task. The iteration within the implementation occurred as students shifted from individual work to teamwork. Before working with their teams, the students individually answered questions about roughness and situations where roughness is important. We provided this opportunity so that students would have an opportunity to think about the task, to become familiar with the context, and to consider their ideas before working with the team. Sometimes one person can dominate the idea devel-

---

[8]MATLAB requires students to learn a basic programming language and write programs to analyze data and do computations.

opment if all students have not had an opportunity to consider theproblem at hand. When the teams began working together after the individual tasks, the individual understandings could be integrated and refined.

The second type of iteration occurred as the students worked on the task. Iteration was necessary as students tested their procedures using the digital images. Within the task asking students to develop a measure of roughness, procedures could be iterated as the students interacted with their teams. In the Nanotechnology task, they could test their procedure using the sample images and the sample images provided examples of characteristics of surfaces that might require encapsulate different aspects of roughness. For instance, surfaces could have a large number of peaks that are about the same height but very close together. A surface could also have a few peaks that are of varying heights and far apart. The surface defined as "more rough" could depend on the context of measurement. In addition, we also asked them to list information that might have been helpful for solving the problem as part of the final product. Further refinement of procedures occurred in later tasks that introduced accepted engineering methods for measuring roughness. The final project in the course required the development of a MATLAB program that would take a data file with the pixel information from a digital image and generate a measure of roughness. The final task required the use of computer programming skills as well as mathematical modeling for the measure of roughness.

To analyze the responses, we used the quality assurance guide developed by Lesh and Clarke (2003) and used in other studies including evaluation of student responses to similarly structured modeling activities (Carmona-Dominguez, 2004; Chamberlin, 2002). The quality assurance guide uses five levels to evaluate student responses holistically based on how well the students met the client's needs. Principally, this includes a focus on the development of a procedure that the client can use and understand in the given situation and extend to other situations. For example, level one is applied to responses which are completely off target. The levels progress upward according to how well the product could be applied in the present situation or generalized to other contexts. Level five solutions are sharable to other situations beyond the immediate context posed by the client. Table 21–2 shows the levels for the quality assurance guide as written in Lesh and Clarke (2003, p. 145).

In order to demonstrate how the quality assurance guide was applied to evaluating student responses, we will discuss two responses and the questions they raise about student work on design tasks. Two responses are shown in Figures 21–3 and 21–4 in order illustrate the differences in the detail of the procedures in the responses. The response in Figure 21–3 received a 2 for being on the right track toward a solution since they mentioned finding equal sections and a set of statistical measures. However, the team did not provide enough information about how big the sections should be, how many sections there should be, or how to interpret information from the statistical analysis of the sample data points. The team should also have specified how to measure "a great difference between average heights of the surface". Using multiple statistics without providing a rationale or an interpretation for the statistics was frequently a response to this and other tasks where the students computed statistics.

The response in Figure 21–4 received a 5 (the highest possible score) according to the quality assurance guide because they explained a procedure in enough detail that someone else could use it for the given images as well as other images that may need to be measured for roughness. In addition, the procedure might be modified in order to measure other characteristics of the surface (e.g., size of peaks).

TABLE 21-2.
Quality Assurance Guide

Performance Level	How Useful is the Product?
1 Requires redirection	The product is on the wrong track. Working longer or harder won't work. The students may require some additional feedback from the teacher.
2 Requires major extensions or revisions	The product is a good start toward meeting the client's needs, but a lot more work is needed to respond to all of the issues.
3 Requires only minor editing	The product is nearly ready to be used. It still needs a few small modifications, additions or refinements.
4 Useful for this specific data given	No changes will be needed to meet the immediate needs of the client.
5 Sharable or reusable	The tool not only works for the immediate situation, but it also would be easy for others to modify and use it in similar situations.

The series of steps to measure the roughness of the sample are:
- Randomly draw lines of random lengths are placed on the image sample.
- The length of each line is measured.
- An interval for how often a measurement of height is determined by the scale of the axis divided by 10.
- At every interval the height of the certain point is determined by the color bar on the right of the image.
- The average height of the points for each line is determined and the average height of the lines for each sample is determined.

For sample A, B and C randomly drawn lines of random lengths would be drawn on the sample. The average height of each line would be determined based on the steps above, and average height per sample would be found. Then the averages of the samples would be taken to determine an average height of the roughness of the gold.

FIGURE 21-3. Using multiple statistics in nanoroughness.

The series of steps that could be used would be to look at the same size sections for all three sections. With these equal sections we would look for the highest points of the surface. After knowing this we would then look to see how frequent this occurs, and if there is a great difference between average heights of the surface.

* For this set of data we would look at 0.5 micrometer x 0.5 micrometer random squares of each data. Looking at each square we see how widely distributed the heights are. We could do various statistics of the data. Such as the deviations of the heights, maxes, mins, ranges, and modes.

FIGURE 21-4. High quality student response to nanoroughness task.

One way to look at the two responses is as "drafts" of a procedure. The level–2 response (Figure 21–3) could have been developed further by the team into something that looked more like the level–5 response in Figure 21–4. However, it did not. The question is why. We do not have observational data about either of the teams, but it is probably a safe assumption that the level–5 team in Figure 21–4 had some discussion about which statistics to compute and how to find a set of sample points. They may have discussed how long the random lines should be. Other teams used lines of equal length and an equal number of lines per sample. Actual engineering methods for computing roughness (and other surface characteristics) use the placement of random lines on an image in order to generate a sample of points to measure. Finding the mean or the standard deviation is also part of some engineering methods. So, the team of students (even though they presumably have little to no experience with nanotechnology) can develop a way of thinking about a procedure for characterizing the roughness of a surface. To return to the earlier question, why did the level–5 group develop their procedure further and the level–2 group did not? Why does the level–2 response look like a first draft of a procedure and the level–5 response look like a final draft of a procedure?

Role of Uncertainty

The first piece of uncertainty in the Nanotechnology task was part of the task context. The students had little to no experience with nanotechnology and it is an emerging field of engineering. A challenge in nanotechnology is the small scale (measurements are taken in nanometers or units less than 10^{-9} meters at the molecular level) so the objects are difficult to see literally and figuratively. There is uncertainty within the field itself as applications, procedures and methods are currently being developed. For instance, the field of Atomic Force Microscopy is still developing methods for generating the images of different types of surfaces. Once the images have been generated, it is still unclear how the information from the images may be used. Medical, genetic, and biological applications may be possible, but there are still unknown applications. So, conveying an emerging, uncertain field to students becomes a chicken-before-the-egg problem. Students should graduate with information about emerging fields of engineering, but there may not be information or problem contexts that are accessible to first-year engineering students if the field is cutting-edge. The context can also feel abstract since the objects cannot be touched or seen with the naked eye. The images are real, but the surfaces they represent may not feel "real" to the students working on the task. We attempted to connect to their previous experiences with rough surfaces, but further analysis would be required to determine how well the students connected with the context. In any case, the context represented a new field for the students and a new type of problem that students may not have thought was engineering before working on the task. As with the tasks described in the previous sections, the students were developing a paper or computer-based design rather than a physical object which may have added to the uncertainty felt by the students. Where a procedure for measuring the roughness of a road could be tested by driving on the road or touching the road, that kind of physical testing was not possible in this context.

A second component of uncertainty was in the task itself and part of the design. The students had a limited set of information for the Nanotechnology task since they had only one sample from each surface. As part of the response, many students mentioned that color images would have been helpful. While color images are available, color images require 3 values per pixel (red, green, and blue) where black-and-white only need one value on the grey-scale in order to relate to a height of the surface which creates a linear relationship between the height of the surface and the pixel value. Using three values would have added significant mathematical complexity to the subsequent programming project to model roughness the students were assigned. This is an example of how a complex problem scenario might be simplified for freshman students. As part of a more complex, capstone project, upper-level undergraduates might be expected to work with color images. However, we had to simplify the available information in order to make the task doable within the time frame and given the mathematical experience of the students. Most freshmen are enrolled in an introductory calculus course. Many have taken an advanced placement calculus course, but we could not assume calculus background or the differential equations background necessary to complete some of the modeling tasks that could have been posed related to nanotechnology.

THEMES AND CONCLUSIONS ACROSS DESIGN-TASK CONTEXTS

There are a number of similarities between the three design contexts presented in this chapter. All three of the design tasks ask students to design something for a client. In addition, iteration and uncertainty play significant roles in the design process for the student-designers and raise questions for engineering educators related to design activity in a learning context. What can the three design tasks tell us about the nature of uncertainty in design activity? To begin to answer this question, we can consider twelve dimensions of task environments for prototypical design activities (e.g., nature of constraints, complexity, interconnectivity of component parts, personalized stopping rules, multiple workable solutions, and temporal sequence of activities) identified by Goel and Pirolli (1992). Goel and Pirolli suggest that the extent to which a task environment fulfills these twelve dimensions captures the extent to which a task is characteristic of design activity. Table 21–3 extends Table 21–1 presented at the beginning of the chapter by adding salient characteristics of the tasks that provide means for comparison across design scenarios. We list a number of characteristics to show both how the tasks are in different contexts but represent similar aspects of engineering.

All of the tasks contain some element of reality. For the Capstone Project, the students are working for a real industry partner. In the other tasks, a fictitious yet realistic client is included. The Playground Design and Nanotechnology tasks are also similar to real work that engineers do: designing playgrounds, designing procedures for measuring characteristics of materials. Playground Design includes a number of variables (e.g., safety, climate, materials) that students need to weigh in order to determine a best possible solution given the particular constraints of time and access to information. In all three tasks, the amount of information available is both a source of uncertainty and an impetus for iterations in the design. The students in the Capstone Project must gather and sort information and data in order to develop a design. The design is revised as they assemble information. In Playground Design and Nanotechnology, the students have a lack of potentially relevant information and have to accommodate that lack of data. For Nanotechnology, the students simultaneously have too much and too little information. Each image contains thousands of pixels that could be used to measure rough-

TABLE 21.3.
Salient Task Features

Task Feature	Capstone Project	Playground Design	Nanoroughness
Nature of problem	Design distribution center and redesign parts of a supply chain	Design a community playground	Design a procedure for measuring surface roughness
Work structure	Team of 4 students	Individual	Team of 4 students
Level of students	Juniors and Seniors	Freshmen and Seniors	Freshmen
Time to complete	2 academic quarters	3 hours	2 hours (laboratory session)
Nature of client	Authentic industry partner	Fictitious community client	Fictitious industry partner
Role and access to information	Undefined, involved many assumptions	Constrained and explicit, although evidence of assumptions	Constrained and explicit
Nature of constraints	Sources include client needs and familiarity with key concepts, requires strategies to manage complexity (e.g., prioritize, reduce, trade-offs)	Sources include client needs and familiarity with task, requires strategies to manage complexity	Sources include client needs and limitations of resources (inability to test working models)
Likelihood of many feasible solutions (open-ended)	High, previous designs available	High, previous designs available	High, previous designs available
Interconnectivity of parts	Can decompose into many parts	Can decompose into many parts	Can decompose into parts
Familiarity of task	New task but builds on prior coursework, likelihood of making rough estimates	Common and tacit experience (outside of school), likelihood of making good estimates	New field and abstract concepts, difficult to make good estimates
Stopping rules	Personalized, judged by client	Personalized, not judged by client but scoring rubric	Personalized, not judged by client but scoring rubric
Uncertainty	At both ends—problem and solution	At both ends—problem and solution	At both ends—problem and solution (or primarily for solution since were asked to use mathematical knowledge)
Motivation to participate and succeed	$1000 competition	Small monetary compensation	Assignment grade

ness. However, the images are black-and-white and there are a limited number of images.

Across the studies there remain questions about how students working on design tasks interact, use knowledge, and develop knowledge as they design. In all three tasks, there is movement from iteration to iteration which hopefully leads to improved design. The mechanisms and events that provoke iteration are still unclear and may be related to characteristics of individuals doing the design work or the design scenario itself. In all three cases, there is uncertainty about how to design and what to design. The problems are all open-ended. There are multiple solutions and procedures. The students' task is to move from the space of all possible solutions to a working solution and justify their response in some way by making assumptions and constraints explicit. The justification may happen only within the team or may be part of the final product. For the Capstone Project, some of their interaction is related to justifying and explaining ideas to each other. As part of the Nanotechnology response, the students were explicitly asked to describe unavailable information that would have helped them develop a solution.

In all three contexts there is a reflexive relationship between uncertainty and iteration. Iteration is used both as a noun and a verb. There are iterations of solutions as students reach points of uncertainty and resolve that uncertainty within the solution. Solutions are also iterated over time. While it may seem obvious to state there is interaction between process and product, in the examples we have described the product drives the development and vice versa. Uncertainty motivates iteration and further iteration generates further uncertainty. As prior complications are resolved, new complications may arise. This can be very frustrating to students who expect to be "done" in one or two iterations, who expect to be clearly told what to do by the client (either fictitious or real), and have limited experience with complex design. As described for the Playground Design task, students vary significantly in terms of time spent on iterations and the number of iterations and that was an individual task. Iterations in a group task will likely include more complex interactions between individuals, information, and the task.

As we consider the challenges of recruitment and retention that engineering education is currently grappling with, it is ever more important that students are quickly exposed to the core of engineering—that students are able to see engineering for what it really is. Through the studies described in this chapter, we would like to suggest that uncertainty, iteration and design are some of the core features of engineering, and therefore are aspects of engineering that students should realize early in their engineering education experiences. If uncertainty and iteration are important aspects of engineering, how can students have opportunities to encounter them before they enter the workplace? We have presented three tasks as examples. All three tasks have implications for instruction, teaching, learning, and development in engineering education. Beyond engineering education, other design fields incorporate similar processes.

REFERENCES

Adams, R. S. (2001). *Cognitive processes in iterative design behavior*. Unpublished doctoral dissertation: University of Washington.

Adams, R. S. (2002). *Understanding design iteration: Representations from an empirical study*. Proceedings of the International Conference of the Design Research Society, September, London.

Adams, R., & Atman, C. J. (1999). *Cognitive processes in iterative design behavior*. Proceedings of the Annual Frontiers in Education Conference, November, San Juan, Puerto Rico.

Adams, R. S., Turns, J., & Atman, C. J. (2003). Educating effective engineering designers: The role of reflective practice. *Design Studies, 24* (3), 275–294.

Archer, B. (1979). Design as a discipline. *Design Studies, 1*(1), 17–24.

Atman, C. J., & Bursic, K. M. (1998). Documenting a process: The use of verbal protocol analysis to study engineering student design. *Journal of Engineering Education, April*, 121–132.

Atman, C. J., & Turns, J. (2001). Studying engineering design learning: Four verbal protocol analysis studies. In M. McCracken, W. Newstetter, & C. Eastman (Eds), *Design learning and knowing*. Mahwah, NJ: Lawrence Erlbaum Associates.

Bryson, M., Bereiter, C., Scardamalia, M., & Joram, E. (1991). Going beyond the problem as given: Problem solving in expert and novice writers. In R. J. S. P. A. Frensch (Ed.), *Complex problem solving: Principles and mechanisms* (pp. 61–84). Hillsdale, NJ: Lawrence Erlbaum Associates.

Cardella, M. E., & Atman, C. J. (2005). A qualitative study of the role of mathematics in engineering capstone design projects: Initial insights. In W. Aung, R.W. King, J. Moscinski, S. Ou, & L. Ruiz (Eds.), *Innovations 2005: World innovations in engineering education and research* (pp. 347–362). Arlington, VA: International Network for Engineering Education and Research.

Carmona-Dominguez, G. (2004). *Designing an assessment tool to describe students' mathematical knowledge*. West Lafayette, IN: Purdue University.

Chamberlin, S. A. (2002). *Analysis of interest during and after model eliciting activities: a comparison of gifted and general population students*. West Lafayette, IN: Purdue University.

Dewey, J. (1938). *Experience and education*. New York: Macmillan.

English, L., & Lesh, R. (2003). Ends-in-view problems. In R. A. Lesh & H. M. Doerr (Eds.), *Beyond constructivism: A models & modeling perspective on mathematics problem solving, learning & teaching* (pp. 297–316). Mahwah, NJ: Lawrence Erlbaum Associates.

Gainsburg, J. (2003). *Abstraction and concreteness in the everyday mathematics of structural engineers*. Paper presented at the annual meeting of the American Education Research Association, Chicago, IL.

Goel, V., & Pirolli, P. (1992). The structure of design spaces. *Cognitive Science, 16*, 395–429.

Hall, R. (1999). Following mathematical practices in design-oriented work. In C. Hoyles, C. Morgan, & G. Woodhouse (Eds.), *Rethinking the mathematics curriculum* (pp. 29–47). Philadelphia: Falmer Press.

Hjalmarson, M. A. (2005). *What is learned about engineering students' mathematical modeling?* Paper presented at the Annual Meeting of the American Educational Research Association, Montreal.

Lesh, R., & Clarke, D. (2000). Formulating operational definitions of desired outcomes of instruction in mathematics and science education. In A. E. Kelly & R. A. Lesh (Eds.), *Handbook of research design in mathematics and science education* (pp. 113–149). Mahwah, NJ: Lawrence Erlbaum Associates.

Lesh, R. A., Hoover, M., Hole, B., Kelly, A. & Post, T. (2000). Principles for developing thought-revealing activities for students and teachers. In A. E. Kelly & R. A. Lesh. (Eds.), *Handbook of research design in mathematics and science education* (pp. 591–646). Mahwah, NJ: Lawrence Erlbaum Associates.

Moore, T., & Diefes-Dux, H. (2004). *Developing model-eliciting activities for undergraduate students based on advanced engineering content*. Proceedings of the 34th ASEE/IEEE Frontiers in Education Conference. Savannah, GA.

National Academy of Engineering. (2004). *The engineer of 2020: Visions of engineering in the new century*. Washington, DC: The National Academies Press.

Schoenfeld, A. H. (1992). Learning to think mathematically: Problem solving, metacognition and sense-making in mathematics. In D. Grouws (Ed.), *Handbook for research on mathematics teaching and learning*, (pp. 334–370), New York: MacMillan.

Schön, D. A. (1993). *The reflective practitioner: How professionals think in action*. New York: Basic Books.

Turns, J., Adams, R. S., Linse, A., & Atman, C. J. (2003). Bridging from research to teaching in undergraduate engineering design education. *International Journal of Engineering Education, 20* (2).

CHAPTER 22

Teacher Development in a Large Urban District and the Impact on Students

Roberta Y. Schorr and Lisa Warner
Rutgers University

Darleen Gearhart and May Samuels
Newark Public Schools

This chapter describes how a modeling perspective, in the context of a multi-tiered research design, is used to strengthen and improve a K–8 systemic reform initiative in mathematics in Newark, the largest urban school district in New Jersey. One of the main goals of the initiative, entitled the Newark Public Schools Systemic Initiative in Mathematics (NPSSIM), is to provide professional development for K–8 teachers to help them encourage the development of powerful ideas in students, particularly those whose abilities and achievements often go unnoticed in settings involving traditional tests, textbooks, and teaching (Lesh, 2001; Schorr, 2003). In this chapter, we will describe the overall approach to the professional development—focusing on one teacher's growth as an example of the process, provide background on the initiative in general, and then conclude by sharing the overall impact of the initiative on students' mathematical achievement.

ASSUMPTIONS ABOUT KNOWLEDGE DEVELOPMENT AND PROFESSIONAL DEVELOPMENT

We recognize that changing teaching practice involves changes in both content and pedagogical content knowledge (see Hill et al., in press; Ball and Bass, 2000). However noble a goal this may seem, it is easier said than done. Schorr (2004), Schorr et al. (2003), and others report that overall, many teachers feel that they *are* teaching in ways that are consistent with the NCTM Standards (2000). Yet, despite their best intentions, there is often little evidence of practices that would encourage student understanding and associated increases in achievement. For example, our research and the research of others (see Simon and Tzur, 1999; Spillane and Zeuli, 1999; Firestone et al., 2004; Schorr, et al., 2003; Schorr and Koellner-Clark, 2003), document that teachers rarely probe students to determine whether their answers make sense; rarely ask students to explain, justify or share their reasoning; and usually focus their teaching on procedural knowledge, without encouraging deeper understanding. Consequently, a key focus of our research on this project is designed to better understand the evolving nature of teachers' knowledge. Our work is grounded in a models and modeling approach to knowledge development.

According to this perspective, knowledge is seen as being organized around situations and experiences. New situations and experiences are mapped into previously existing internal descriptive or explanatory systems (models). The models that are used to make sense of a situation are not static rather; they can evolve over time, as the need arises. So, when learners look at something for the first time, or in a different way, the models that they use to make sense of the situations (e.g., to make predictions that guide actions) are likely to be shallow, distorted, or less sophisticated versions of later models. Consequently, new models for, in this case, teaching mathematics, do not emerge instantaneously; rather they develop over time and in stages. The professional development design used in this initiative, therefore, is deliberately designed to help all learners (whether they be researchers, teachers, or administrators) revise, refine, extend, test and share their evolving models for teaching and learning mathematics over extended periods of time.

A main approach used in the professional development process involves a multi-layered research design involving interacting teams of teachers, administrators, undergraduate and graduate students, as well as University researchers who collaborate as "co-investigators" during year-long sequences of activities aimed at helping participants to revise, test, refine and share their approaches to teaching and learning mathematics—all for the purpose of helping students to develop deeper and higher-order understandings (Lesh, 2001; Schorr and Lesh, 2003). In this process, no clear lines are drawn between researchers and practitioners. In fact, in many cases, researchers function as practitioners, and practitioners function as researchers, and it is through this process that knowledge development is enhanced.

Consistent with our models and modeling approach, we believe knowledge development is far more cyclic and interactive in nature than is suggested by one-way transmissions in which teachers ask questions and researchers or teacher educators answer them. We contend that "telling" teachers about their students' mathematical thinking or pedagogical practices that are associated with the Standards (NCTM, 2000), or even "modeling" the behavior *for* them is no more effective than "telling" students about a complex mathematical idea or "modeling" a solution process *for* them and then expecting the students to "understand" the mathematical concepts involved and be able to apply them to new or unique situations (Schorr and Lesh, 2003). Indeed, teachers may change specific behaviors or teaching strategies, while still missing key ideas and understandings about their students' ways of thinking and their own pedagogical practices (Schorr et al., 2003; Spillane and Zeuli, 1999). Consequently, when considering professional development, parallel assumptions regarding students' knowledge acquisition and learning must also be applied to teachers' knowledge acquisition and learning. These assumptions are as follows (see also Schorr and Lesh, 2003):

- Students' knowledge includes not only skills, processes, and attitudes, but also includes models for describing and explaining mathematically rich problem solving experiences;

- Teachers' knowledge should include not only skills, practices, and dispositions, but also should include models for describing and explaining mathematical teaching and learning;
- Students should learn through meaningful problem solving experiences;
- Teachers should learn through personally meaningful problem solving experiences.

Further, we maintain that the teaching practices that we care about involve complex and interrelated knowledge and understandings. This is true for all individuals regardless of whether they are students (K–12, undergraduates, or graduates), teachers, administrators, teacher educators, or researchers. Therefore, we do not establish, a priori, a pre-conceived notion regarding the single "best" type of teacher (or student) behavior. All teachers (indeed, all individuals) have complex profiles, which enable them to be effective in some ways or places under some conditions, while being less effective in others. A key point that must be emphasized is that all teachers, no matter the level, should continue to develop, since there can be no final or fixed state of expertise or excellence. Indeed, as teachers develop new knowledge, they notice new things about themselves and the others that they interact with-particularly their students. This, in turn, causes them to revise their approaches to teaching and learning mathematics (Schorr, 2004; Schorr and Lesh, 2003; Koellner-Clark and Lesh, 2003). As this happens, their students begin to change, thus resulting in further changes in the teacher. This particular aspect will be documented in greater detail as we examine the practice of one teacher.

BACKGROUND

The school district that is the subject of this chapter is the largest district in the state of New Jersey, and serves a population characterized by, amongst other things, high poverty (its combined community wealth is in the bottom 10 percent of communities in the state and 77 percent of students are from families whose income qualifies them for free or reduced lunch programs), high student mobility (40 percent either enter or leave their school after the school year has begun) and poor student achievement on local and state assessments (approximately 85 percent of Newark students were classified as not competent or minimally competent in mathematics before implementation of the initiative). Eighty-three percent of Newark residents are African American or Hispanic (according to the 2000 U.S. Census) and 89 percent of the children in school are from these families.

The initiative, which began in October 2002, has several goals. They can be summarized as follows:

- To provide professional development for elementary and middle grade teachers;
- To institutionalize standards-based curricula, in every mathematics classroom in Newark's elementary and middle schools[1];
- To strengthen administrative support of standards-based math instruction;
- To build capacity among the existing cadre of mathematics resource teachers (MRT's) to support and facilitate standards-based mathematics instruction;

---

[1] Everyday Math in grades Kindergarten through five, and Connected Math in grades six through eight.

- Parental outreach so that parents can understand and support the changes that will take place when standards-based approaches and materials are used;
- The development of After-School Centers which will serve as "laboratories" for collaboration between University partners, prospective teachers, graduate students, teachers, MRT's, administrators, students, and parents. As part of the After-School Center, middle school children from local Newark schools attend sessions (generally speaking, they occur every other week) to participate in mathematical problem solving sessions. During these sessions, teachers have the opportunity to try out new instructional methods and pedagogic techniques; they also have the opportunity to closely observe the students' thinking. After the sessions, University researchers help teachers as they implement these ideas in the context of their own classrooms; and
- Using data and feedback as a way to inform the project so that constant improvement can take place.

As noted in the second bullet, a key goal of the NPSSIM is the institutionalization of mathematics materials that have been evaluated as being consistent with the goals of the National Council of Teachers of Mathematics (2000), and shown to be more effective than commercial materials (UCSMP, 2005; NRC, 2004). While the materials may be seen as having the potential to positively impact students, we note that their use, in and of itself, does not necessarily lead to changes in instruction or student achievement. Rather, long held beliefs, a lack of knowledge (both in content and pedagogy), and embedded traditions often result in new materials being used in old ways—ways that diminish the chances for increases in student achievement (NCTM, 1989). Further, as noted in the NRC (2004) report, "...positive results are enhanced when accompanied by adequate professional development and the use of pedagogical methods consistent with those indicated by the curricula" (p. 159). Therefore, a key component of this project is meaningful professional development.

Before continuing, as indicated in bullets 3, 4, and 5, we believe that teachers are not the only ones whose actions and beliefs have a strong influence on what goes on in mathematics classrooms. Other influential individuals include parents, administrators, curriculum developers, resource teachers, etc. whose knowledge needs are no less important than teachers. To this end, the project also offers professional development to administrators, and to the teacher leaders (many of whom participate in the project described in this chapter). It is not the purpose of this chapter to discuss, in detail, all of the above-mentioned bullets, but rather to focus on the professional development that is provided for teachers, particularly in the context of the After-School Centers (see bullets 1 and 6).

PROFESSIONAL DEVELOPMENT COMPONENTS

One component of the professional development program involves weekly meetings at Rutgers University. As mentioned above, these meetings are not simply a series of seminars or workshops, rather they involve interactions among students from local schools, undergraduate students at the University, teachers, administrators, and researchers working together to consider mathematical content, pedagogical content knowledge (see Hill et al., in press; Ball and Bass, 2000) and the ways in which students build understanding of mathematical ideas. The sessions are highly interactive, and make wide use of what we refer to as "thought revealing activities" (see Schorr, 2004, Schorr and Lesh, 2003). These problems are designed to employ contexts in which students can use their sense-making abilities to solve complex problems in a collaborative atmosphere. The problems activities that are chosen are specifically designed to simulate a "case study" approach to instruction (Schorr and Lesh, 2003 and Lesh et al., 2002). Problem activities of this type are designed to produce models for constructing, describing, explaining, manipulating, predicting, and controlling complex systems. These activities also have another important characteristic: as students solve these problem activities, they tend to produce a trail of documentation that reveals important aspects about the nature of the constructs or models that develop. The specific design characteristics of these activities encourage problem solvers to produce products that are not simply answers to specific questions, rather they involve constructions, descriptions, or explanations, that reveal many aspects of the thought process that go into the final solution. In this way, teachers, administrators, researchers etc. can focus on the work that the students produce in an effort to deepen their own understanding of the content, and the ways in which children learn the content, build representations of the content, and formulate justifications.

The work that is done essentially allows researchers, teachers, and teacher educators to collaboratively investigate the following:

- To understand how to create a mathematics classroom that promotes understanding.
- To identify students' mathematical thinking in order to make better instructional decisions.
- To identify how problem solving can be used to effectively teach mathematical concepts.
- To identify the interrelatedness of the skills and concepts that makes up the domain of middle school mathematics (including rational number concepts; ratio and proportion; etc.).
- To collaborate with colleagues to better understand the mathematical foundation of rational numbers.
- To grapple with case studies to better understand how students think mathematically.
- To reflect on their own teaching through videotape analysis and debriefing with the class.
- To reflect about their own mathematical experiences through journal writing, articulate what they know through discussion and interviews and to make their mathematical knowledge their own.

Several overarching themes permeate all sessions. One involves the development of a classroom atmosphere that is conducive to student understanding and involves the maintenance of high cognitive demand in the implementation of all activities (see Stein et al., 2000). Good tasks are important, however they must be implemented in classroom environments in which students are encouraged to formulate conjectures, test the conjectures, and defend and justify solutions, in the context of an inquiry-oriented learning approach. Such instructional environments have been shown to be effective with inner city students (Schorr, 2000; Schorr, 2003; Campbell, 1995; Silver and Stein, 1996; NCTM, 2000).

A second and closely related theme involves the development of powerful mathematical affect (see Goldin, this volume). As noted by Goldin, affect is much more complex than simply considering student attitudes toward mathematics (positive or negative), and the problems associated with negative emotional feelings such as anxiety or fear. *Mathematically powerful* affect (i.e., the affect that enables individuals to do mathematics powerfully) is not the same thing as positive affect. Powerful affect means that often, students experience *both* positive feelings about mathematics (e.g., curiosity, enjoyment, elation in relation to mathematical insight, pride, and satisfaction) and ambivalent or negative feelings (e.g., annoyance, impatience, frustration, anxiety, nervousness, fear). However, the negative feelings occur in safe contexts, so that the students (and their teachers) can benefit from the experiences. Thus, frustration with a difficult problem leads to anticipation of learning something new, and increased pride of achievement when the problem is solved. Our work therefore often focuses on how teachers deal with student frustration, annoyance, and anxiety in positive and productive ways (Goldin, 2002; Goldin et al., 2005).

In particular, as part of our After School Learning Center (ASLC), researchers, teachers, and teacher educators observe/interact with students during the first 60 minutes of the session, and when the students leave, they discuss their own solutions, and the solutions of the students, the mathematical ideas present in the problem, related and deeper mathematical ideas, implications for teaching such ideas, and most importantly, how they saw the students solve the problems. They then implement the ideas in their own classrooms, and bring in student solutions to discuss with their peers, noting in particular, what they observed, what questions they asked, and other key issues related to the implementation of the problem. As part of this process, teachers often generate "observation" lists, or question lists to be used to help them or one of the colleagues implement the activity in the future. These lists serve as a window into their ideas about what they consider to be the most important features to notice (including mathematical ideas) as their students are working in class. Therefore they serve both as a reflection tool, and as a source of documentation of change over time.

In addition to work in ASLC sessions, Rutgers University researchers and mathematics specialists often accompany teachers as they implement ideas in the context of their own classrooms. During these visits, they discuss mathematical ideas that may be elicited, implementation strategies and issues, ways in which to consider the development of a classroom culture in which proof, justification, sense making, and high cognitive demand could be elicited and maintained, etc. After implementation, the teacher and researchers/mathematics educators discuss the ideas stated above.

AN EXAMPLE: FILOMINA

In order to better understand the professional development process, we will highlight the development of one teacher, Filomina[2].

Filomina is a seventh grade teacher who participated in the courses and the ASLC, and received direct assistance in her classroom (both from the second author, a Rutgers University researcher, and several undergraduate student-mentors). She confesses that when she initially attended the professional development sessions (before attending the ASLC where students were actually present

---

[2] The name has been changed to preserve the anonymity of the teacher.

during the course), she didn't agree with many of the ideas that were discussed. She often nodded her head in agreement, but privately said, "yeah right." She noted that several of her colleagues had spoken of changes in practice, but this was not particularly convincing to her, nor were the videos of her colleagues classrooms in which their students were defending and justifying their solutions, and doing the kinds of things that the teachers and researchers had spoken about. She would often solve the problems that were implemented in the course, and talk about the strategies that she had used, but felt that *her* students could not do them in the ways that she had observed. Filomina was convinced that her students were too far behind, mathematically, would become too frustrated and quit, and lacked the perseverance needed to do the kinds of things that she had observed in the videos. She began to change her mind when she began to observe the students in the ASLC. In that context, Filomina could actually work with the students as they solved the problems, and watch them as they grappled with solutions, revised their work, revisited their thinking and shared it with their peers and the other participants in the ASLC.

Filomina then challenged one of the researchers to come in to her afternoon classroom and help her to implement the thought-revealing problems in the way that she had observed in the ASLC and in the videotapes. Indeed, the researcher and several Rutgers University undergraduate students visited Filomina' afternoon class. Filomina agreed to have the sessions videotaped, even when the researcher was not present. During each session, two cameras captured different views of the group work, class presentations and associated student-student and student-teacher interactions. In addition, Filomina, the researcher and the undergraduate students (who actually videotaped all sessions) took careful field notes (which included description and interpretation) immediately after each session.

During the first few sessions, it was often noted that Filomina stopped students' thought process by asking them questions that were not related to their ideas, but rather to *her* ideas. For example, she didn't consider the representations and notations that they had chosen or used, and didn't actually "listen" to their ways of thinking about the mathematics. Rather, students shared their solutions, but not their ideas. This was reflected in her observation forms, where her notes indicate that she was primarily focused on group dynamics and how well students performed operations and procedures. For example, Filomina had implemented a task involving probability with the second author. As part of her observation guide, she wrote the following:

> This problem should be accompanied by a step by step guide for teachers who will introduce this problem to another class. The guide should include the skills that will be addressed such as fractions, decimals, percent and ratio, and probability (including tree diagram). It should also include the questions you asked when students derived the percents, etc. Teachers should have had prior lessons on changing decimals to percents, to fractions, to ratio. I don't think this is the problem to use when doing that. I would rather focus on the probability aspect of the problem.

She continued by discussing the particulars of her own students, "right or wrong, students were trying. They were focused and engaged. They wanted to...write on the transparencies. When it came to the discussions, they all wanted to present. Some more hesitant than others, but all who were not able to do so were disappointed. The level of commitment was exceptional."

Filomina shared these ideas with her peers and the researchers, and in reflecting on them, she began to notice that she was not paying much attention to the mathematical ideas of the students, or the their ways of thinking about the problem, rather

her ideas were more focused on the surface characteristics of the mathematics and classroom dynamics taking place. Despite her limited progress, Filomina began to notice differences between the mathematical activity of her afternoon class and morning class (where she continued to teach as usual). For example, she noticed that her afternoon students were able to talk about their ideas with greater fluency, and were more willing to justify their solutions, both in the context of small group and full class presentations. Filomina also noticed differences in the two class's performance on tests:

> When I gave an assessment to both the morning and the afternoon classes, I was amazed at the disparity in the way they were able to answer the questions. The task [that they were asked to solve] was patterned after a patterning task [that she had done with both groups]. The students were asked to create stages and predict the higher stages. At this point, students have only worked with 2 shapes. The afternoon class, which was allowed to discuss more, was able to apply their knowledge of stages to the other different shapes even the ones [shapes] they have not worked with in the past. The morning class...was generally lost. They asked many more questions. The afternoon class was more independent and thoughtful.

Filomina began to notice that the majority of her morning class was only able to solve problems that had already been modeled in class. If the problem was changed, even slightly, they were unable to solve it. She said that both of the classes "began the year on an even plane." As evidence she noted that at the beginning of the school year, both classes had the same number of students scoring at the highest, middle and lower end of the state assessment. In particular, she noted that both classes had the same number of students scoring 1's (highest score on the state test) and 3's (lowest score). In fact, when Filomina tested these students at the beginning of the school year (prior to any project implementation), the scores were essentially the same. During the month immediately prior to project implementation, the morning class had actually scored higher on the assessments contained in the Connected Math books (average score for morning class was 44 percent; and the average for the afternoon class was 39 percent). Seven weeks later, the two classes were tested again (again, using identical assessments from the Connected Math book). The average test score for the morning class was now 45 percent and the average test score for the afternoon class was 55 percent.

Filomina noted that little by little, the gap in test scores between the two classes increased. She stated, "Something must be working better in the afternoon class." It was at this time that Filomina decided to implement the same types of practices she had been using with her morning students. In particular, she stated that during "the next few days, I was more open to discussions with both classes. I was looking for wrong answers to spark further discussion. I was asking questions."

As she continued to reflect on her own behaviors, she stated, "There are days when I revert back to my old ways. Giving more answers than questions. Each time I do, I notice the difference in the students' behavior and response. They are always less responsive when I offer too much information. They appear more confused when I give them an algorithm. They produce very little when the tasks are low-level and repetitive. Students deserve more than a lecture from the teacher. They deserve more than a repetitious task. They rise up to the occasion when they are allowed and expected." Filomina began to become convinced that *her* students could do the kinds of things that she had observed in the videos and in the ASLC.

Further, Filomina experienced something that would prove very valuable. She was working on problem involving combinatorics (see Maher, 1998) with several other teachers and undergraduate students during one of the ASLC sessions. As part of her reflection, she noted,

> One of the teachers next to me said, 'We have to organize our towers. The way you are doing it, there is no organization. You are just choosing one color and then another.' Before he said that, I was working very hard on trying to organize my towers. He just could not understand my way of organizing. As a teacher, I may see that some students are 'not Organizing' things at first glance. I know now, that I have to investigate to find out if a student has a different technique of organizing than what I perceive to be disorganization.

Filomina's classmate questioned her method for solving the problem and wouldn't back down. He didn't understand her way of organizing and said that she was incorrect. When her method was questioned, Filomina said that she realized that just because we listen to students' ideas, it doesn't mean that we always understand them.

Over the next few weeks, Filomina made a genuine effort to "listen" to her students' answers, probe them for justifications, and refrain from simply telling them how to solve a problem when they reached an impasse. Her progress was noted by one of the Rutgers University students: "She didn't lead the students to believe that they were correct or incorrect. Filomina's questions appeared to prompt them to convince her that their solution made sense." Similarly, another Rutgers University student wrote, "I would hear comments from students saying Ms E. can you please just tell us the answer...." The students were now beginning to realize that their teacher is no longer providing the solutions for them, even when the mathematics education researcher is not in the room. Rather than tell students the answer, or provide a hint, Filomina asked more "Why" types of questions like: "Do you think that Anthony's formula works?" "Does yours work?" "Why is your answer different than Anthony's?" Filomina was now asking the students to justify their solutions while simultaneously encouraging them to look at the relationship between their solutions, which stimulated student-to-student questions, and the linking of representations. When she talks about asking questions, she now means clarifying questions, questions that asked students to describe their thinking, and examine questions as opposed to recall types of questions (see Heibert and Wearne, 1993 for a more complete description of the different types of questions that teachers tend to ask).

Over the course of the next few months, Filomina continued to change her perspectives and practices. She noted:

> ...I began actively listening and asking important questions. The more I listened, the more I learned about the students. I went from the teacher who showed the students the way or the solution, to one who listens. When I started listening, I was able to follow the students' thought processes and uncover misconceptions. My students are now about 4 months into the program and I am not the only one asking questions in the class anymore. I choose the first presenter and what happens next depends on that presentation. Most of the time the questions I want to ask gets asked by the students from the audience themselves. When I ask a question and one of the students start to ask a question, I consciously stop to allow the student to ask his or her question. If you ask me now what questions I'd ask my class on Monday when I go back to work, I won't be able to tell you exactly what they are. I'd have to listen to the students' conversations or the questions they ask each other and I'd have to look at their work. But one thing is certain; my students will have time to think the problem through. They may not all be working on the same aspect of the problem and that's o.k. They wouldn't have to beat the clock to answer many questions. They would be able to take their time. And I'll find time to

listen. My definition of a teacher has now changed. I'm now taking a different role in the classroom. My role now is to help create a non-judgmental environment where student ideas are valued and the best way to do that is to model a positive behavior. My role is to provide direction. My role is to listen some more. And on days that I am able to do such, I know those days, I've done my job.

Filomina's observation forms began to focus more and more on her students' mathematical ideas. For example, in referring to a task involving geometric concepts, she began to note the many different ways that students might approach the problem of finding the area of a trapezoidal region:

Students were looking for the formula to find the area of a trapezoid. M created rectangles by piecing the trapezoid. F later came up with the idea of adding triangles to the sides to create a rectangle, then subtracting the area of the added piece. O insisted that he had an easier formula. When he shared his formula, it wasn't as simple to me as he insisted it was. The technique that N's group came up with was to create a square/rectangle that was a little bit longer than the top, but a little bit smaller than the base. The excess part of the trapezoid from the base, they say has the same area that the excess of the rectangle was on top. They were counting the one unit squares within the squares to get the area.

Filomina is still grappling with ways in which to express the students' mathematical ideas with words. She also actively reflects on the questions that she asks, and their impact on student questions and student understanding. For example, she wrote,

A asked if this works for all triangles. Ar said it depends on the triangle. I think some of the students in this class are now internalizing the type questioning that teachers ask and they are now confident to ask the same questions. [For example] S asked why it was necessary to add the rectangle. She asked how you can find the area of a triangle when there is a rectangle. Ar, J, and Jo were trying to explain to her and she was not convinced. B's question was why Ar needed to add two parts when there is only one triangle. He said that if you remove the rectangle, there is only one triangle, and not two. Ar answered that the triangle was divided into 2 separate parts. P, Ar and S added a rectangle from the height of the isosceles to one side of the triangle, creating a smaller half rectangle as compared to J and Jo's. They then multiplied the length and width (without identifying it as the altitude and half of the base). When I asked why they did not divide by two, they said that they did not need to. They also explained that the other half of the triangle, if flipped, would fit into the rectangle. Some did not understand that and needed clarification. At some point, the questioning got heated, there were several students in front and they were discussing on their own. They did not need my assistance. P's group named their half of the base as 19, when the whole base was 42. I questioned why that was 19, Pa said it was supposed to be 21. When I asked him why, he said he multiplied 42 by .5 and got 21. I questioned why he did that and he explained that .5 is half. I asked then if he could use that technique with other numbers and get half. He said yes, and so did several other students. I said, check it with 100, 24, and other numbers. More and more seemed to agree.

Filomina attributes her growth to all of the different aspects of her involvement including the classroom visits by the researcher and the opportunities to share ideas, watch students, and work with colleagues and researchers as part of the ASLC. Recently, she noted:
Teaching the way I did before I:

1. Forced the students to have a conversation with me that only I decided was important and that was dominated by me.

2. Not only controlled and monopolized the questions that will be asked in the classroom, I also controlled the students' answers by asking leading questions and by asking them in a tone of voice that would make them answer the answer I want to hear.

As noted before, the quality of the students' work increased. Test scores also increased: Prior to project implementation: 44 percent test average in her morning class and a 39 percent test average in her afternoon class. By May, the test average went up to 68 percent in the morning class and 72 percent in the afternoon class.

Filomina, in reflecting on her own growth, felt that she had gone through four different stages. They appear below, in her words:

> I identified 4 stages that I went through. The first was that which was shown in the earlier tapes where I would go to one group and then leave shortly. At this point in time, I thought my job was to make sure all the groups have a task. I would pose one short question or problem, such as find the area of this rectangle, and then head to the next group. Aside from just assigning the task, I didn't know what to ask. The second phase was when I started asking questions. At this point, I thought my job was to ask as many questions as I can. I would dominate the questioning. Even if students ask questions, I would still insist that my question be asked. The third phase was when I started actively listening. I started to encourage student-to-student questioning. But at this point I found myself still asking too many questions, that the students often stop me because I would interject in the middle of a good student to student discussion. The fourth stage is the stage I'm working on. I'm taking a more backseat role. I continue to listen this time for longer periods of time; I encourage student-to-student questioning, and I try ask[ing] only questions that allow students to build on their ideas instead of introducing other topics in my questioning.

The particular activities that were chosen for these sessions provided an opportunity for students to use their concrete, intuitive, and natural sense-making capabilities. They could literally "enter" the mathematical task in a place that was comfortable for them—that is, they could use their prior experiences and real world understandings to begin the problem while simultaneously being challenged to reveal, test, refine, revise, and extend important aspects of their ways of thinking. The activities that were chosen not only contributed to learning, but also simultaneously produced trails of documentation that revealed important aspects about the nature of the mathematical constructs that developed. In this way, the teacher was able to continuously adapt, modify and extend her own thinking about the mathematical engagement, discourse, justification, and sense-making that was going on, and consequently make informed decisions about how to help the students. She was also able to share these aspects of her thinking with the other participants in the Center, thereby allowing them to be co-investigators in her work.

OVERALL DISTRICT RESULTS

Filomina was not unique in her progress. Rather, she is one example of a growing network of teachers, teacher educators, researchers and supervisors who join together to help students learn conceptually deep mathematics. Results thus far indicate that indeed, Newark students are achieving at significantly higher levels. For example, at the fourth grade level in 1999 (before implementation of the NPSSIM), there was no statistically significant difference between NPS schools who

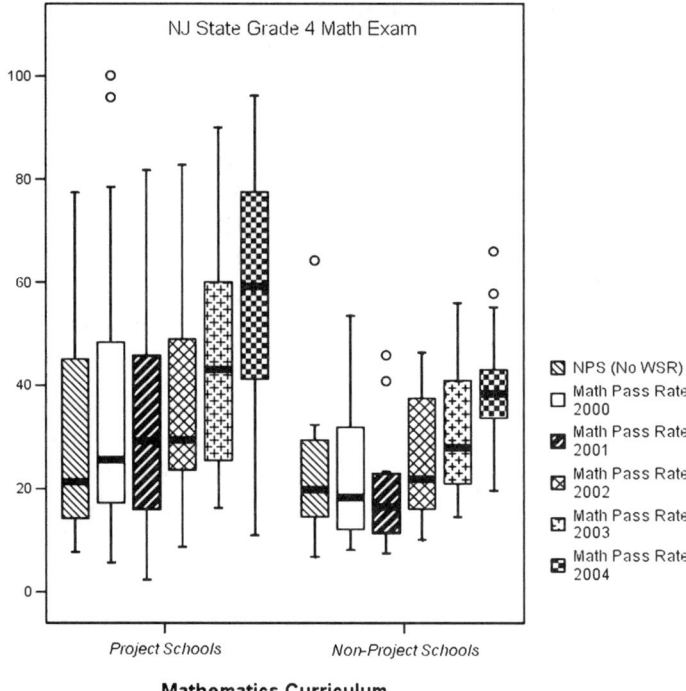

FIGURE 22–1. N.J. State Grade 4 Math Exam

were participating in the program and those who were not[3] (see Figure 22–1). In subsequent years, the non-project schools did not improve at the same rate as schools involved in the NPSSIM Project. By 2002, the differences between the groups of schools were statistically significant, i.e., schools involved in the NPSSIM Project dramatically improved and schools not involved showed little improvement, if any (see Figure 22–1). In 2004, at the request of both principals and teacher, all schools joined the NPSSIM Project.

In grade 4, initial equivalence/differences across groups can be measured (and accounted for) using students' reading/writing achievement scores (LAL). Reading/writing achievement scores have tended to be highly correlated to mathematics achievement scores in Newark, and are not likely to be affected by NPSSIM; therefore, serve as an ideal covariate. In 2004, the rate of change in reading/writing scores (9.3% increase) for the New Jersey Assessment of Skills and Knowledge of grade 4 (NJASK4) from 2003 was not significantly different than the rate of change in mathematics scores (11.1 percent increase). In 2005, the rate of change in reading/writing achievement scores (2.0 percent decrease) for the NJASK 4 was significantly different than the rate of change in mathematics achievement scores (5.2 percent increase). (See Figure 22–2). Considering the lack of improvement of reading/writing achievement scores and the significant gains made by mathematics achievement scores, this preliminary data suggests that the rise in standardized test scores can be attributed to the NPSSIM Project.

In comparison to other school districts in the state of New Jersey, NPSSIM has had a significant impact on grade 4 and grade 8 NPS students, as measured by New

---

[3]The non-project schools were participating in a "Whole School Reform Model" which had prescribed materials, and at the time, they could not adopt the new materials which were part of the project.

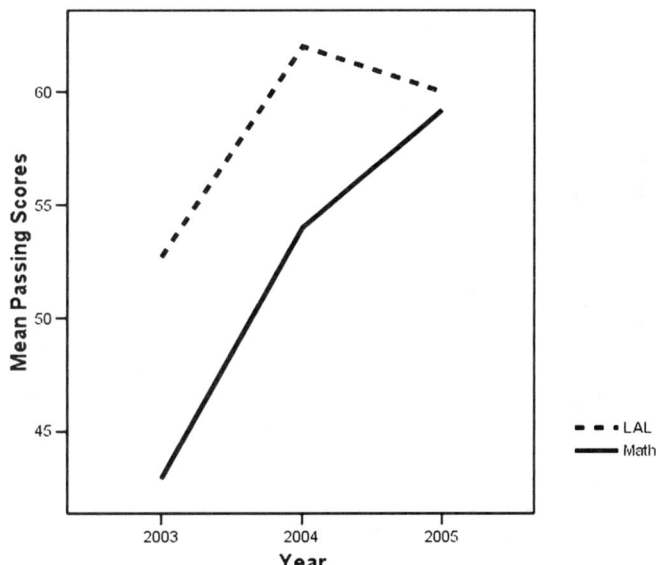

FIGURE 22–2. Comparison of Grade 4 results in math & language arts

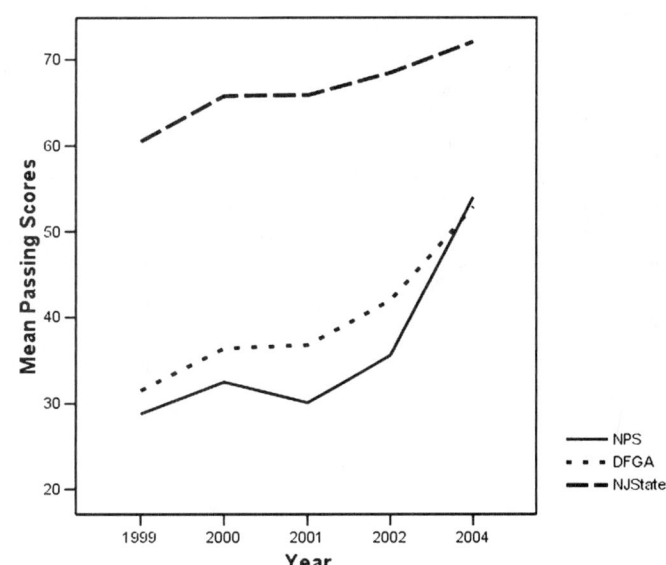

FIGURE 22–3. Comparison of scores across state, Grade 4

Jersey's state-mandated NJASK4 (Grade 4) and GEPA (Grade 8) assessments. On the graph below (see NJ State Grade 4, Grade 8, and Grade 11 Math Exams), the blue line represents the mean score for the state of New Jersey; the green line represents the mean score for District Factor Group A (factor groups are based primarily on district's socioeconomic status); and the red line represents the mean score of Newark Public School students. (Note that Newark Public Schools is classified as District Factor Group A and therefore, included in the factor group's data.)

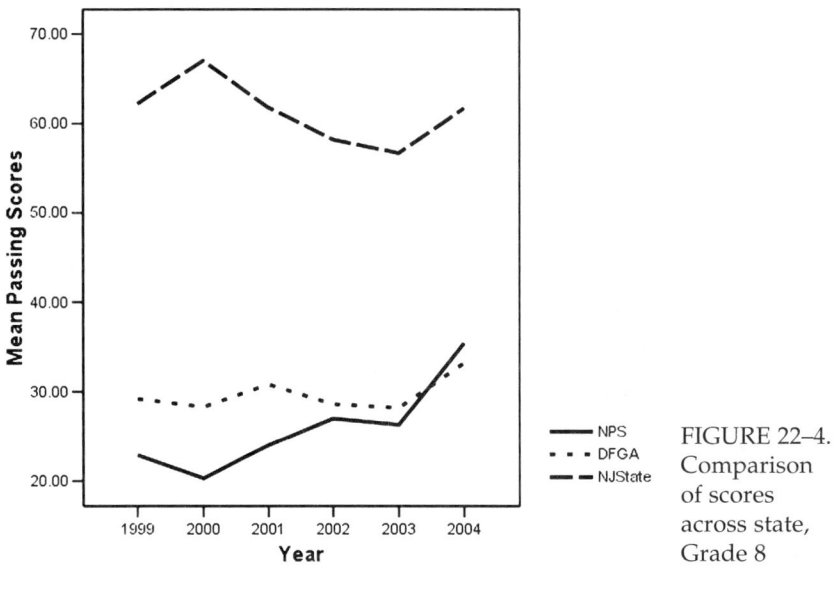

FIGURE 22–4. Comparison of scores across state, Grade 8

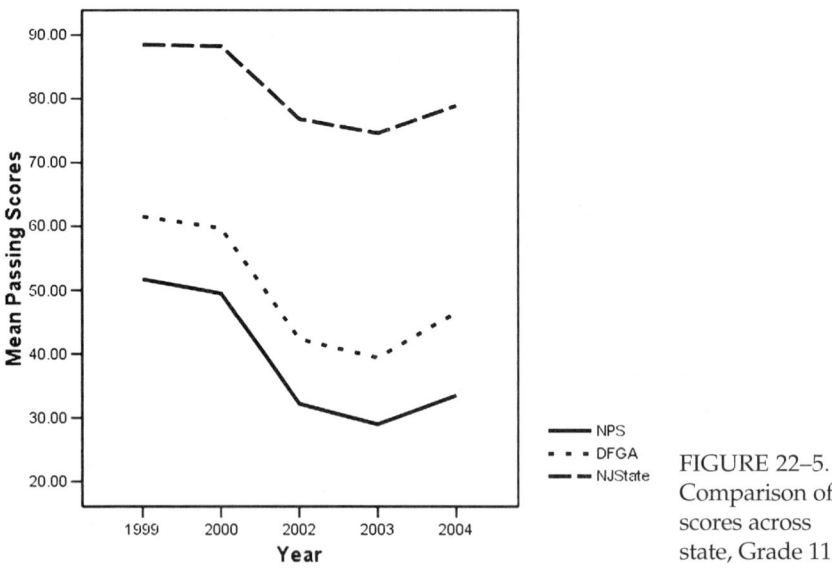

FIGURE 22–5. Comparison of scores across state, Grade 11

Grades 4 and 8 teachers received standards-based professional development in 2001, 2002, and 2003 in pedagogy and content through the NPSSIM Project funded by the National Science Foundation. In SY2004–05, standards-based curriculum was implemented at these grade levels and subsequently, the rate of increase for Newark Public School students surpassed both District Factor Group A as well as the state of New Jersey.

Grade 11 NPS teachers received comparable amounts of professional development and were neither using standards-based programs nor involved in the

NPSSIM Project (See NJ State Exams HSPA below). Considering the lack of improvement of grade 11 students and the significant gains made by grades 4 and 8, this preliminary data further suggests that the rise in standardized test scores was the result of the NPSSIM Project.

CONCLUSIONS

Our work appears to indicate three critical things. First, the data indicate that the project has had a significant and highly positive impact on students' mathematical achievement across the entire district. Students, particularly those whose abilities and achievements often go unnoticed in urban settings, are now performing at higher levels—and this is occurring across the entire district, not just in a few targeted classrooms. Although the mathematics education research literature has many accounts of teachers' practices there is little evidence of this having an impact on students across an entire urban district. Of course, a deep analysis of one teacher or a group of teachers is very useful for understanding many of the psychological, mathematical, and/or pedagogical dimensions relating to how and why teachers do the types of things that they do in classrooms. It is also useful for analyzing teachers' beliefs, knowledge, and attitudes concerning mathematics, student learning, and the teaching and learning of mathematics. However, while extremely valuable, such research is not designed to generalize about the effects that this may have on teachers throughout an entire district, particularly in an urban district under state takeover. While we recognize the limitations of using test score data in order to document growth in students, the case of Filomina provides a backdrop for considering the types of changes that the teachers are making using this approach to professional development, and the type of growth that students are experiencing in their classrooms.

The second, and closely related conclusion is that as teachers change and develop, their students also change and develop. The data presented, in conjunction with the case study of Filomina, demonstrates that as teachers see new things in their students, better understand their students' ways of thinking, and build deeper content knowledge, their students improve as well. The study of Filomina shows that teachers do not need to be taken away from their teaching to improve—rather they need to learn to do what they do better in the context of their actual teaching environments. That is, taking teachers out of their classrooms to engage in activities like writing curriculum may not produce the same results as having them involved in activities that take advantage of their own classroom practice.

Finally, the approach to professional development that is outlined in this paper is not at all based upon the passive transmission of information. Teachers, researchers, supervisors, students, and administrators are all active participants in this project. Telling teachers, researchers, and teacher educators about better ways to teach or about their students' mathematical thinking, or even "modeling" the desired behaviors or activities *for* them is no more effective than "telling" students about an idea or "modeling" a solution *for* them and then expecting them to "understand" the mathematical concepts involved and be able to apply them to new or unique situations. Teachers', teacher educators, researchers, supervisors, etc. all must constantly build knowledge that includes not only skills, practices, and dispositions, but also models for describing and explaining mathematical teaching and learning; and this *must* be done through personally meaningful problem solving experiences.

ACKNOWLEDGMENTS

The project is funded by grants from the National Science Foundation grant numbers 0138806 and ESI-0333753. The opinions expressed are those of the authors and not necessarily those of the funding agencies, Rutgers University, or the Newark Public Schools.

REFERENCES

Ball, D. L., & Bass, H. (2000). Interweaving content and pedagogy in teaching and learning to teach: Knowing and using mathematics. In J. Boaler (Ed.), Multiple perspectives on the teaching and learning of mathematics (pp. 83–104). Westport, CT: Ablex.

Campbell, P. (1995). Project IMPACT: Increasing mathematics power for all children and teachers. Phase I, final report. College Park, MD: Center for Mathematics Education, University of Maryland.

Firestone, W. A., Schorr, R. Y., Monfils, L. A. (2004). *The ambiguity of teaching to the test*. Mahwah, NJ: Lawrence Erlbaum Associates.

Goldin, G. A. (2002). Affect, meta-affect, and mathematical belief structures. In G. C. Leder, E. Pehkonen, & G. Törner (Eds.), *Beliefs: A hidden variable in mathematics education?* (pp. 59–72). Dordrecht, the Netherlands: Kluwer Academic.

Goldin, G. A., Richardson, G. Schorr, R.Y., Shtelen, V. (2005). *The affective dimension of innovative urban mathematics programs: An exploratory investigation.* A research proposal submitted to MetroMath: The Center for Mathematics in America's Cities.

Heibert, J., and Wearne, D. (1993). Instructional tasks, classroom discourse, and students' learning in second grade arithmetic. *American Educational Research Journal*, 30, 393-425.

Hill, H., Rowan, B., & Ball, D. (in press). Effects of teachers' mathematical knowledge for teaching on student achievement. American Educational Research Journal.

Kelly, A., & Lesh, R. (Eds.). (2000). The Handbook of Research Design in Mathematics and Science Education. Mahwah, NJ: Lawrence Erlbaum Associates.

Lesh, R. (2003). Models and modeling in mathematics education. Monograph for *International Journal for Mathematical Thinking & Learning*. Mahwah, NJ: Lawrence Erlbaum Associates.

Maher, C. A. (1998). *Communication and constructivist teaching. Constructivism in mathematics education.* Lund, Sweden: Studenlitteratur.

National Council of Teachers of Mathematics. *Curriculum and Evaluation Standards for School Mathematics.* Reston, VA: NCTM, 1989.

National Council of Teachers of Mathematics. (2000). *Principles and standards for school mathematics*. Reston, VA: Author.

National Research Council, Mathematical Sciences Education Board. (1989) *Everybody Counts: A Report to the Nation on the Future of Mathematics Education.* Washington D.C.: National Academy Press.

Schorr, R. Y. (2000). Impact at the student level. *Journal of Mathematical Behavior 19*, 209–231.

Schorr, R. Y. (2003). Motion, speed, and other ideas that "should be put in books." *Journal of Mathematical Behavior 22*(4), 467–479.

Schorr, R. Y., (2004). Helping teachers develop new conceptualizations about the teaching and learning of mathematics. *AMTE Monograph I: The Work of Mathematics Teacher Educators.* (pp. 212–230).

Schorr, R. Y., Firestone, W., Monfils, L. A. (2003). State testing and mathematics teaching in New Jersey: the effects of a test without other supports. *Journal for Research in Mathematics Education 34*(5) 373–405.

Schorr, R. Y., Koellner-Clark, K. (2003). Using a modeling approach to consider the ways in which teachers consider new ways to teach mathematics. *Mathematical Thinking and Learning: An International Journal 5*(2) 191–210.

Schorr, R. Y., & Lesh, R. (2003). A modeling approach to providing teacher development. In R. Lesh & H. Doerr (Eds.), *Beyond constructivism: A model and modeling perspective on teaching, learning, and problem solving in mathematics education.* (pp. 141–157). Mahwah, NJ: Lawrence Erlbaum.

Silver, E. A., & Stein, M. K. (1996). The QUASAR project: The "Revolution of the possible" in mathematics instructional reform in urban middle schools. *Urban Education 30*(January): 476–521.

Simon, M. A., & Tzur, R. (1999). Exploring the teacher's perspective from the researchers' perspectives: generating accounts of mathematics teachers' practice. *Journal for Research in Mathematics Education 30*, 252–264.

Spillane, J. P., & Zeuli, J. S. (1999). Reform and teaching: Exploring patterns of practice in the context of national and state mathematics reforms. *Educational Evaluation and Policy Analysis, 21*(1) 1–127.

Stein, K. S., Smith, M. S., Henningsen, M. A., & Silver, E. A. (2000). *Implementing standards-base mathematics instruction: A casebook for professional development.* New York: Teachers College Press.

Stigler, J. W., & Hiebert, J. (1997). Understanding and improving classroom mathematics instruction. *Phi Delta Kappan, 79*(1), 14–21.

Stigler, J. W., & Hiebert, J. (1999). *The teaching gap: Best ideas from the world's teachers for improving education in the classroom.* New York: The Free Press.

CHAPTER 23

Directions for Future Research

Richard Lesh
Indiana University

Eric Hamilton
U.S. Air Force Academy

Jim Kaput
University of Massachusetts, Dartmouth

This book began with the three-part observation that, as we enter the 21st century: (a) dramatic changes have occurred in the kinds of situations where some kind of "mathematical thinking" is needed for success beyond school, (b) equally large changes have occurred in the kinds of mathematical understandings and abilities that are needed for success in the preceding situations, and (c) corresponding changes have occurred in the kinds of people who are most sought-after in job interviews—especially in future-oriented fields that are becoming increasingly heavy users of mathematics, science, and technology.

Chapters throughout this book have described details about the nature of many of the preceding changes. New fields of mathematical inquiry are emerging as important; and, new levels and types of abilities are replacing old notions of basic skills. Examples of the former include discrete mathematics and the mathematics of complex systems; and, examples of the latter often emphasize capabilities associated with description and explanation—even more than traditional abilities associated with computation and deduction.

A related observation is that, in virtually every field where ethnographic studies have investigated similarities and differences between experts and novices, results have shown that experts not only DO things differently but they also SEE things differently. That is, the models that they use to make sense of their experiences tend to be qualitatively different than those used by novices; and, for both experts and novices, competence appears to be far more situated, piecemeal, distributed, and socially influenced than traditional theories of learning have assumed.

One consequence of the preceding trends is that some of the most important goals of mathematics instruction probably should involve the development of powerful models and modeling capabilities for interpreting the kind of mathematically interesting systems that increasingly impact the lives of most people in a technology-based age of information. Therefore, a productive agenda research should (a) help to clarify the nature of the most important among these models and modeling capabilities, and (b) help to design effective learning environments where students can develop these relevant constructs and capabilities.

Throughout this book, the challenge was given for authors to go beyond describing conclusions from our past research, and instead to use results from our past research to describe promising directions for future research related to the questions we've raised about foundations for the future in mathematics education. Therefore, it is appropriate for this final chapter to focus not only on promising topics for research but also on promising research methodologies.

It is reasonable to expect that the kind of research and theory development that will be most appropriate for investigating the questions and issues emphasized in this book might be quite different than research on problem solving which assumes that the nature of mathematics and problem solving have not changed in the past millennium. For example, when the former changes are recognized, researchers probably need to be suspicious of any methodology which begins with the assumption that there is no need to question the researchers' own preconceived notions about the nature of mathematics, and about the ways that mathematics should be useful beyond school. Consider:

a. *Consensus Building*: One way that the mathematics education community has attempted to answer questions about the nature of foundation level mathematics has been through consensus-building activities aimed at the development of standards for curriculum and assessment. Typically, these efforts enlist "stakeholders" who mainly included practitioners representing schools and school teachers, or mathematics professors representing traditional university departments. Little input tends to be sought from people who represent hyphenated fields lying at the intersections of traditional disciplines—or future-oriented fields that are heavy users of mathematics in the 21st century. Consequently, results from these consensus-building efforts tend to focus on making incremental improvements in traditional curriculum materials—rather than on taking a fresh look at foundations for the future.

According to the views of authors in this book, when we investigate *what's needed for success beyond school,* it is important to avoid becoming so preoccupied with the goal of avoiding failure in school so that no appropriate attention is given to questions about what's needed to prepare for success beyond mathematics classrooms. Similarly, it is important to avoid becoming so preoccupied with low-level facts and skills that we neglect to emphasize more complex constructs and conceptual systems. In particular, it is important to avoid false assumptions about prerequisite relationships between basic facts or skills and more complex achievements. For example: *Do "basic skills" need to be mastered (completely) before students can be introduced meaningfully to more complex ideas and experiences?* Answers to such questions tend to be obvious outside of schools—in fields ranging from art, to athletics, to cooking, to carpentry. Both "basic skills" and "complex understandings" are important; and, development is parallel and interactive. For example, cooks and carpenters do not need to master the names and skills associated with every tool at Sears before they can begin to cook meals and craft furniture; and, athletic coaches who are strong proponents of "fundamentals" seldom enjoy long careers if they neglect scrimmages.

b. *Ethnographic Observations*: A second way that educators have investigated questions about what's needed for success beyond school is by observing people "thinking mathematically" in everyday situations. Sometimes, such studies compare "experts" with "novices" who are working in fields such as engineering, agriculture, medicine, or business management—where "mathematical thinking" often is critical for success. Several chapters in this book emphasize ethnographic approaches. Such investigations often have been exceedingly productive and illuminating. Nonetheless, even when extraordinary precautions are taken by exceedingly insightful researchers, such observational studies also tend to have some significant shortcomings. For example, we must be skeptical of observations which depend heavily on preconceived notions about *where* to observe (in grocery stores? carpentry shops? car dealerships? engineering firms? Internet cafés?), *whom* to observe (street vendors? shoppers? farmers? cooks? engineers? baseball fans?), *when* to observe (when they're estimating sizes? calculating with numbers? minimizing routes? describing, explaining, or predicting the behaviors of complex systems?), and *what* to count as "mathematical thinking" (e.g., planning, monitoring, assessing, explaining, justifying steps during multi-step projects, or deciding what information to collect about specific decision-making issues). Too often, in ethnographic studies, close examinations of underlying assumptions expose unwarranted prejudices about what it means to "think mathematically"—and about the nature of "real life" situations in which mathematics is useful. For example, in most fields where mathematics is useful, an important part of expertise consists of developing routines that reduce large classes of tasks to situations that no longer are problematic. Therefore, situations that once were problematic often become little more than exercises in following rules; and, situations that once required significant types of mathematical thinking may not require much thinking at all.

c. What are the most important characteristics that distinguish mathematics from other domains of knowledge? Our own prejudices are similar to those expressed by Lynn Steen in the book *On the Shoulders of Giants* (1990). That is: *Mathematics is the science and language of pattern.* In fields that range from mathematics to the physical sciences, to the life sciences, to the social sciences, the deepest and most high quality thinking that goes on tends to focus on developing, discerning, and investigating fundamental properties of patterns and regularities beneath the surface of things. In physics, these pattern and regularities involve physical properties of relevant systems. In biology, they involve biological properties; in economics, they involve economic properties; and, in mathematics, they involve structural properties of the relevant systems.[1]

Not everybody shares the preceding opinions, of course. For example, information processing theories treat all knowledge as if it were reducible to lists of condition-action rules; and, even more restricted views underlie the kind of psychometric theories that are used by most testing programs. Therefore, when the goals of research focus on questions about the nature of mathematics and mathematical problem solving, it is sensible to be skeptical of such theories—and accompanying research methodologies which do not require researchers to express → test → revise their own preconceived notions about such issues.

d. *Multi-Tier Design Experiments*: To investigate *what's needed for success beyond school*, research on *models* and *modeling* has tended to emphasize the use *multi-tier*

---

[1] Pure scientists develop and investigate constructs and conceptual systems for their own sakes. Whereas, applied scientists develop and investigate constructs and conceptual systems that are intended to be useful for creating or making sense of other systems.

A Three-Tiered Design Experiment

Tier #3: Researcher Level	Researchers express → test → revise their ways of thinking about students' and teachers' ways of thinking. The thought-revealing artifacts (models or conceptual tools) that they develop often are intended to elicit, document, or assess teachers' and/or students' ways of thinking.
Tier #2: Teacher Level	Teachers express → test → revise their ways of thinking about students' ways of thinking. The thought-revealing artifacts (models or conceptual tools) that they develop often are sharable and re-useable tools (such as observation forms or guidelines for assessing of students' responses) which describe, explain, or predict students' behaviors.
Tier #1: Student Level	Students express>test>revise their ways of thinking by working on thought-revealing activities which are simulations of "real life" situations where important new types of "mathematical thinking" are needed for success beyond school.

design experiments[2] (Lesh, 2002) in which diverse teams of experts work together in semester-long sequences of bi-weekly meetings in which they collaborated to design thought-revealing activities which (they believed) to be high fidelity simulations of "real life" situations where "mathematical thinking" is needed for success in their fields. Experts in such studies commonly include mixed teams of teachers, mathematicians, psychologists, and professors or professionals in fields that are heavy users of mathematics, professors in relevant schools and departments); and, the thought-revealing activities that these teams developed generally are field tested at diverse sites—where the designers also develop tools to document and assess the mathematical understandings and abilities that they actually observe to be most critical for success in such problem solving activities. In this way, the designers themselves repeatedly express, test, and revise their own current ways of thinking about the nature of "mathematical thinking" beyond school. So, their conceptions tend to change over time; and, as changes occur their activities automatically generate auditable trails of documentation that reveal significant information about the nature of changes that occur.

We often refer to such studies as *evolving expert studies*—because the results often represent significant extensions or revisions in the thinking of each of the participants who were involved. Such methodologies respect the opinions of diverse groups of stakeholders whose opinions should be considered. On the other hand, nobody (including especially the researchers) is considered to have privileged access to the truth. Everybody's prejudices need to be subject to examination and possible revision.

We refer to our investigations as *multi-tier* because a series of *thought-revealing activities* for *students* often provides the context for a series of *thought-revealing activities*

---

[2] Details about scientific principles governing multi-tier design experiments have been described in the Handbook of Research Design in Mathematics, Science, and Technology Education (Kelly & Lesh, 2000) and in our chapter in the International Handbook of Research in Mathematics Education (English, 2002)—as well as in a recent special issue of the Educational Researcher (Kelly, 2003). Such research designs are based on approaches that are well established in design sciences, such as engineering, where: (i) the goals of projects typically involve the development of complex artifacts, but (ii) the underlying design is one of the most important components of the product that is produced.

for *teachers* (or other evolving experts) which in turn provide the context for a series of *thought-revealing activities* for *researchers* (Lesh, 2002). Finally, we refer to our investigations as *design experiments* because, at each tier, participants express>test>revise their current ways of thinking during the process of designing complex artifacts (mathematical models, or conceptual tools) for meeting specific purposes.

For the preceding kind of *three-tiered design experiments*, each tier can be thought of as *a longitudinal development study in a conceptually enriched environment* (Lesh, 2002). That is, a goal is to go beyond studies of *typical* development in *natural* environments to also focus on *induced* development within *carefully controlled* environments. For example, at the student-level, *three-tier design experiments* have proven to be especially useful for studying the nature of students' developing knowledge about fractions, quotients, ratios, rates, and proportions (or other ideas in algebra, geometry, or calculus) which seldom evolve beyond primitive levels in "natural" environments that are not artificially enriched.

REFERENCES

Lesh, R. (2002). Research design in mathematics education: Focusing on design experiments. In L. English (Ed.), *International handbook of research design in mathematics education.* Mahwah, NJ: Lawrence Erlbaum Associates.

Steen, L. (1990). Pattern. In L. A. Steen (Ed.), *On the shoulders of giants.* Washington, DC: National Academy Press.

Author Index

A
Abdul Rahim, 306
Abrahamson, L., 181, *191*
Acioly, N. M., 79, *97*
Adamic, 229, *239*
Adams, R., 416, *428*
Adams, R. S., 416, 419, 428, *429*
Adams, V. M., 283, *296*
Adelson, B., 355, *362*
Akin, O., 99, 115, *123*
Albert, 237, *239*
Aleksandrov, A., 135, 139, *153*
Alvarez, 227, *243*
Anderson, 227, 229, *230*
Anderson, 54, *55*
Anderton, F., 100, 101, *123*
Andrade, 385, *391*
Andrade, R., *241*
Anthony, K. H., 100, 101, 116, *123*
Archer, B., 415, *428*
Ares, N., 239, 243, 244, 367, 369, 370, 372, 374, 378, 379, 383, 384, 385, *390*
Atman, C. J., 415, 416, 417, 419, 428, *429*
Au, 385, *390*
Au, K., *239*
Axtell, 229, *239*

B
Bak, 229, *239*
Ball, D., 431, 434, *446*
Ball, D. L., 274, 279, 431, 434, *445*
Barab, 371, *390*
Barab, S., *239*
Barabasi, 237, *239*
Barron, B., *239*
Bass, H., 431, 434, *445*
Bates, E., 148, *153*
Beer, R., 398, *408*
Beers, J., *97*
Behr, M., 274, 277, 279, *280*
Bellugi, U., *153*
Bereiter, C., 419, *428*
Bessot, A., 9, *34*

Beth, E., *347*
Biehl, L. C., 219, *222*
Bishop, L., 8, *34*
Bliss, J., *170*
Blythe, T., 99, *124*
Bodilly, S. J., *311*
Boohan, *170*
Boroditsky, L., 133, *153*
Borovoy, R., 239, 379, *390*
Bouwen, J. E., 100, *124*
Bowers, 227, *239*
Bowker, G. C., 16, *34*
Bowman, K., viii, x, 315, *347*
Branki, N., 99, *123*
Bransford, 374, *390*
Bransford, J., 239, 359, *362*
Bransford, J. D., 248, *261*
Briggs, 379, *392*
Briggs, W. L., *241*
Brousseau, G., 278, *279*
Brown, A., 238, 359, *362*
Brown, A. L., 102, 113, 121, *123*
Brown, S., 269, *279*
Brownell, W. A., 273, *279*
Bryson, M., 419, *428*
Bursic, K. M., 417, *428*

C
Callahan, 369, *390*
Callahan, P., *239*
Camazine, 228, *240*
Campbell, J. R., *239*
Campbell, P., 435, *445*
Campbell, 371, *390*
Campione, J. C., 240, *123*
Cancho, 229, *240*
Carambo, C., 92, *97*
Cardella, M. E., *428*
Carmona, 339
Carmona, G., 240, 369, *394*
Carmona-Dominguez, G., *429*
Carnine, 226, 227, 243, 374, *391*
Carnine, D., *241*

Carraher, D., 242
Case, R., 246, *261*
Casti, 226, *228*
Chafee, R., 101, *123*
Chamberlin, S. A., 423, *429*
Chapman, 370, *390*
Chapman, M., *239*
Chazan, D., 271, *279*
Chi, 227, *240*
Chiel, H., *408*
Chrostowski, *312*
Cienki, A., 148, *153*
Civil, 385, *391*
Civil, M., 340
Clark, 228, *240*
Clarke, D., 356, 362
Clayton, M., 9, *34*
Cobb, 227, 278, *240*
Cobb, P., 101, 123, 309, *311*
Cohen, D. K., 274, *279*
Cole, 229, *240*
Colella, 379, 390
Colella, V., *240*
Collins, R., 89, 91, *96*
Confrey, J., 246, *261*
Conley, D. T., *311*
Cooper, R., *311*
Corbett, H. D., *311*
Cossentino, J., 99, 122, *123*
Courant, R., 136, *153*
Coy, M., 80, *96*
Coyne, R., 99, *123*
Craig, D. L., 100, *123*
Cramer, K., 293, *296*
Crismond, D., 99, *124*
Crowe, N. A., 100, *123*
Csikszentmihalyi, M., 355, *362*
Curphy, J., *362*
Curtis, C. L., 122, *124*

D

Dai, D. Y., 284, *295*
Damasio, A. R., 79, 89, 91, 92, 93, 95, *96*
Darling Hammond, L., 301, *311*
Davies, R., 99, 119, *123*
Davis, 225, *240*
Davis, R. B., 273, *279*
Davis, S., *241*
Davis, S. M., 375, *391*
Dawson, J. A., *311*
DeBellis, V. A., 215, 222, *223*
De Corte, E., *283, 296*
del Rio, 227, *244*
Derry, 227, *240*
de Sessa, A., *311*
Desimone, L., *311*
Dewey, J., 102, 123, 240, 304, 311, 372, 410, *429*

Dick, F., 148, *153*
Diefes-Dux, H., viii, x, 193, 200, 347, 429,
  Diehl, 379, *391*
Diehl, E., *240*
Diodati, 229, *240*
diSessa, 227
diSessa, A. A., 55, 102, 123, 246, 247, 248, 249, 250, 251, 252, 253, 256, *261*
Dochy, 302, *312*
Doerr, 37, *56*
Doerr, H., 193, 200, 203, 206, 208, 279, 280, 315, 319, 325, 334, 339, 346, 347, *390*
Doerr, H. M., 57, 76, 283, 284, 292, 293, 296, 396, *408*
Dolmans Diana, H. J. M., 303, *312*
Douglas, L., 9, *35*
Drushel, R., 398, *408*
Dubinsky, E., 269, *279*
Dutton, T. A., 100, *123*

E

Ebbesen, 229, *244*
Edwards, D., 201, 203, *208*
Edwards, L., 135, 142, *154*
Eisenhart, M., 74, 75
Elmesky, R., 92, *97*
Empson, 228
Empson, S. B., 291, *295*
Engeström, Y., 8, 15, 33, 34, 35, 89, *96*
English, L., 410, *429*
Ennis, 225, *240*
Eraut, M., 14, *34*
Erickson, J., 99, 102, *123*
Erlwanger, S., 277, *279*
Ernest, 227, 240
Ernest, P., 269, *279*
Essegbey, 148
Esterly, 256
Evans, J., 283, *295*

F

Fashola, O. S., 301, *313*
Fauconnier, G., 131, *153*
Feltovich, 227, *240*
Fey, J., 179, *191*
Fielding, M., *312*
Firestone, W., 431, 432, *446*
Firestone, W. A., 381, *445*
Fischbein, E., 272, *279*
Flemming, D., 100, *123*
Follman, D., 193, *200*
Forman, E., 358, *362*
Fox-Keller, E., 84, *96*
Franzblau, D., 216, *223*
Frederickson, M. P., 100, 101, *123*
Freire, 369, *391*
Freire, A. M. A., *240*
Freire, P., 240, 369

AUTHOR INDEX

Freudenthal, H., 135, 140, 144, *153*
Fullan, M. G., *312*

G

Gabaix, 238, 240
Gagne, R., 247, *261*
Gainsburg, J., 38, 39, 55, 410, 415, *429*
Galbraith, P., 353, *362*
Garden, 302, *312*
Gardner, 371, *391*
Gardner, H., *240*
Gee, J., 302, *312*
Geertz, C., 101, 104, *123*
Gentner, D., 133, *153*
Gibbs, R., 147, *153*
Gijbels, D., 302, *312*
Gillespie, 256
Ginnett, R., 360, *362*
Giordano, F. R., 202, *208*
Glaser, 227, 240
Gleick, 381, 391
Gleick, J., *241*
Glew, R. H., 303, *312*
Goel, V., 426, *429*
Goetting, M. M., 271, *279*
Goh, C. T., 304, *312*
Goldin, G. A., 282, 283, 285, 287, 288, 289, 290, 292, 295, 436, *445*
Goldin-Meadow, S., 147, 148, *153*
Goldschmidt, G., 99, *124*
Goldstein, B., 57, 69, 73, *75*
Gomez-Chacon, I. M., 283, *295*
Gonzales, P., 76, *241*
Gonzalez, 385, *391*
Gonzalez, G., 302, *312*
Gonzalez, N., *240*
Goodman, 388, *391*
Goodman, N., *240*
Goodwin, C., 70, 75, 85, *96*
Goodwin, M. H., 85, *96*
Goos, M., 353, *362*
Gopinathan, 302, 304, *313*
Gopinathan, S., 306, *312*
Gray, J., 99, 122, *124*
Greenberg, 385, *391*
Greenberg, J. B., *240*
Greene, B. A., 355, *362*
Greenfield, T., 21, *34*
Greeno, J. G., 57, 75, 99, 122, *124*
Griesemer, J., 16, 35, 68, *76*
Guerra, M. R., 102, 121, *124*
Guile, D., 2, 6, 15, 17, *34*
Gunther, *312*

H

Hadamard, J., 355, *362*
Hall, R., 9, 34, 74, 75, 75, 99, 124, 410, *429*
Hall, R. P., 57, *75*

Hamson, M., 201, 203, *208*
Hancock, C., *239*
Hannula, M. S., 283, *295*
Harel, 339
Harel, G., 38, 54, 55, 264, 267, 268, 269, 270, 274, 275, 279, 279, *280*
Hartman, H., 90, *96*
Harvey, M., 360, *362*
Hawkins, R., 99, *124*
Hegedus, S., 181, 188, 190, 191, 240, 243, 378, *390*
Heger, M., 284, 291, *296*
Heibert, 439
Henderson, D., *153*
Henningsen, M. A., 435, *446*
Hernstein, 371
Hernstein, R., *240*
Herrenkohl, L. R., 102, 121, *124*
Hersh, R., 143, *153*
Hewitt, M., 119, *124*
Heylighen, A., 100, *124*
Hickok, G., *153*
Hiebert, 410, *447*
Hiebert, 227, *241*
Hiebert, J., 5, 6, 273, 280, *446*
Hill, H., 431, 434, *446*
Hills, 230, 237, *241*
Hirsch, C., 216
Hjalmarson, M. A., 422, *429*
Hmelo, C. E., 99, 122, *124*
Hmelo, 225, *241*
Hofer, B., 249, *261*
Holbrook, J., 99, 122, *124*
Hole, 396, *429*
Hole, B., 193, 200, 240, 369, 372, 376, 390, 422, *429*
Holland, 381, *391*
Holland, J., 228, *240*
Holton, D. L., 99, 122, *124*
Holzkamp, 90, 91, *96*
Hombo, C. M., *239*
Hoover, 396, *429*
Hoover, M., 193, 200, 240, 369, 372, 376, 390, 422, *429*
Hoyles, C., 2, 6, 9, 11, 12, 13, 17, 34, 35, 79, 96, 193, *200*
Huberman, 229, *241*
Hughes, 237, *243*
Hughes, R., 360, *362*
Hurford, 225, *241*
Hurford, A., 243, 244, 367, 369, 370, 372, 374, 378, 379, 383, *390*
Hurtt, S. W., 100, *123*
Hutchins, E., 62, *76*

I

Imbrie, P. K., 193, *200*
Ineke, H. A. P., 303, *312*

Ioannides, 229, 238, *240, 241*
Iverson, J., 148, *153*

J
Jackson, A. A. D. G., 301, *312*
Jacobsen, 225, *241*
Jacobson, C., 99, 122, *124*
Jacoby, S., 76, *241*
James, 371, 372, 373, *391*
James, W., 238, *240*
Jansson, D., 99, *124*
Jaques, J., 148, *153*
Johnson, 302, *312*
Johnson, D. A. K., 302, *312*
Johnson, H., *347*
Johnson, M., 131, 132, 133, *153*
Johnston, 229, *241*
Johnstone, T., 85, *96*
Joiner, B., 19, *34*
Jonassen, D. H., 302, *312*
Joram, E., 419, *428*

K
Kameenui, 374, *391*
Kameenui, E., *240*
Kaput, J., 127, 143, 181, 183, 184, 188, 190, 191, 240, 242, 378, *390*
Kaput, J. J., 179, 191, 295, 369, 374, *390*
Kauffman, 381, *391*
Kauffman, S. A., *240*
Kaur, B., 302, *312*
Kedem, I., 272, *279*
Kelly, 190, *200*, 307, *312*, 381, *391*, 396, *429*
Kelly, A., 193, 200, 240, 307, 312, 346, 369, 372, 376, 390, 422, *429*, *446*
Kelly, D. L., 302, *312*
Kelly, K., *240*
Keltner, B., *311*
Kendon, A., 147, 148, *153*
Kenney, M., *216*
Kent, P., 2, 6, 9, 11, 12, 13, 17, *34*
Kerns, S., 3, *6*
Kieran, C., 358, *362*
Kirshner, 371, *390*
Kirshner, D., *239*
Kita, 148
Klima, E., *153*
Koellner-Clark, K., 431, 433, *446*
Kolmogorov, A. N., 135, 139, *153*
Kolodner, J. L., 99, 122, *124*
Konold, C., 18, 19, *35*
Kornhaber, M. L., *240*
Kramer, E., 141, *153*
Kvan, T., 100, *124*

L
Ladson-Billings, 385, *391*
Ladson-Billings, G., *240*

Lakoff, G., 128, 131, 132, 133, 139, 140, 142, 143, 144, *153*, *154*
Larsson, A., *240*
Larsson, 384, *391*
Latour, B., 69, *76*
Lave, 54, *55*, 372, *391*
Lave, J., 79, *96*, 102, *124*, 177, *191*, 227, *240*
Lavrent'ev, M. A., 135, 139, *153*
Leder, G. C., *295*
Ledewitz, S., 100, *124*
Lee, V., 301, *312*
Lehrer, A., *311*
Lehrer, R., 57, *76*, 99, 102, 122, *123, 124*
Leo, E., 306, *312*
Leont'ev, A. N., 89, *96*
Lesh, 158, 190, 204, 205, 307, 323
Lesh, R., 37, 38, 54, *55*, 57, *76*, 193, *200*, 203, 206, *208*, 240, 274, 279, *279, 280*, 283, 284, 291, 292, 293, *296*, 315, 319, 325, 334, 339, 343, 345, 346, *346, 347*, 369, 372, 376, *390*, 396, *408*, 410, 423, *429*, 431, 432, 433, 435, *446*, 449, 450, *451*
Lesh, R. A., 422, *429*
Levin, B., *312*
Lewis, C., 360, *362*
Lim, 306
Lim, T. M., *312*
Linse, A., *429*
Liu, 306
Lobato, J., 357, *362*
Loeb, A., 99, 100, *124*
Logan, 229, 242

M
Ma, L., 274, *280*
Mabogunje, A., *240*
Macedo, P., *240*
Mack, A., *241*
Maclean, R., 304, *312*
Maher, 438, *446*
Maki, D. P., 201, 202, *209*
Malamud, et al., 229, *242*
Malmivuori, M.-L., 283, *296*
Mandelbrot, 229, *242*
Mardiana, A. B., 306, *312*
Martin, F., *239*
Martin, M. O., 302, *312*
Martin, W., 359, *363*
Matos, J. F., 135, 142, *154*
Maurer, S., 216, *223*
Maurice-Naville, 370, *391*
Maurice-Naville, D., *240*
Mayberry, R., 148, *153*
Mayer, 227, *242*
Mazzeo, J., *240*
McClain, 227, 239
McCrone, S. S., 358, *362*
McCullough, M., 102, 115, *124*

McDonald, M., *239*
McKnight, Y., 92, *97*
McLeod, D. B., 283, 288, *296*
McNeill, D., 147, 148, *153*
Mellar, H. *170*
Michaels, S., 102, 121, *124*
Middleton, 350, 353, 356
Middleton, J. A., 284, 291, *296*
Mien, M., 302, *312*
Miles, M. B., *312*
Miller, R. B., 355, *362*
Milne, A., *242*
Mitchell, W. J., 99, 102, 115, *124*
Moll, 372, 385, *391*
Moll, L., *240*
Moll, L. C., *240*
Molyneux-Hodgson, S., 11, 12, 13, 17, *34*
Monfils, L. A., 431, 432, 445, 446
Montangero, 370, *391*
Montangero, J., *240*
Moore, A., *239*
Moore, T., 193, *200*, 422, *429*
Mortimore, J., 306, *312*
Mortimore, P., 306, *312*
Moss, 385, *392*
Moss, B. J., *240*
Motz, B., 133, *154*
Mullis, I. V. S., 302, *312*
Mullis, M., 302, *312*
Murphy, G., 147, *154*
Murray, 371, *391*
Murray, C., *240*
Myers, D. G., 358, *362*
Myers, K., 306, *312*
Mylander, C., *153*

N

Narayanan, S., 131, *154*
Nardi, B., 8, *35*
Nava, 229, *241*
Nemirovsky, R., 253, *261*
Neuckermans, H., 100, *124*
Neves, 229, *242*
Newmann, F., 301, *312*
Nichols, E., 277, *279*
Nicolopoulou, 369
Nicolopoulou, A., *241*
Nonaka, I., 360, *362*
Noss, R., 2, *6*, 9, 17, *34, 35*, 79, *96*, 193, *200*
Novicevic, M. M., 360, *362*
Nuñes, 372
Nuñes, T., *243*
Nuñez, R., 128, 131, 133, 134, 135, 139, 140, 142, 143, 144, 145, 148, 153, *154*

O

Ochs, E., *76, 243*
Ochs, 384, *392*

O'Connor, M. C., 102, 121, *124*
O'Connor, 302, *312*
Ogborn, J. *170*
Op't Eynde, P., 283, *296*
Osterkamp, U., 90, *96*
Overman, 229

P

Pacey, 378, *392*
Pacey, A., *243*
Papert, 369, *392*
Papert, S., 168, *170, 241*
Pea, R., 181, 184, *191*
Pehkonen, E., *295*
Peirce, 372, *392*
Peirce, C. S., *243*, 378, *392*
Penuel, B., 181, 184, *191*
Pereira-Mendoza, 302
Perkins, D., 99, *124*
Perlwitz, 227, *240*
Petrosino, A., *239*
Pfannkuch, M., 18, 24, *35*
Philippou, G., 283, *295*
Piaget, 227, 228, *243*
Piaget, J., *347*
Picker, S. H., 219, *223*
Pintrich, P., 249, *261*
Pirolli, P., 426, *429*
Pittam, J., 88, *96*
Polanyi, M., 358, *362*
Pollatsek, A., 18, 19, *35*
Polya, G., 343, *347*
Post, 396, *429*
Post, T., 193, *200, 240*, 274, *280*, 293, *296*, 369, 372, 376, *390*, 422, *429*
Pozzi, S., 9, *34, 35*, 79, *96*, 193, *200*
Prigogine, 228, *243*
Puntambekar, S., 99, 122, *124*
Purnell, *311*

R

Raizen, S. M., 7, *35*
Ramsdan, P., 302, *312*
Raup, 229, *243*
Reder, 227, *239*
Reed, 237, *243*
Reich, R. B., 7, *35*
Reichardt, R., *311*
Reiner, 227, *243*
Resnick, 227, 372, 379, *243, 392*
Resnick, L. B., 225, *241*
Resnick, M., 181, 182, *191, 239*
Richardson, 436, *445*
Ricoeur, P., 92, 93, *96*
Ridgway, J., 9, *34*
Riley, K., *312*
Robbins, H., 136, *153*
Robert, A., 135, *154*
Roberts, F., 216, *223*

Roberts, F. S., 202, *209*
Rogoff, 372, *392*
Rogoff, B., *243*
Roschelle, J., 181, 184, *191, 243*, 246, *261, 390*
Rosenstein, J. G., 215, 216, 219, *222, 223*
Roth, W.-M., 79, 80, 82, 83, 89, 92, *96, 97*, 283, *296*
Rowan, B., 431, 434, *446*
Rowe, P. G., 99, 119, *124*
Rubin, 229, *243*

S

Sancar, F. H., 100, *124*
Saxe, 372, *392*
Saxe, G. B., 79, *97, 243*
Scardamalia, M., 419, *428*
Schademan, A., *239*, 369, *390*
Schauble, L., 57, *76*, 311
Scherer, K. R., 85, 88, *96*
Schliemann, A., *243*
Schliemann, A. D., 79, *97*
Schoenfeld, 372, *392*
Schoenfeld, A., 267, 268, *280*
Schoenfeld, A. H., *243*, 343, *347*, 414, 415, *429*
Schon, D. A., 99, 100, 101, 102, 115, 117, 118, *124*, 418, *429*
Schooler, 229, *239*
Schorr, R. Y., 431, 432, 433, 435, 436, *445, 446*
Schuyler, G., *311*
Schwartz, 375, 383, *390*
Schwartz, D., *239*
Schwartz, D. L., 248, *261*
Schwartz, J. L., 163, *243*
Scribner, S., 79, *97*
Selting, M., 85, *97*
Senge, 369, 371, 379, 381, *392*
Senge, P., *312*
Senge, P. M., *243*
Senker, J., 360, *363*
Sfard, A., 17, *35*
Shaffer, D. W., 99, 122, *123, 124, 125*, 179, *191*
Sharpe, L., 306, *312*
Sharpe, T. 302, 304, *313*
Sherin, B., 251, *261*
Shtelen, *no initial*, 436, *445*
Silbert, 226, 227, *243*
Silbert, 374, *392*
Silbert, J., 225, *243*
Silver, E. A., 345, *347*, 435, *446*
Silverman, B., *239*
Simmons, G. F., 139, 140, *154*
Simms, A., *313*
Simmt, 225, *240*
Simon, 227, *244*
Simon, H. A., 99, 101, 118, *125*
Simon, M., 274, *280*
Simon, M. A., 431, *446*

Sinclair, N., 283, *296*
Skinner, 226, 243
Slavin, 371, 392
Slavin, R. E., *243*, 301, *313*
Slotta, 227, *243*
Smith, 225, *244*
Smith, J., 9, *35*
Smith, J. P., 246, *261*
Smith, M. S., 435, *446*
Smith, 302, *312*
Sole, 229, *240*
Soloway, 379
Sowder, L., 264, 267, 268, 270, 275, *280*
Spillane, J. P., 431, 432, *446*
Star, S. L., 8, 16, *34, 35*, 68, *76*
Steen, L., 449, *451*
Stein, 226, 227, *243*
Stein, K. S., 435, *446*
Stein, M., 225, *241*
Stein, M. K., 435, *446*
Sternberg, R., 90, *96*, 358, *362*
Sternberg, R. J., 284, *295*
Stevens, R., 9, *34*, 74, *76*, 99, *124*
Stevens, R. R., 99, *125*
Stewart, I., 141, *154*
Stigler, 410, *447*
Stigler, J. W., *446*
Stoll, L., 306, *312*
Stor, 379, *392*
Stor, M., *243*
Strauss, A., 8, *35*
Stroup, W., 182, *191*, 225, 228, *243, 244*, 367, 369, 370, 371, 372, 374, 375, 376, 378, 379, 381, 383, 387, *390*
Stroup, W. M., 230, *239, 243, 244*, 247, *261*
Sweetser, E., 131, 133, 134, 148, *154*

T

Takeuchi, H., 360, *362*
Tall, D., 135, *154*
Talmy, L., 131, 145, *154*
Tan, H., 306, *313*
Tan, J., 306, *313*
Tan, S., 306, *313*
Tan, 305, 310, *313*
Teuscher, 133
Tharman, *313*
Thelen, 225, *244*
Thelen, E., 148, *153*
Thompson, M., 201, 202, *209*
Tobin, K., 92, *97*
Tompsett, C. *170*
Törland, P., *240*
Törner, G., *295*
Torraco, R., 38, *56*
Torralba, A., 74, *76*
Torralba, T., 9, *34*
Turner, J. H., 89, 90, *97*

Turner, M., 131, *153*
Turns, J., 416, 419, *428, 429*
Tzur, R., 431, *446*

U

Uluoglu, B., 100, *125*
Underwood, 227, *240*

V

Vahey, P., *390*
Van den Bossche, P., 302, *312*
Van der Vleuten, 303, *312*
Vemuri, S., *239*
Verschaffel, L., 283, *296*
Vidakovica, D., 359, *363*
Vinner, S., 135, *154*
von Glasersfeld, E., 370
Vye, N., *239*
Vygotsky, 380, 383, 385, 386, 388, *393*
Vygotsky, L., 227, *244*, 302, *313*
Vygotsky, L. S., 101, 102, 117, *125, 244*, 347

W

Wagner, J. F., 54, *55*, 248, *261*
Wagner, R., 358, *363*
Wake, G., 9, *35*
Wake, G. D., 9, *35*
Wake, W. K., *240*
Waldrop, 381, *393*
Waldrop, M., *244*
Wallace, M., *313*
Wankat, P. C., 405, *408*
Wearne, 439, *446*
Weir, M. D., 202, *208*
Weiss, R. S., 104, *125*
Wenger, 372, *391*
Wenger, E., 102, *124*, 177, *191*, 227, *240*
Wenzel, 229, *243*
Wertsch, 368, 383, 384, *393*
Wertsch, J. V., 102, *125*, 227, *244*

White, D. Y., 358, *363*
Wijnen, H.F. 303, *312*
Wilber, D. M., *313*
Wild, C. J., 18, 24, *35*
Wilensky, U., 168, *171*, 181, 182, *191*, 225, 228, 230, *243, 244*, 376, 378, 379, 381, 387, *390*
Williams, J. S., 9, *35*
Wilson, B., 163, *171*
Wiske, M. S., 301, *313*
Wittrock, M. C., *244*, 368, *390*
Wixted, 229, *244*
Wolf, A., 11, 12, 13, 17, *34*
Wolfhagen Cees, P. M., 303, *312*
Wolters, C., 355, *363*
Wynard, H. F. W., 303, *312*
Wynne, B., 74, *76*

Y

Yaeger-Dror, M., 85, *96*
Yakubinskii, 368, 383, 384
Yakubinskii, L. P., *244*
Yerushalmy, 383, *392*
Yerushalmy, M., 163, *171*, 243
Yoon, 339
Yoon, C., *55*
Young, M., 15, 17, *34*

Z

Zan, R., 283, *295*
Zawojewski, 204, 205, *209*
Zawojewski, J., 315, *347*
Zawojewski, J. S., viii, x, 193, *200*, 293, *296*, 343, 345, *347*
Zech, L., *239*
Zeuli, J. S., 431, 432, *446*
Zevenbergen, R., 9, *35*
Zimring, C., 100, *123*
Zipf, 229, *244*
Zuboff, S., 7, *35*

Subject Index

A

Abductive reasoning, 338
A BET (Accreditation Board for Engineering and Technology), 397
Abilities, modeling, 202–203
    cultivating, 203–204
Abstraction, situated, 15, 17
Abstractness, levels of, 256
Actions
    emotion and, 89–90
    identity and, 92
Activity
    emotions and, 89–90
    sensible fabrics of, 247, 255
    understanding and, 100–101
Activity theory, 79
Adaptive learning, 369
Additive model, 251
Advanced Placement
    in calculus, 212
    discrete mathematics and, 222
Aesthetics, 377–378
Affect, 281–294
    affective competencies, 289
    affective pathways, 288–289
    affective structures, 289–292
    domains of, 288
    global, 289
    local, 288–289
    mathematical, 436
    meta-affect and, 287–288
    modeling and, 293–294
    as representational, 285–286
    as system of communication, 286–287
    theoretical trends and, 282–285
Affective competencies, 289
Affective pathways, 288–289
Affective structures, 289
    essential, 289–292
After-School Learning Centers, in Newark, 434, 436
Agency, interactivity and, 178–179
Agentsheets, 163
Aggregating task, 323–335

Algebra, 250
    calculus and, 213–215, 246
    rush to, 214
Ambiguity
    in design, 409, 415–416, 420
    in problem-solving, 359
American Society for Engineering Education, 3
Animations, 163
Apperceptual mass, 368, 383, 384
Architectural design. *See also* Oxford Studio
    architectural ideas, 119–120
    classroom mathematics and, 122–123
    design iterations and, 4, 101–102
    design studios, 101–103
Arena, 414, 415
Artifact-creating play, 370
Artifacts, 323
    development of, 308
    interactive networks and, 375
Ascertaining, 267
Aspectual schemas, 131
AspenPlus, 404–407
Attitudes, 288
    student, 281–282
Axiomatic proof scheme, 274
Axioms/axiomatic methods, 129, 137
    Least Upper Bound, 138

B

Baby O, 149
Balance, systemics and, 248, 249
Balanced Assessment in Mathematics project, 168–169
Bank deposit model, 310
Bar charts, 12
Basic skills, 448
    holistic conceptual systems and, 341–342
Beer Game, The, 379
Behaviorist model of learning, 226, 227
Beliefs, 288
    about mathematics, 266, 268, 278
Between-concept systems, 337–339
"Beyond Cognitivism," 284–285

Beyond Constructivism, 283
Biologists, divisions between, 60–74
Black box issue, 406–407
Boundary objects, 15, 16
    individual and collective knowledge and, 16–17
    meaning of variation at, 17–19
    SPC chart as, 19–20, 32
B spread, 150
Butterfly effect, 381

C

Cabri Geometre, 163
Calculator Based Ranger (CBR), 185
Calculators, 340
Calculus
    accessibility in elementary school, 246–247
    Advanced Placement, 212
    motion metaphors and, 127
    rush to, 211–215
Calculus, 139
Capstone design project, 411–415, 426–427
Carbi, 178
Cartesian coordinate system, 320
CAS. *See* Computer Algebras System
Case-focused analysis, 104
Case studies, 10, 299, 435
CBR. *See* Calculator Based Ranger
Center for Learning & Technology, viii
Centre for Research in Pedagogy and Practice, 306
Change, time-scales of, 180–181
Characters, of mental acts, 265
ChemCad, 405
Chemical Engineering courses, use of simulators in, 403–407
Classroom atmosphere, learning and, 435
Classroom connectivity/networks, 181–183
    changes in mathematical problem solving and, 183–190
    generative design and, 367, 368, 370–371, 376, 378
    need for research on, 182
    pedagogical implications of, 188–190
    questions for investigating, 182–183
    role-playing and, 376
Clay Mathematics Institute, 219
Co-construction, 385
Codes, 216, 217
Cognition, 134
    affect and, 285–286
    embodied, 144–151
    emotions and, 79–80, 89–90
    gesture as, 147–151
    situated, 54, 157, 302
Cognitive ecologies, viable and generative, 249

Cognitive linguistics, 131, 132
Cognitive schemas, development of, 239
Cognitive science of mathematics, 130
    case study, 134–144
    embodied cognition, 144–151
    mathematics curriculum and, 130–134
Cognitive simplicities, 246, 247, 252, 253, 254
Cognitive strategies, switching, 237
Cognitivism, limitations of, 284–285
Coho salmon case study, 77–83
Collaborative learning, 302
Collective knowledge, boundary objects and, 16–17
College
    calculus courses and admission to, 212–213
    effect of high school calculus in, 212–214
Combinatorics, 216, 217
Commercial simulators, 403–407
Common cause variation, 19, 21, 22, 25, 28, 31
Communication
    affect as system of, 286–287
    power law distribution in, 229
    technology and, 175–177
    in techno-mathematical knowledge, 15, 16, 33
Communication technology affordance, 181–183
Communities
    learning, 371–372
    of practice, 177–183
Community norms of practice, 102
Company manager, role in statistical process control, 22–26
Competence, 289
    problem solving and, 344
Complementarity, systemics and, 248, 249
Complexity theory, 155, 225
    participatory simulations and, 379
Complex systems, 235–236
    learning as, 225, 226–228
    mathematics of, 447
Composite beam case study, 39–45
Computational kinematics, 250
Computational literacies, 250
Computational technologies, 176
Computer Algebras System (CAS), 178, 180
Computer languages, 404
Computer programming as modeling, 251, 395–408
    using commercial simulators, 403–407
    writing programs, 397–403
Concept-then-word problem math curriculum, 4
Conceptual blends, 131
Conceptual development, 334

SUBJECT INDEX

Conceptualizing, 335
Conceptual mappings, 132–134
Conceptual metaphors, 131, 132-133, 144, 146–147, 152
Conceptual metonymy, 131, 144
Conceptual model, 106
Conceptual shifts, modeling cycles and, 206–207, 208
Conceptual tools, 321
Conjecture, 267
Consensual intuitive models, 257
Consensus building
    reflection tools and, 352
    regarding curriculum and assessment standards, 448
Consequential Task, 102
Conservation biologists, 1
    modeling among, 57–76
Constructivism, 265, 370
    limitations of, 283–284
Constructivist learning, 227, 302
Contact, 128–129
Context-specific mathematical modeling, 2–3, 5
Continuity of functions, 138–144
    ε-δ definition of, 139–140, 142–144, 152
    natural continuity, 139, 140–144
Continuity of trigonometric functions, 136
Continuous functions, 135, 136
Control limits, 25, 28, 31
Cooperation, classroom-based learning and, 370
Cooperative learning, 302
Coupled iterations, 419, 421
Creativity, design and, 409
Critical path analysis, 219
Critical thinking, encouraging in schools, 304
Cross-factor analysis, 9
Crowdedness, measures of, 169
Cultural diversity, generative design and, 383–384. *See also* Diversity
Cumulativity, 246–248
Curriculum, Oxford Design, 105–106, 116–118. *See also* Mathematics curriculum

D

Databases, 2
    fish hatchery, 82
    storage and manipulation of data in, 175
Dead metaphors, 146–147
Decision-making
    emotions and, 84
    model building and, 194, 195, 196–197
    motivation and, 91
    power law distribution in, 230–238
    scale invariance in classroom-situated, 226
    simulations and, 323–325
Definitions, mathematical, 129, 137, 142–143
Description, theory development and, 245
Descriptive modeling, 82
Descriptive reflections, 359
Descriptive self-models, 293
Descriptive strategies, 343–345
Design
    architectural. See Architectural design; Oxford Studio
    constraints on, 409
    generative. See Generative design
    iteration in, 4, 101–102, 369, 378, 409–411, 415–421, 426–428
    uncertainty in, 411–415, 421–428
Design activities, 377–378
Designed experiences, 297
Design studios, 100
    case study. See Oxford Studio
    as learning environment, 102–103
    pedagogical activities in, 102–103
    research on, 101–102
Design tasks, 299
    role-playing and, 179
Desk crits (crits), 101, 109–110, 117-118, 120
Developing-first-then-formalize methodologies, 326–327, 330–332
Devolution, 15, 16
Dewey, John, 315, 319, 320, 372
Diagrams, 106, 393
Dialogic speech, 383–384
Didactical contracts, 278
Direct instruction, 374
Disc-ness, measures of, 169
Discontinuity of trigonometric functions, 136
Discrete mathematics, 155, 214, 215–217, 447
    defining, 215–216
    in K-8 curriculum, 219–221
    in K-12 curriculum, 217–221
Discrete Mathematics Across the Curriculum K-12, 216
Discrete Mathematics in the Schools, 216, 219
Discretization task, 47-52
Disembedded background work, 8
Display technologies, 176
Diversity. *See also* Generative design
    generative design in support of, 383–384, 388–389
    learning and, 372–373
    of resources, 385
DNR-based instruction, 263–279
    determinants, 265–268
    duality principle of, 264, 271–273, 276
    example, 276–278
    goal of, 268

instructional principle of, 270–271
necessity principle of, 264, 273–275
premises of, 264–265
problem-solving approaches, 266–267
proof schemes, 267–268
repeated-reasoning principle of, 264, 275–276
ways of understanding and ways of thinking and, 268–270
DNR determinants, 264–265
Doing, 335
Domain general environments, 163
Domain particular knowledge, 249
Domain specific environments, 163
Duality principle, 264, 271–273, 276
Dynamical systems theory, 253, 256
Dynamic Geometry, 174, 178
Dynamic Mathematics, 178
Dynamic schemas, 131
Dynamic systems, participatory simulations and, 379

E

Ecole des Beaux Arts, 101
Education, 127
    role of models in, 161–169
Educational Researcher, 307
Effective flange, 40
Electronic gesture, socially situating, 385–388
Elevations, 108
Elicit or tell dilemma, 357
Embodied cognition, 144–151
Emergence, 387
Emergent structure, classroom-based learning and, 370
Emotions, 288
    action and, 89–90
    activity and, 89–90
    cognition and, 89–90
    ethnography of, at work, 84–89
    identity and, 92, 93
    modeling and, 77-80, 86–89, 95
    operations and, 89–90
Empirical proof scheme, 271–272, 278
End-in-view, 320
Endpoints, in generative activities, 373–378
Engagement, 292
    levels of, in model-eliciting activities, 351, 355–356
Engineering. *See also* Structural engineering
    computer programming as modeling in, 395–408
    defined, 409
    iteration in design in, 409–411, 415–421, 426–428
    mathematical modeling in, 421–426
    robotic design, 396–403

    uncertainty in design in, 411–415, 421, 426–428
    use of simulators in, 403-407
Epistemic complexity of mathematics, 174
Epistemological knowledge, 249
Epistemology, 250
    of design, 103, 119–120, 121, 122
    meta-knowledge and, 255–256
Equality, 266
Equation, solution to, 276
Equilibration, 265, 371
    thermal, 257–260
    tick model of, 254
Escher's World project, 122
Ethnographies, 1, 2–3. *See also* Architectural design; Factory statistical process control; Fish hatchery; Habitat conservation planning; Structural engineering
    of emotions at work, 84–89
    of mathematical thinking, 448–449
    methods, 9–11
Ethnomathematics, 79
Euclidean point, 130–131
Evolving expert studies, 450
Excel, 326, 329, 413, 415, 422
"Exciting Sack Race," 185
Expertise, 342
Expert studies, evolving, 450
Explanation, model building and, 194, 195, 198–199
Exploratory environments, 163
Exponential growth processes, random stopping of, 237
Externalization, in problem solving, 358–359, 360

F

Fabrics of activity, sensible, 247, 255
Facilitating Communities of Learners (FCL) curriculum, 102
Fact, 267
Factory statistical process control, 1, 7–34
    boundary objects, 15–30
    company manager and, 22–26
    company statistician and, 20–22
    conceptualizing techno-mathematical knowledge, 15–30
    individual and collective knowledge and, 16–17
    information technology and modern workplace, 11–15
    invisibility of knowledge at work, 8–9
    meanings of variation, 17-19
    methods, 9–11
    SPC chart as boundary object, 19–20
    team leader and, 26–30
    theoretical framework, 30–34

Fathom, 178, 326, 327, 330
FCL. *See* Facilitating Communities of Learners
Feedback, design cycle and, 116–118, 120, 122
Feigenbaum Bifurcation Diagram, 178
Fictive motion, 131, 145–146, 152
Field, 137
Fifth Discipline, The, 379
Fish hatchery, mathematical modeling in, 1, 77-95
    emotions and actions in theory, 89–90
    ethnography of emotions at work, 84–89
    identity and, 92–94
    implications for teaching mathematical modeling, 94–95
    motivation and, 90–92
    salmon case study, 77–83
Five handshape, 149
Flow, 292
    in model-eliciting activities, 355–360
Flowchart, 203
Formalisms, mathematical cognition and, 3
Fourth Generation Management, 18
Fractal phenomena, 229
Fractions, 276–277
Function-based models, 164
    lumped variable, 165–166
    spatially distributed, 164–165
Functions, 136
    continuity of, 138–144
    continuous, 135, 136
    discontinuous, 136
    oscillating, 144–145

G

Game theory, 155
Gaussian distributions, 235
Gender Equity in Engineering Project, viii
Generative design, 367–389
    characteristics of, 369-370
    design activities, 375–378
    generative teaching and learning and, 368–369
    modeling, 375–377
    moving from individualism to learning communities, 371–372
    multiple pathways and agreed upon endpoints, 374–378
    nominally generative activities, 373–374
    pathways, 382–383
    role of network technology in, 370–371
    role-playing activities, 382–386
    social significance of, 383–388
    toward framework of participation and representational truth-making, 372–373

Generative learning, 368–369, 378
Generative rules, models and simulations and, 163
Generative teaching, 368–369, 378
Generative words, 369
Genetic modeling languages, 251
Genscope, 163
Geometer's Sketchpad, 163
Geometric Supposer, 163, 383
Geometry, Dynamic, 174, 178
Geometry in Design curriculum, 122
Gestures, 286, 339, 340
    phases of, 148
    role in connected classroom, 189, 190
    socially situating, 385–388
Gesture studies, 147–151, 152
Girder case study, 39–45
Global affect, 289
Globalization, educational reform and impact of, 307
Goodness of fit, 169
Graphs, 250, 252, 259
    boundary objects and, 16
    transformable, 330–331, 333, 342, 343
    vertex-edge, 216, 217, 218
Gridlock simulation, 379–381, 386–387
Group design. *See* Generative design
Group functioning, in model-eliciting activities, 351, 353–355, 365–366
Group-level gestures, 386–388
Group modeling, 350
Group personality, 354–355
Group problem solving, affect and, 287
Growth with preferential attachment theory, 237
Guest crits, 110, 116, 120

H

Habitat conservation plan (HCP), 58
Habitat conservation planning, modeling in, 57–76
    creating different organizational futures through, 73–74
    difficulties in SAC meetings, 59–67
    performing differences in scientific practice through, 69–70
    professional vision and, 70–73
Half-cycle modeling problems, 321
Handbook of Research Design in Mathematics, Science, and Technology Education, 311
Handshapes, 149–150
Hands-on work, 308
HCP. *See* Habitat conservation plan
Heuristics, 266–267, 278
Higher-order conceptual systems, problem solving and, 344–345

SUBJECT INDEX

Holistic conceptual systems
    basic skills and, 341–342
    higher-order systems and, 342–346
Holmes, Oliver Wendell, 316, 319
Horizontal systemics, 248–253
How to Solve It, 343
HubNet architecture, 379, 382
Human capital, globalization and, 303, 304
Hyper-richness hypothesis, 256, 257–259
HySys, 405

I

Iconic roles, 378
Identity, 92–94, 281
    community norms and, 102
    mathematical self-, 289, 290–291, 294
    modeling and, 93–94, 95
    work and, 78, 79–80
Implicit knowledge, in problem solving, 361–362
Indexing a metonymical gesture, 149
Indiana University, viii
Individualism, moving to learning communities, 371–372
Individualistic theories of learning, 226–228
Individual knowledge, boundary objects and, 16–17
Individual roles, in model-eliciting activities, 351, 353–354, 363–365
Individual thinking, community norms and, 102
Industrial engineering project, 411–415, 420
Inferential organization of mathematics, 131, 132–134, 152
Infinite series, 135, 136, 137
    limits of, 144
Information, school mathematics and, 217
Information age, multi-media mathematics and, 156
Information processing perspectives on learning, 226–227
Information technology, workplace and, 11–15
Infrastructurality, necessity *vs.* convenience as factor in, 178
Infrastructural representations, 250–252
Infrastructure
    changes in institutional, 180–181
    meanings of, 177
Inner speech, 380
Innovation & Enterprise, 304
In-person services, 7
Institutional infrastructures, changes in, 180–181
Instructional activities, 297–299
Instructional principle, 270–271
Integrated Program, in Singapore, 305
Integrity, 281, 282, 289, 290
Intellectual need of learner, 263, 269, 274–275
Interactive Physics, 163
Interactivity, agency and, 178–179
International Handbook of Research in Mathematics Education, 307
Internet, 175
Interpretation, in modeling, 206
Interpreting, 335
Interviews, 9
Intimacy, 281, 282, 289, 291–292
Intuitive ideas, mathematics education and, 142–143
Intuitive knowledge, role in instruction, 246
Iteration
    coupled, 419, 421
    in design, 4, 101–102, 369, 378, 409–411, 415–421, 426–428
    discrete mathematics and, 216–217
    role in modeling, 422–425
Iterative analyses, 9
Iterative development cycles, 157
Iterative web diagrams, 417, 418

J

James, William, 316, 319-320, 339
Judgments, 271
Jury(ies), 101, 110

K

Kellogg School of Management, 323
Key Performance Indicators (KPIs), 11–12, 19
Knowledge. *See also* Mathematical knowledge
    collective, 16–17
    epistemological vs. domain particular, 249
    implicit, 361–362, 364
    naive, 252
    pragmatic vs. theoretical, 17
    prior knowledge principle, 359–360
    relevant, in modeling, 321
    tacit, 359–358, 360
    techno-mathematical, 14–15
    vertical systemics and, 246
    at work, invisibility of, 8–9
Knowledge development
    assumptions about, 431–433
    as premise of DNR-based instruction, 265, 268–269, 270
Knowledge economy, contributors to, 7
Knowledge systems, 249
KPIs. *See* Key Performance Indicators

L

Land managers, 58, 70–71
Landscape, 145

SUBJECT INDEX

Leadership Program in Discrete Mathematics, 220–221
Learners, intellectual need of, 263, 269, 274–275. *See also* Students
Learning
    adaptive, 369
    behaviorist model of, 226, 227
    collaborative, 302
    as complex adaptive system, 225, 226–229
    constructivist, 227, 302
    cooperative, 302
    design, 101–102
    diversity and, 372–373
    generative, 368–369, 378
    problem-based, 95, 302–303
    problem solving and, 38, 45, 53–54
    process, 304
    questioning and, 91
    student, 270–271, 272–273
    systems-theoretic approaches to, 228–229
    theories of, 226–229
    by using models vs. by making models, 168–169
Learning by Design project, 122
Learning communities, 371–372
Learning environments, as coherent systems, 102–103
Learning Technology Center at Vanderbilt University, 368–369
Learning to learn, 180–181
Learning transfer, 53–54
Least Upper Bound axiom, 138
Limits, 135–137
Limit statements, 127
Linear theories of learning, 226–228
Line-between-two-sensors design, 401–402
Line following task, 400–402
Linguistic diversity, generative design and, 383–384
Link-It, 163
Listing factors, 203, 205
Literacies, workplace, 13–15
Literacy communities, 177–183
Lizard habitat. *See* Habitat conservation planning
Local affect, 288–289
Local concept development, 37–38, 54, 293
Logic
    fabrics of activity vs., 247
    mathematics and, 128
Log-log distribution, 233, 236
Log-normal distribution, 232
Lumped variable function-based models, 165–166

M

Making Mathematics Engaging, 222
Manipulation, model building and, 194, 195, 199–200
Manning, Peyton, 345
Maple, 175
Massachusetts Institute of Technology, 100
Master Plan for Information Technology in Education, 304
Matematika, ee soderzhanie metody i znachenie, 135, 139
Materials, design studio, 114–115
Math anxiety, 287
Mathcad, 404
Mathematica, 175, 404
Mathematical affect, 283, 436. *See also* Affect
Mathematical cognition, 3. *See also* Cognition
    problem solving and, 5
Mathematical idea analysis, 131–132
Mathematical induction, 269
Mathematical integrity, 281, 282, 289, 290
Mathematical intimacy, 281, 282, 289, 291–292
Mathematical knowledge. *See also* Knowledge
    basic skills and systemic understanding and, 341–342
    between-concept systems and, 337–339
    higher-order ideas and systemic understanding and, 342–346
    interacting representational systems and, 341, 349
    invisibility of, 9
    used in workplaces, 9, 79
    within-concept systems and, 335–337
Mathematical performances, in wireless classroom, 185
Mathematical powerful affect, development of, 436
Mathematical representations, use of at work, 79
Mathematical self-identity, 289, 290–291, 294
Mathematical thinking, 157–158, 318
    observational studies of, 448–449
    in the workplace, 1, 2–3
Mathematics
    architectural problem solving and, 121–123
    assumptions about, 318–319
    beliefs about, 266, 268, 278
    changes in concepts of, 155–158
    cognitive science of. *See* Cognitive science of mathematics
    conceptions of, 128–130, 151
    discrete, 214, 215–217, 217–221, 447
Dynamic, 178
    epistemic complexity of, 174

as human activity, 129–130, 151
human cognition and, 151–152
inferential organization of, 132–134, 152
making practical, 315–316, 317
making practice, 316, 317
nature of, 131
patterns and, 449
promised usefulness of, 95
pure, 3, 137–138
rationality of, 282–283
real-life situations requiring, 316–318
representational advances in, 179–180
romance of, 128–130
thinking mathematically, 157–158, 318
Mathematics, Its Contents, Methods and Meaning, 135, 139
Mathematics curriculum
discrete mathematics as part of, 217–221
goals of, 158
need for changes in, 3–6
preparation for future and, 247–248
research agenda for, 447–451
rush to calculus and, 211–215
Mathematics education, 127–130
intuitive ideas and, 142–143
managing complexity of, 176–177
nature of curriculum in, 130–134
technology becoming infrastructural, 173–191
Mathematics epistemology premise, of DNR-based instruction, 265
Mathematics resource teachers, 433
Mathematizing reality, 337, 346
MATLAB, 404, 422
Matrices, 216
MBL. *See* Microcomputer Based Laboratory
Mead, George Herbert, 316, 319
Meaning, fostering shared, 384–385
MEAs. *See* Model-eliciting activities
Measurement
fractional, 276
model building and, 194, 195–196
Media, role in design process, 102
Mental acts, 265, 278
Meta-affect, 282, 287–288
Metacognitive functions, problem-solving strategies and, 346
Meta-knowledge, epistemology and, 255–256
Metaphorical thinking, gestures and, 148
Metaphors, 127, 146–147, 340
Meta-representational competence (MRC), 252–253
Microcomputer Based Laboratory (MBL), 175
Middle School Math through Applications Program, 122
Millennium Problems, 219
Misconceptions hypothesis, 256
Misconceptions studies, 246
Model-building environments, 163

Model development sequences, 293
Model-eliciting activities (MEAs), 37–38, 54–55, 193, 291, 316, 317, 318, 319
aspects of, 351
development of relevant skills, 342
engagement in, 351, 355–356
group functioning in, 351, 354–355, 365–366
individual roles in, 351, 353–354, 363–365
metaphors used during, 340
reflection tools and, 349
roles assumed during, 350, 351, 353–354, 363–365
solutions to, 348–339
strategies in, 351, 356–357
Volleyball Problem, 323–335, 336–337
Model-exploration activities, 316, 317, 330
Modeling cycles, 293
Modeling languages, 251
Modellus, 163
Model of dropped object, 251
Models/modeling, 4, 245
abilities in, cultivating, 201–208
affect and, 281, 293–294
anticipating use of, 75
architectural, 106–107, 112, 114–115
bank deposit, 310
of coho salmon and work, 77–83
with commercial simulators, 403–407
computer programming as, 395–408
conceptions of, 320–322
conceptual, 106
consensual intuitive, 257
containing disagreement, 67–69, 74
context-specific, 2–3, 5
creating different organizational futures through, 73–74
cycles of, 206–207
decision making and, 194, 195, 196–197
descriptive, 82
design research and, 308
discrete mathematics and, 218
emotions and, 77–80, 86–89, 95
in engineering, 421–426
explanation and, 194, 195, 198–199
function-based, 164–166
as goal of mathematical curriculum, 158–159
group, 350
in habitat conservation planning, 57–76
homogeneity of, 259–260
identity and, 79–80, 93–94, 95
learning by using vs. learning by making, 168–169
manipulation and, 194, 195, 199–200
measurement and, 194, 195–196
motivation and, 79–80, 90–92, 94–95
multiple pathways and endpoints and, 375–377

normative, 257, 258, 259, 260
parts of, 320
performing differences in scientific practice through, 69–70
personal, 349-351
perspective, 318–320
phase shifts during, 356
prediction and, 194, 195, 198
prescriptive, 82
problem solving and, 345, 346
professional vision and, 70–73, 74
purpose of, 162
real-life situations and, 315, 317
reasons for building, 193–200
reflecting on experience of, 204–205
reflection tools for, 207-208
replication and, 194, 195, 197-198
role in education, 161–169
role of iteration in, 422–425
role of uncertainty in, 425–426
role-playing and, 379
in scientific advisory committee meetings, 59–67
simulations vs., 162–163
in structural engineering, 38–52
structure-based, 166–168
student, 322
tension over model construction and deconstruction, 65–67
theoretical foundations of, 319-320
understanding underlying, 45, 52, 53
using satellite imagery to model habitat, 60–65
writing and drawing abilities and, 340–341
Monologic speech, 383–384
Motion
 fictive, 145–146
 origin of in mathematical ideas, 144
Motion metaphors, in calculus, 127
Motivation
 engagement and, 355
 identity and, 92, 94
 modeling and, 79–80, 90–92, 94–95
MRC. *See* Meta-representational competence
MSHCP. *See* Multiple-species habitat conservation plan
Multiple-species habitat conservation plan (MSHCP), 57-58, 73–74
 narrative timeline, 59, 68
Multiplicative model, 251
Multi-stage projects, mathematical thinking and, 316, 318
Multi-tier design experiments, 307–311, 432, 449–450

N

Naive knowledge, 252
Nanotechnology, 422–425, 426–428
National Council of Teachers of Mathematics (NCTM), 216, 434

National Science Foundation, 220, 444
Natural continuity, 139, 140–144
Navigating through Discrete Mathematics, 222
Navigations Series, 222
NCTM. *See* National Council of Teachers of Mathematics
Necessity principle, 264, 273–275
 -ness tasks, 168–169
Networks, 216. *See also* Classroom connectivity/networks
Network technologies, 176
Newark (NJ) Public Schools Systemic Initiative in Mathematics (NPSSIM), 431–435
 background, 433–434
 district results, 441–444
 example, 436–441
 goals of, 433-434
 professional development components, 434–436
Newtonian Sandbox, 163
Nominally generative activities, 373–374
Non-referential symbolic proof scheme, 266, 277, 278
Non-spatial structure-based models, 167-168
Normative model, 257, 258, 259, 260
Norms, 278
Novice-expert perspectives on learning, 227
NPSSIM. *See* Newark (NJ) Public Schools Systemic Initiative in Mathematics
"Numbers are locations in space," 146
NUS High School for Mathematics and Science, 305, 306

O

Obstacle-avoidance task, 399–400
One-over-f noise, 229
On the Shoulders of Giants, 449
Operations, emotions and, 89–90
Optimization, 381
Organizational future, modeling and, 73–74
Oscillating function, 144–145
Oscillation, 257
 tick model of, 254
Ownership, 291
Oxford Studio, 100
 alternation of design and feedback in, 117-119
 analysis of, 120–123
 assignments, 106–109
 case study, 103–123
 crits and design work, 109–110
 curriculum, 105–106, 116–118
 design cycles, 116–117
 epistemology, 119–120
 interdependence of structures, 115–116
 lessons learned in, 111–113
 materials, 114–115
 personnel, 113–114
 physical setting, 104
 presentation, 110–111

space in, 104–105, 113
time in, 105, 113

P

Parental outreach, 434
Pareto power law distribution, 229
Participant structures, 102
Participatory simulations, 379-381
Partition, 276
Part-whole, 276
Pathways
    exploring, 382–383
    in generative activities, 373–378
Patterns, 253–260
    case of thermal equilibration, 257–260
    general sketch, 253–256
    mathematics and, 449
    prototype rubric, 255
    systems-theoretical perspective and, 229, 238
Pedagogical activities, of studio design, 102–103
Pedagogically organized assistance, 175
Peirce, Charles Sanders, 316, 319, 372, 378
Peoples-Molecules activity, 371
Perfect Gas Law, 167–168
Performance criteria, determining, 13–14
Personal models of modeling, 349-351
Persuading, 267
Phase shifts, during modeling, 356
Pinups, 110, 116, 118, 120
Pitch, in prosody, 85–89
Plan of building, 107, 108
Play
    artifact-creating, 370
    problem solving and, 355–356
    space-creating, 374
Playground design, 415–421, 426–427
Point-slope forms, 187–188
Power law distribution, 229
    in decision making, 230–238
Power laws
    structural analysis of emergence of, 237-238
    systems that do and do not exhibit, 236–237
Pragmatic knowledge, theoretical *vs.*, 17
Pragmatism, 372
    model-eliciting activities and, 338–339
Pragmatists, 316, 319–320
Precedence, in design, 106
Predator-prey phenomena, models of, 164–165
Prediction, model building and, 194, 195, 198
Premises, of DNR-based instruction, 264–265
Preparation, of gesture, 148
Prescriptive modeling, 82
Prescriptive reflections, 359
Prescriptive self-models, 293
Prescriptive strategies, 343–345, 362

Presentation, in design studio, 110–111
Preservation of closeness, 142
Priming techniques, 147
Principles and Standards for School Mathematics (PSSM), 216, 217, 220, 221
Prior knowledge principle, 359-360
Probability, 214, 215
Problematic situations, 321
Problem-based learning, 95, 302–303
Problem-solvers, 321, 322, 341
Problem-solving
    activities, vii, 1–6
    defined, 273, 321
    externalization and, 358–359
    knowledge of adult, 37
    learning and, 38, 45, 53–54
    mathematical cognition and, 5
    prescriptive strategies in, 362
    strategies, 266–267, 343–345, 356–357
    tacit or implicit knowledge in, 357–358, 360
    use of reflection tools in. See Reflection tools
    using teams, 2
Process learning, 304
Products, 341
    of mental acts, 265, 278
    in modeling, 321
Professional development, 431–445
    assumptions about, 431–433
    components of, 434–436
    district results example, 441–444
    in DNR-based instruction, 276–278
    educational reform and, 307, 310
    knowledge development and, 436–441
    in Singapore, 306
    teacher example, 436–441
Professional vision, modeling and, 70–73
Program, for building, 107
Pro/II, 405
Project brief, 105–106
Proof-eliciting problems, 279
Proof schemes, 137, 266, 265–268
    axiomatic, 274
    defined, 267
    empirical, 271–272, 278
Prosodic features of talking, 85–89
Proving act, 267–268, 272
PSSM. *See Principles and Standards for School Mathematics*
Pumping/resonance, 257
Purdue University, viii
Pure mathematics, 3, 137–138
Purposes, impact on construct development, 337–338

Q

Quadratic formula, 277–278
Quality assurance guide, 423, 424
Quantifiers, continuity concept and, 140
Quarter-cycle modeling problems, 317

Quasi-induction, 269
Questions/questioning
    in connected vs. nonconnected classrooms, 189–190
    learning and, 91

R

Radical constructivism, 370
Randomness, 257
Ranking task, 323–335
Rationality of mathematics, 282–283
Rational numbers, 138, 277
Real-life situations
    mathematizing, 335–346
    needing mathematical thinking, 316–318
Real numbers, 137–138
Reason/reasoning
    abductive, 338
    mathematics and, 128
    quality of pathways and, 382
Recall techniques, 205
Recursion, 216–217
Reflection-in-action, 102
Reflection tools, 207–208, 346, 389–362, 363–366
    engagement and, 351, 355–356
    examples of, 351–352
    Group Functioning Reflection Tool, 351, 354–355, 365–366
    Individual Roles Reflection Tool, 351, 353–354, 363–365
    prior knowledge principle and, 359–360
    problem-solving strategies and, 351, 356–357
    promise of, 357–359
    research directions, 360–361
    technical challenges in designing, 352–353
Regulatory biologists, 58, 60–72
Re-mediation, 246–247, 254
Repeated-reasoning principle, 264, 275–276
Replication, model building and, 194, 195, 197–198
Representation, affect and, 285–286
Representation affordances, technology and, 174–177
Representational advances, 179–180
Representational infrastructures, changes in, 180–181
Representational misconceptions, 252
Representational systems, 339–341
Representational truth, 373, 388
Research agenda
    mathematics instruction, 447–451
    reflection tools, 360–361
    Singapore educational reform, 307–311
Research design, multi-tier, 307–311, 428, 449–450
Researchers
    effects of classroom connectivity on, 182

models of students' models, 322
Resources, diversity of, 385
Retraction, of gesture, 148
Reviews, 101, 110, 113–114, 116, 118, 120
Robotic design, 392–399
Role-playing
    generative design and, 378–382
    network-mediated, 376
Romance of mathematics, 128–130
Routine producers, 7

S

SAC. *See* Scientific advisory committee
Scale drawings, 107, 108, 115
Scale invariance, in classroom-situated decision making, 226
Scales, creating comparable, 331–332
Schema-theoretic perspectives on learning, 227
School Mathematics & Science Center, viii
Scientific advisory committee (SAC), 57–59
Scientific knowledge, 249
Sections, 108
Seeing, 335
Seeking open space task, 402–403
Self-identity, mathematical, 289, 290–291, 294
Self-monitoring, iteration and, 421
Semiotic acts, 292
Servo motor, 402
Sharpness, measures of, 169
Shear walls, 46
SimCalc MathWorlds software project, 183–188
Simulations, 254, 297, 299, 316. *See also* Model-eliciting activities
    computer program design process, 396–403
    foraging, 230–235
    gridlock, 379-381, 386–387
    models vs., 162–163
    participatory, 379-381
    purpose of, 162
Simulators, developing models using commercial, 403–407
Simultaneous equations, 187–188
Singapore educational reforms, 301–311
    education in Singapore, 304–306
    future challenges, 306–307
    niche schools, 305–306
    problem-based learning and, 302–303
    research agenda, 307–311
Singapore Ministry of Education Workplan Seminar, 304
Singapore Polytechnic, 106
Singapore Sports School, 305, 306
Site visits, 9, 10
Situated abstraction, 15, 17
Situated cognition, 54, 157, 302
Situations, need for model building and, 194–200

Six Sigma, 18
Sketchpad, 178
Skills
  hierarchical decomposition of, 247
  meaning acquired by, 341–342, 343
Smoothness, measures of, 169
Social choice, mathematics and, 217
Social context, for modeling and design, 381
Socially situating electronic and physical gesture, 385–398
Social norms, solutions and, 378
Social significance, of generative design, 383–388
Solutions processes, 321
Sonar sensor, 402
Source-path-goal schema, 141–142
Space, in design studio, 104–105, 113
Space-creating play, 374
Spatially distributed function-based models, 164–165
Spatial structure-based models, 166–167
SPC. *See under* Statistical process control
Special cause variation, 19, 22, 28
Specialized School for the Arts, 305, 306
Speech
  gestures and, 147–148
  monologic *vs.* dialogic, 383–384
  prosodic features of, 85–89
Sphere, 128
Spreadsheets, 2, 340
  boundary objects and, 16
  data display and, 24
  fish hatchery, 81–82
  for function-based models, 164
  to monitor factory performance, 11–12
  non-spatial structure-based models and, 168
  structural engineering, 46–52
Sprites, 165
Squareness, measures of, 169–170
"Staggered Races," 186–187, 189
StarLogo, 163, 165
Statistical process control (SPC), 18
Statistical process control (SPC) chart, 19–34
  company manager and, 22–26
  company statistician and, 20–22
  role of, 31–34
  team leader and, 26–30, 32
Statistical variation in work context. *See* Factory statistical process control
Statistician, SPC chart and company, 20–22
Statistics, 18, 214, 215
STELLA, 163, 165
Strawberry Vale Elementary School, 106
Stroke, 148
Structural engineering, problem solving in, 1, 37–55
  educational implications of, 54–55
  girder case study, 39–45
  learning of structural engineers, 53–54
  modeling and, 38–39
  nature of, 52–53
  tiedown case study, 46–52
Structural engineers, characteristics of successful, 4
Structure, mathematics as study of, 318
Structure-based models, 166–168
  non-spatial, 167–168
  spatial, 166–167
Structures d'ensemble, 370
Student learning, 270–271, 272–273
  defined, 270
Students
  attitudes of, 281–282
  effects of classroom connectivity on, 182
  intellectual needs of, 263, 269, 274–275
  mathematizing real-life situations, 335–346
  reflecting on modeling experiences, 204–205
Subjectivity premise, of DNR-based instruction, 265, 276
Surface procedures, 102–103
Surface structures, 103, 113, 115–116
Symbol analysts, 7–8, 33
Synergy, systemics and, 249
Systematic counting, 216
Systematicity, 246
Systemics, 246
  horizontal, 248–253
  vertical, 246–248
Systemic thinking
  basic skills and, 341–342
  between-concept systems and, 337–339
  higher-order ideas and, 332, 346
  interacting representational systems and, 339-341
  within-concept systems and, 335–337
Systems
  power law distribution in, 229
  that exhibit power law relationships, 236–237
Systems-theoretic approaches to learning, 228–229
Systems theory, 155
  dynamical, 253, 256
System-visualization software, 2

T

Tacit Dimension, The, 358
Tacit knowledge, in problem solving, 357–358
Tapered O, 150
Target line, 25
Tarski's World, 163
Taxonomy of Generative Activity Design...Classroom Networks, 389–390
Teachers
  addressing students as learners, 273–274
  benefits of discrete mathematics for, 220

effects of classroom connectivity on, 182, 183
mathematics resource, 433
practices in connected vs. nonconnected classrooms, 188–190
professional development for. See Professional development
role in encouraging modeling, 201, 203–205
Teaching
developing-first-then-formalize methodologies, 326–327, 330–332
for future learning, 248
generative, 368–369, 378
teach-first-then-apply methodologies, 326–327, 338
Teaching action, 270–271, 272–273
Teaching premise, of DNR-based instruction, 265
Teach Less Learn More initiative, 304–305
Team leader, role in statistical process control, 26–30, 32
Teams, as problem-solving entities, 2
Technological Resonance and Generative Activity, 389
Technology
becoming infrastructural in mathematics education, 173–191
communication technology affordance, 181–183
computational, 176
display, 176
experience of, 177
human partnership with, 179
increasing agency of tools, 179
network, 176
regularities in changes in distribution of skill across tool and user, 179–180
representational affordances of, 174–177
technology-driven evolution of technology-user communities, 177–178
types of hardware, 176
Technology-based tools, 340
Technology-user communities, 177–178, 180
Techno-mathematical knowledge, 14–15
conceptualizing, 15–30
Techno-mathematical literacies (TmL), 2, 14–15
Testing-and-revising, reflection tools and, 352
Theoretical knowledge, vs. pragmatic, 17
Theory development, description and, 245
Thermal equilibration, 257–260
Thick description, 104
Thinking Schools, Learning Nation, 304
Thinking Tags, 379
Third International Mathematics and Science Study-Repeat (TIMSS-R), 302
Third International Mathematics and Science Study (TIMSS), 215, 302, 304
Thought-revealing experiments, 450
Three-tiered design experiment, 450
Tick model, 250–251, 254
Tiedown case study, 46–52
Time
continuity and, 142
decision making and, 231–235, 238
in design studio, 105, 113
"Time events are things in unidimensional space," 132
"Time passing is motion of an object," 132–133
"Time passing is motion over a landscape," 133
Time-scales of change, 180–181
Timing, reflection tools and, 352
TIMSS. *See under* Third International Mathematics and Science Study
TinkerPlots, 326, 327, 330, 333, 342, 343
Tipping-point, 257
TmL. *See* Techno-mathematical literacies
Toolforthought, 179
Tools
computer, 404–407
conceptual, 321
distribution of skill across user and, 179–180
increasing agency of, 179
reflection. *See* Reflection tools
representational media, 339
technology-based, 340
Towers of meta-affect, 287
Trajector, 145–146
Transformable graphs, 330–331, 333, 342, 343
Traveling Salesman Problem, 216, 217
Trust, 292
Truth
conceptual metaphors and, 133–134
pragmatic notions of, 373, 376
representational, 388
Tutoring, one-on-one, 374
"2020 Engineer," 3
2001: A Space Odyssey, 128

U

Uncertainty
in design, 409–415, 421, 426–428
role in modeling, 425–426
Understanding, activity and, 100–101
Unit fraction, 274
Universality, of speech-accompanying gestures, 148
University of California at Berkeley, 212
UNIX, 422
Upper control limit, 25, 28
Use-patterns, 175

V

Validation of analyses, 9
Validity, manipulation of, 73–74
Values, 288
Variation
    common cause, 19, 21, 22, 25, 28, 31
    meanings of, at boundaries, 17–19
    special cause, 19, 22, 28
    statistical process control and, 18, 19, 21
Vectors, 247, 250
Vertex-edge graphs, 216, 217, 218
Vertical systemics, 246–248
Volleyball Problem, 323–325, 336–337, 338
    difficulties and insights emerging from solutions to, 326–335
    relevant skills needed for, 342
Vulnerability, 292

W

Ways of thinking, 206, 263–264, 265–268, 268–270, 278
Ways of understanding, 263–264, 265–266, 267–270, 278
Weighted averages, 326, 327, 336
Weighted sums, 326, 336
What Is Mathematics, 136
Where Mathematics Comes From, 128, 131, 143

Within-concept systems, 335–337
Word problems, 317
Work
    disembedded background, 8
    ethnography of emotions at, 84–89
    identity and, 78, 79–80
    invisibility of knowledge at, 8–9
    outcomes, employees distinguishing, 20–21
Workforce, impact of globalization on, 303, 304
Work groups, innovation generation and, 360
Working practices, types of, 7
Workplace
    information technology and, 11–15
    mathematical knowledge used in, 9
    mathematical thinking in, 1, 2–3
    new literacies required in, 13–15
Worldmaker, 163
Written symbolic statements, 339–340

Z

Zipf distribution, 229, 232–233
Zone of proximal development, 101
Zone of proximal knowledge, 265